Novel Compound Semiconductor Nanowires

Novel Compound Semiconductor Nanowires

Materials, Devices, and Applications

edited by

Fumitaro Ishikawa
Irina A. Buyanova

PAN STANFORD PUBLISHING

Published by

Pan Stanford Publishing Pte. Ltd.
Penthouse Level, Suntec Tower 3
8 Temasek Boulevard
Singapore 038988

Email: editorial@panstanford.com
Web: www.panstanford.com

British Library Cataloguing-in-Publication Data
A catalogue record for this book is available from the British Library.

Novel Compound Semiconductor Nanowires: Materials, Devices, and Applications

ISBN 978-981-4745-76-5 (Hardcover)
ISBN 978-1-315-36440-7 (eBook)

Printed in the USA

Contents

Foreword

As a researcher working on semiconductor nanostructures for about 50 years, it is my great pleasure to write a foreword for this book, which has been edited by Drs F. Ishikawa and I. Buyanova to cover recent research accomplishments on the epitaxial growth, physics, and device applications of semiconductor nanowire structures.

As is well known, most of key semiconductor devices, such as field-effect transistors (FETs) and laser diodes (LDs) make use of very thin layers as their core parts, since the control of channel conductance in FETs by gate voltage and that of optical gains in LDs by injected carriers can be efficiently achieved only if the channel layer of FETs and the active layer of LDs are sufficiently thin. These core layers, however, cannot be made too thin, since the quantum confinement of carriers in such layers leads to an excessive rise of the carrier energy and possibly weakens the carrier confinement.

As a result, these layers are formed with the thickness of a few to 10 nm by employing advanced epitaxy and other semiconductor techniques so that carriers are well confined and move freely only along the layer, while their motion normal to the layer is quantized. The two-dimensional (2D) nature of these carriers has induced a variety of important consequences, such as the formation of a series of 2D sub-bands with a step-like density of states, and the enhancement of excitonic effects. Consequently, various new devices, such as resonant tunneling diodes and inter-sub-band photodetectors and lasers, have been realized and new phenomena, such as quantum Hall effect, have been discovered.

While the thinning of these core layers has allowed the progress of advanced devices and the exploration of new physics of 2D carriers, the width reduction of such layers or films to form wire structures, such as narrow FET channels, has been done mainly in the field of LSIs to shrink devices for higher integration and to reduce the current and power consumption. It is also noted that the use of nanowire channels reduces the short-channel effect of FETs, as the gate around the wire acts more effectively than that in planar FETs. Although this down-scaling has been widely done, the typical width of FETs is set still at 100 nm or above to keep enough current drive capabilities.

Possibilities of squeezing the wire width down to 10 nm to confine electrons quantum mechanically and to use such 1D electrons for possible device applications had not been discussed until 1975, when I analyzed electron transport in coupled nanowire or planar superlattice structures. In 1980, potentials of quantum nanowire FETs were studied also by me. In 1982, Arakawa and I proposed and studied the possible use of quantum wires and quantum dots as gain media of semiconductor lasers.

Although such nanowire structures could not be formed in those early years, several groups started exploratory works to develop various methods that have enabled the formation of nanowire structures. The attempts are categorized into the three groups:

(a) top-down approach, based on the lithographic film patterning and the passivation
(b) bottom-up approach, based on the wire growth on self-assembled nanoparticles
(c) hybrid approach, which combine various patterning and selective growth processes

This book covers key results on the epitaxial growth of nanowires that belong to either group (b) or group (c).

Thanks to the progress of these fabrication methods, the formation of nanowires has been greatly facilitated, though there are still a lot of problems to be taken care of. As a consequence, a large number of studies have been made to clarify the physics and chemistry of various nanowire structures and to disclose unexplored potentials of nanowire-based devices, such as nanowire LEDs and solar cells.

I wish to close this foreword by wishing that this book promotes more research activities and accelerates the progress in the field.

Hiroyuki Sakaki
Professor Emeritus, University of Tokyo
President, Toyota Technological Institute
Nagoya, Japan
September 2017

Preface

The area of nanoscale science and technology is currently gaining increasing attention because of intriguing fundamental physics at the nanoscale, as well as potential applications of the low-dimensional structures in nanoscale electronics, optics, energy storage, and biology. The fabrication of these structures from crystalline semiconductors is now possible with a high degree of complexity, due to impressive developments in epitaxial growth techniques. Molecular beam epitaxy and metalorganic vapor phase epitaxy approaches are among the representative techniques utilized for the growth of low-dimensional semiconductor heterostructures with an atomic structural accuracy and precision. High-quality, single-crystalline III-V one-dimensional (1D) nanowire heterostructures are an example, promising for the next-generation nanoscale photonic and electronic devices, such as highly efficient nanowire light-emitting diodes, lasers and solar cells, as well as high-speed transistors. The key advantages of employing III-V semiconductor materials (i.e., nitride-, phosphide-, and arsenide-related heterostructure systems) for the realization of these devices include a high carrier mobility and superior optical properties. Furthermore, the developed epitaxial growth techniques enable electronic carrier control through the formation of quantum structures and precise doping, which can be introduced into a nanowire system. Most recently, it also became possible to fabricate III-V nanowires from highly mismatched alloys formed from III-V compounds with a large miscibility gap, or, alternatively, by the formation of hybrid heterostructures between a semiconductor and another material system based on, for instance, magnetic half metals and oxides. This book reviews the recent progress of such novel III-V semiconductor nanowires, covering a wide range of aspects from the epitaxial growth to the device applications. The prospects of such advanced 1D structures for nanoscience and nanotechnology are also discussed.

The book is organized as follows. A general overview of the area of one-dimensional structures is provided in Chapters 1 and 2. Specifically, Chapter 1 reviews the nanowire heterostructures for electrical and optical applications, focusing on the basic concepts and growth methods and the challenges of controlling the structure and the composition, and providing examples of selected nanowire devices. Chapter 2 describes the historical progress of molecular beam epitaxy of nitride nanocolumns and related nanocolumn emitters. The authors of this chapter have pioneered the growth of high-quality, metastable InGaN nanocolumns and related devices, which has opened the possibility of their applications in optical devices.

Chapters 3–9 review the properties of nanowires derived from novel strategic materials fabricated by epitaxial growth techniques.

Chapters 3–5 describe the highly mismatched dilute nitride and dilute bismide systems, which allow increased tunability in the band gap energy and lattice constants compared with conventional III-V semiconductor compounds and alloys. In Chapter 3, the structural and optical properties of GaNP nanowires are discussed along with their potential for future optoelectronic applications. In addition to significant improvements in the radiative efficiency, defect engineering via alloying with nitrogen is shown to be advantageous for the realization of polarized nano light sources and also for improving energy harvesting. In Chapter 4, structural and optical properties as well as the possible applications of GaNAs nanowires are reviewed. It is shown that alloying with nitrogen leads to the passivation of the nanowire surface and the formation of embedded quantum dot-like emitters. The existing research on dilute III-V-Bi nanowires, particularly dilute GaAsBi nanowires, is reviewed in Chapter 5. These novel alloys are of potential importance as infrared emitters with suppressed intrinsic non-radiative recombination losses at high temperature.

Chapters 6 and 7 are devoted to the characterization of nanowires based on novel spintronic materials. Chapter 6 describes experimental results from hybrid structures of ferromagnetic MnAs and non-magnetic III-V compounds and demonstrates their potential for magneto–nanoelectronic or spintronic device applications. In Chapter 7, the results from the GaAs–Fe_3Si semiconductor-ferromagnetic hybrid nanowires are presented. A high Curie temperature is a prerequisite for the application of these materials in spintronic devices, and the binary Heusler alloy Fe_3Si is a promising material in this sense.

Chapters 8 and 9 discuss hybrid nanowires consisting of GaAs and oxides. In Chapter 8, the synthesis of GaAs/$AlGaO_x$ nanowires combining molecular beam epitaxy and subsequent wet oxidation is presented. The characteristic broad visible light emission from the $AlGaO_x$ materials is demonstrated. Chapter 9 describes the growth and properties of GaAs/$SrTiO_3$ core–shell nanowires providing strategies for the synthesis of these high-quality hybrid structures. It demonstrates the feasibility of monolithical integration between the monocrystalline epitaxial shell of functional oxides and the nanowire, promising for the development of heterostructures that couple light emission/absorption and piezoelectricity or ferroelectricity.

The recent progress in device applications of nanowires is reviewed in Chapters 10–14. An overview of growth and applications of GaN nanowires together with a brief history of the related discoveries is given in Chapter 10. Both top-down and bottom-up approaches for the formation of nanowire arrays are presented. Additionally, the applications of these materials as quantum light emitters and nano-transistors are discussed. In Chapter 11, the potential of InP-related light-emitting devices is reviewed based on the in-depth investigations of the nanowire growth and device fabrication. The growth and the future prospect of InP/InAs nanowires for applications in field-effect transistors are described in Chapter 12. The achieved reproducible low-temperature synthesis of radial InP/InAs structures grown with top-down patterning and the demonstrated use of these structures as field-effect transistors highlight the potential of these materials in beyond-CMOS technologies. Chapter 13 describes the growth of InGaN nanowires and their applications in green light-emitting diodes, in lasers with a photonic crystal cavity, and also in photovoltaic devices. Chapter 14, the last chapter, provides a comprehensive overview of the historical development and the state-of-the-art of semiconductor nanowire transistors and solar cells. The emphasis is on the progress of III-V nanowires grown via metalorganic vapor phase epitaxy and their applications. This chapter also summarizes the most representative achievements in nanowire device applications for a wide range of III-V and group-IV materials and oxides.

We would like to express our gratitude to all contributors for their willingness to share with the readers their experience and in-depth insights into the exciting area of semiconductor nanowires. We hope that the comprehensive collection of the review articles on the current status of novel nanowire materials and systems provided in the book will stimulate further research efforts in this exciting field. F. I acknowledges the transfer of the experience and growth techniques at the initial stage of the nanowire study from Masahito Yamaguchi (who sadly passed away in 2013) and Yoshio Honda.

We are grateful to Stanford Chong and Arvind Kanswal of Pan Stanford Publishing for the invitation of this book editing and continuous support during the preparation.

Fumitaro Ishikawa, Matsuyama
Irina A. Buyanova, Linköping
September 2017

PART I

OVERVIEW

Chapter 1

Epitaxial Heterostructure Nanowires

Nari Jeon and Lincoln J. Lauhon

Department of Materials Science and Engineering, Northwestern University, Evanston, Illinois 60208, USA

lauhon@northwestern.edu

Semiconductor heterostructures are the basis for modern solid-state electronic and optoelectronic devices, including field-effect transistors, lasers, and light-emitting diodes. Epitaxial nanowire heterostructures enable crystal growers to transcend limitations of conventional thin-film heterostructures to create new opportunities for device engineering. This chapter provides a brief introduction to opportunities and challenges provided by nanowire heterostructures. Key concepts are illustrated with selected examples, primarily of III-V materials. We first introduce growth methods and classes of heterostructures and then identify targets for the control of structure and composition. We then describe the principles that govern structure–property relationships, provide recent examples of nanowire electrical and optical devices, and highlight the fundamental properties that currently limit performance. The control of interface structure and complete modeling of device properties are noted as important outstanding challenges.

Novel Compound Semiconductor Nanowires: Materials, Devices, and Applications
Edited by Fumitaro Ishikawa and Irina Buyanova

ISBN 978-981-4745-76-5 (Hardcover), 978-1-315-36440-7 (eBook)
www.panstanford.com

1.1 Introduction: Transcending Limits of Epitaxial Thin-Film Heterostructures

Semiconductor heterostructures are the basis for modern solid-state electronic and optoelectronic devices, including field-effect transistors, lasers, and light-emitting diodes. Heterostructures enable the engineering of energy band discontinuities and built-in potentials to control charge carrier injection, confinement, transport, and recombination. Charge carrier transport may occur along heterointerfaces, as in field-effect transistors, or across heterointerfaces, as in lasers and light-emitting diodes. Heterostructures can be formed by heteroepitaxy, a process in which a one crystalline material is grown on another crystalline material with a well-defined orientation between the two crystals. Beyond forming active junctions, thin film heteroepitaxy is also used to create large area substrates of high quality when bulk crystals are either unavailable or prohibitively expensive. For example, a rather small number of crystalline semiconductor substrates (e.g., Si, GaAs, sapphire) are of sufficiently high quality and low cost to support large-scale device fabrication, and they are available with limited crystal orientations. Substrates of semiconductor alloys and/or unusual crystal orientations are generally grown as thin films on commercially available substrates, provided the substrates are compatible in terms of thermodynamic stability, crystal structures, and lattice constants. When appropriate combinations are identified, heteroepitaxy can also be exploited for device integration. Practical operation of many electronic and optoelectronic devices often requires integration of multiple components of different functions. Heteroepitaxy can provide an approach by which the multiple components are combined economically.

Key challenges in heteroepitaxial film growth include the control of growth morphology, stress, strain, defects, and polarity. When mismatches of lattice constants and/or thermal expansion coefficients between the thin film and substrate are substantial, dislocations form to relax strain at the hetero-interfaces. In general, structural defects, including dislocations, antiphase domains or boundaries, point defects, and precipitates are detrimental to the performance and reliability of devices by forming midgap states that act as charge carrier traps and scattering/recombination centers. The mitigation of defect formation and the reduction of defect density are key goals of heteroepitaxy, particularly in forming the active region of devices. Approaches to suppress formation of defects include nano-patterning, compositional grading, and buffer layer growth [1–3]. However, these methods may provide incomplete suppression of dislocation formation or involve complicated processing. Epitaxial nanowire heterostructures enable crystal growers to transcend some limitations of conventional thin-film heterostructures to open up new opportunities in device engineering. This chapter provides a brief introduction to the opportunities and challenges created by nanowire heterostructures, including the basic structures and concepts. In addition, we provide some perspective on structure–property relationships using select examples taken primarily from the literature on group III-V nanowires.

Epitaxial heterostructure nanowires [4] combine two or more segments of different composition and/or doping while maintaining epitaxial relationships between the segments [4, 5]. In contrast to the conventional planar heterostructures, epitaxial heterostructure nanowires can accommodate significant interfacial strain at nanowire–substrate interfaces and heterojunction interfaces without defect formation due to their small lateral dimensions. The minimization of defect induced scattering and recombination improves the control of injection, separation, and collection of charge carriers, enhancing device performance. Moreover, the efficient strain accommodation in epitaxial heterostructure nanowires widens the range of materials that can be combined into heterostructures, including the range of substrates on which nanowires can grow. By enabling a wider range of material combinations, nanowires provide greater flexibility in selecting bandgaps and band offsets. The monolithic growth of III-V nanowires on Si substrates in particular provides the opportunity to integrate optoelectronic devices with Si-based complementary metal-oxide semiconductor (CMOS) technology [6]. Additionally, epitaxial nanowire growth methods can be used to create combined radial/axial heterostructures and mixed dimensional heterostructures such as dot-in-a-nanowire, which can provide new or improved functionality.

This chapter is not a comprehensive review of epitaxial nanowire heterostructures. Rather, it provides a perspective on structure–property relationships in epitaxial nanowire heterostructures by illustrating basic concepts with selected examples, primarily from III-V materials. Accordingly, Section 1.2 introduces basic concepts in the growth and structure of epitaxial heterostructure nanowires. Section 1.3 outlines goals for controlling structure and composition. Section 1.4 describes relationships between structures and properties and provides select recent examples of nanowire devices with an emphasis on electrical and optical devices.

1.2 Target Heterostructures and Growth Methods

Heterostructures may be formed within nanowires or between a nanowire and a growth substrate. We first consider the epitaxial growth of nanowires on a substrate. In homoepitaxy, nanowires are grown on a substrate of the same composition and phase. The crystal structure, lattice constants, and orientation are maintained across the nanowire substrate boundary without the formation of strain fields at the interface. In heteroepitaxy, nanowires are grown on a substrate of different material, typically with some degree of lattice mismatch. The finite and small cross section of nanowires implies that the total interfacial strain energy can be simply controlled with nanowire diameter. Therefore, the total interfacial strain energy can be held below the threshold for nucleation of a dislocation. Furthermore, any dislocations that cannot propagate parallel to the growth direction eventually terminate. In van der Waals epitaxy [7], nanowires are grown on a substrate that does not form covalent bonds with the nanowire. Interactions are sufficiently weak to prevent dislocation formation, but strong enough to provide some degree of orientation. Epitaxial heterostructures can also be formed within individual nanowires as axial heterostructures, radial heterostructures, and combinations thereof. In axial heterostructures, heterointerfaces are perpendicular to the nanowire growth axis. When the material is deposited epitaxially on the surface of a nanowire, a radial heterostructure is formed. Complex heterostructures combining both radial and axial heterostructures can be fabricated by modifying growth conditions to favor one growth mode or the other.

Nanowires are the result of highly anisotropic crystal growth driven by preferential nucleation on a single facet or set of facets. There are many ways to achieve preferential nucleation, but the vapor–liquid–solid (VLS) growth mechanism is perhaps the most often used. Precursor molecules in the vapor phase are chemisorbed and dissociated at the surface of a liquid alloy droplet, and crystal growth occurs at the solid–liquid interface with the crystal diameter defined by the size of the droplet. Metals that alloy with semiconductors to form deep eutectics, such as Au, are used to seed VLS growth. However, the incorporation of metal impurities into the growing nanowires is a concern for many applications. For example, Au impurities in Si nanowires form mid-gap impurity levels and act as charge carrier traps and recombination centers [8]. VLS growth also encompasses "self-catalyzed" processes, such as the growth of GaAs nanowires from a Ga droplet, which avoid the use of an impurity species. Furthermore, nanowires can be grown in the absence of seed particles if growth on a particular facet is strongly preferred. In all cases, the locations at which nanowires grow can be influenced by patterning. For growths without a seed particle, the growth location can be controlled by selective area epitaxy, which involves creating openings in a substrate masked by a barrier layer. Nanowires grow epitaxially from the openings, while the barrier layer promotes precursor diffusion and inhibits nucleation. Selective area epitaxy is typically implemented using metalorganic chemical vapor deposition (MOCVD) or molecular beam epitaxy (MBE). In MOCVD, metalorganic precursors are preferentially decomposed on the open surface with respect to mask, and undecomposed precursors diffuse to the open areas. In MBE, molecules and atoms have higher sticking coefficients on the open surface compared to the mask layers, and adatoms impinging on the masked area diffuse to the mask opening.

1.3 Controlling Structure and Composition: Achievements and Challenges

The major goals of nanowire heterostructure growth are to control position, orientation, and crystallinity. The position is typically controlled by patterning the growth substrate, as mentioned above. The orientation is established by some form of epitaxy, including standard epitaxy, van der Waals epitaxy, and graphoepitaxy. Finally, most nanowire growth processes produce crystalline material, but the defect type and density is strongly influenced by both the substrate and growth conditions, as for thin film growth. This section briefly reviews the role of epitaxy in controlling position, orientation, and crystallinity.

1.3.1 Epitaxy on Substrates

In heteroepitaxial growth of nanowires, as with thin films, the combination of substrate and nanowire is typically chosen to avoid the formation of misfit dislocations due to lattice mismatch between the nanowires and substrate. More specifically, there is in principle a critical nanowire diameter above which plastic deformation occurs via the formation of misfit dislocations at the interface of a nanowire and a substrate for a given lattice mismatch [9, 10]. While strain is shared by a nanowire and a substrate, elastic relaxation occurs primarily along the growth axis of the nanowire [11]. Tomioka et al. reported that InAs and GaAs nanowires grown on Si (111) substrate have misfit dislocations at the nanowire–substrate interface with a periodicity corresponding to the values expected from lattice mismatch [12, 13].

Another important consideration for heteroepitaxial growth is the preferred growth direction, which is influenced by the substrate as well as the seed particle (if used) and growth conditions. For the integration of nanowire arrays into devices, the growth of nanowire arrays oriented in a common direction, usually perpendicular to the substrate, is desired. Most nanowires prefer to grow along <111> directions, which favor substrates that are lattice matched to the selected nanowire (111) planes.

However, the extent to which a single growth direction dominates depends on both the substrate orientation and polarity, if present. For example, when III-V nanowires that prefer to grow along <111>B directions are grown on non-polar Si(111) substrates, nanowires grown along the <111> direction normal to the substrate coexist with nanowires grown along three equivalent off-normal <111> directions. Surface treatments can be used to influence the predominant growth direction by, for example, modifying the Si surface reconstruction [12, 13]. The lattice mismatch between nanowire and substrate can also influence the growth direction. Bakkers et al. showed that the percentage of non-vertical nanowires increases as the lattice mismatch between a nanowire and a substrate increases [14]. The mismatch can also induce kinking, in which the nanowire changes direction, to relieve strain. Under this circumstance, a compositionally graded interface was shown to be effective in eliminating defect formation and kinking [14]. The composition of the seed catalyst was also observed to influence the preferred growth direction [15, 16]. As can be inferred from these examples, the degree of control over nanowire orientation depends greatly on the specific system of interest. Finally, it is possible to grow oriented arrays of nanowires or even more complex arrangements with the nanowires parallel to the substrate through a process of epitaxy [17] and/or graphoepitaxy [18].

1.3.2 Axial and Radial Heterostructures

Heterostructures within nanowires can be formed either along (axial) or perpendicular (radial) to the nanowire growth direction. Goals for heterostructure formation include interface abruptness, crystalline perfection, and uniformity of composition. Axial heterostructures are formed by switching sources or precursors while maintaining conditions that prefer axial growth. Atomically sharp heterointerfaces are generally preferred but may be difficult to achieve when growing with a seed particle due to the so-called "reservoir" effect, in which one or more species that have accumulated in the catalyst continue to be incorporated in the nanowire even after the vapor phase source of that species has been eliminated. In such a case, the abruptness of the interface depends on the solubilities of the elements in the catalyst. Heterostructures formed by switching group III elements tend to have longer interface widths than those formed by switching group V elements due to higher solubilities of group III elements in the catalyst [19]. Various growth schemes have proved to be effective in suppressing the reservoir effect and enhancing abruptness of the heterojunctions, including changes of the catalyst particle phase from a liquid to a solid, varying the catalyst composition, and changing the size of the catalyst [20, 21]. It should be noted that misfit dislocations can be formed at the interfaces regardless of the abruptness of the heterointerfaces [22, 23]. In many cases, misfit dislocations may be avoided in axial heterostructures even for systems of large lattice mismatch such as InAs-InP (3.5%) [24] and GaP-GaAs (4%) [25] axial heterostructure nanowires due to efficient radial strain relaxation.

Radial heterostructure or core–shell nanowires also provide more effective strain relaxation than the corresponding planar heterostructures, enabling the epitaxial growth of Si on Ge at a 4% lattice mismatch [26, 27]. For compound semiconductors, ternary alloy shells can introduce band offsets at the core–shell interface while also controlling or minimizing strain. A homogeneous alloy distribution is generally desired to avoid potential fluctuations, which can partially deplete charge carriers and act as charge scattering centers. However, element-dependent variations in incorporation kinetics among the various exposed facets, in addition to the influence of surface curvature on elements of different diffusivities, can lead to nonuniformities in the composition of the shell [28–30]. The evolution of nanowire sidewall facets is determined by the surface energy anisotropy for a given growth condition and the relative diffusion rates of adatoms on the sidewall surfaces. For example, GaAs nanowires of zinc blende crystal structure exhibit a truncated hexagonal cross section, with a sidewall composed of large {110} facets and nanosize {112} facets [28, 31, 32]. Chemical potentials of different facets and surface diffusivities of different alloying elements are known to cause alloy segregation on the {112} nanofacets [28, 31, 32]. The formation of misfit dislocations also becomes favorable when the lattice mismatch between the core and the shell is large. Line edge dislocations parallel to the nanowire axis and dislocation loops with dislocation lines perpendicular to the nanowire axis relieve strain along radial and axial directions, respectively [10, 33]. It was found that the strain in InAs-GaAs core–shell nanowires is partially relaxed by forming paired line dislocations and loop edge dislocations [34].

1.3.3 Impurity Doping

The general goals for impurity doping are to selectively dope portions of nanowire heterostructures and to limit undesirable diffusion of dopants. Here we consider intentional and unintentional doping during growth (in situ). One possible source of unintentional doping is the diffusion of atoms from the growth substrate into the nanowire [35]. In VLS growth, for example, metal catalysts may dissolve some of the substrate, and substrate atoms can thereby become incorporated into the nanowires [36]. Diffusion from the substrate can also occur in catalyst-free growth when the growth temperature is sufficiently high. Such unintentional doping forms a doping gradient along the nanowire axis with the highest doping concentration near the nanowire base. This can be beneficial in cases where a low-resistance contact between the nanowire and substrate is needed. When this is undesirable, the growth of a low-temperature buffer layer can help prevent unintended doping [37].

In situ doping of nanowires at the growth interface presents new challenges compared to doping in conventional planar heterostructures due to the possibility of multiple dopant incorporation pathways with different incorporation kinetics, which can lead to an inhomogeneous dopant distribution [38]. Nanowire growth is by definition anisotropic, and the dopant incorporation rate relative to that of the semiconductor elements may vary between different interfaces. Dopant incorporation efficiency and spatial uniformity will therefore depend on the dopant precursors, growth methods, and growth conditions. Furthermore, the introduction of dopant precursors during growth can change the axial and radial growth rates [39] and nanowire nucleation density [40]. In general, and particularly for shell growth in core-shell nanowires, dopants may modify the surface energy of side-facets, as manifest in modified aspect ratios of nanowires [41]. Epitaxial heterostructures are used to implement modulation doping, in which carriers ionized from dopants in a large-gap material are transferred to a small-gap material to form a high-mobility, conducting channel. Modulation doping has been implemented in group IV [42, 43] and III-V core–shell nanowires [44–46], as discussed further.

1.3.4 Strain Engineering

Strain can be used to modify band degeneracies and offsets in semiconductor heterostructures and is an important degree of freedom in engineering electrical and optical properties. The general goals are to control offsets while avoiding defects. The potential to impose high-strain states at nanowire heterointerfaces without the formation of defects broadens the window of strain-induced band structure engineering. Strain can be imposed at the heterointerfaces between segments of different compositions or different crystal structures within a nanowire, as well as at the interface between a nanowire and a substrate. Several theoretical and experimental studies have explored the feasibility of strain engineering in heterostructure nanowires. Wölz et al. showed that strain states of GaN-InGaN axial heterostructure nanowires can be tuned via a ratio of quantum well thickness and superlattice period [47]. Copple et al. anticipated from first-principles calculations that tensile and compressive strain can induce the bandgap of wurtzite GaAs nanowires to change from indirect to direct [48]. The distribution of strain fields in heterostructure nanowires and the critical dimension above which plastic relaxation occurs have been calculated as a function of lattice mismatch and geometry, including the diameter and cross-sectional shape [49–51]. The critical dimensions are determined by comparing the energies of the system when the first dislocation is introduced at the interface and when the interface is coherently strained. In comparison with numerous theoretical studies, there are fewer experimental studies in which the distribution of strain has been directly measured [52, 53]. Recently, X-ray diffraction using a nanofocused beam has been developed to map strain distributions when combined with simulation of elastic strain relaxation by finite element methods [54, 55].

1.4 Controlling Properties for Emerging Applications

The degree to which the properties of nanowire heterostructures can be controlled depends on the degree to which the structure can be controlled. The previous section described goals for controlling structure, and this section discusses the properties that one would like to control through structure. Many properties are probed by measurements of devices, so below we illustrate a few key structure–property relationships and give some perspectives on how they impact device performance. Readers are referred elsewhere for reviews focusing on nanowire heterostructure devices and best-in-class performance [4, 56].

1.4.1 Effects of Size and Shape

The electronic and optical properties of nanowires differ from those of corresponding bulk materials by virtue of nanowire size and shape. Confinement of electrons/holes, phonons, and photons is prominent in nanowires when the dimensions become comparable to the respective wavelengths. At the shortest length scales, radial electron confinement (in two dimensions) leads to the formation of 1-D sub-bands, and exciton confinement modifies photoluminescence spectra. 1D sub-bands have been observed in nanowire field-effect transistors at low temperatures [57, 58]. Due to the geometrical anisotropy, the changes in conduction and valence bands depend on the growth direction [59]. Engineering of confinement potentials at the nanoscale enables devices that exploit quantum phenomena, including tunneling. For example, epitaxial heterostructure nanowires have been configured as tunnel field-effect transistors (TFET). TFETs can achieve a subthreshold slope (S) lower than the theoretical limit of a Si MOSFET (60 mV/dec) by exploiting interband tunneling of carriers. Narrow, direct bandgap III-V heterostructures or III-V/Si heterostructures have been explored for TFETs due to large bandgap and low band-to-band tunneling efficiency of Si [60, 61]. The optimization of nanowire heterostructure TFETs requires further improvements of heterojunction abruptness, doping profile, and defect mitigation to produce higher on-currents, lower off-currents, and lower subthreshold slope.

While the nanowire surface generically provides a confinement potential, the electronic properties are also modified by the condition of surfaces and interfaces, including dangling bonds, reconstructions, and passivation layers, including native oxides. The Fermi level can be pinned at the surface by interface states, and near surface band bending redistributes charge carriers. Therefore, surface and interface passivation is critical to enhance electrical properties by avoiding surface-mediated depletion of charge carriers, surface recombination, and short carrier lifetimes.

Modeling and simulation are essential to the interpretation of measurements and the design of devices influenced by finite size effects on band structure, charge carrier distribution, and optical interactions. For example, self-consistent solutions of the Schrödinger and Poisson equations have been used to determine the distribution of charge carriers for a given geometry and composition, including diameter, cross section shape, and doping [46, 62, 63]. From the calculated distribution of carrier densities, the photocurrent spectrum was calculated by solving the Boltzmann transport equation [64]. A comparison of simulations with and without photoexcitation can reveal pathways of charge carrier injection, separation, and collection [64]. The distribution of currents can be also simulated at the level of nanowire device by building an electrical network composed of circuits equivalent to different parts of nanowire device with multiple resistances accounting for different segments in nanowire heterostructures [65]. Photoluminescence and photoluminescence excitation spectra can be also simulated from the spatial distribution of carrier density [66]. Comparison of the simulated and measured spectra can help unveil the origins of spectral features, for example, states of electrons and holes that recombine. Electromagnetic simulations can be combined with the modeling of electronic properties described above to provide a more comprehensive understanding of optoelectronic properties. For example, finite-difference time-domain simulations are commonly used to understand optical modes of nanowire cavities, as well as the distribution of light reflection/absorption and photocarrier generation [67].

Size-dependent optical properties arise from the dielectric mismatch between the nanowire and its surroundings. The resulting size-dependent and polarization-dependent scattering and absorption can be qualitatively discussed in terms of optical modes in weak and strongly confining regimes. Strong 2D confinement of optical modes guided along the nanowire axis and high end-facet reflectivity enable nanowires to serve as a gain medium and optical cavity for lasing. In core–shell nanowires, the core can serve as a gain medium and shell as cavity, or vice versa [68, 69]. GaN-AlGaN core–shell nanowires were used in an early demonstration of photon confinement [68], which can be exploited to lower lasing thresholds, reduce the temperature dependence of the thresholds, and narrow the spectral linewidth [70]. Axial heterostructure growth enables the embedding of quantum dots in nanowire waveguides, which are useful as single or entangled photon sources [71–73]. Tapering can be used to make the light extraction more efficient [74]. The site-specific growth of nanowires in an array combined with deterministic growth of QDs within the nanowires in terms of size provides a platform for a photon source with high-fidelity entanglement [73].

1.4.2 Carrier Concentration and Mobility

Carrier concentrations and mobilities are influenced by bulk (interior) defects, surface and interface states, impurities, and intentional dopants. The degree of control over these factors in nanowire heterostructures depends on the details of the material system in growth method, but approaches to minimize the impact of defects and exploit opportunities to enhance properties are common between different material systems. For example, the charge carrier mobility depends on the effective mass and scattering rates; strain induced by epitaxial heterostructures can be used to tune the effective mass, and the general goal is to minimize scattering through maximizing structural perfection in the conducting channel. In radial heterostructure nanowires for which the shell bandgap is larger than that of the core, the energy band offsets (assuming a type I heterojunction) can be used to confine charge carriers to the core regions and reduce carrier scattering from the nanowire surface, though the carrier mobility in heterostructure nanowires can still be limited by scattering by defects, impurities, and interfaces, as well as polytypism. The goal of optimizing mobility has motivated many efforts to optimize the growth conditions of the shell. For example, Joyce et al. reported that deposition of AlGaAs shell on GaAs core under optimized condition improved the mobility to 65% of the mobility of bulk GaAs [75].

To control carrier concentration, substitutional doping of the "bulk" of a nanowire during and after synthesis has been demonstrated, but high doping levels reduce carrier mobility due to impurity scattering [76]. In this context, modulation doping of a heteroepitaxial shell is of interest because it enables engineering of high extrinsic carrier concentrations while preserving carrier mobility. For example, modulation doped III-V core–shell nanowires have been reported by several research groups [45, 62, 77]. Morkötter et al. demonstrated high doping efficiency by directly measuring Si dopant concentration ($\sim 6 \times 10^{19}$ atoms/cm^3) using atom probe tomography, and comparing it with the free carrier concentration derived from transport measurements [46]. Boland et al. measured the carrier concentration ($1.10 \pm 0.06 \times 10^{16}$/cm^3) and mobility ($2200 \pm 300$ cm^2/Vs at room temperature) in δ-doped GaAs-AlGaAs core–shell nanowires using optical pump terahertz probe spectroscopy and were able to increase the dopant concentration without degradation of mobility [62]. Despite these efforts, mobilities in heterostructure nanowires remain inferior to those of the corresponding bulk materials, and the relative contributions of interface, defect, and impurity scattering require further study. However, nanowire heterostructure field-effect transistors showing high electron mobility can still achieve very good transconductance values and electrostatic gate control. For example, Morkötter et al. measured a field-effect mobility of $4 - 5 \times 10^3$ cm^2/Vs at 4.2 K from Si delta-doped GaAs-AlGaAs core–shell nanowires, along with a subthreshold slope close to the theoretical limit (60 mV/dec) and excellent dc current saturation [46]. Tomioka et al. reported InGaAs-InP-InAlAs-InGaAs core-multi-shell nanowires with field-effect mobilities higher than typical Si MOSFETs, as well as enhanced on-state current and transconductance [44]. The wrap-gate geometry, where a standing nanowire is surrounded by gate, was proposed to be a main reason for excellent electrostatic gate control.

1.4.3 Lifetime and Quantum Efficiency

The lifetime of excess charge carriers created by electrical biasing or optical excitation strongly influences performance of many semiconductor devices. Here we provide a few examples involving GaAs nanowires, considering several factors that reduce minority carrier lifetime. For reference, well-passivated GaAs has minority carrier lifetime on the order of few microseconds [78]. In heterostructure nanowires, carrier lifetime is typically dominated by nonradiative recombination at the heterointerface and surfaces due to the large surface area to volume ratio. There have been many efforts to enhance lifetime using confining shell layers in core–shell nanowires to keep nonequilibrium carriers away from the nanowire surface. Jiang et al. reported that carrier lifetime in GaAs-AlGaAs core–shell nanowires increases with the shell thickness up to a critical value because the probability of carriers tunneling out of the core is reduced [79]. However, a prolonged shell growth time reduced the minority carrier lifetime, which was attributed to interdiffusion at the GaAs-AlGaAs interface. Controlled doping of shell layers can also enhance carrier lifetime by modifying band bending to reduce surface recombination [80]. As noted for carrier mobility, carrier lifetime is also sensitive to the presence of planar defects and impurities. The elimination of twin boundaries via a two-temperature growth method led to a threefold increase of minority carrier lifetime relative to the nanowires with twins [81]. In spite of these efforts, the longest minority carrier lifetime in GaAs-AlGaAs core–shell nanowires is in the nanosecond regime [79–81]. This is still short compared to "surface-effect-free" GaAs epilayers with a GaAs-AlGaAs heterointerface, which have a microsecond minority carrier lifetime [82].

Recombination of minority carriers reduces both the open-circuit voltage and short-circuit current and therefore the efficiency of solar cells. The internal quantum efficiency (IQE) of a solar cell, for example, is the fraction of photogenerated charge carriers that is collected as current. Nanowire heterostructures offer new opportunities to improve efficiencies by separating pathways of photon absorption and charger carrier collection [83]. In principle, radial heterojunction nanowires can reduce the impact of surface recombination by separating the active junction from the nanowire surface, and by maximizing absorption in the core. However, the quality of the interface may still limit the minority carrier lifetime [83], so optimizing heterostructures and passivating surfaces are still key areas for improvement [67, 84–86]. A recent comparison of axial and radial heterostructures suggests that axial heterostructures are generally more sensitive to surface recombination, yet may achieve high open-circuit voltages over a wider range of doping concentrations [87]. In addition, axial heterostructures provide the freedom to stack multiple, lattice-mismatched materials, allowing construction of multi-junction solar cells with optimal bandgap combinations and large open–circuit voltages [88]. For comparison, reported efficiencies of III-V nanowire solar cells have reached 11.4% for GaAs axial heterostructures [88] and 19.6% for InP core–shell heterostructures [86]. The efficiency might be further increased by optimization of contacts for efficient carrier extraction as well as high transparency, optimization of array architecture design, and realization of highly uniform nanowire distribution.

A long minority carrier lifetime is also important for devices that convert electrical energy to light. The IQE of an LED, for example, is the fraction of excitons in the active region that combine radiatively. The emissive IQE can be extracted from the dependence of the integrated PL emission intensity with excitation power, and is typically low in nanowire heterostructures due primarily to surface recombination [89]. Hence, the surface passivation and/or confinement away from the surface is important to enhancing IQE. The PL intensity of InAs-InAsP core–shell nanowires increased ~100 times compared to that of unpassivated InAs nanowires when the thickness and P content of InAsP shell was optimized. Pseudomorphically strained, high-quality heterointerfaces were proposed to suppress surface states and confine charge carries in core, thereby leading to the enhancement of PL intensity. Along with surface passivation and high crystal quality, controlling strain is important in polar materials because the interfacial strains may create electric fields that separate electrons and holes by the strain-induced quantum-confined Stark effect [90, 91]. The PL intensity of InGaN-GaN dot-in-a-wire nanowire LED structures was improved by deposition of an AlGaN shell, which improved radial confinement and reduced nonradiative surface recombination [92], leading to a

100-fold increase in output intensity compared to those without an AlGaAs shell [92]. InGaAs-InGaP radial heterostructures of ultralow lasing threshold exhibited an IQE that is 16.8% higher than those of GaAs- or AlGaAs-passivated radial structures, likely due to higher band offset providing better carrier confinement [93]. Other factors limiting the performance of nanowire array LEDs and lasers include a low density of emitting nanowires in an array and poor carrier injection efficiency. In III-N LEDs, including nanowire LEDs, electrons exhibit a relatively uniform distribution across the entire active region, but the lower mobility holes tend to not reach the active region before recombining [94–96]. The hole injection efficiency in nanowire LEDs can be improved by using tunnel junctions [96] or enhancing doping efficiency [97]. Increasing the uniformity of carrier distribution decreases Auger recombination and electron overflow, thereby improving the optical emission efficiency at high injection levels. Lastly, for the optoelectronic nanowire array devices, including solar cells, lasers, and LEDs described above, optimization of junction design at a single-nanowire (e.g., diameter, length, and doping) level and array level (e.g., spacing between nanowires), achievement of nanowire uniformity at a wafer scale, as well as optimized contact fabrication are also important areas of research [98, 99].

1.5 Concluding Remarks

This chapter provides a concise review of nanowire heterostructures focusing on (1) basic concepts and growth methods; (2) challenges of controlling structure and composition; and (3) goals for electrical and optical properties with select examples of nanowire devices. While many advantages of the nanowire geometry have been demonstrated at the single-nanowire device level, in practice many challenges remain, particularly associated with the control of interface quality. Advanced materials engineering approaches such as modulation doping have been demonstrated, but further improvements in the control of carrier concentration and the achievement of mobilities beyond existing heterostructures are needed. Theoretical studies and modeling of the strain distribution and resulting band structure modification show the promise of strain engineering in defect free epitaxial nanowire heterostructures, which is just beginning to be exploited. Direct measurement of strain fields at the single-nanowire level, along with advances in uniformity of nanowire arrays, will support efforts to correlate strain, composition, and structure, providing a foundation for strain engineering. The high surface area-to -volume ratio of nanowires presents opportunities and challenges. While nanowire arrays can increase the effective interface area of heterojunctions on a planar substrate, which is useful for carrier injection and collection, the effective density of surface states also increases, which can lead to recombination, band bending, and the redistribution of carriers. The abruptness and perfection of heterointerfaces strongly influence electrical and optical properties, including mobility, carrier concentration and lifetime, and quantum efficiency. Hence, the optimization of surface passivation and heterojunction growth strategies remain important areas of research as in planar heterostructures, but with the additional challenge of many different facets or interface orientations being present. Modeling and simulation have been very useful in exploring photon and charge carrier confinement in the nanowire geometry, including the prediction of band profiles, carrier distributions, and transport/recombination, thereby assisting the interpretation of experimentally observed electrical and optical properties in design of improved devices. Modeling of device behavior both at the single-nanowire level and in arrays of nanowires, together with improvements in uniformity of structure and composition, is essential to realize the benefits of epitaxial heterostructure nanowires at a relevant scale to impact technology.

Acknowledgments

L. J. L. and N. J. acknowledge NSF DMR-1308654 for support of this work.

References

1. Nakajima, K., Ujihara, T., Miyashita, S., Sazaki, G., Effects of misfit dislocations and AlN buffer layer on the GaInN/GaN phase diagram of the growth mode. *Journal of Applied Physics* 2001, 89, 146–153.

2. Bertoli, B., Suarez, E. N., Ayers, J. E., Jain, F. C., Misfit dislocation density and strain relaxation in graded semiconductor heterostructures with arbitrary composition profiles. *Journal of Applied Physics* 2009, 106, 073519.

3. Niu, G., Capellini, G., Schubert, M. A., Niermann, T., Zaumseil, P., Katzer, J., Krause, H.-M., Skibitzki, O., Lehmann, M., Xie, Y.-H., von Känel, H., Schroeder, T., Dislocation-free Ge nano-crystals via pattern independent selective Ge heteroepitaxy on Si nano-tip wafers. *Scientific Reports* 2016, 6, 22709.

4. Hyun, J. K., Zhang, S., Lauhon, L. J., Nanowire heterostructures. *Annual Review of Materials Research* 2013, 43, 451–479.

5. Lauhon, L. J., Gudiksen, M. S., Lieber, C. M., Semiconductor nanowire heterostructures. *Philosophical Transactions of the Royal Society of London A: Mathematical, Physical and Engineering Sciences* 2004, 362, 1247–1260.

6. Deleonibus, S., *Intelligent Integrated Systems: Devices, Technologies, and Architectures*. 2014, p. 1–499.

7. Koma, A., Papers presented at the International Workshop on Science and Technology of Thin Films for the 21st Century, Evanston, IL, USA, July 28-August 2, 1991 Van der Waals epitaxy—a new epitaxial growth method for a highly lattice-mismatched system. *Thin Solid Films* 1992, 216, 72–76.

8. Bullis, W. M., Properties of gold in silicon. *Solid-State Electronics* 1966, 9, 143–168.

9. Chuang, L. C., Moewe, M., Chase, C., Kobayashi, N. P., Chang-Hasnain, C., Crankshaw, S., Critical diameter for III-V nanowires grown on lattice-mismatched substrates. *Applied Physics Letters* 2007, 90, 043115.

10. Karen, L. K., Misfit dislocations in nanowire heterostructures. *Semiconductor Science and Technology* 2010, 25, 024006.

11. Stanchu, H., Kladko, V., Kuchuk, A. V., Safriuk, N., Belyaev, A., Wierzbicka, A., Sobanska, M., Klosek, K., Zytkiewicz, Z. R., High-resolution X-ray diffraction analysis of strain distribution in GaN nanowires on Si(111) substrate. *Nanoscale Research Letters* 2015, 10, 51.

12. Tomioka, K., Motohisa, J., Hara, S., Fukui, T., Control of InAs nanowire growth directions on Si. *Nano Letters* 2008, 8, 3475–3480.

13. Tomioka, K., Tanaka, T., Hara, S., Hiruma, K., Fukui, T., III-V Nanowires on Si substrate: Selective-area growth and device applications. *IEEE Journal of Selected Topics in Quantum Electronics* 2011, 17, 1112–1129.

14. Bakkers, E. P. A. M., Borgström, M. T., Verheijen, M. A., Epitaxial growth of III-V nanowires on group IV substrates. *MRS Bulletin* 2007, 32, 117–122.

15. Man Suk, S., Jae Hun, J., Yong, K., Wang, Y., Zou, J., Joyce, H. J., Gao, Q., Tan, H. H., Jagadish, C., Vertically standing Ge nanowires on GaAs(110) substrates. *Nanotechnology* 2008, 19, 125602.

16. Fortuna, S. A., Li, X., Metal-catalyzed semiconductor nanowires: A review on the control of growth directions. *Semiconductor Science and Technology* 2010, 25, 024005.

17. Fortuna, S. A., Wen, J. G., Chun, I. S., Li, X. L., Planar GaAs nanowires on GaAs (100) substrates: Self-aligned, nearly twin-defect free, and transfer-printable. *Nano Letters* 2008, 8, 4421–4427.

18. Tsivion, D., Schvartzman, M., Popovitz-Biro, R., von Huth, P., Joselevich, E., Guided growth of millimeter-long horizontal nanowires with controlled orientations. *Science* 2011, 333, 1003–1007.

19. Dubrovskii, V. G., Sibirev, N. V., Factors influencing the interfacial abruptness in axial III–V nanowire heterostructures. *Crystal Growth & Design* 2016, 16, 2019–2023.

20. Priante, G., Glas, F., Patriarche, G., Pantzas, K., Oehler, F., Harmand, J.-C., Sharpening the interfaces of axial heterostructures in self-catalyzed AlGaAs nanowires: Experiment and theory. *Nano Letters* 2016, 16, 1917–1924.

21. Wen, C.-Y., Reuter, M. C., Bruley, J., Tersoff, J., Kodambaka, S., Stach, E. A., Ross, F. M., Formation of compositionally abrupt axial heterojunctions in silicon-germanium nanowires. *Science* 2009, 326, 1247–1250.

22. Furtmayr, F., Teubert, J., Becker, P., Conesa-Boj, S., Morante, J. R., Chernikov, A., Schäfer, S., Chatterjee, S., Arbiol, J., Eickhoff, M., Carrier confinement in $GaN/Al_xGa_{1-x}N$ nanowire heterostructures $(0 < x < 1)$. *Physical Review B* 2011, 84, 205303.

23. Pratyush Das, K., Heinz, S., Mikael, T. B., Lynne, M. G., Chris, B., John, B., Cedric, D. B., Heike, R., Selective area growth of III–V nanowires and their heterostructures on silicon in a nanotube template: Towards monolithic integration of nano-devices. *Nanotechnology* 2013, 24, 225304.

24. Magnus, W. L., Jakob, B. W., Mathias, W., Paul, H., Linus, E. F., Lars, S., Wallenberg, L. R., Strain mapping in free-standing heterostructured wurtzite InAs/InP nanowires. *Nanotechnology* 2007, 18, 015504.

25. Verheijen, M. A., Algra, R. E., Borgström, M. T., Immink, G., Sourty, E., van Enckevort, W. J. P., Vlieg, E., Bakkers, E. P. A. M., Three-dimensional morphology of GaP–GaAs nanowires revealed by transmission electron microscopy tomography. *Nano Letters* 2007, 7, 3051–3055.

26. Xiang, J., Lu, W., Hu, Y., Wu, Y., Yan, H., Lieber, C. M., Ge/Si nanowire heterostructures as high-performance field-effect transistors. *Nature* 2006, 441, 489–493.

27. Lauhon, L. J., Gudiksen, M. S., Wang, C. L., Lieber, C. M., Epitaxial core-shell and core-multishell nanowire heterostructures. *Nature* 2002, 420, 57–61.

28. Zheng, C., Wong-Leung, J., Gao, Q., Tan, H. H., Jagadish, C., Etheridge, J., Polarity-driven 3-fold symmetry of GaAs/AlGaAs core multishell nanowires. *Nano Letters* 2013, 13, 3742–3748.

29. Rudolph, D., Funk, S., Döblinger, M., Morkötter, S., Hertenberger, S., Schweickert, L., Becker, J., Matich, S., Bichler, M., Spirkoska, D., Zardo, I., Finley, J. J., Abstreiter, G., Koblmüller, G., Spontaneous alloy composition ordering in GaAs-AlGaAs core–shell nanowires. *Nano Letters* 2013, 13, 1522–1527.

30. Heiss, M., Fontana, Y., Gustafsson, A., Wust, G., Magen, C., O'Regan, D. D., Luo, J. W., Ketterer, B., Conesa-Boj, S., Kuhlmann, A. V., Houel, J., Russo-Averchi, E., Morante, J. R., Cantoni, M., Marzari, N., Arbiol, J., Zunger, A., Warburton, R. J., Fontcuberta i Morral, A., Self-assembled quantum dots in a nanowire system for quantum photonics. *Nature Materials* 2013, 12, 439–444.

31. Sköld, N., Wagner, J. B., Karlsson, G., Hernán, T., Seifert, W., Pistol, M.-E., Samuelson, L., Phase segregation in AlInP shells on GaAs nanowires. *Nano Letters* 2006, 6, 2743–2747.

32. Zhang, Y., Sanchez, A. M., Wu, J., Aagesen, M., Holm, J. V., Beanland, R., Ward, T., Liu, H., Polarity-driven quasi-3-fold composition symmetry of self-catalyzed III–V–V ternary core–shell nanowires. *Nano Letters* 2015, 15, 3128–3133.

33. Raychaudhuri, S., Yu, E. T., Calculation of critical dimensions for wurtzite and cubic zinc blende coaxial nanowire heterostructures. *Journal of Vacuum Science & Technology B* 2006, 24, 2053–2059.

34. Popovitz-Biro, R., Kretinin, A., Von Huth, P., Shtrikman, H., InAs/GaAs core–shell nanowires. *Crystal Growth & Design* 2011, 11, 3858–3865.

35. Dhaka, V., Haggren, T., Jussila, H., Jiang, H., Kauppinen, E., Huhtio, T., Sopanen, M., Lipsanen, H., High quality GaAs nanowires grown on glass substrates. *Nano Letters* 2012, 12, 1912–1918.

36. Han, H., Huang, Z., Lee, W., Metal-assisted chemical etching of silicon and nanotechnology applications. *Nano Today* 2014, 9, 271–304.

37. Huang, H., Ren, X., Ye, X., Guo, J., Wang, Q., Yang, Y., Cai, S., Huang, Y., Growth of stacking-faults-free zinc blende GaAs nanowires on Si substrate by using AlGaAs/GaAs buffer layers. *Nano Letters* 2010, 10, 64–68.

38. Dufouleur, J., Colombo, C., Garma, T., Ketterer, B., Uccelli, E., Nicotra, M., Fontcuberta i Morral, A., P-Doping mechanisms in catalyst-free gallium arsenide nanowires. *Nano Letters* 2010, 10, 1734–1740.

39. Tutuc, E., Appenzeller, J., Reuter, M. C., Guha, S., Realization of a linear germanium nanowire p–n junction. *Nano Letters* 2006, 6, 2070–2074.

40. Limbach, F., Caterino, R., Gotschke, T., Stoica, T., Calarco, R., Geelhaar, L., Riechert, H., The influence of Mg doping on the nucleation of self-induced GaN nanowires. *AIP Advances* 2012, 2, 012157.

41. Wirths, S., Weis, K., Winden, A., Sladek, K., Volk, C., Alagha, S., Weirich, T. E., von der Ahe, M., Hardtdegen, H., Lüth, H., Demarina, N., Grützmacher, D., Schäpers, T., Effect of Si-doping on InAs nanowire transport and morphology. *Journal of Applied Physics* 2011, 110, 053709.

42. Dillen, D. C., Kim, K., Liu, E.-S., Tutuc, E., Radial modulation doping in core-shell nanowires. *Nature Nano* 2014, 9, 116–120.

43. Yang, C., Zhong, Z., Lieber, C. M., Encoding electronic properties by synthesis of axial modulation-doped silicon nanowires. *Science* 2005, 310, 1304–1307.

44. Tomioka, K., Yoshimura, M., Fukui, T., A III-V nanowire channel on silicon for high-performance vertical transistors. *Nature* 2012, 488, 189–192.

45. Spirkoska, D., Fontcuberta i Morral, A., Dufouleur, J., Xie, Q., Abstreiter, G., Free standing modulation doped core–shell GaAs/AlGaAs hetero-nanowires. *Physica Status Solidi (RRL)–Rapid Research Letters* 2011, 5, 353–355.

46. Morkötter, S., Jeon, N., Rudolph, D., Loitsch, B., Spirkoska, D., Hoffmann, E., Döblinger, M., Matich, S., Finley, J. J., Lauhon, L. J., Abstreiter, G., Koblmüller, G., Demonstration of confined electron gas and steep-slope behavior in delta-doped GaAs-AlGaAs core–shell nanowire transistors. *Nano Letters* 2015, 15, 3295–3302.

47. Wölz, M., Ramsteiner, M., Kaganer, V. M., Brandt, O., Geelhaar, L., Riechert, H., Strain engineering of nanowire multi-quantum well demonstrated by Raman spectroscopy. *Nano Letters* 2013, 13, 4053–4059.

48. Copple, A., Ralston, N., Peng, X., Engineering direct-indirect band gap transition in wurtzite GaAs nanowires through size and uniaxial strain. *Applied Physics Letters* 2012, 100, 193108.

49. Trammell, T. E., Zhang, X., Li, Y., Chen, L.-Q., Dickey, E. C., Equilibrium strain-energy analysis of coherently strained core–shell nanowires. *Journal of Crystal Growth* 2008, 310, 3084–3092.

50. Ertekin, E., Greaney, P. A., Chrzan, D. C., Sands, T. D., Equilibrium limits of coherency in strained nanowire heterostructures. *Journal of Applied Physics* 2005, 97, 114325.

51. Raychaudhuri, S., Yu, E. T., Critical dimensions in coherently strained coaxial nanowire heterostructures. *Journal of Applied Physics* 2006, 99, 114308.

52. Taraci, J. L., Hÿtch, M. J., Clement, T., Peralta, P., McCartney, M. R., Jeff, D., Picraux, S. T., Strain mapping in nanowires. *Nanotechnology* 2005, 16, 2365.

53. Wen, C.-Y., Reuter, M. C., Su, D., Stach, E. A., Ross, F. M., Strain and stability of ultrathin Ge layers in Si/Ge/Si axial heterojunction nanowires. *Nano Letters* 2015, 15, 1654–1659.

54. Stankevič, T., Hilner, E., Seiboth, F., Ciechonski, R., Vescovi, G., Kryliouk, O., Johansson, U., Samuelson, L., Wellenreuther, G., Falkenberg, G., Feidenhans'l, R., Mikkelsen, A., Fast strain mapping of nanowire light-emitting diodes using nanofocused X-ray beams. *ACS Nano* 2015, 9, 6978–6984.

55. Stankevič, T., Dzhigaev, D., Bi, Z., Rose, M., Shabalin, A., Reinhardt, J., Mikkelsen, A., Samuelson, L., Falkenberg, G., Vartanyants, I. A., Feidenhans'l, R., Strain mapping in an InGaN/GaN nanowire using a nano-focused x-ray beam. *Applied Physics Letters* 2015, 107, 103101.

56. Joyce, H. J., Gao, Q., Hoe Tan, H., Jagadish, C., Kim, Y., Zou, J., Smith, L. M., Jackson, H. E., Yarrison-Rice, J. M., Parkinson, P., Johnston, M. B., III–V semiconductor nanowires for optoelectronic device applications. *Progress in Quantum Electronics* 2011, 35, 23–75.

57. Ford, A. C., Kumar, S. B., Kapadia, R., Guo, J., Javey, A., Observation of degenerate one-dimensional sub-bands in cylindrical InAs nanowires. *Nano Letters* 2012, 12, 1340–1343.

58. Tian, Y., Sakr, M. R., Kinder, J. M., Liang, D., MacDonald, M. J., Qiu, R. L. J., Gao, H.-J., Gao, X. P. A., One-dimensional quantum confinement effect modulated thermoelectric properties in InAs nanowires. *Nano Letters* 2012, 12, 6492–6497.

59. Scheel, H., Reich, S., Thomsen, C., Electronic band structure of high-index silicon nanowires. *Physica Status Solidi* (b) 2005, 242, 2474–2479.

60. Ganjipour, B., Wallentin, J., Borgström, M. T., Samuelson, L., Thelander, C., Tunnel field-effect transistors based on InP-GaAs heterostructure nanowires. *ACS Nano* 2012, 6, 3109–3113.

61. Tomioka, K., Yoshimura, M., Fukui, T. In Steep-slope tunnel field-effect transistors using III-V nanowire/Si heterojunction, *VLSI Technology (VLSIT)*, 2012 Symposium on, June 12–14, 2012, pp. 47–48.

62. Boland, J. L., Conesa-Boj, S., Parkinson, P., Tütüncüoglu, G., Matteini, F., Rüffer, D., Casadei, A., Amaduzzi, F., Jabeen, F., Davies, C. L., Joyce, H. J., Herz, L. M., Fontcuberta i Morral, A., Johnston, M. B., Modulation doping of GaAs/AlGaAs core–shell nanowires with effective defect passivation and high electron mobility. *Nano Letters* 2015, 15, 1336–1342.

63. Jadczak, J., Plochocka, P., Mitioglu, A., Breslavetz, I., Royo, M., Bertoni, A., Goldoni, G., Smolenski, T., Kossacki, P., Kretinin, A., Shtrikman, H., Maude, D. K., Unintentional high-density p-type modulation doping of a GaAs/AlAs core–multishell nanowire. *Nano Letters* 2014, 14, 2807–2814.

64. Dai, X., Zhang, S., Wang, Z., Adamo, G., Liu, H., Huang, Y., Couteau, C., Soci, C., GaAs/AlGaAs nanowire photodetector. *Nano Letters* 2014, 14, 2688–2693.

65. Tchernycheva, M., Lavenus, P., Zhang, H., Babichev, A. V., Jacopin, G., Shahmohammadi, M., Julien, F. H., Ciechonski, R., Vescovi, G., Kryliouk, O., InGaN/GaN core–shell single nanowire light emitting diodes with graphene-based P-contact. *Nano Letters* 2014, 14, 2456–2465.

66. Fickenscher, M., Shi, T., Jackson, H. E., Smith, L. M., Yarrison-Rice, J. M., Zheng, C., Miller, P., Etheridge, J., Wong, B. M., Gao, Q., Deshpande, S., Tan, H. H., Jagadish, C., Optical, structural, and numerical investigations of GaAs/AlGaAs core–multishell nanowire quantum well tubes. *Nano Letters* 2013, 13, 1016–1022.

67. Mariani, G., Zhou, Z., Scofield, A., Huffaker, D. L., Direct-bandgap epitaxial core–multishell nanopillar photovoltaics featuring subwavelength optical concentrators. *Nano Letters* 2013, 13, 1632–1637.

68. Choi, H.-J., Johnson, J. C., He, R., Lee, S.-K., Kim, F., Pauzauskie, P., Goldberger, J., Saykally, R. J., Yang, P., Self-organized GaN quantum wire UV lasers. *The Journal of Physical Chemistry B* 2003, 107, 8721–8725.

69. Qian, F., Li, Y., Gradecak, S., Park, H.-G., Dong, Y., Ding, Y., Wang, Z. L., Lieber, C. M., Multi-quantum-well nanowire heterostructures for wavelength-controlled lasers. *Nature Materials* 2008, 7, 701–706.

70. Yan, R., Gargas, D., Yang, P., Nanowire photonics. *Nature Photonics* 2009, 3, 569–576.

71. Kremer, P. E., Dada, A. C., Kumar, P., Ma, Y., Kumar, S., Clarke, E., Gerardot, B. D., Strain-tunable quantum dot embedded in a nanowire antenna. *Physical Review B* 2014, 90, 201408.

72. Huber, T., Predojević, A., Khoshnegar, M., Dalacu, D., Poole, P. J., Majedi, H., Weihs, G., Polarization entangled photons from quantum dots embedded in nanowires. *Nano Letters* 2014, 14, 7107–7114.

73. Versteegh, M. A. M., Reimer, M. E., Jöns, K. D., Dalacu, D., Poole, P. J., Gulinatti, A., Giudice, A., Zwiller, V., Observation of strongly entangled photon pairs from a nanowire quantum dot. *Nature Communication* 2014, 5, 5298.

74. Claudon, J., Bleuse, J., Malik, N. S., Bazin, M., Jaffrennou, P., Gregersen, N., Sauvan, C., Lalanne, P., Gerard, J.-M., A highly efficient single-photon source based on a quantum dot in a photonic nanowire. *Nature Photonics* 2010, 4, 174–177.

75. Joyce, H. J., Parkinson, P., Jiang, N., Docherty, C. J., Gao, Q., Tan, H. H., Jagadish, C., Herz, L. M., Johnston, M. B., Electron mobilities approaching bulk limits in "surface-free" GaAs nanowires. *Nano Letters* 2014, 14, 5989–5994.

76. Ketterer, B., Uccelli, E., Fontcuberta i Morral, A., Mobility and carrier density in p-type GaAs nanowires measured by transmission Raman spectroscopy. *Nanoscale* 2012, 4, 1789–1793.

77. Sladek, K., Klinger, V., Wensorra, J., Akabori, M., Hardtdegen, H., Grützmacher, D., MOVPE of *n*-doped GaAs and modulation doped GaAs/AlGaAs nanowires. *Journal of Crystal Growth* 2010, 312, 635–640.

78. Ahrenkiel, R. K., Measurement of minority-carrier lifetime by time-resolved photoluminescence. *Solid-State Electronics* 1992, 35, 239–250.

79. Jiang, N., Gao, Q., Parkinson, P., Wong-Leung, J., Mokkapati, S., Breuer, S., Tan, H. H., Zheng, C. L., Etheridge, J., Jagadish, C., Enhanced minority carrier lifetimes in GaAs/AlGaAs core–shell nanowires through shell growth optimization. *Nano Letters* 2013, 13, 5135–5140.

80. Boland, J. L., Casadei, A., Tütüncüoglu, G., Matteini, F., Davies, C. L., Jabeen, F., Joyce, H. J., Herz, L. M., Fontcuberta i Morral, A., Johnston, M. B., Increased photoconductivity lifetime in GaAs nanowires by controlled *n*-type and *p*-type doping. *ACS Nano* 2016, 10, 4219–4227.

81. Parkinson, P., Joyce, H. J., Gao, Q., Tan, H. H., Zhang, X., Zou, J., Jagadish, C., Herz, L. M., Johnston, M. B., Carrier lifetime and mobility enhancement in nearly defect-free core–shell nanowires measured using time-resolved terahertz spectroscopy. *Nano Letters* 2009, 9, 3349–3353.

82. Nelson, R. J., Sobers, R. G., Minority-carrier lifetimes and internal quantum efficiency of surface-free GaAs. *Journal of Applied Physics* 1978, 49, 6103–6108.

83. Kayes, B. M., Atwater, H. A., Lewis, N. S., Comparison of the device physics principles of planar and radial p-n junction nanorod solar cells. *Journal of Applied Physics* 2005, 97, 114302.

84. Mariani, G., Scofield, A. C., Hung, C.-H., Huffaker, D. L., GaAs nanopillar-array solar cells employing in situ surface passivation. *Nature Communication* 2013, 4, 1497.

85. Sandhu, S., Fan, S., Current–voltage enhancement of a single coaxial nanowire solar cell. *ACS Photonics* 2015, 2, 1698–1704.

86. Ko, W. S., Tran, T.-T. D., Bhattacharya, I., Ng, K. W., Sun, H., Chang-Hasnain, C., Illumination angle insensitive single indium phosphide tapered nanopillar solar cell. *Nano Letters* 2015, 15, 4961–4967.

87. Yao, M., Huang, N., Cong, S., Chi, C.-Y., Seyedi, M. A., Lin, Y.-T., Cao, Y., Povinelli, M. L., Dapkus, P. D., Zhou, C., GaAs nanowire array solar cells with axial p–i–n junctions. *Nano Letters* 2014, 14, 3293–3303.

88. Yao, M., Cong, S., Arab, S., Huang, N., Povinelli, M. L., Cronin, S. B., Dapkus, P. D., Zhou, C., Tandem solar cells using GaAs nanowires on Si: Design, fabrication, and observation of voltage addition. *Nano Letters* 2015, 15, 7217–7224.

89. Li, K., Ng, K. W., Tran, T.-T. D., Sun, H., Lu, F., Chang-Hasnain, C. J., Wurtzite-phased InP micropillars grown on silicon with low surface recombination velocity. *Nano Letters* 2015, 15, 7189–7198.

90. Chang, C. H., Chen, L. Y., Huang, L. C., Wang, Y. T., Lu, T. C., Huang, J. J., Effects of strains and defects on the internal quantum efficiency of InGaN/GaN nanorod light emitting diodes. *IEEE Journal of Quantum Electronics* 2012, 48, 551–556.

91. Sun, Y. H., Cheng, Y. W., Wang, S. C., Huang, Y. Y., Chang, C. H., Yang, S. C., Chen, L. Y., Ke, M. Y., Li, C. K., Wu, Y. R., Huang, J., Optical properties of the partially strain relaxed InGaN/GaN light-emitting diodes induced by p-type GaN surface texturing. *IEEE Electron Device Letters* 2011, 32, 182–184.

92. Nguyen, H. P. T., Zhang, S., Connie, A. T., Kibria, M. G., Wang, Q., Shih, I., Mi, Z., Breaking the carrier injection bottleneck of phosphor-free nanowire white light-emitting diodes. *Nano Letters* 2013, 13, 5437–5442.

93. Sun, H., Ren, F., Ng, K. W., Tran, T.-T. D., Li, K., Chang-Hasnain, C. J., Nanopillar lasers directly grown on silicon with heterostructure surface passivation. *ACS Nano* 2014, 8, 6833–6839.

94. Han, S.-H., Lee, D.-Y., Lee, S.-J., Cho, C.-Y., Kwon, M.-K., Lee, S. P., Noh, D. Y., Kim, D.-J., Kim, Y. C., Park, S.-J., Effect of electron blocking layer on efficiency droop in InGaN/GaN multiple quantum well light-emitting diodes. *Applied Physics Letters* 2009, 94, 231123.

95. Wang, C. H., Ke, C. C., Lee, C. Y., Chang, S. P., Chang, W. T., Li, J. C., Li, Z. Y., Yang, H. C., Kuo, H. C., Lu, T. C., Wang, S. C., Hole injection and efficiency droop improvement in InGaN/GaN light-emitting diodes by band-engineered electron blocking layer. *Applied Physics Letters* 2010, 97, 261103.

96. Sadaf, S. M., Ra, Y. H., Nguyen, H. P. T., Djavid, M., Mi, Z., Alternating-current InGaN/GaN tunnel junction nanowire white-light emitting diodes. *Nano Letters* 2015, 15, 6696–6701.

97. Zhao, S., Connie, A. T., Dastjerdi, M. H. T., Kong, X. H., Wang, Q., Djavid, M., Sadaf, S., Liu, X. D., Shih, I., Guo, H., Mi, Z., Aluminum nitride nanowire light emitting diodes: Breaking the fundamental bottleneck of deep ultraviolet light sources. *Scientific Reports* 2015, 5, 8332.

98. LaPierre, R. R., Chia, A. C. E., Gibson, S. J., Haapamaki, C. M., Boulanger, J., Yee, R., Kuyanov, P., Zhang, J., Tajik, N., Jewell, N., Rahman, K. M. A., III–V nanowire photovoltaics: Review of design for high efficiency. *Physica Status Solidi (RRL)–Rapid Research Letters* 2013, 7, 815–830.

99. Yunyan, Z., Jiang, W., Martin, A., Huiyun, L., III–V nanowires and nanowire optoelectronic devices. *Journal of Physics D: Applied Physics* 2015, 48, 463001.

Chapter 2

Molecular Beam Epitaxial Growth of GaN Nanocolumns and Related Nanocolumn Emitters

Katsumi Kishino[a] and Hiroto Sekiguchi[b]

[a]*Sophia University, 7-1 Kioicho Chiyodaku, Tokyo 102-8554,* Japan
[b]*Toyohashi University of Technology, 1-1 Hibarigaoka, Tempaku-cho, Toyohashi, Aichi 441-8580, Japan*

kishino@sophia.ac.jp, sekiguchi@ee.tut.ac.jp

GaN nanocolumns, which are one-dimensional nanocrystals with diameters of less than several hundred nanometers, have been self-organized by radio-frequency-plasma-assisted molecular beam epitaxy (RF-MBE) and subsequently extensively studied. In this chapter, studies on GaN-based nanocolumns performed at Sophia University are systematically reviewed. These studies have addressed the first self-organization of nanocolumns, visible-light-emitting InGaN-based nanocolumns, the nanocrystal effects of such nanocolumns, controlled growth of regularly arranged nanocolumns by MBE, and application of GaN nanocolumns to nanocolumn light emitters. Achievements in research on GaN-based nanocolumns by other groups are described in the subsequent chapters.

Novel Compound Semiconductor Nanowires: Materials, Devices, and Applications
Edited by Fumitaro Ishikawa and Irina Buyanova

ISBN 978-981-4745-76-5 (Hardcover), 978-1-315-36440-7 (eBook)
www.panstanford.com

2.1 Introduction

GaN-based nanocolumns, which have high crystalline quality because of their dislocation-free nature, high light extraction efficiency, and strain relaxation, have attracted considerable attention among researchers. They are also referred to as nanowires, nanorods, and nanopillars [1]. GaN nanocolumns were first self-organized on (0001) sapphire substrates, and GaN/AlGaN nano-heterojunctions with GaN multiple quantum disks (MQDs) have been integrated in self-organized GaN nanocolumns. GaN nanocolumns have also been self-organized on (111) Si substrates. Using self-organized nanocolumns, light-emitting diodes (LEDs) on Si have been operated in the blue to red and ultraviolet regions. Random and spontaneous nucleation in the self-organization process of nanocolumns introduces fluctuations in their diameter and position, resulting in variation in the In composition among InGaN-based nanocolumns. Thus, we have frequently observed near-field images of nanocolumn LEDs with multicolor emissions, in which spots of various colors, including red, green, blue, and yellow, have been observed in the vicinity of each other within a micrometer-scale area.

A technique for selective-area growth (SAG) of nanocolumns on GaN templates has been developed to achieve precise control in the size and position of nanocolumns, resulting in uniform arrays of GaN nanocolumns. A novel technology to control the emission color of triangular-lattice nanocolumn arrays from blue to red based on nanocolumn parameters (diameter and period) has also been developed. This technology has been used to demonstrate monolithic integration of InGaN-based LEDs with different emission colors. Green-light nanocolumn LEDs based on InGaN-based nanocolumn arrays have been fabricated. The periodic arrangement of nanocolumns induces a photonic crystal effect. Light diffraction at the band edge of the nanocolumn photonic crystal provides highly directional beam radiation from InGaN-based LEDs. Nanocolumn photonic crystals have also been used in surface-emitting lasers based on a two-dimensional distributed feedback (2D-DFB) scheme, resulting in optically pumped blue to green lasers.

In this chapter, the nanocolumn technologies described above and developed over the last two decades are described.

2.2 Self-Organized GaN Nanocolumns

2.2.1 First Self-Organization of GaN Nanocolumns

GaN nanocolumns were self-organized on c-Al_2O_3 substrates for the first time by means of radio-frequency-plasma-assisted molecular beam epitaxy (RF-MBE). The experimental results were reported at the *23rd International Symposium on Compound Semiconductors* in 1996 [2] and subsequently published in the April 15 issue of the *Japanese Journal of Applied Physics* in 1997 [3] and the *Journal of Crystal Growth* in 1998 [4]. These types of nanocolumns stand up on the substrate. The synthesis of tangled GaN nanorods (nanowires) by means of a carbon nanotube-confined reaction was reported in the August 29 issue in 1997 [5], and the fabrication of self-organized GaN nanocolumns fabricated on (111) Si substrates was reported in 1998 [6]. The first achievements in the development of self-organized GaN nanocolumns, described in full in refs. 3 and 4, are summarized in this chapter.

The growth process began by heating an Al_2O_3 substrate to 900°C in a vacuum for 30 min for the purpose of thermal cleaning. This was followed by surface nitridation at 500°C at a nitrogen flow rate (Q_{N2}) of 2 sccm and a radio-frequency (RF) input power (P_{in}) of 350 W. Subsequently, an AlN buffer layer (BL) with a thickness ($d_{AlN\text{-}BL}$) of 2–5 nm was deposited, with a streaky reflection high-energy electron diffraction (RHEED) pattern overlapping a spotty wurtzite pattern. After the substrate temperature was raised to 800°C, GaN nanocolumns were grown by changing Q_{N2} from 1 to 3 sccm (P_{in} = 450 W) at a constant Ga supply and Ga effusion cell temperature (T_{Ga}) of 1122°C (Experiment 1) [3] and by changing T_{Ga} from 1044 to 1170°C at a constant Q_{N2} of 6 sccm (Experiment 2) [4]. Figure 2.1a shows top-view and cross-sectional scanning electronic microscopy (SEM) images of a GaN nanocolumn grown in Experiment 1. Uniform upstanding one-dimensional GaN nanocrystals (GaN nanocolumns) were self-assembled without any patterning or etching of the underlying substrate.

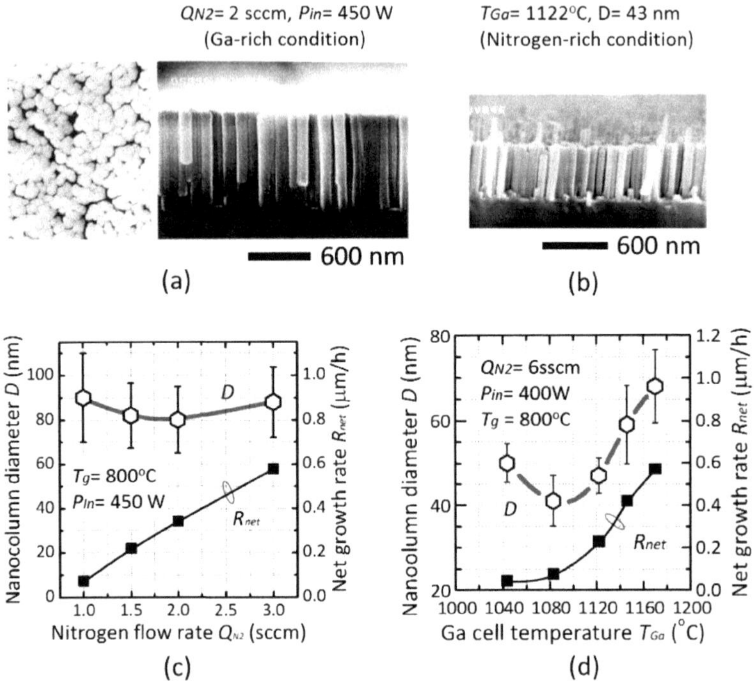

Figure 2.1 (a) Top-view and cross-sectional SEM images of self-assembled GaN nanocolumns grown under Ga-rich conditions, (b) bird's-eye-view SEM image of self-assembled GaN nanocolumns under nitrogen-rich conditions, (c) nanocolumn diameter and net growth rate versus nitrogen flow rate (Ga-rich growth conditions, $d_{\text{AlN-BL}}$ = 4.9 nm), and (d) nanocolumn diameter and net growth rate versus Ga cell temperature (nitrogen-rich conditions) (after refs. 3 and 4).

GaN nanocolumns were obtained for all of the samples in Experiments 1 and 2. The net growth rate (R_{net}), which corresponds to the total volume of GaN nanocolumns grown, was defined by the following equation:

$$R_{\text{net}} = D_{\text{NC}} \times R \times \pi\left(\frac{D}{2}\right)^2, \quad (2.1)$$

where D_{NC} is the nanocolumn density, R is the growth rate determined from the nanocolumn height, and D is the average nanocolumn diameter. The R_{net} values determined in Experiments 1 and 2 are shown in Figs. 2.1c,d to increase monotonically with increasing Q_{N2} and T_{Ga}, respectively. These trends indicate that the growth rate was limited by the nitrogen supply, i.e., that growth occurred under Ga-rich conditions in Experiment 1 [3]. In contrast, in Experiment 2, the growth was performed under nitrogen-rich conditions. The substrate temperature was sufficiently high that excess Ga adatoms were re-evaporated from the growth surface. The GaN film was grown at T_g = 760°C. In this experiment, the Ga beam flux was calibrated with respect to the growth rates of the GaN film, which were 1.38 and 0.75 μm/h for T_{Ga} = 1120 and 1080°C, respectively. From the value of R_{net} (0.34 μm/h) at 2 sccm, as shown in Fig. 2.1c, it was estimated that approximately 75% of the Ga flux supplied at T_{Ga} = 1122°C was desorbed in the growth of nanocolumns under Ga-rich conditions.

In Experiment 1, the nanocolumn diameter varied over a relatively large range of 80–90 nm for a Q_{N2} range of 1.0–3.0 sccm. However, with lowering of P_{in} from 450 to 250 W at 2 sccm, which decreased the generation of atomic nitrogen, D decreased monotonically from 80 to 53 nm [3], probably because of suppression of lateral growth of the nanocolumns. Note that the nanocolumn density (D_{NC}) and growth rate (R) also decreased monotonically.

For the nitrogen-rich growth of nanocolumns conducted in Experiment 2 [4], the D value was minimized at 40–45 nm at T_{Ga} = 1083°C. Figure 2.2 shows D_{NC} as a function of R_{net}. The nanocolumn density was proportional to the density of the nucleation sites in the initial growth. At a high net growth rate, the GaN islands come into contact with each other, side by side, because of the increased nucleation density, resulting in saturation in D_{NC}, as shown in Fig. 2.2. A circular area occupied by one nanocolumn was calculated from D_{NC} to be 82–89 nm in diameter. In this region, randomness in the nucleation position produces coalescence between the GaN islands, increasing the nanocolumn diameter. At the same time, when R_{net} is low, the nanocolumns grow independently, and D_{NC} increases monotonically with increasing R_{net}, as shown in Fig. 2.2. The abrupt decrease that occurred in D_{NC} in Experiment 2 above an R_{net} of approximately 0.25 was caused by coalescence of the GaN nanocolumns.

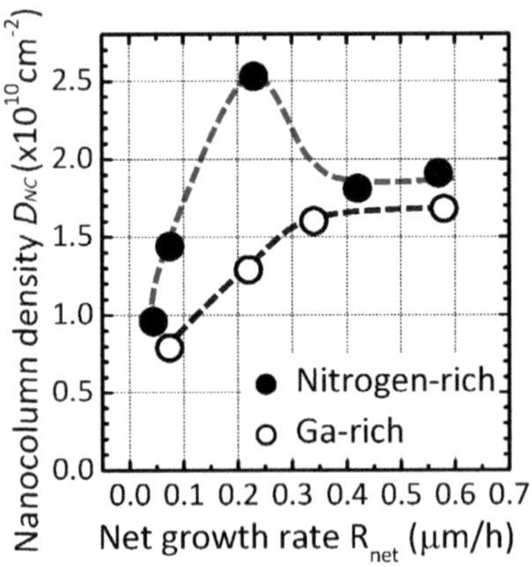

Figure 2.2 Nanocolumn density as a function of net growth rate.

It was noted in ref. 3 that the AlN-BL plays an important role in the growth of GaN nanocolumns. The growth rate of the AlN was 175 nm/h [3]. Nanocolumns were obtained for a growth time of 40 s, but the morphology of the GaN was somewhat film-like for a thin AlN-BL grown for 10 s. The role of the AlN buffer for the nanocolumn growth is discussed in the next section.

Based on the nanocolumn growth experiment described above, GaN (6 nm)/$Al_{0.18}Ga_{0.82}N$ (9 nm) MQDs were integrated into the top region of the self-organized GaN nanocolumns at a height of 320 nm, as shown in Fig. 2.3a [4]. GaN disks with a diameter of 46 nm and a thickness of 6 nm were prepared. The photoluminescence (PL) spectra of the MQD and GaN nanocolumn samples were evaluated (see Fig. 2.3b). The PL peak wavelength of the MQD sample exhibited a blueshift, and the spectrum was broadened as a result of overlapping with the spectrum from the underlying GaN nanocolumn. This experiment was the first trial of GaN-based nano-heterojunction fabrication.

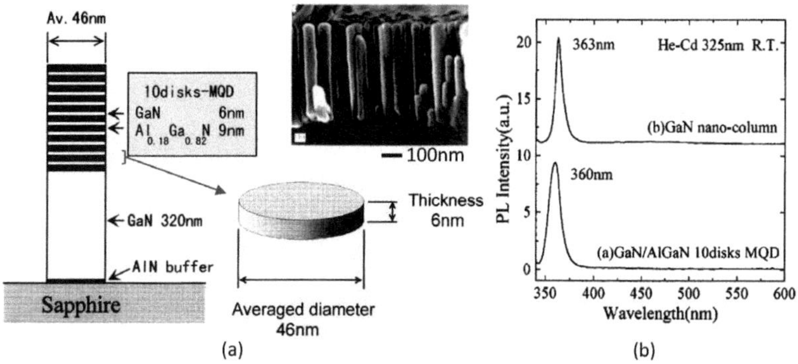

Figure 2.3 (a) Schematic illustration of 10 GaN/$Al_{0.18}Ga_{0.82}N$ MQDs and cross-sectional SEM image of the MQD structure, and (b) PL spectra of the MQDs and GaN nanocolumns (after ref. 4).

2.2.2 Growth of GaN Nanocolumns on (0001) Sapphire Substrates with AlN Buffer Layers (AlN-BLs)

In this section, the RF-MBE growth behavior of GaN nanocolumns on (0001) sapphire as a function of $d_{\text{AlN-BL}}$ is discussed in the context of ref. 7. The value of $d_{\text{AlN-BL}}$ was varied from 1.8 to 8.2 nm [7]. The surface of sapphire substrate was thermally cleaned at 950°C. This was followed by surface nitridation at ~760°C under irradiation of plasma nitrogen. AlN-BLs were grown at approximately 850°C by alternately supplying Al and nitrogen beams, i.e., in migration-enhanced epitaxy (MEE) mode. The AlN-BL thickness was controlled by the total Al irradiation time during the MEE process. GaN nanocolumns were grown at approximately 950°C for 1 h under a Ga-beam-equivalent pressure (Ga^{BEP}) of 4.1×10^{-4} Pa at $Q_{N2} = 1.6$ sccm and $P_{in} = 450$ W. Figure 2.4a shows a cross-sectional transmission electron microscopy (TEM) bright-field image of GaN nanocolumns grown on a 3.9 nm-thick AlN-BL. No threading dislocations were observed inside the nanocolumns with the flat tops. The nanocolumn diameter, which did not change from the bottom to the top, was typically 50 nm. The AlN-BL was observed at the bottom of the nanocolumns (see Fig. 2.4a). Figure 2.4b shows that GaN nanocolumns grew on the edges of the AlN islands. The boundary between the nanocolumn and the substrate surface was flat;

therefore, the nanocolumn grew directly on the sapphire, not on the AlN islands.

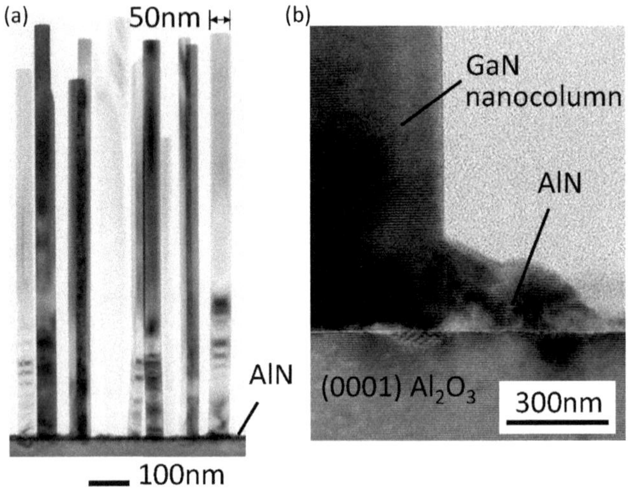

Figure 2.4 (a) Cross-sectional TEM image of GaN nanocolumns on (0001) sapphire substrate with AlN-BL, and (b) magnified TEM image at the bottom part of nanocolumn. The nanocolumn was grown at the side edge of an AlN island (after ref. 7).

We examined top-view SEM images of GaN nanocolumns for different $d_{\mathrm{AlN\text{-}BL}}$ values to evaluate D and D_{NC} as a function of $d_{\mathrm{AlN\text{-}BL}}$. The $d_{\mathrm{AlN\text{-}BL}}$ values fell into three groups within the range of 1 to 3 (see Fig. 2.5d). When the AlN buffer layer was thin, i.e., less than 3.2 nm (group 1), AlN islands were sparsely grown, providing wide and irregular spaces between the islands. The growth of GaN nanocolumns started from the edges of the AlN islands, increasing their diameters, and resulting in nanocolumns with a variety of complex polygonal and network shapes (see Fig. 2.5a). As $d_{\mathrm{AlN\text{-}BL}}$ increased, D decreased from 150 to 80 nm, and D_{NC} increased. For group 2, the nanocolumns were partially placed individually, and D remained in the range of 80–90 nm. Room-temperature PL spectra of GaN nanocolumns with various $d_{\mathrm{AlN\text{-}BL}}$ values were compared with those of GaN bulk crystal with a dislocation density of 6–8 × 10^6 cm^{-2} that was grown by hydride vapor phase epitaxy (HVPE). The maximum PL intensity for group 2, which occurred at $d_{\mathrm{AL\text{-}BL}}$ = 4.6 nm, was four times stronger than that of the HVPE-GaN. The full width at half maximum (FWHM) of the PL spectrum was approximately 48 meV for groups 1 and 2, although it increased dramatically in the group 3 region with increasing $d_{AlN\text{-}BL}$. For group 3, the spaces between the AlN islands became very small, which resulted in suppressed nucleation of the nanocolumns, and thus, both D and D_{NC} decreased monotonically.

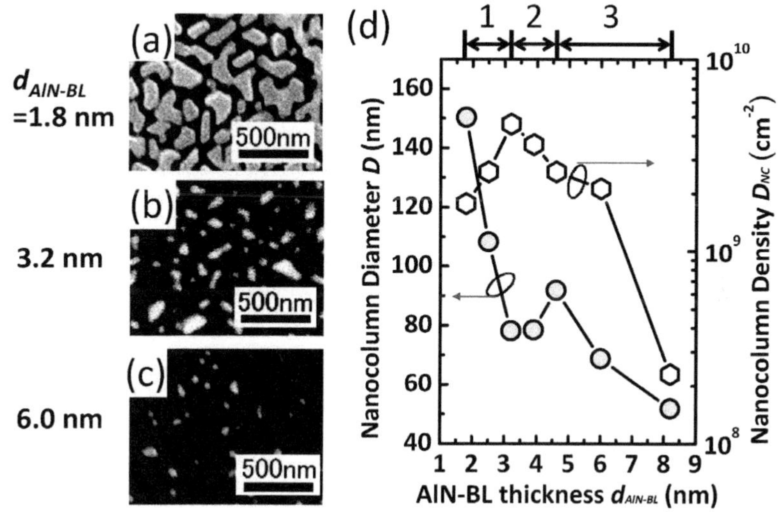

Figure 2.5 Surface SEM images of GaN nanocolumns grown on sapphire by RF-MBE for AlN thicknesses of (a) 1.8 nm, (b) 3.2 nm, and (c) 6.0 nm; and (d) nanocolumn diameter and density as a function of $d_{AlN\text{-}BL}$ (after ref. 7).

2.2.3 AlN-BL Effect on Crystal Quality of Self-Assembled GaN Nanocolumns on (111) Si Substrates

GaN nanocolumns were self-organized on (111) Si substrates by RF-MBE [8]. The Si substrate was first thermally cleaned at 570°C in a vacuum. Subsequently, at the same temperature, a Ga beam was supplied to the Si surface for 25 s to make Ga droplets, and the Ga droplets were nitrided for 60 s under the irradiation of RF-plasma-excited active nitrogen (Q_{N2} = 0.4 sccm, P_{in} = 450 W). This process was expected to produce dot-like GaN nucleation sites. The substrate temperature was raised to 885–935°C, and GaN nanocolumns were grown for 60 min at a Ga^{BEP} of 5.7 × 10^{-4} Pa, Q_{N2} = 3.0 sccm, and P_{in} = 450 W. The maximum PL intensity of the GaN nanocolumns, which occurred at 906°C, was 1.8 times stronger than that of the HVPE-GaN bulk crystal.

Film-like GaN was grown at a low temperature of T_g < 885°C. The growth temperature range of 885–935°C was extremely high for the growth of GaN by RF-MBE. As T_g increased, Ga adatoms were substantially desorbed from the growth surface, changing the growth stoichiometry into a nitrogen-rich condition, and the average diameter of nanocolumns decreased. The trend observed for D was similar to that observed in a previous experiment (see Fig. 2.1d), in which the Ga supply decreased as T_{Ga} decreased. Based on SEM observations of grown GaN nanocolumns, D was determined as a function of T_g, as shown in Fig. 2.6a. We observed that D decreased dramatically with increasing T_g and reached a minimum of 40 nm at 922°C. This minimum value for D was similar to that shown in Fig. 2.1d. However, in this experiment, the thin nanocolumns were slanted (see Fig. 2.6c) and had a low nanocolumn density. It is noteworthy that high-density, upstanding, thin (D = 40–50 nm) nanocolumns were prepared on sapphire with an AlN-BL (see Figs. 2.1b,d), which suggests that the use of an AlN-BL was effective in improving nanocolumn growth on Si.

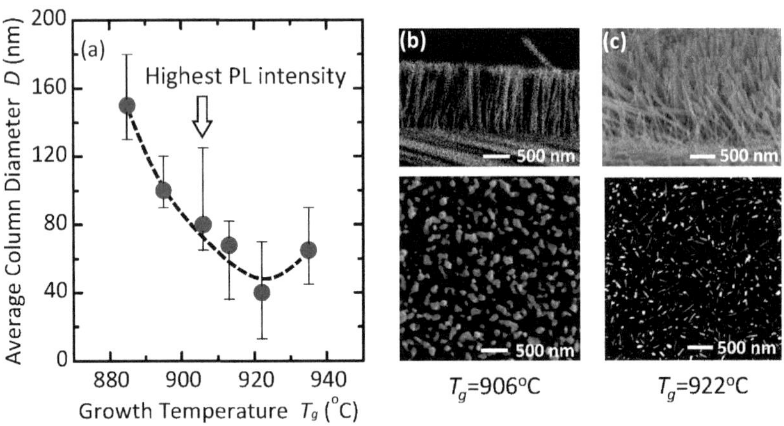

Figure 2.6 (a) Average nanocolumn diameter of GaN nanocolumns on (111) Si substrates as a function of growth temperature, and cross-sectional and top-view SEM images of GaN nanocolumns at (b) T_g = 906°C and (c) T_g = 922°C (after ref. 8).

GaN nanocolumns were prepared on n-(111) Si substrates with or without AlN-BLs (not published), where GaN dots were not utilized in this experiment for the case without AlN-BL. In the growth, the substrate surfaces were thermally cleaned at approximately 960°C. For some samples, AlN-BLs were prepared on Si at approximately 800°C using MEE mode and GaN nanocolumns were grown on them at various T_g from 915 to 970°C. Figures 2.7a,b show examples of top-view and bird's-eye-view SEM images of GaN nanocolumns with and without AlN-BLs, i.e., α (T_g = 920°C) and β (T_g = 915°C), respectively, and Fig. 2.7c shows D_{NC} as a function of T_g. When GaN nanocolumns were grown directly on Si without AlN-BLs and GaN dots, D_{NC} decreased dramatically, from 6.4×10^9 to 1.7×10^8 cm^{-2}, for a slight increase in T_g of 5°C, from 915 to 920°C, and the nanocolumns were slanted (see Fig. 2.7b). In contrast, the nanocolumns grown on the AlN-BL stood straight up (see Fig. 2.7a) and were nearly uniform in diameter (D ~ 70 nm) from the bottom to the top, and the sensitivity of D to T_g was low. The nanocolumn density was nearly constant, within 0.7 to 1×10^{10} cm^{-2}, as shown in Fig. 2.7c. As the GaN nanocolumns were nucleated at the edges of the islands of the AlN-BL, the D_{NC} was determined based on the surface morphology of the AlN-BL, resulting in the independence of D and D_{NC}.

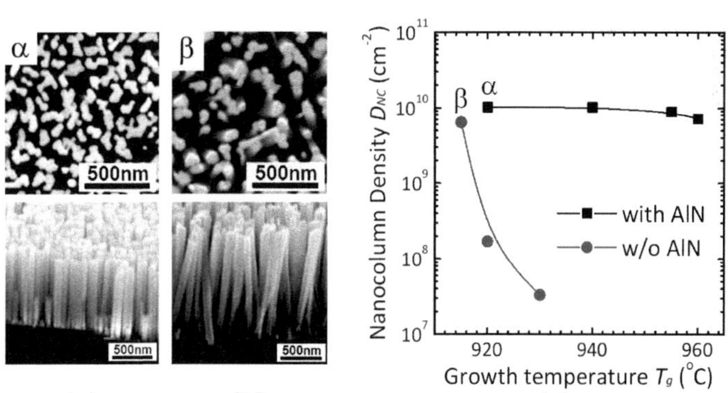

Figure 2.7 (a) Top-view and bird's-eye-view SEM images of GaN nanocolumns on Si with and without AlN-BLs, i.e., α and β, respectively, and (b) nanocolumn density as a function of growth temperature.

The PL spectra of the GaN nanocolumns and HVPE-GaN bulk crystal with a dislocation density of 6 to 8 × 10^6 cm^{-2} (the reference sample) were evaluated at room temperature (RT), as shown in Fig. 2.8. The PL peak intensities of the nanocolumns were normalized by that of the HVPE-GaN. We noticed that the PL peak intensity increased monotonically with increasing T_g and reached a maximum at 960°C. The nucleation temperature of the GaN nanocolumns was increased substantially by the use of the AlN-BL, and high-temperature growth improved the crystal quality of the nanocolumns.

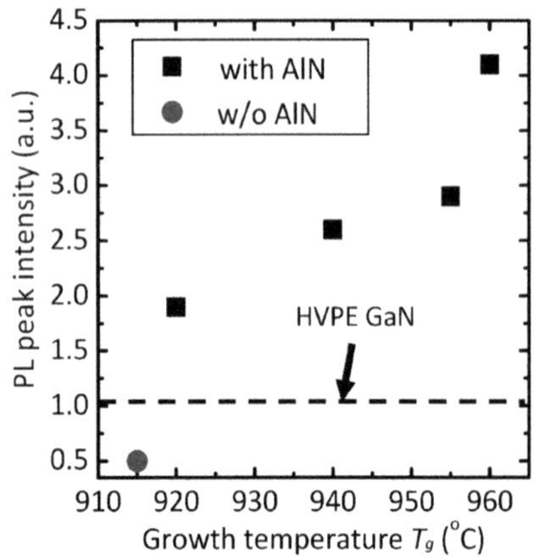

Figure 2.8 PL peak intensity as a function of growth temperature for GaN nanocolumns grown with and without AlN-BLs.

2.3 Self-Organized Nanocolumn LEDs

2.3.1 PL Emission Characteristics of InGaN Quantum-Disk Nanocolumns

Eight-pair InGaN/GaN multiple-quantum wells (MQWs) were grown at the tops of GaN nanocolumns and were capped by approximately 30 nm-thick GaN, as shown in Fig. 2.9a [9]. The InGaN QWs were sandwiched at the top and bottom surfaces by GaN barriers, and the sides of those were limited by the nanocolumn diameter, resulting in InGaN quantum disk configurations referred to as InGaN multi-quantum disks (MQDs). Carriers were confined in the MQDs. The InGaN MQDs were grown at low temperatures, in the range of 670–700°C, to incorporate In into the crystal efficiently. The thicknesses of the InGaN QD and GaN barriers were 3 and 5 nm, respectively. The change in T_g and the In/Ga beam flux ratio resulted in InGaN MQD samples emitting in the wavelength range of 436 to 614 nm, as shown in Fig. 2.9b. Wide emission spectra were observed because of the randomness of the nanocolumn self-assembly process. For example, the spectral FWHM was 77.5 nm for the 583 nm-peak-wavelength spectrum.

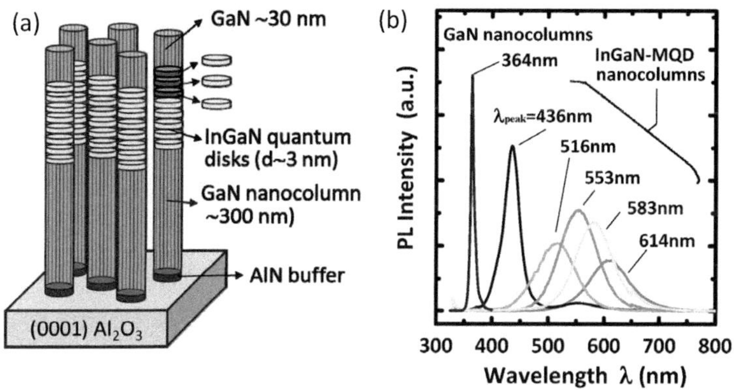

Figure 2.9 (a) Schematic diagram of InGaN/GaN multi-quantum-disk (MQD) nanocolumns and (b) emission spectra of the InGaN MQDs [9].

Figure 2.10a illustrates the dependence of the PL integrated intensity of the InGaN-MQD nanocolumns on the peak wavelength. The intensities were normalized to unity at the maximum intensity observed in Fig. 2.9b, i.e., at the peak intensity at 516 nm. These results were reported at the *6th International Conference on Nitride Semiconductors* [9]. As the figure shows, the normalized intensity increased from that of ultraviolet-emission GaN to that peaking at 516 nm and then decreased with increasing wavelength. At a red emission wavelength of 614 nm, the intensity decreased to 36% of that at a blue emission wavelength of 410 nm. The external quantum efficiency (EQE) of the InGaN LEDs, plotted using closed circles in Fig. 2.10a, was normalized by the maximum value of EQE at 486 nm. The EQE of the InGaN-film LEDs decreased dramatically with increasing wavelength from blue to green to red. At red (630 nm), the EQE is a hundredth of that at blue, because of the fundamental properties of InGaN QWs.

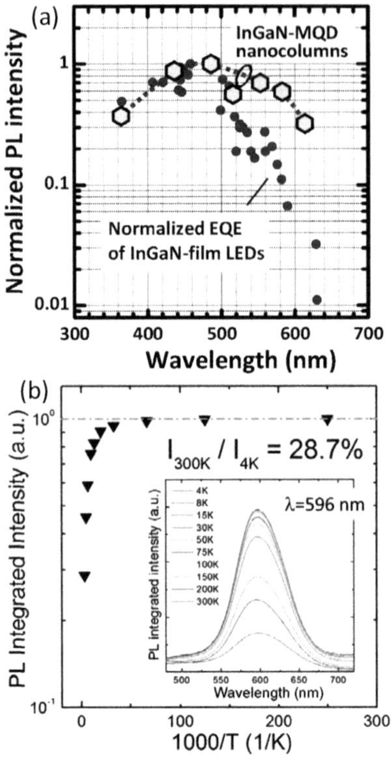

Figure 2.10 (a) Dependence of PL integrated intensity of InGaN-MQD nanocolumns on peak wavelength and (b) Arrhenius plot of InGaN SQD nanocolumns. The inset shows the PL spectra as functions of temperature [10].

Amber-emission (587 nm-wavelength) InGaN single-quantum-disk (SQD) nanocolumns were grown on (0001) Al_2O_3 substrates using an AlN-BL approximately 3 nm-thick to evaluate its internal quantum efficiency (*IQE*) [10]. At the top of the GaN nanocolumns (D = 60–100 nm and H = 1.5 μm), a 5 nm-thick InGaN SQD was prepared and was capped by 30 nm-thick GaN. The temperature dependence of the PL spectrum was evaluated over a range of 4–300 K under excitation by a 405 nm semiconductor–laser light. An Arrhenius plot of the PL-integrated intensity was prepared to estimate the *IQE*, as shown in Fig. 2.10b. The ratio of the intensity at 4 K to that at 300 K was 28.7% at an excitation power density of 0.55 kW/cm^2 [10]. This value corresponds to the *IQE*, with the low-temperature *IQE* assumed to be unity. Recently, the

IQE of amber (590–600 nm) InGaN-based ordered nanocolumns has been evaluated by a more rigorous method and found to be 17% to 22% [11]. This method for evaluating the *IQE* more accurately involved defining the PL efficiency by dividing the integrated intensity by the excitation power density and measuring the excitation power dependence of the PL efficiency at 4 and 300 K. The *IQE* was calculated as the ratio between the 300 K and 4 K PL efficiencies, maximized at each excitation power [12]. Further details are provided in refs. 11 and 12.

The *IQE* of InGaN QWs is known to decrease considerably with increasing wavelength because of the increase in the piezoelectric field and threading dislocation in the QW. GaN nanocolumns are expected to be applied in InGaN-based emitters to overcome this situation. GaN nanocolumns exhibit nanocrystal effects that contribute to improved luminous efficiency in the following ways:

(1) *Dislocation filtering effect,*
(2) *Suppressed defect generation,*
(3) *Decreased lattice strain*
(4) *Highly efficient light extraction,* and
(5) *Suppressed In fluctuation.*

The experimental results shown in Fig. 2.10a motivated us to conduct research on InGaN-based nanocolumn emitters. One of the challenges was how to create electrodes on the tops of *pn*-junction nanocolumns with diameters of several tenths of a nm. In 2004, InGaN-based nanocolumn LEDs were demonstrated [13, 14].

2.3.2 Champagne-Glass InGaN-Based Nanocolumn LEDs

Figure 2.11a shows a schematic diagram of InGaN-nanocolumn LEDs fabricated on n-type (111) Si substrates by RF-MBE [13]. The LED crystals were grown by the following procedure. First, GaN dots, which act as initial nucleation islands, were formed on the Si surface through nitridation of Ga-droplets, as described above. Si-doped n-type GaN nanocolumns were then grown on them at a temperature of approximately 900°C. The nanocolumn height (*H*) range was 0.7–1.5 μm, *D* was typically in the range of 80–100 nm, and D_{NC} was in the range of 1–2 × 10^{10} cm^{-2}. On the tops of the nanocolumns, un-doped GaN (10 nm), InGaN

(2 nm)/GaN (3 nm) eight-pair or two-pair MQDs, un-doped GaN (10 nm), and Mg-doped p-type GaN (typically 0.6 μm) were grown at low temperatures, in the range of T_g = 680–750°C. The low-temperature growth increased the incorporation of the high-vapor-pressure elements of In and Mg into the crystal. Interestingly, lowering the growth temperature with Mg doping enhanced lateral growth of the nanocolumns during the growth of the p-type GaN, creating a continuous film at the top of the LED crystals (see Fig. 2.11a).

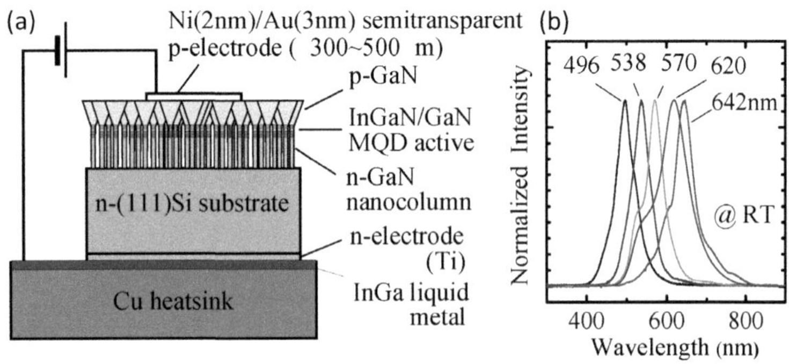

Figure 2.11 (a) Schematic diagram and (b) electroluminescence spectra of champagne-glass InGaN/GaN-MQD nanocolumn LEDs (after ref. 13).

The cross-sectional configuration of the nanocolumn crystal looked like a bundle of champagne glasses, as shown in Fig. 2.12. We therefore referred to LEDs of this type as "*Champagne-glass nanocolumn LEDs*" [8]. This structure offered the advantage of easy device processing; thus, a p-type electrode can be formed on the surface by the conventional method. Ni (2 nm)/Au (3 nm) semitransparent electrodes were formed on the as-grown surface of the LED crystal. Then, the LED was mounted on a Cu heat sink in the epi-side-up configuration with Ga-In liquid metal, as shown in Fig. 2.11a. The TEM image in Fig. 2.12b shows that no clear contrasts related to threading dislocations were observed in the p-type GaN cladding region, at least for each nanocolumn. New dislocations are not generated during the lateral growth of crystals. Thus, the corresponding underlying n-type nanocolumn should be dislocation-free.

The LED characteristics were evaluated under the direct-current injection at RT. The turn-on voltage range was 2.5–3.0 V, and the applied voltage was 5 V at a forward current of 10 mA. The doping density of n-type GaN nanocolumns was designed to be 1 × 10^{18} cm^{-3}, and capacitance–voltage measurements that were carried out for the diodes showed the net acceptor concentration (N_A-N_D) to be approximately 1 × 10^{17} cm^{-3} [13]. The electroluminescence (EL) spectra of the champagne-glass nanocolumn LED are shown in Fig. 2.11b. Visible emissions at wavelengths from 496 to 642 nm were observed.

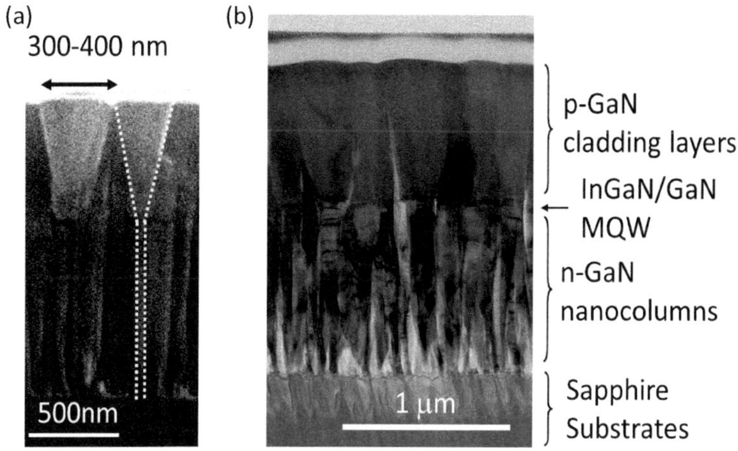

Figure 2.12 Cross-sectional SEM (a) and TEM (b) images of champagne-glass nanocolumn LED crystals.

2.3.3 Micro-EL Property of InGaN/GaN Champagne-Glass Nanocolumn LEDs

Champagne-glass InGaN/GaN-MQD nanocolumn LEDs occasionally operated in multi-color [8], although the emission could be controlled to be more uniform, as shown in Fig. 2.11b. An example of a near-field image of multi-color nanocolumn LEDs is shown in Fig. 2.13. The luminescence was observed through the semitransparent electrode under the microscope. Luminescence spots with various colors from blue to red were observed. The spot size was typically less than 2 μm. The spectrum of the nanocolumn LED was evaluated by focusing on emission areas of 500 and 2 μm diameters, i.e., by means of macroscopic- and microscopic-electroluminescence (macro-EL and micro-EL) measurements, as

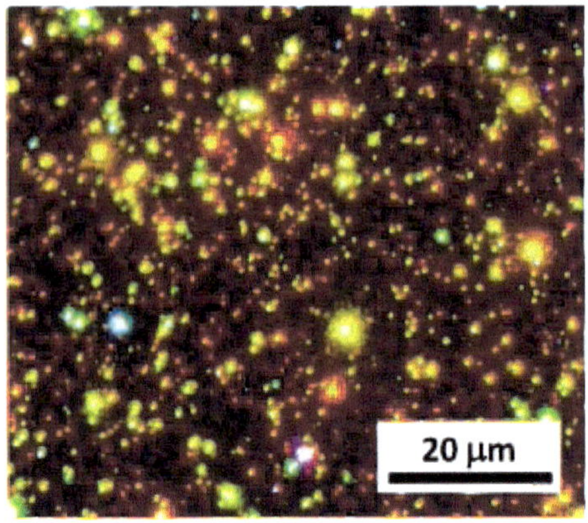

Figure 2.13 Near-field-emission image of multi-color emission champagne-glass InGaN MQD (eight-pair) nanocolumn LEDs.

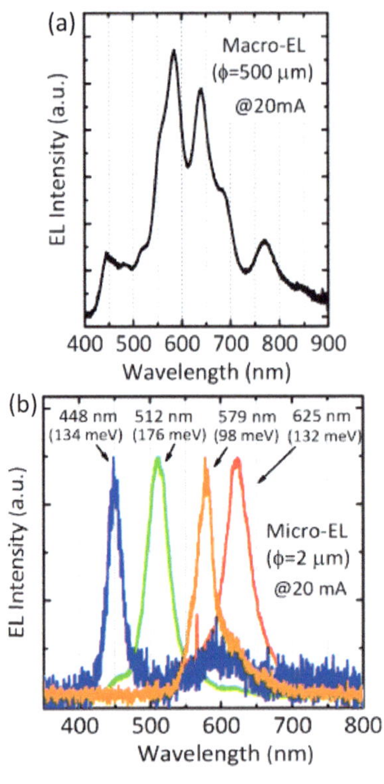

Figure 2.14 (a) Macroscopic and (b) microscopic electroluminescence spectra of multi-color nanocolumn LEDs observed for emission areas with diameters of 500 and 2 μm.

shown in Fig. 2.14. A macro-EL spectrum of nanocolumn LEDs distributed across wavelengths from 450 to 780 nm is shown in Fig. 2.14a. The spectrum for each luminescence spot in Fig. 2.13 can be measured by micro-EL measurement, as shown in Fig. 2.14b. Narrow emission spectra with wavelengths from 448 to 625 nm were distinguished from multi-color emission, and their FWHMs are shown in the figure. The FWHMs were extremely small, e.g., ~98 meV at 579 nm ($\Delta\lambda$ = 26.5 nm) and 132 meV at 625 nm ($\Delta\lambda$ = 42 nm).

For each luminescence spot, the peak emission wavelength was measured and plotted as a function of the injection current, as shown in Fig. 2.15a. In conventional InGaN-based LEDs, a spectral blueshift occurs with increasing injection current as a result of the quantum-confined Stark effect (QCSE), the screening of the polarization-induced electric field by carrier injection, and the state-filling effect [15]. The total blueshift observed in Fig. 2.15a was in the range of 5–10 nm, which is smaller than the blueshift of conventional c-plane LEDs (20–47 nm) [16–17] and reasonably close to the values for semi-polar and/or non-polar green LEDs (7–10 nm) [18, 19]. An example of the EL spectra examined in this experiment is shown for the amber emission spot in Fig. 2.15b. The EL-integrated intensity increased linearly with the injection current. The diameters of self-organized nanocolumns were typically in the range of 50 to 100 nm. Thus, this suggests that the internal field is suppressed in InGaN/GaN-based nanocolumns with diameters less than 100 nm.

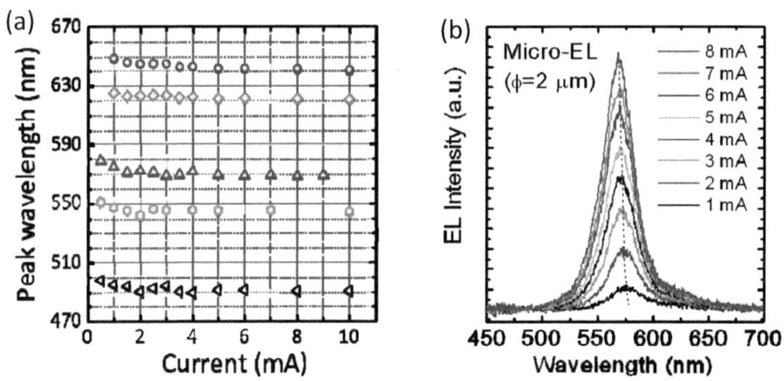

Figure 2.15 (a) Peak wavelength as a function of injection current for each emission spots in the multi-color emission, and (b) EL spectra of an amber emission spot for various currents from 1 to 8 mA.

2.3.4 Micro-PL Measurement of a Single InGaN-SQD Nanocolumn and Strain Relaxation Effect in Nanocolumns

In the case of micro-EL measurement, an emission area with a diameter of 2 μm still includes several hundred nanocolumns. The PL properties of individual nanocolumns were evaluated using the following method [20]. A self-organized array of InGaN/GaN nanocolumns without p-GaN on the top was removed using a mechanical lift-off technique. The nanocolumns removed were dispersed in ethanol, and the solution was dropped onto a quartz plate to be dried. The individual nanocolumns were characterized by means of micro-PL measurement. The self-organized nanocolumn array consisted of a 1.1 μm-thick underlying GaN nanocolumn, a 3 nm InGaN SQD, and a 55 nm GaN capping layer. The diameters of the nanocolumns were in the range of 40 to 120 nm, with a mean value of approximately 80 nm.

Figure 2.16a shows the PL spectra of a single nanocolumn under a photoexcitation energy density (I_{ex}) of 6.8 (spectrum α in the figure) and 680 μJ/cm^2 (β) at 13 K [20]. A frequency-doubled Al_2O_3:Ti laser was employed to selectively excite the InGaN SQD at an excitation wavelength of 400 nm, a pulse width of 2 ps, and a repetition rate of 80 MHz. The PL signal that was detected was integrated over the entire time domain. The photogenerated carrier (exciton) density (N_e) for spectra α and β were approximately 7×10^{17} and 7×10^{19} cm^{-3}, respectively. The main PL peak for α was ~2.339 eV, and the PL linewidth was approximately 20 meV. For the high-excitation case (β), the main PL band broadened slightly, but the blueshift of the peak energy was small (5 meV). Compared with a conventional InGaN/GaN single-quantum well film emitting at 500 nm, this shift is quite small. The blueshift for a conventional film is typically 100 meV under the same photoexcitation conditions [20]. The blueshift is induced by the screening of the piezoelectric field and the band filling of the localized states, as described above. Therefore, the results of this experiment suggest that both carrier localization and the piezoelectric polarization field are suppressed in InGaN/GaN nanocolumns. The sharp PL lines at 2.538 and 2.606 eV appeared under high photoexcitation. The reason for this is discussed in ref. 19.

The compressive strain in a 3 nm-thick $In_{0.2}Ga_{0.8}N$ SQD, calculated as a function of the nanocolumn diameter using an analytical method [21], is shown in Fig. 2.16b. InGaN SQD sandwiched by GaN at the top and bottom experienced compressive strain. The nanocolumn sidewall was unrestrained, and therefore the compressive strain decreased toward the sidewall, producing a strain distribution with a maximum at the center of the SQD [22]. As Fig. 2.16b shows, the average compressive strain in the SQD decreased rapidly in nanocolumns with diameters less than approximately 100 nm. Thus, in a thin nanocolumn with $D < 100$ nm, the piezoelectric field is effectively suppressed by the nanocolumn free-end effect.

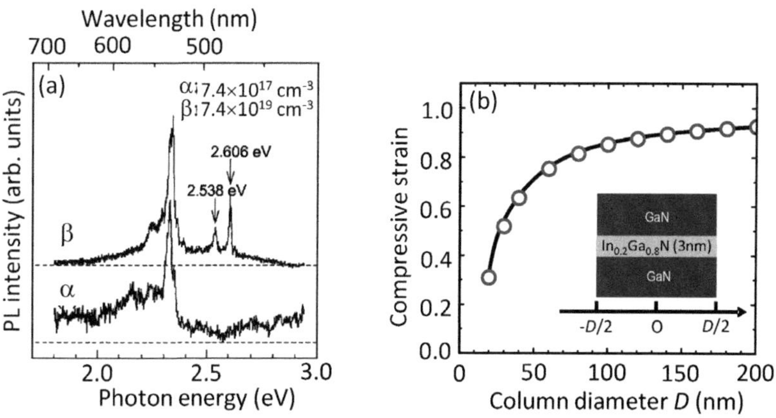

Figure 2.16 (a) PL spectra of a single InGaN/GaN nanocolumn at 13 K under α: I_{ex} = 6.8 μJ/cm^2 and under β: I_{ex} = 680 μJ/cm^2 and the corresponding photo-generated excitation densities (after [20]), and (b) calculated compressive strain averaged along the radial direction of 3 nm-thick $In_{0.2}Ga_{0.8}N$ SQD as a function of nanocolumn diameter. The calculation is performed by applying the method reported in ref. [21] to the InGaN/GaN system.

2.3.5 Flip-Chip InGaN/GaN Nanocolumn LED with One-Micron-Diameter Injection Area

Nanocolumn LEDs consist of a huge number of GaN-based nanocolumns, spontaneously nucleated during their self-organization. Thus, the positions and diameters of nanocolumns tend to be distributed randomly. Multi-color emission was therefore observed, as described above, with spots of various colors and diameters of 0.5–2 μm randomly scattered from blue to red in the length range of several tenths of micrometer. The emission spectrum for each color spot was evaluated using micro-EL measurement with a spatial resolution of approximately 1 μm, and a very narrow emission spectrum with an FWHM of 26.5 nm was observed at a wavelength of 579 nm (see Section 2.3.3). This value is small in comparison to the reported value of 50 nm at 594 nm for conventional InGaN-based LEDs [23]. Thus, a small area of current injection into the nanocolumn LEDs may reveal inherent emission characteristics of the nanocolumns. The development of the flip-chip technology of InGaN-based nanocolumn LEDs on a 50 μm-thick Au plate, illustrated in Fig. 2.17, was reported at the *36th International Symposium on Compound Semiconductors* (ISCS 2009) [24].

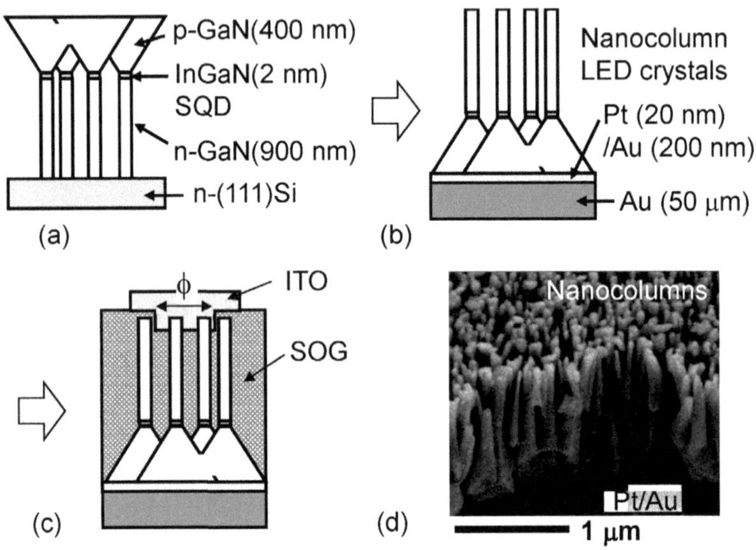

Figure 2.17 (a–c) Flip-chip of InGaN SQD nanocolumn LED crystal on a gold plate and (d) Bird's-eye-view SEM image of nanocolumn LEDs after etching Si substrates [24].

The results of that research are summarized in this section. Si substrates were removed from LED crystals, and a micro-size indium tin oxide (ITO) electrode with a diameter of ϕ was prepared on the n-type GaN nanocolumns (see Fig. 2.17c).

This configuration enabled us to excite a small number of self-organized InGaN-based nanocolumns.

InGaN SQD nanocolumn LED wafers, each of which consisted of a Si-doped n-GaN cladding layer (900 nm), an InGaN single-quantum disk (SQD) (1.5 nm), and an Mg-doped p-GaN cladding layer (400 nm), were prepared on n-type (111) Si substrates. As the first step in the flip-chip process, Pt/Au (20 nm/200 nm) electrodes were deposited on the p-GaN surface by electron beam evaporation. This was followed by electrolytic plating of a 50 μm-thick Au film. The Si was then chemically etched off using a HF/HNO_3/H_2O (6:3:2) solution, thereby transferring the nanocolumn LED structure onto the thick Au film. A bird's-eye-view SEM image of the etched-off structure is shown in Fig. 2.17d. Note that the Si substrate was removed completely and that each n-sided GaN nanocolumn was observed separately. Subsequently, spin-on glass (SOG) was embedded in the spaces between the nanocolumns, and circular apertures (ϕ = 11, 2.3, and 1.2 μm) were opened on them by means of electron beam lithography (EBL) and chemical etching in buffered HF, exposing n-sided nanocolumn tops. Finally, transparent 150 nm-thick ITO electrodes 100 μm × 50 μm in size were deposited on the circular apertures by spattering to inject the current.

The current was injected into the InGaN SQDs through the individually separated n-sided GaN nanocolumns. We were thus able to access a limited number of InGaN SQDs precisely to characterize their emission properties. In contrast, when a micron-size electrode is prepared on the p-side top surface of nanocolumn LEDs on Si, the current spread in the p-GaN cladding layer prohibits access to the desired InGaN SQD nanocolumns. The flip-chip process is the key to being able to evaluate the inherent properties of nanocolumn emission.

The EL characteristics of small-electrode InGaN/GaN nanocolumn LEDs were evaluated at room temperature under direct-current injection conditions. Figure 2.18 shows the near-field emission images (a)–(c) and EL spectra (d) of nanocolumn LEDs for emission windows of ϕ = 11, 2.4, and 1.2 μm, respectively. In the 11 μm-diameter emission area, multiple color spots approximately 1 μm in diameter, typically blue and green, were observed. Fewer color spots were observed for ϕ = 2.4 μm: just two blue spots and one green spot. Reducing the emission window area made the spectrum narrower. For ϕ = 1.2 μm, a single green spot (507 nm in wavelength) was observed, and the current was injected precisely into a single emission-color spot to clarify its emission properties [24]. Note that an extremely narrow emission spectrum was maintained for the whole range of applied voltages. The EL-FWHM at 4.0 V was 13 nm (ΔE = 63 meV), and the peak wavelength was 507 nm, as shown in Fig. 2.19. For another device, a spectrum with an EL-FWHM of 19 nm was observed at a 533 nm peak wavelength. Small EL-FWHM values observed in Figs. 2.14b and 2.19 could suggest suppressed In fluctuation in nanocolumns.

The three emission areas (i.e., ϕ = 11, 2.4, and 1.2 μm) were formed on the same ITO electrode, operating in a parallel conduction, so the injection is described as the applied voltage. However, the inset shows that the integrated EL intensity increased linearly above a voltage of 3 V (note that the turn-on voltage was 2.5 V) [24]. This means that a monotonic increase in the injection current resulted in the spectral change shown in Fig. 2.19. The peak wavelength shifted very little, from 505 to 507 nm, with increasing voltage from 3.0 to 4.0 V, which reflects the inherent nature of self-organized thin nanocolumns.

In this experiment, nanocolumn LED structures with 1 × 1 cm^2 square area were transferred onto a thick Au film, and the transfer of large-area wafers of quarter of 2 inch substrates was demonstrated. This technology for bonding wafers to a thick metal and removal of Si is expected to contribute to the development of flexible InGaN-based visible-light-emitting devices. The removal of the Si substrate eliminates light absorption by the Si and improves the light extraction efficiency as a result of the emergence of top-faced needle-like n-sided GaN, both of which contribute to the improved emission efficiency of nanocolumn LEDs.

Although this technology has not received much attention since this first trial, flip-chip bonding of white-light self-organized nanocolumn LEDs to copper substrates was recently demonstrated [25]. The technology for fabricating flip-chip InGaN/GaN well-ordered nanocolumn-array LEDs has also been developed [26, 27]. This technology employs Ti-mask selective area growth [28, 29] of GaN nanocolumns on a sputter-deposited AlN/Si (111) substrate.

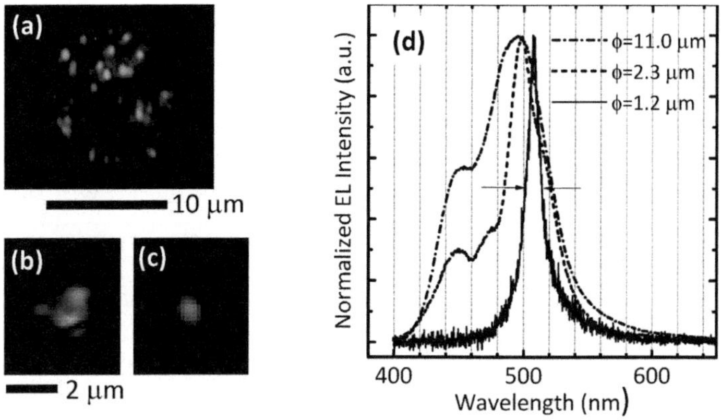

Figure 2.18 Near-field emission images of nanocolumn LEDs for emission window diameters of 11 (a), 2.4 (b), and 1.2 μm (c), and (d) EL spectra from these emission windows [24].

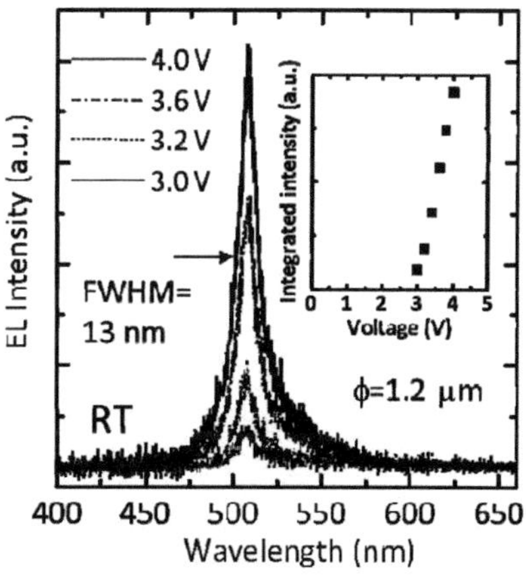

Figure 2.19 EL spectra of the ϕ = 1.2 μm emission window for different applied voltages [24]. The inset is integrated EL intensity vs. applied voltage.

2.3.6 Ultraviolet Nanocolumn LEDs on Si Using Underlying GaN Nanocolumn Template

Self-organized nanocolumns have also been utilized to fabricate ultraviolet LEDs on Si substrates. Figure 2.20 shows the device structure and spectra of ultraviolet (UV) nanocolumn LEDs, first demonstrated in 2008 [30, 31]. A GaN/$Al_{0.25}Ga_{0.75}N$ three-pair

MQD was integrated into the nanocolumns and cladded by p- and n-type AlGaN. In this experiment, GaN nanocolumn templates with heights of 150–200 nm were employed to enhance the nucleation of AlGaN nanocrystals on them, which was essential to forming the nanocolumn structure of the AlGaN. The lateral growth of nanocolumns was enhanced during the growth of the p-type AlGaN, and a p-type GaN contact layer was grown that bridged neighboring nanocolumns and resulted in a continuous film surface. Thin Ni/Au (2 nm/3–10 nm) semitransparent circular electrodes with diameters of 500 μm were prepared on the surface by electron beam evaporation through a metal mask. Because a p-GaN contact absorbs UV light, a p-AlGaN contact should be utilized instead.

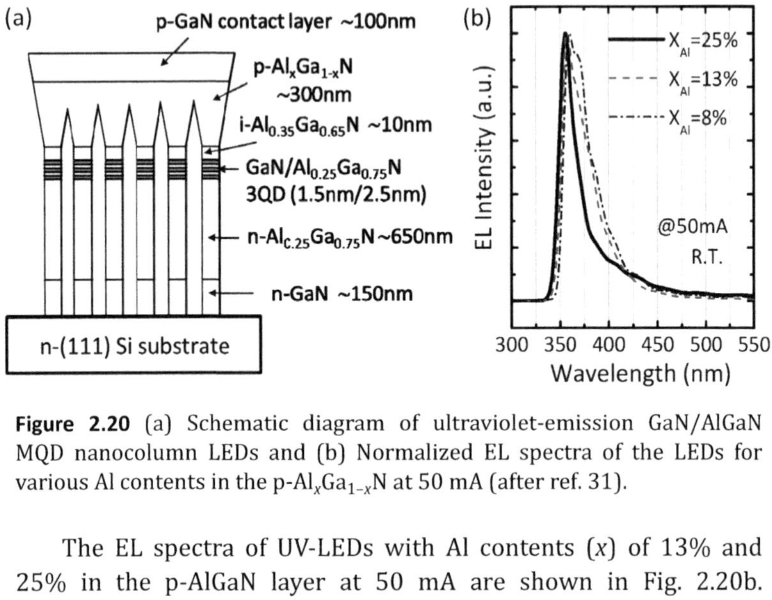

Figure 2.20 (a) Schematic diagram of ultraviolet-emission GaN/AlGaN MQD nanocolumn LEDs and (b) Normalized EL spectra of the LEDs for various Al contents in the p-Al$_x$Ga$_{1-x}$N at 50 mA (after ref. 31).

The EL spectra of UV-LEDs with Al contents (x) of 13% and 25% in the p-AlGaN layer at 50 mA are shown in Fig. 2.20b. The UV emissions at a peak wavelength of 354 nm were observed. The spectral FWHM was approximately 300 meV for the LED device with x = 8.8 and 13%, but it decreased to 200 meV for x = 25%. A high Al content has the effect of suppressing electron overflow from the active region to the p-cladding layer. Electron overflow probably caused emission of a donor–acceptor pair (DAP) in the p-cladding layer, inducing spectral broadening at a low Al content.

Self-organized nanocolumns (nanowires) for UV LEDs did not receive much subsequent attention. Very recently, pn-junction self-organized-AlN-nanowire LEDs emitting at 210 nm were fabricated on Si [32, 33]. Direct deposition of 15 nm Ni/10 nm Au metal was employed to form a p-metal semitransparent electrode on the top surfaces of independently formed nanowires [33]. However, the spacing between nanowires typically varies within a range of less than 30 nm, so it may be difficult to bridge all of the nanocolumns in the region. For both devices, the underlying GaN nanocolumn templates were used for the growth of AlGaN nanocolumns, as shown in Fig. 2.20a. The use of that template was essential to growing high-Al-content AlGaN nanocolumns, because high sticking of Al on the surface prohibits formation of the column configuration. On the GaN nano-template, AlGaN is preferentially nucleated on the tops of GaN nanocolumns, and the beam shadowing effect assists the growth of Al(Ga)N nanocolumns, as discussed in detail in ref. 34. For nano-templates approximately 100 nm in height, the beam shadowing effect functions effectively for spacings of less than 40 nm [34].

The development of UV nanocolumn emitters for practical use faces the following challenges:

(1) *Eliminating high light absorption of Si,*
(2) *Increasing light extraction efficiency,*
(3) *Lowering p-type contact resistance for AlGaN, and*
(4) *Bridging nanocolumns with a thin semi-transparent electrode.*

The flip-chip bonding and Si substrate removal techniques described in the previous section can be used to address the first and second of these challenges. With the flip-chip configuration, we can deposit a thick metal electrode on the nanocolumn top surface, which addresses the third challenge. The removal of Si using a strong acid solution is problematic, because careful and precise etching control is necessary to avoid damaging the nanocolumn region [26, 27].

2.3.7 Self-Organization of GaN Nanocolumns on Graphene/SiO_2/Si

Self-organization of high-density and upstanding GaN nanocolumns on multilayer graphene (MLG)/SiO_2 by RF-MBE was successfully accomplished using a thin AlN buffer layer, as shown in Fig. 2.21a,b [35]. MLG/SiO_2 substrates were prepared by the transfer of chemical vapor deposition (CVD) graphene onto thermally oxidized SiO_2/Si (100) substrates. TEM observations revealed no threading dislocations, stacking faults, or twinning defects in the self-organized nanocolumns (see Fig. 2.21c). The PL peak intensities of the self-organized GaN nanocolumns were 2.0 to 2.6 times higher than that of a GaN substrate grown by hydride vapor phase epitaxy. Furthermore, no yellow luminescence or zinc blende (ZB)-phase GaN emission was observed from the nanocolumns. An InGaN/GaN MQW and p-type GaN were grown on n-type GaN nanocolumns on MLG, and a single-peak PL emission was observed at a wavelength of 533 nm, as shown in Fig. 2.21d. High-density nitride p-i-n nanocolumns were fabricated on SiO_2/Si using the transferred MLG interlayer, suggesting the possibility of developing visible nanocolumn LEDs on graphene/SiO_2. Recently, this technology for growing nanocolumns on MLG was applied to the growth of GaN nanocolumns on single- and double-layer graphene/SiO_2 (fused silica).

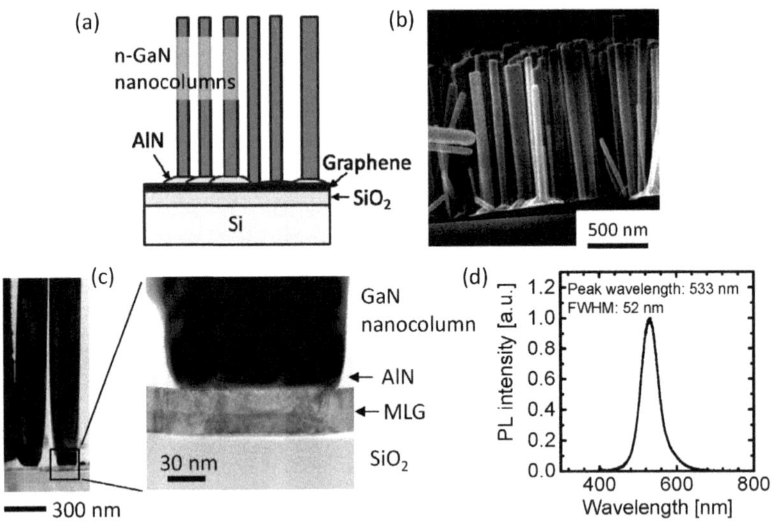

Figure 2.21 Schematic diagram (a), cross-sectional SEM view (b), and cross-sectional TEM images (c) of GaN nanocolumns grown on multilayer graphene (MLG)/SiO_2/Si substrate, and (d) photoluminescence spectrum of InGaN/GaN nanocolumns on MLG (after ref. 35).

Graphenes are thin, flexible, two-dimensional film with high optical transparency, good electrical and thermal conductivities, and good mechanical strength. Recently, graphene has been used as a crystal nucleation film for the growth of various semiconductors (InGaN-based LED [36], InAs [37], and GaAs [38] nanowires) on foreign substrates, and GaN nanocolumns have been grown on graphene/SiO_2 [35]. Graphene can be transferred easily to another functional substrate [39]. Success in growing crystalline GaN nanocrystals on graphene will expand the range of substrates for GaN-based compounds and contribute to the development of nano-devices such as LEDs, laser diodes (LDs), and solar cells, including flexible visible-range devices.

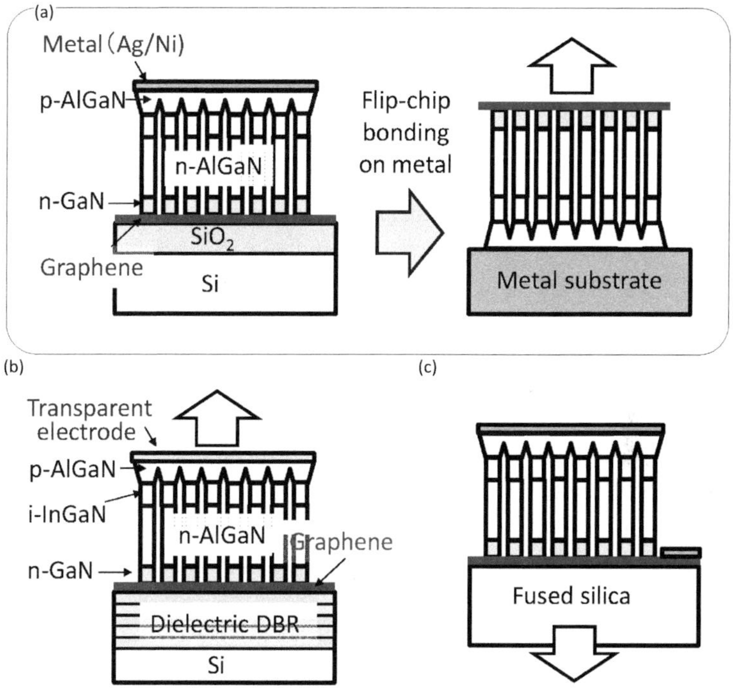

Figure 2.22 Future prospects for GaN-based nano-device structures based on graphene technology. A thin AlN buffer contributes to the growth of upstanding GaN nanocolumns on graphene [35].

Figure 2.22 illustrates schematically the future prospects for GaN-based nano-devices based on graphene technology. This technology will be used to prepare AlGaN-based nanocolumn

LED crystals on SiO_2/Si (see Fig. 2.22a). Etching the intermediate layer of SiO_2 between nanocrystals and Si permits lift-off of the Si, enabling flip-chip bonding of the nanocolumn LEDs. Thus, it is expected that the challenges facing the development of UV nanocolumn LEDs, described in the previous section, can be addressed.

If nanocolumn LEDs are fabricated on a dielectric distributed Bragg reflector (DBR) as shown in Fig. 2.22b, the light extraction upward is enhanced. A highly efficient InGaN-based visible LED can therefore be fabricated. With an additional DBR on the top, this device configuration will also contribute to the fabrication of a vertical surface-emitting laser.

Devices can be grown on a fused silica substrate with low light absorption at wavelengths down to the UV-C region, as shown in Fig. 2.22c. A transparent graphene with a low sheet resistance extracts the current laterally, and a thick metal electrode at the p-side surface functions as a reflection mirror. The deep UV-C light can be radiated from the backs of silica substrates, through some surface structures that enhance the light extraction from there, such as a patterned indented surface.

2.3.8 Effect of Be-Doping on InGaN/GaN Nanocolumn LED Structure

The effect of Be-doping on GaN nanocolumns, which was investigated using an InGaN/GaN nanocolumn LED structure grown by RF-MBE [40], is described in this section. Be was doped into the upper cladding region of nanocolumn LEDs, as shown in Fig. 2.23a. Si-doped n-type GaN nanocolumns (900 nm in thickness) were grown at 940°C on n-(111) Si substrates under N-rich conditions, followed by the growth of an InGaN SQD active region (5 nm). Be-doped GaN upper cladding layers (680 nm) with a thickness of ~550 nm were then grown at 770°C.

Be is expected to be a p-type dopant of GaN with a shallow accepter level, but p-type conductivity was not achieved in this experiment. The doping characteristics were evaluated by means of capacitance–voltage (C–V) measurement. All of the samples exhibited n-type conduction in the Be-doped layer. For a Be cell temperature of 920°C, a net donor concentration (N_D-N_A) of 5.4 × 10^{14} cm^{-3} was estimated. The low N_D-N_A value is speculated to be attributable to compensation for the residual n-type carrier by the Be doping. For p-type conduction, the growth conditions should be optimized. Co-doping with Mg may be helpful in improving the device performance, because Be doping enhances the smoothness of the top surface.

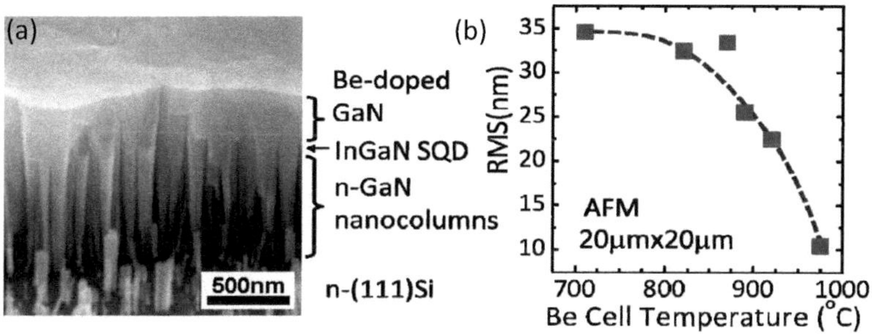

Figure 2.23 (a) Bird's-eye-view SEM images of nanocolumn LEDs with Be-doping in upper cladding region. The Be cell temperature was 920°C. (b) RMS surface roughness of nanocolumn LED structures with Be-doped GaN upper cladding region as a function of Be cell temperature (after ref. 40).

A remarkable enhancement in the lateral growth of the GaN nanocolumns was achieved by Be doping. The root mean square (RMS) surface roughness was reduced from 35 to 10 nm as the Be cell temperature increased from 710 to 975°C, as shown in Fig. 2.23b. The surface of the Be-doped region became smoother and more continuous with increasing Be doping concentration.

Free-standing GaN layers have been overgrown on self-assembled GaN nanocolumns [41], and GaN crystals have been overgrown on a Be-doped coalesced GaN nanocolumn layer. Interestingly, under some conditions, the overgrown GaN included large micrometer-scale hexagonal columnar crystals (microcolumns) [42] whose optical properties were comparable to those of HVPE-grown GaN bulk crystals.

2.4 Dislocation Suppression Effects of Nanocolumns

2.4.1 Dislocation Filtering Effect of GaN Nanocolumns as a Function of Nanocolumn Diameter

Dislocation filtering is one of the key effects exerted by nanocolumns. For nanocolumns on a GaN template/sapphire, the propagation of dislocations through nanocolumns was reported to be suppressed for nanocolumns with small diameters of less than approximately 300 nm [29, 43]. Complete filtering of the underlying dislocations was demonstrated for very thin nanocolumns with diameters of less than 50–80 nm [44]. However, the dependence of the dislocation filtering effect on D was not examined in these studies.

This section discusses a systematic investigation of the effect of nanocolumns on dislocation propagation behavior, illustrated in Fig. 2.24a, as a function of the nanocolumn diameter [45]. GaN nanocolumns with different diameters were grown on a 1.6 μm-thick GaN template prepared on Si substrates using an AlN/GaN superlattice (SL) buffer. The template had a threading dislocation density of approximately 10^{11} cm^{-2}. The average number of dislocations calculated to occur in a 200 nm-diameter area of nanocolumns was 35, but amazingly, cross-sectional TEM images showed that many of the dislocations were effectively filtered out within a region up to approximately 0.3 μm in height from the bottom of the nanocolumn (see Fig. 2.24b).

The following dislocation filtering mechanism in nanocolumns is described briefly in ref. 45. The bending of dislocations occurs when they reach oblique crystal facets during the growth of crystals, as discussed elsewhere for epitaxial lateral overgrowth (ELO) of GaN [46, 47]. A similar phenomenon is considered to occur in nanocolumns because the top crystal facets of nanocolumns tend to protrude from the Ti-mask openings at the initial stage of growth [29]. A narrow nanocolumn possesses a pyramidal top, whereas, for a nanocolumn of a larger diameter, a flat surface appears at the top, with an oblique facet at the edge. Thus, the bending of dislocations in nanocolumns is considered to begin to occur at some small diameter. In contrast, for a wide nanocolumn, the dislocations are filtered in the area close to the sidewall, but the dislocations near the centers of the nanocolumn propagate without bending (see Fig. 2.4d in ref. 45). The strain distribution in the crystal in the radial direction of the nanocolumn caused by the existence of the sidewall may be another phenomenon related to the dislocation filtering effect. The strain is considered to have a considerable effect in the case of narrow nanocolumns, i.e., for those with diameters less than ~100 nm (see Fig. 2.16b). The increased crystal strain induced by the inclusion of dislocations probably has the effect of filtering the dislocations. Further research is necessary to clarify the dislocation filtering effect.

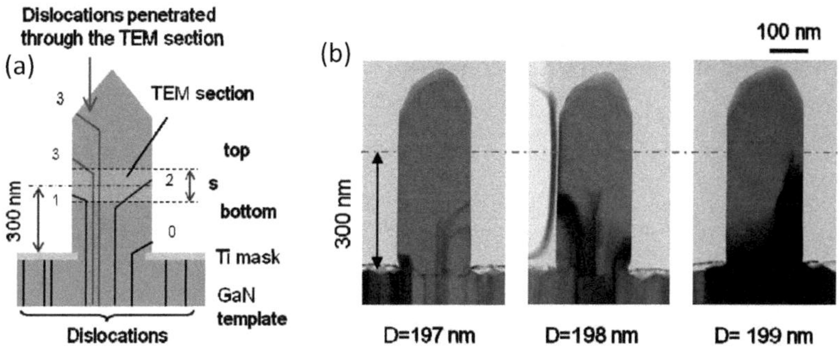

Figure 2.24 (a) Schematic diagram of behavior of dislocations propagating in a TEM section. (b) Cross-sectional bright-field STEM images of GaN nanocolumns with diameters of approximately 200 nm, observed by thinning the nanocolumns to ~100 nm in the direction of the *a*-axis (after ref. 45).

Plan-view TEM observation of nanocolumns was performed by slicing the nanocolumns horizontally to a thickness of s at a height of ~300 nm [45]. The behaviors of the dislocations propagating in the TEM section are described based on the discussion in ref. 45. As shown in Fig. 2.24a, these dislocations can be classified into four categories: Nos. 0, 1, 2, and 3, in which the dislocations were filtered out at the bottom of the nanocolumn (No. 0), bent in the bottom region of the TEM section and then terminated at the sidewall (1), reached the sidewall at an intermediate height (2), or penetrated through the TEM section (3). Figures 2.25a,b show magnified plan-view bright-field scanning transmission electron microscopy (STEM) images of nanocolumns with diameters of 254 and 205 nm, respectively.

They were observed before and after thinning from the bottom side. The estimated thicknesses were s = 150 and 100 nm. A number of dislocations in category No. 1 appeared in the TEM section for s = 150 nm that were not present for s = 100 nm, i.e., after the thickness of the TEM section was reduced by 50 nm. The dislocations in category No. 2 were terminated at the sidewall inside the TEM section. This experiment demonstrated that the dislocations were effectively filtered out at the bottom of the nanocolumns during their growth. The dislocations in category No. 3 propagated through the TEM section to the upper part of the nanocolumns, but the dislocations in category No. 3 with a lateral trajectory eventually terminated at the sidewall in the region above the TEM section as the growth of the nanocolumns proceeded.

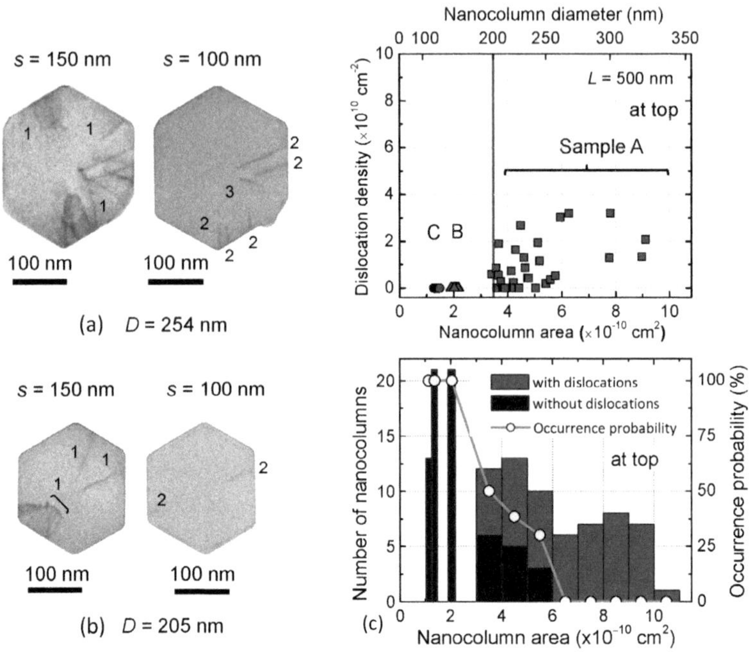

Figure 2.25 Magnified plan-view TEM images of nanocolumns with diameters of (a) D = 254 nm and (b) D = 205 nm for s = 150 and 100 nm, and (c) dislocation densities and occurrence probabilities of dislocation-free nanocolumns at the bottom and top of the TEM section, respectively (after ref. 45).

By analyzing magnified plan-view STEM images for s = 100 nm, the dislocation propagation behavior of a total of 79 nanocolumns with diameters in three different ranges: A (D = 198–324 nm), B (D = 150–159 nm), and C (D = 120–130 nm). Among the dislocations in category Nos. 1–3, only category No. 3 dislocations existed at the top of the TEM section and propagated to the upper part of the nanocolumns. The category No. 3 dislocations were counted, and the dislocation densities at the top of the TEM section were plotted as a function of the nanocolumn area A_{nc}, as shown in the upper figure in Fig. 2.25c. Assuming a regular hexagonal shape, the equivalent diameter D was calculated using the formula $A_{nc} = \sqrt{3/2D^2}$, as shown on the upper x-axis. For nanocolumns with D > 200 nm, dislocations propagated to the upper part of the nanocolumns above the TEM section, but the occurrence probability of dislocation-free nanocolumns at the top started to increase at $A_{nc} = 6 \times 10^{-10}$ cm^2 ($D \sim 250$ nm), as shown in the lower figure in Fig. 2.25c. When D was less than approximately 200 nm, the dislocation density at the top and bottom of the TEM section decreased dramatically, and thus, the nanocolumns effectively filtered the dislocations before they reached a height of approximately 300 nm. All 55 of the nanocolumns with D < 160 nm that were analyzed were dislocation-free at the top of TEM section, as shown in the lower histogram in Fig. 2.25c. Narrow nanocolumns with D < 160 nm are therefore suitable for application to nanoemitters.

2.4.2 Critical Thickness of InGaN/GaN Nanocolumns as a Function of Nanocolumn Diameter

In addition to reducing threading dislocations (TDs), it is crucial to avoid introducing misfit dislocations (MDs) into InGaN/GaN heterojunctions to obtain high emission efficiencies for InGaN-based optical devices. When the thickness of InGaN is greater than the critical layer thickness h_c, MDs are generated at the hetero-interface between InGaN and GaN by lattice relaxation, degrading the crystal quality. For a planar InGaN/GaN heterostructure, h_c decreases dramatically with increasing the In composition. For example, h_c < 10 nm for $In_{0.3}Ga_{0.7}N$/GaN. On the other hand, for InGaN/GaN nanocolumns, the free-end effect at the side wall decreases the lattice strain at the heteroboundary [21] (see Fig. 2.16b). That is, the stress tensors σ_{zz} and σ_{rz} become zero in the cylindrical coordinate system (r, ϕ, z) at the lateral end of the InGaN/GaN heterostructure in the nanocolumn, which introduces a smaller total strain energy into the nanocolumn, especially for a thin nanocolumn, compared with that of a planar material system. This increases the h_c value. Models for calculating h_c in a nanocolumn have been presented [49–52]. Ertekin et al. presented the calculation of h_c by a variational method [49], and Glas demonstrated the radius (diameter) dependence of h_c in cylindrical heterostructures [50]. Figure 2.26a shows h_c values of InGaN with various In compositions on GaN nanocolumns [48], calculated using the method described in refs. 50 and 51. When the column diameter is less than a certain critical value, h_c diverges to infinity.

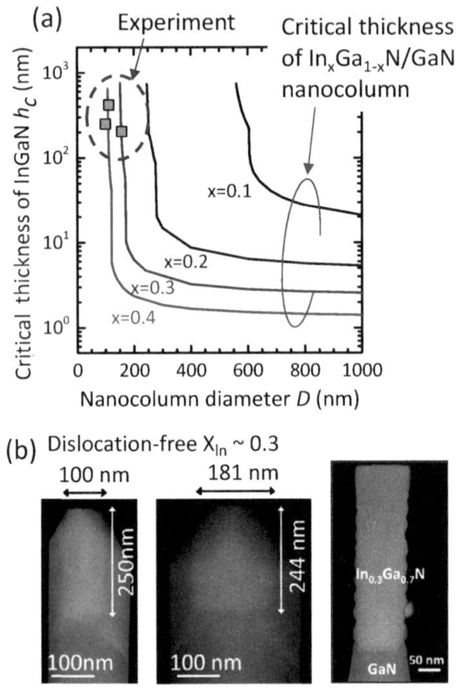

Figure 2.26 (a) Calculated critical thickness h_c of InGaN for InGaN/GaN nanocolumns, and (b) cross-sectional TEM images of InGaN grown on GaN nanocolumns.

To confirm the theoretical prediction, ~400 nm-thick $In_{0.3}Ga_{0.7}N$ nanocolumns were grown on GaN nanocolumns with heights of 500–600 nm. The nanocolumns were grown at the mask openings by the selective area growth technique using a Ti-mask [28, 29]. The Ga^{BEP}, nitrogen flow rate, and growth temperature were 3.0×10^{-4} Pa, 1.0 sccm, and 910°C, respectively [48]. The InGaN nanocolumns were grown using an In^{BEP}, Ga^{BEP}, nitrogen flow rate, and growth temperature of 3.5×10^{-4} Pa, 2.5×10^{-4} Pa, 1.65 sccm, and 720°C, respectively. Figure 2.26b shows high-angle annular dark-field (HAADF)-STEM images of $In_{0.3}Ga_{0.7}N$ GaN nanocolumns in which no MDs were observed. Thus, an extremely large critical thickness of InGaN, $h_c > 400$ nm, was observed in the nanocolumn system, although the calculated h_c for planar $In_{0.3}Ga_{0.7}N$/GaN heterojunction films is less than 10 nm. From Fig. 2.27b, we observe that two different InGaN structures were prepared by one-time crystal growth on the same substrate by changing the diameter of the underlying GaN nanocolumns; that is, on a thin GaN nanocolumn, an InGaN nanocolumn with almost the same diameter was grown axially (axial InGaN), while InGaN-InGaN core–shell structures were spontaneously formed (core–shell InGaN) on the thick GaN nanocolumns [48].

Figure 2.27a shows the horizontal distributions of the In composition, as analyzed by energy dispersive X-ray (EDX) analysis, and the HAADF-STEM images for the axial and core-shell InGaN on the underlying GaN nanocolumns with $D_{GaN} = 130$ and 248 nm, respectively. The circles and squares correspond to the axial and core–shell InGaN, respectively. For the axial InGaN, the In composition varied within the range of ± ~4% both vertically and horizontally, with an average value of ~34%. For the core–shell InGaN, the $In_{0.3}Ga_{0.7}N$ core and $In_{0.1}Ga_{0.9}N$ shell were self-assembled. Figure 2.27b shows the diameter of the InGaN core (D_{InGaN}) as a function of the diameter of the underlying GaN nanocolumn (D_{GaN}) for InGaN/InGaN core–shell structures. However, for $D_{GaN} \leq$ 120–130 nm, axial InGaN nanocolumns were grown, and D_{InGaN} increased linearly with $D_{InGaN} \simeq D_{GaN}$. Then, for $D_{GaN} >$ 120–130 nm, the InGaN structure switched to core–shell InGaN, and D_{InGaN} became almost constant at $D_0 \simeq 120$ nm. Thus, in the larger D_{GaN} range, D_{InGaN} was restricted to be less than

D_{GaN}, as a result of the minimization of the total strain energy during the crystal growth.

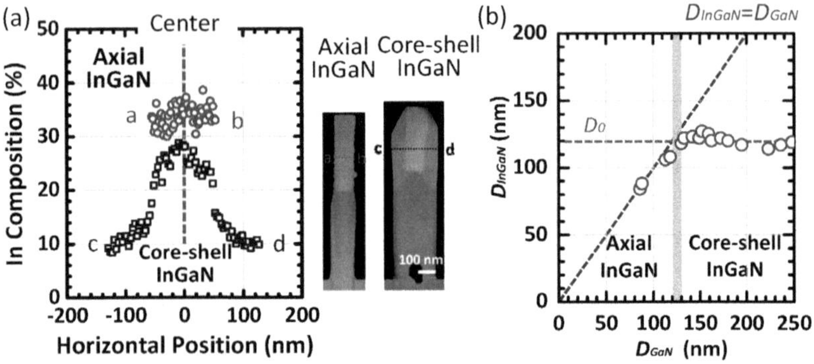

Figure 2.27 (a) Horizontal distributions of In composition analyzed by EDX measurement and HAADF-STEM images for axial InGaN (circles) and core–shell InGaN (squares), respectively, and (b) diameter of InGaN core (D_{InGaN}) as a function of the diameter of the underlying GaN nanocolumn (D_{GaN}) (after ref. 48).

2.5 Selective-Area Growth of GaN Nanocolumns

2.5.1 Development of SAG of GaN Nanocolumns

SAG of GaN nanocolumns using nano-mask patterns has contributed to the homogenization of the size and composition of GaN nanocolumns. Studies of SAG of GaN nanocolumns by RF-MBE [53, 54] and metal-organic vapor phase epitaxy (MOVPE) were reported in 2006 [55]. At that time, SAG of GaN, but not nanocrystals, was well developed for HVPE and MOVPE, employing SiO_2 SAG masking [56, 57]. Using a pulsed MOCVD process, a uniform array of GaN nanorods has been prepared on SiN_x mask-patterned substrates through SAG [55]. SAG of GaN micro-crystals by metal-organic molecular beam epitaxy (MO-MBE) using triethyl gallium [58] and by gas-source MBE using an ammonia source [59], where the gas-phase supply of ingredients contributes to SAG, has been reported. However, SAG by standard metal-source RF-MBE has been difficult [8, 60, 61], with the growth of GaN on the mask surface not completely suppressed.

The first breakthrough in the complete SAG of GaN nanocolumns on Si was achieved in growing GaN at a high temperature (966°C) by RF-MBE on nitridated Al nanodot patterns [62]. GaN nanocrystals were grown at the edge of the nitridated Al (thus, AlN) nanodots, whereas no growth of GaN occurred on the Si surface between the nanodots. However, the nanocolumn shape was insufficiently controlled because of the deformation of the Al patterns before nitridation. We have also reported on the use of the Ti-mask technique for SAG of GaN on Si [63]. The self-assembly of GaN nanocolumns was suppressed on the Ti mask, and a single nanocolumn was aligned along the one-dimensional mask opening, which had a width of 120 nm. However, it was difficult to control the nanocolumn size by the width of the mask opening because thin GaN nanocolumns were independently self-organized on the Si surface in the mask opening when the diameter of the opening was larger than approximately 100 nm.

The use of nitride-based nucleation layers was shown to be effective in achieving uniform nucleation of nanocolumns in the mask opening. Thus, on GaN templates, uniform arrays of GaN nanocolumns have been prepared using Ti-mask SAG by RF-MBE [28, 29]. SAG of nanocolumns on Si has also been achieved using AlN buffer layers covered with patterned SiN_x [64], SiO_2 [65], and Ti [26] masks. This section summarizes research on SAG of GaN nanocolumns on Si and GaN templates.

2.5.2 Selective-Area Growth of GaN Nanocolumns on Si Using Nitridated Al Metal Pattern

When we found that the nucleation of nanocolumns occurred at higher temperatures on AlN-BL, as shown in Fig. 2.7, we immediately understood how a patterned AlN on Si contributed to the development of SAG of nanocolumns on Si. However, in advance of the patterning, it is necessary to prepare an AlN film using a deposition technique such as RF-MBE, MOCVD, or high-temperature sputtering. Therefore, patterned pre-deposited Al metal layers were utilized to demonstrate SAG [60, 62]. Al can be deposited easily using an evaporation coating. Patterned Al layers were nitrided under the irradiation of RF-plasma-excited atomic nitrogen, producing AlN crystals, which functioned as nucleation sites for the SAG [60]. SAG on Si was performed using Al-disk patterns, as described in ref. 60, published in 2007.

The patterned thin Al metal layers were fabricated on the substrate surfaces through photolithography, evaporation coating of Al, and lift-off processes. Figure 2.28a shows a top-view atomic force microscope (AFM) image of the patterned Al layer, in which Al disks with a diameter of 2.8 μm and the thickness of 18 nm are arranged in a square lattice with a period of 4 μm. The substrates were loaded into the MBE chamber, and surface nitridation was performed at 860°C for 15 min. The nitridation temperature was the same as the T_g of AlN-BLs for the growth of ordinary GaN nanocolumns. After nitridation of the patterned Al metal layers, GaN nanocolumns were grown on them at 960°C for 60 min under nitrogen-rich conditions [7].

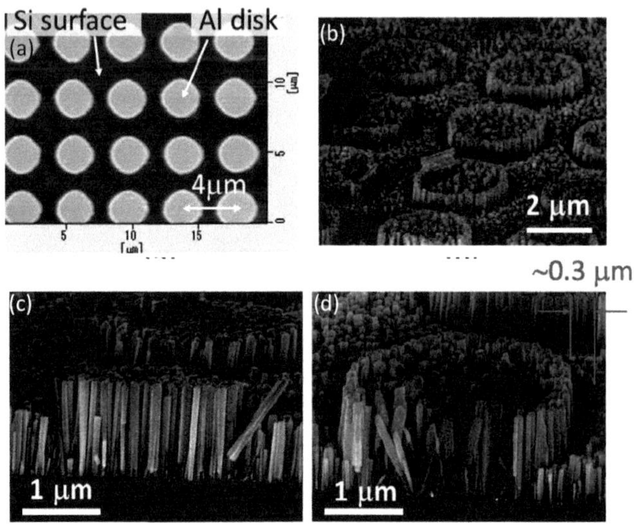

Figure 2.28 (a) AFM image of Al disk patterns on Si (diameter: 2.8 μm), (b) low-magnification bird's-eye-view SEM image of grown GaN nanocolumns on Al disks with a thickness of 18 nm, (c) and (d) high-magnification images of GaN nanocolumns grown on Al disks with thicknesses of 15 and 23 nm, respectively (after ref. 60).

Figures 2.28b,d shows bird's-eye-view SEM images of GaN nanocolumns on a Si substrate with 15 and 23 nm-thick Al disks, respectively. The nucleation of GaN on the nitrided Al disk area was enhanced, and the Al-disk patterns were transcribed to the GaN nanocolumns. Taller GaN nanocolumns, with a height of 1.2 μm, were observed on the nitrided Al disk areas than on the Si surface areas between the disks. It is noteworthy that the growth of nanocolumns was substantially suppressed in the ring area close to the Al disk edges (see Fig. 2.28b,d). The width of the ring was approximately 0.3 μm. In this region, the impinging Ga adatoms diffused on the substrate surface were absorbed to GaN nanocolumns on the Al disk areas, which were first nucleated at the initial growth stage. The crystal nucleation was delayed on the Si surface.

Upstanding nanocolumns perpendicular to the substrate were grown on the nitrided Al disks with thicknesses from 7 to 15 nm, as shown in Fig. 2.28c. This indicates that the pre-deposited nitrided Al layers functioned in the same way as the MBE-grown AlN films. For Al disk thicknesses greater than 18 nm, the GaN nanocolumns were densely produced along the contour right outside the circular Al patterns (see Fig. 2.28d). In fact, the inner diameter of the circularly arranged nanocolumns was nearly equal to the diameter of the underlying Al disk. Thus, the nanocolumns were grown at the edge of the nitrided Al disks.

On the basis of this experiment, we demonstrated the position control of GaN nanocolumn growth using Al nanodot patterns [62], as shown in Fig. 2.29. The nanohole patterns were prepared on a 2 inch n-Si(111) substrate by EBL, followed by the evaporation coating of Al thin film on them. The liftoff process left Al nanodots 85 nm in diameter and 20 nm in thickness, arranged in a triangular lattice with a 300 nm period. Each Al nanopattern consisted of 109 Al nanodots that formed a circular area with a diameter of approximately 3 μm. After the Al nanodots were nitrided to convert them to AlN nanodots, GaN nanocolumns were grown on them at T_g from 941 to 970°C. Figures 2.29b,d show SEM images of the GaN nanocolumns at 961°C. The tubular nanocolumns were selectively grown on the nitride Al nanodot patterns. The average air hole diameter of the nanotube was 69 nm ± 12%. On the Si surface outside the nanopatterns, spontaneous random nucleation occurred at 961°C, as shown in Figs. 2.29b,d. With increasing T_g, self-assembly of nanocolumns on the Si surface was suppressed, and at 966°C, no spontaneous nucleation of GaN occurred on the Si surface (see Fig. 2.29c,e). Meanwhile, the nitrided Al nanodots functioned as nucleation sites for nanocolumn growth [62]. The difference in the crystal nucleation temperature between the nitrided Al and the Si surface contributed to the success of the SAG. At 970°C, however, no nanocolumn growth occurred, even on the nanopatterns.

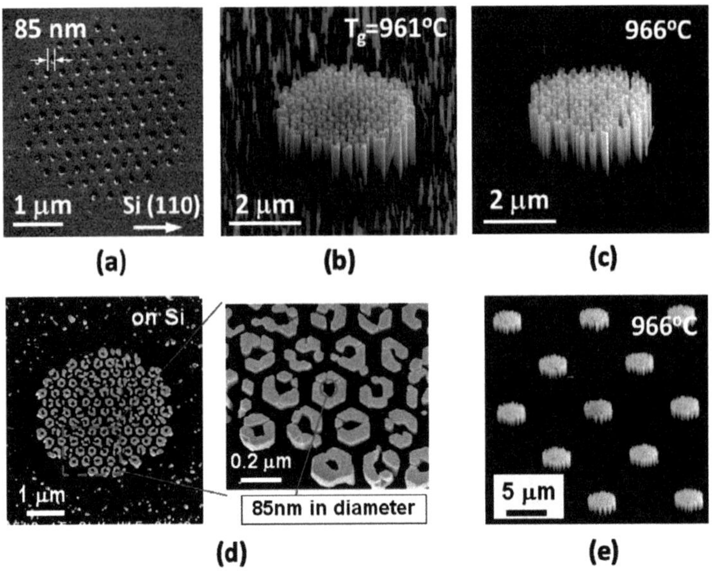

Figure 2.29 (a) Top-view SEM image of Al nano-pattern with triangular lattice of Al nanodots with 300 nm period, 85 nm diameter and 20 nm thickness, (b) and (c) Bird's-eye-view SEM images of GaN nanocolumns grown on the Al nanopattern at T_g = 961 and 966°C, respectively, (d) High-magnified top-view SEM image of the sample of (b), where GaN nanocolumns were grown at the edge of each nitrided Al nanodot, forming nanotube-like structures and (e) Low-magnification bird's-eye-view SEM image of the sample of (c) (after ref. 62).

2.5.3 Ti-Mask SAG of GaN Nanocolumns on GaN Template/Sapphire Substrates

Ti [28, 29] or TiO_2 [66] films with thicknesses of 5–10 nm, which were prepared on MOCVD-grown (0001) GaN templates, were utilized for the SAG mask of GaN nanocolumns. Then, triangular or square lattice arrays were prepared on the mask films by etching down into the underlying GaN to depths of 10–50 nm. GaN nanocolumns were grown on the nano-patterns by RF-MBE, as schematically shown in Fig. 2.30a. The growth was initiated by the surface nitridation of the Ti mask at 400°C, and the subsequent SAG of GaN nanocolumns occurred at the mask openings at a T_g of approximately 900°C. The growth of GaN tends to be initiated at the side face of the nanohole in a GaN template crystal. In fact, for a large nanohole diameter, the nucleation at the nanohole side face resulted in the formation of GaN nanorings (nanotubes), owing to insufficient coalescence at the nanohole center [67]; this tended to occur for deep nanoholes, probably at depths greater than 50 nm. When the nanohole diameter is small, the crystal quickly fills up the hole and grows out of the hole (Fig. 2.30b). The lateral growth of the nanocolumn proceeds outward along the *a*-axis because of the large lateral growth rate along the *a*-axis, finally forming stable hexagonal side facets (*m*-faces) of the nanocolumns (see Fig. 2.30c). The GaN nanocolumns then grow preferentially along the *c*-axis, because the lateral growth rate along the *m*-axis is extremely low.

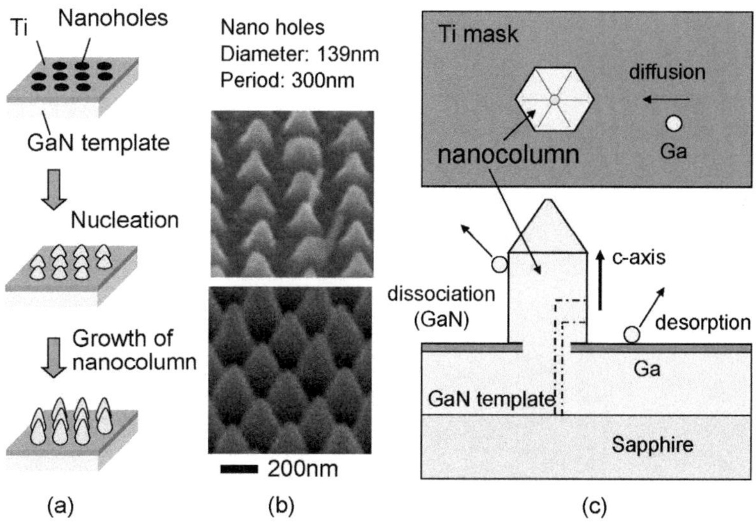

Figure 2.30 (a) Procedure of Ti-mask SAG, (b) crystal nucleation, and (c) schematic of growth mechanism of SAG [29].

Figure 2.30c illustrates the SAG mechanism schematically. Key elements of SAG are the surface passivity of the nitrided

Ti mask, the diffusion and desorption of Ga adatoms, and the dissociation of GaN. The growth temperature range of high-quality GaN thin films is 710–750°C [68, 69], but the SAG of GaN was achieved at 900°C [28, 29], which is an unusually high growth temperature for RF-MBE. At a sufficiently high temperature, the diffusion and desorption of Ga on the mask surface should be sufficiently enhanced. The diffusion length of Ga was estimated to be increased from 92 to 495 nm as T_g increased from 900 to 925°C [28]. The accelerated diffusion contributes to SAG at a high T_g. The desorption of Ga from the surface of the nitrided Ti mask is another key factor in understanding SAG.

The critical temperature for SAG was between 880 and 900°C, as shown in Fig. 2.31. At the lower temperature of 880°C, no SAG of GaN nanocolumns was observed. Crystal nucleation occurred in the spaces between the nanoholes, where Ga was considered to be insufficiently desorbed from the nitrided Ti-mask surface because of the suppressed surface diffusion of Ga. In the temperature range from 900 to 925°C, SAG of GaN was achieved [28]. With increasing T_g, the growth rate of the nanocolumns decreased from 1.1 to 0.57 μm/h, which indicated accelerated desorption of Ga from the nitrided Ti mask surface. Random and spontaneous nucleation on the nitrided Ti mask area outside the nanohole patterns was sufficiently suppressed at the high temperature, contributing to SAG through the enhancement of both the diffusion and desorption of Ga. However, the increase in T_g above 900°C induced an increase in the shape inhomogeneity of the nanocolumns and a decrease in column diameter (see Fig. 2.31d).

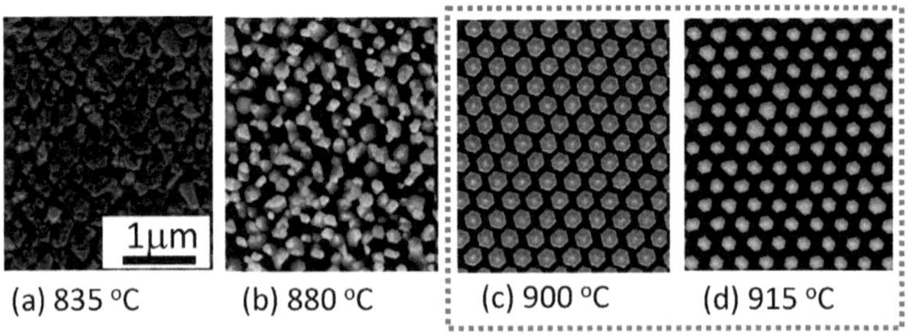

Figure 2.31 Top-view SEM images of GaN nanocolumns with a 600 nm-period triangular lattice, grown at 835 to 915°C with a Q_{N2} of 3.5 sccm.

The decreased Q_{N2} at 900°C contributed to the improved SAG, because a decrease in Q_{N2} accelerates the diffusion and desorption of Ga on/from the nitrided Ti-mask surface, which suppress spontaneous nucleation on the nitrided Ti mask. Figure 2.32 shows bird's-eye-view and top-view SEM images of GaN nanocolumns grown at a T_g of 900°C with decreasing Q_{N2} from 3.5 to 0.5 sccm, where the triangular-lattice nanocolumns were arranged with a lattice constant of 600 nm. The samples with Q_{N2} of 3.5, 2.0, and 1 sccm were grown for 3 h, and the sample with Q_{N2} of 0.5 sccm was grown for 2.5 h. At Q_{N2} = 3.5 nm, for example, a uniform nanocolumn array, in which the GaN nanocolumns of 3.2 μm height were precisely arranged at each nanohole position, was successfully demonstrated.

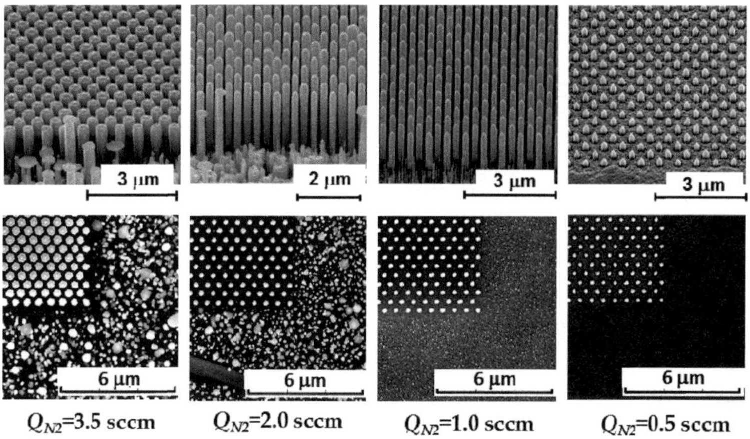

Figure 2.32 Bird's-eye view and top-view SEM images of GaN nanocolumn arrays with 600 nm-period triangular lattice grown with different nitrogen flow rates from Q_{N2} = 3.5 to 0.5 sccm at a T_g of 900°C (after ref. 29).

The lower top-view SEM images show the random and spontaneous crystal nucleation of GaN on the nitrided Ti-mask surface outside the nanohole patterns. A large number of self-organized nanocolumns were observed outside the nanohole patterns at Q_{N2} = 3.5 and 2.0 sccm. It is noteworthy that the spontaneous nucleation was well suppressed, even for Q_{N2} = 3.5 sccm, in the neighborhood of the nanohole pattern, in which the Ga atoms diffused toward the nanohole pattern and were absorbed.

The Ga diffusion length on the nitride Ti-mask was evaluated as a function of Q_{N2} [29], and the average diffusion length increased from 92 to 187 nm as Q_{N2} decreased from 3.5 to 1.0 sccm, as shown in Fig. 2.33a. When the spacing between the nanoholes was less than double the diffusion length, i.e., approximately 200 nm for Q_{N2} = 3.5 sccm, no self-assembled nanocolumns were grown there, even at the high nitrogen supply rate of 3.5 sccm. Note that spontaneous nucleation was completely suppressed at Q_{N2} = 0.5 sccm, probably because substantial desorption of Ga annihilated the nucleation.

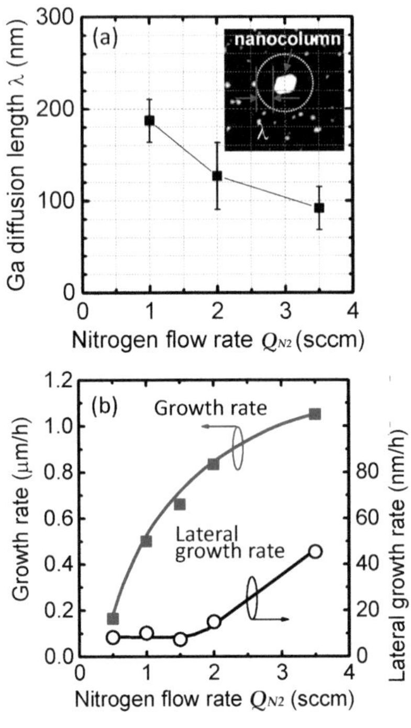

Figure 2.33 (a) Ga diffusion length as a function of nitrogen flow rate (Q_{N2}). The inset shows the neighboring no-nucleation area of a single nanocolumn, the width d of which corresponds to the diffusion length. (b) Growth rate and lateral growth rate of GaN nanocolumns as a function of Q_{N2} (after ref. 29).

As QN_2 increased from 0.5 to 3.5 sccm, the growth rate of nanocolumns along the c-axis increased monotonically from 0.15 to 1.05 μm/h (see Fig. 2.34b), demonstrating that SAG occurred under the nitrogen supply limitation scheme. This is understood as follows: desorption of Ga occurs rapidly at 900°C, but the increased nitrogen supply suppresses the desorption, bringing about an increase in the growth rate. The lateral growth rate did not change at Q_{N2} values less than 1.5 sccm (see Fig. 2.33b); it was sufficiently low, approximately 8 nm/h, although the growth rate along the *c*-axis increased. The ratio between the vertical (*c*-axis) and lateral (*m*-axis) growth rates was very high, 89, and therefore the mask patterns were well transcribed to the grown GaN crystal. The lateral growth rate increased above 2 sccm, and a sufficient supply of nitrogen at 3.5 sccm enhanced the incorporation of Ga and N into the sidewalls of the nanocolumns and caused a jump in the lateral growth rate to 45 nm/h. However, the growth rate was also high, 1.05 μm/h, which was approximately double that at Q_{N2} of 1 sccm.

2.6 Emission Color Control and Multicolor Integrated Nanocolumn LEDs

2.6.1 Emission Color Control of InGaN-Based Nanocolumn Arrays by the Nanocolumn Structural Parameters

As described in Section 2.3.3, self-organized InGaN/GaN nanocolumns whose diameter and position were randomly distributed were observed to exhibit a multi-color emission image that included various color spots from blue to red [8]. The results of an experiment conducted in 2007 strongly suggest that precise control of the nanocolumn size and position permits emission color control. This was demonstrated in 2010 [70] by fabricating uniform arrays of InGaN/GaN nanocolumns using Ti-mask SAG [29]. The emission color of the nanocolumn arrays is controlled by the structural parameters of the nanocolumns (their diameter, lattice constant, and height), as shown in Fig. 2.34. In this section, the emission wavelength changes of regularly arranged nanocolumns as a function of the diameter and period are described, following the description in ref. 70.

In the experiment, triangular-lattice-arranged nanohole patterns with a single lattice constant of 400 nm and diameters from 50 to 167 nm were prepared on a Ti-film/GaN template/sapphire substrate. The nanohole patterns, with a size of 20 × 20 μm^2, were arranged in a square lattice in close contact with each other. Thus, the growth conditions of the neighboring nanocolumn arrays were considered to be the same, although RF-MBE frequently exhibits a distribution in the growth temperature on the same substrate. By means of Ti-mask SAG, GaN nanocolumn arrays were grown at 900°C at a nitrogen flow rate of 2.0 sccm, followed by growth of three-period InGaN/GaN QWs with design thicknesses of 3 and 9 nm for the InGaN well and GaN barrier, respectively. Each nanocolumn had a pyramidal top with six-fold side faces, assigned to be {1–101} semipolar crystal planes. The nanocolumn diameter was precisely controlled for the different nanocolumn arrays to be within the range of 95 to 270 nm, as shown in Fig. 2.34. An emission color change from blue to red was noted with increasing nanocolumn diameter between the neighboring nanocolumn arrays on the same substrate. Near-field emission images of nanocolumn arrays with diameters from 143 to 270 nm were observed under 200 μW He-Cd laser light irradiation at room temperature.

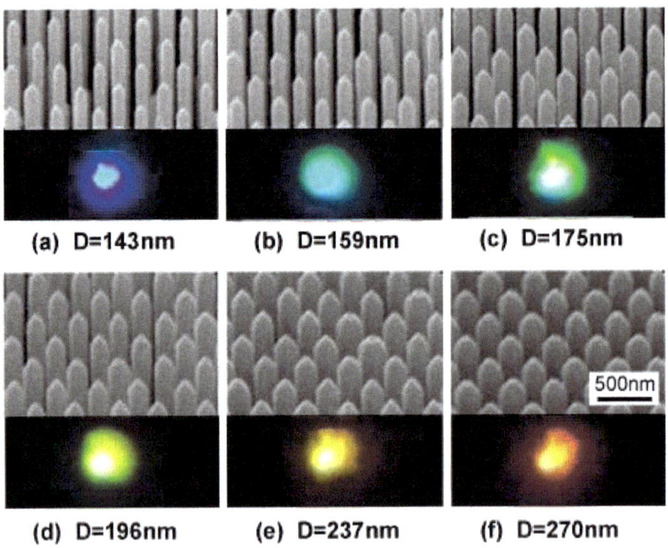

Figure 2.34 Triangular-lattice uniform arrays of InGaN/GaN nanocolumns with different nanocolumn diameters and a constant lattice constant of 400 nm and their PL near-field emission images (after ref. 70).

Figure 2.35 shows PL spectra evaluated for these nanocolumn arrays under InGaN-based semiconductor laser excitation at an excitation power density of 15.2 kW/cm^2. For the samples with diameters less than 166 nm, double-peak emissions were observed. The shorter-wavelength-side peak that appeared at approximately 480 nm became dominant under high excitation conditions, resulting in a distinct blue emission. On the other hand, the PL wavelength for the longer-wavelength-side peak shifted monotonically from 513 to 632 nm with increasing diameter from 137 to 270 nm (see Fig. 2.35b). Therefore, the peak wavelength shifted from 479 to 632 nm. The possible reasons for the wavelength shift induced by increasing the nanocolumn diameter are the piezoelectric effect and compositional change of the InGaN MQWs. If the thickness and In composition of InGaN MQWs change with increasing nanocolumn diameter, the crystal strain induces different piezoelectric fields in the MQWs and shift the emission wavelength as a result of the quantum-confined Stark effect (QCSE). Therefore, the QCSE probably caused the wavelength shift. It was observed from cathodoluminescence (CL) that these samples did not have InGaN QD structures. Semipolar plane InGaN/GaN MQWs are well known to provide a low piezoelectric field [71]; thus, the emission wavelength shift is mainly attributable to the difference in the level of In incorporation in the InGaN MQWs.

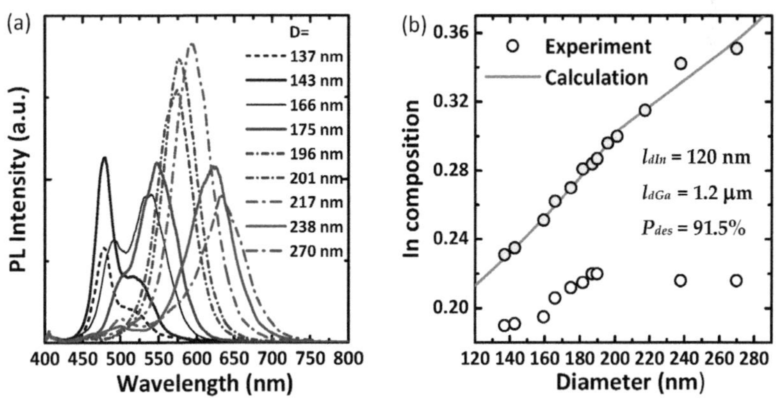

Figure 2.35 (a) Emission spectra of InGaN/GaN nanocolumn arrays with different nanocolumn diameters and (b) In composition (X_{In}) of the InGaN quantum well as a function of the diameter (after ref. 70).

To clarify this, a point analysis by TEM EDX was performed at the centers of the side faces of the pyramids for three samples with diameters of 95, 138, and 175 nm, as shown in Fig. 2.36. Note that the electron beam penetrated the nanocolumn, generating the EDX signals of Ga and In. Thin InGaN layers covered the side faces of the pyramidal GaN core, and thus, the EDX signal of Ga for a wider nanocolumn was intensified with respect to that of In, owing to the larger volume of GaN. Nevertheless, the ratio of the EDX signal intensity of In to that of Ga increased from 0.12 to 0.24 and 0.29 with increasing diameter. This clearly indicated that the In composition of InGaN increased for wider nanocolumns.

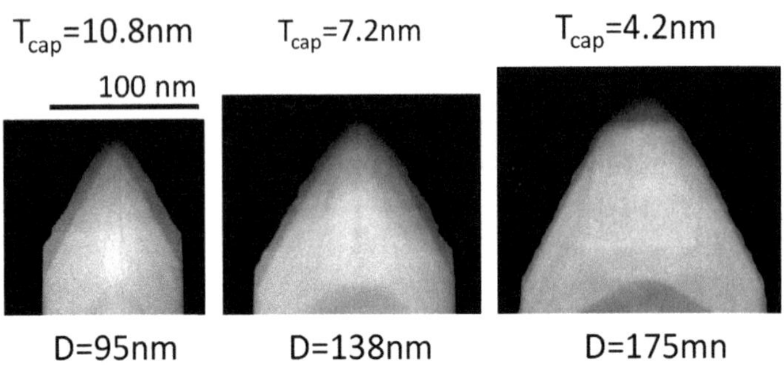

Figure 2.36 Cross-sectional TEM images of InGaN/GaN MQW on the top of GaN nanocolumns with the diameters of 95, 138, and 175 nm. With increasing *D*, the thickness of GaN cap-layer decreased from 10.8 to 4.2 nm, evincing the depression in Ga diffusion amount to the nanocolumn top.

The In composition (X_{In}) of InGaN QWs corresponding to the PL peak wavelength, which was estimated by comparison with the transition energy obtained by solving the Schrödinger equation, considering the spontaneous and piezoelectric field, increased monotonically from 0.23 to 0.35 with increasing diameter, as shown in Fig. 2.35b. The growth mechanism of the increased In composition is explained here (see Fig. 2.37). The growth of InGaN at the nanocolumn top occurs as a result of the ingredient supply by two pathways, that is, the impingement of Ga and In atomic beams directly on the nanocolumn top and the diffusion of adatoms into the nanocolumn top from the nanocolumn sidewall. One nanocolumn is surrounded by other nanocolumns, as shown in Fig. 2.37c; thus, in the RF-MBE, the adjacent nanocolumns function as shadowing incident Ga and In atomic beams on the sidewall surface (m-plane) of the nanocolumn of interest. The supplied Ga and In adatoms diffuse on the sidewall surfaces and are incorporated into the InGaN layer at the nanocolumn top. Therefore, the In composition is calculated as a function of the nanocolumn diameter using this model.

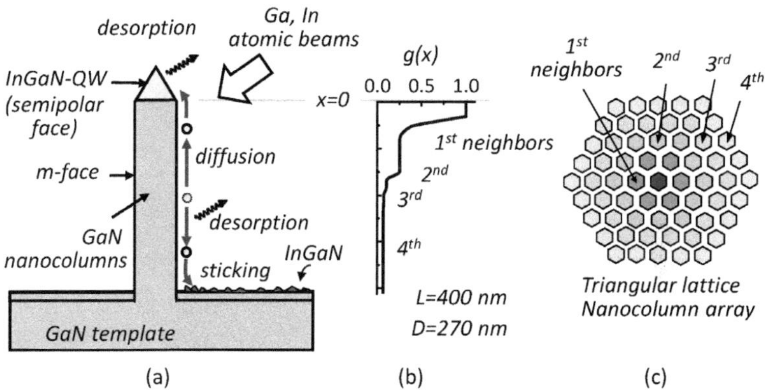

Figure 2.37 (a) Growth mechanism of InGaN at the nanocolumn top, (b) atomic beam distributions $g(x)$ along the nanocolumn axes of Ga and In impinging on the nanocolumn sidewall (m-faces), and (c) spatial nanocolumn arrangement: one nanocolumn is surrounded by other nanocolumns.

First, the atomic beam distributions along the nanocolumn axes of Ga and In impinging on the nanocolumn sidewall (m-faces), $g(x)$, are calculated by considering the shadow effect of the first to fourth neighboring nanocolumns (see Fig. 2.37b). Then, using the atomic beam distributions, the following diffusion equation for Ga and In adatoms on the m-faces is solved to determine the diffusive fluxes of Ga and In from the sidewall to the nanocolumn top.

$$\frac{\partial p(x)}{\partial t} = g(x) - \frac{p(x)}{\tau} + D\frac{\partial^2 p(x)}{\partial x^2}, \; l_{\mathrm{d}} = \sqrt{D\tau}, \tag{2.2}$$

where $p(x)$, D, and l_{d} are the atomic beam distribution along the nanocolumn axes in the static state, the surface diffusion constant, and the diffusion length on the sidewall, respectively. Assuming that adatoms are absorbed in the Ti mask surface and nanocolumn tops or desorbed when they reach the bottoms of the nanocolumns or the tops of the *m*-faces, the boundary condition is given by

$$p(0)=0, p(H)=0, \tag{2.3}$$

where H is the nanocolumn height. By solving Eq. 2.2 using the boundary conditions of Eq. 2.3, the flow rate of adatoms from the nanocolumn sidewalls to the nanocolumn tops is obtained from the following equation:

$$J\Big|_{x=0} = -D\frac{\partial p(x)}{\partial x}\Big|_{x=0} = -\frac{\int_0^H \left(e^{(H-x)/l_d} - e^{-(H-x)/l_d}\right)\cdot g(x)dx}{e^{H/l_d} - e^{-H/l_d}} \tag{2.4}$$

The desorption probability of In from the nanocolumn top is assumed for the total amount of directly impinged and diffused In, whereas no desorption probability of Ga is assumed. In this calculation, the diffusion lengths of In and Ga adatoms (l_{di}, i = In, Ga), and the desorption probability of In are considered to be fitting parameters. The theoretical curve of the In composition is calculated over the entire nanocolumn diameter range using the following equation:

$$X_{\text{In}} = \frac{Q_{\text{In}} \times (1-P_{\text{des}})}{Q_{\text{In}} \times (1-P_{\text{des}}) + Q_{\text{Ga}}}, \tag{2.5}$$

where Q_{In}, Q_{Ga}, and P_{des} are the quantities of In and Ga adatoms supplied at the pyramidal tops and the desorption probability of In, respectively. The calculated result fits the experimental data well, as shown in Fig. 2.35b, when the diffusion lengths of the In and Ga adatoms are 120 nm and 1.2 μm, respectively, and the desorption probability of In is 91.5%. Although there have been few reports on the diffusion length of III-group adatoms on m-plane GaN, the diffusion length of Ga can be expected to be larger than that of In, owing to the rapid desorption of In from the m-plane. The desorption of In should be appreciably accelerated at the high growth temperature of 650°C for InGaN.

In contrast, Ga desorption is suppressed at such a low temperature because GaN is typically grown at a temperature higher than 700°C [68, 69]. From this discussion, we can conclude that Ga adatoms diffuse preferentially on the sidewall and are absorbed into the nanocolumn top but that the number of Ga beams impinging on the sidewall decreases with increasing nanocolumn diameter because of the shadow effect. Thus, the decreased supply of Ga adatoms to the top brings about an increase in the In composition of the InGaN wells. From Fig. 2.36, we notice that the thickness of GaN cap-layer decreased from 10.8 to 4.2 nm with increasing *D*, which phenomenon evinced the depression in Ga diffusion amount to the nanocolumn top.

According to this model, we can easily predict that the increased array period for a constant nanocolumn diameter will induce a blueshift in the nanocolumn emissions. In an experiment with a different series of samples of 210 nm-diameter nanocolumns, the emission wavelength decreased from 508 to 480 nm as the array period increased from 0.4 to 1.0 μm. For these samples, the dependence of the In composition on the nanocolumn period was calculated using the same diffusion length, resulting in a good fit of the calculation curve to the experimental data, as shown in Fig. 2.38. Thus, the phenomenon indicating that the In composition of InGaN changes with the nanocolumn diameter and lattice constant is explained by this model.

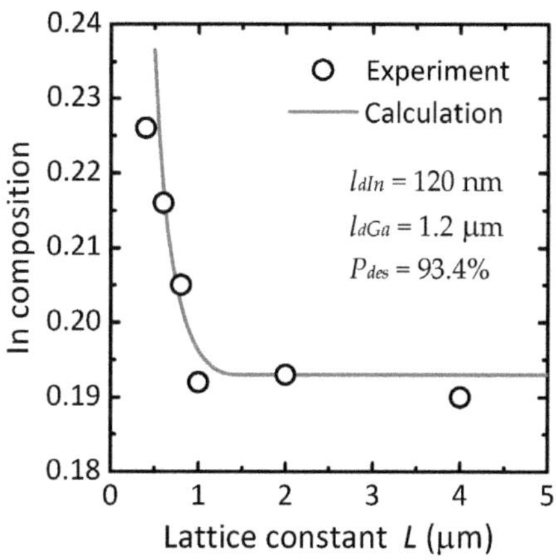

Figure 2.38 In composition of InGaN/GaN nanocolumn arrays as a function of lattice constant (after ref. 70).

However, the PL peak wavelength of the InGaN nanocolumn array moved toward the shorter-wavelength side for diameters greater than 300 nm (for example, 573 nm for D = 349 nm), where c-faces appeared at the nanocolumn top and the change in the mechanism probably changed the wavelength.

2.6.2 Monolithic Integration of Nanocolumn LEDs with Different Emission Colors

On the basis of the emission-color control scheme described above, InGaN-based light emitters have been developed on the basis of a novel concept, in which microscale nanocolumn LEDs with various emission colors, such as red, green, blue, and yellow (RGBY), have been monolithically integrated [72, 73]. This new generation of LEDs will enable the application of LEDs to full-color displays and solid-state lighting by providing high functionality and high-color rendering. In this section, the development of monolithic integration of four-color (blue, sky-blue, green, and yellow) InGaN-based nanocolumn LEDs [73] is described.

Four micro-LEDs (LEDs 1 to 4) with a 65 μm-diameter circular emission area were arrayed in a 2 × 2 square lattice, as shown in Fig. 2.39a–c. Different integrated nanocolumn LED units were prepared on an MOCVD-grown GaN template (3.5 μm) on c-face sapphire by means of Ti-mask SAG via RF-MBE [28, 29]. Each integrated LED unit was grown on four underlying Ti-mask triangular-lattice nanohole patterns, each with a 150 × 150 μm^2 area, with the nanohole diameter being varied for a fixed lattice constant L. For the unit corresponding to Fig. 2.39c, for example, the nanohole diameters were 90, 120, 150, and 210 nm for LEDs 1 to 4, respectively, and L was 300 nm. Si-doped n-type GaN nanocolumns, five pairs of InGaN (3 nm)/GaN (12 nm) MQWs, a Mg-doped p-type GaN cladding layer (approximately 300 nm), and a Mg-doped p-type InGaN contact layer (~50 nm) were grown at substrate heater temperatures of 1160, 860, 965, and 900°C, respectively. The nanocolumn height was approximately 0.9 μm. From observation of InGaN/GaN MQW nanocolumns grown on a patterned substrate designed in the same way, the nanocolumn diameters at the positions of the InGaN/GaN MQWs (in Fig. 2.39c) were estimated to be 119, 145, 188, and 231 nm. With increasing diameter, the emission wavelength shifted toward the longer-wavelength side, in this case, from blue to yellow. An ITO film with a 0.3 μm thickness was deposited on the top surfaces of the nanocolumn arrays to form transparent p-type electrodes. After the crystal growth, the sidewalls of the nanocolumns were coated with an Al_2O_3 thin film (30 nm) by atomic layer deposition, which passivated the surface and suppressed the leakage of current on the sidewall. Then, the nanocolumns were embedded with SOG, and the p-type electrode process was performed. Figure 2.39b shows a top view of the electrode arrangement in the fabricated integrated nanocolumn LED unit. LEDs 1–4, separated by a distance of 190 μm, were prepared in a circular SOG island 1 mm in diameter. The p-type electrode of each LED consisted of a 65 μm-diameter circular ITO emission window and a square p-type electrode pad. An n-type electrode was prepared outside the circular SOG island.

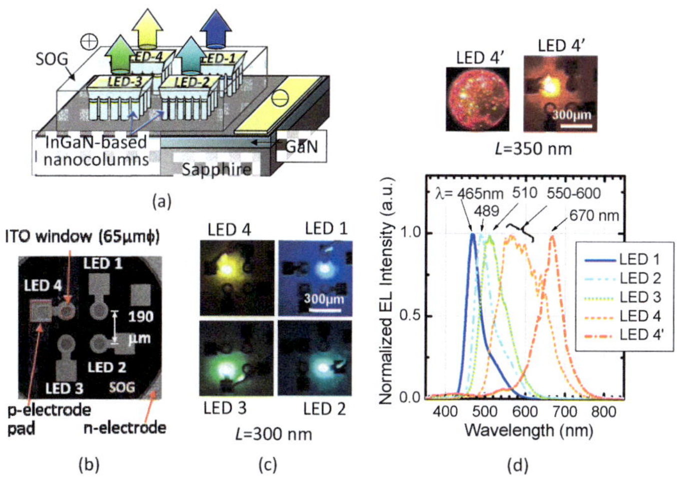

Figure 2.39 (a) Schematic diagram of integrated nanocolumn LED units (LEDs 1-4), (b) top-view of the electrode arrangement of the unit, (c) emission images of an integrated nanocolumn LED, and (d) emission spectra from LEDs 1-4 and LED 4′ and the red emission image of LED 4′ (after ref. 73).

The emission spectra of the integrated nanocolumn LED (units A and B) were evaluated under a DC injection current at room temperature, as shown in Fig. 2.39d. Figure 2.39c shows emission images of the four-color integrated nanocolumn LED (unit A) at an injection current of 10 mA. Blue, sky-blue, green, and yellow lights were emitted from LEDs 1–4, respectively. This emission image clearly indicates that nanocolumn LEDs emitting four different colors were integrated in the vicinity of each other. The peak wavelengths and FHWMs of LED 1–4 (unit A) were (465 nm, 40 nm), (489 nm, 49 nm), (510 nm, 72 nm), and (570 nm, 134 nm), respectively. The red emission image shown in the superior part of Fig. 2.39d indicates that of LED 4′ of the neighboring LED unit (unit B) having a spectrum with a peak wavelength of 670 nm. In this case, the use of a large lattice constant of 350 nm resulted in a larger nanocolumn diameter (260 nm) at the position of the InGaN/GaN MQWs, which increased the emission wavelength [70]. The ability to combine different lattice constants in nanocolumn LED units would be useful in the fabrication of RGBY-color-integrated nanocolumn LEDs.

2.7 LEDs Based on Uniform Nanocolumn Arrays

2.7.1 Green-Light Nanocolumn LEDs with Triangular-Lattice Uniform Arrays of InGaN-Based Nanocolumns

This section describes the preparation of green-light nanocolumn LEDs with emission wavelengths from 515 to 550 nm from InGaN-based nanocolumn arrays, as schematically shown in Fig. 2.40a [66]. The first demonstration of green-light nanocolumn-array LEDs was reported in 2010 [74]. Figure 2.40b shows bird's-eye-view and top-view SEM images of a nanocolumn LED crystal. For the fabrication of the LED crystal, regularly arranged GaN nanocolumns were grown on MOCVD-grown GaN templates (3.5 μm) on *c*-face sapphire substrates (330 μm) by RF-MBE, using of 10 nm-thick titanium oxide (TiO_2) films for the SAG. It was demonstrated that TiO_2 films can function effectively as an alternative to the Ti mask ordinarily used in the SAG of GaN nanocolumns [28, 29]. GaN nanocolumn arrays with lattice constants of 245–350 nm, with integrated InGaN/GaN MQWs, were prepared, and a built-in core–shell structure was self-assembled, embedding InGaN/GaN MQWs in the cores of nanocolumns, with the heterostructure arranged axially along the nanocolumn axis (see Fig. 2.40c).

The nanocolumn LED crystals were prepared as illustrated in Fig. 2.40a. The GaN nanocolumns were grown at approximately 900°C to satisfy the SAG conditions [29]. In contrast, the InGaN (1.5 nm)/GaN (1.5 nm) SL (25 pairs) buffer, InGaN (3 nm)/GaN (12 nm) MQW (three pairs) active layers, and p-type cladding layers were grown at a lower temperature of approximately 650°C to incorporate In and Mg into the crystal. The low-temperature growth and Mg doping enhanced the lateral growth of GaN [13], increasing the nanocolumn diameter during the growth of the p-type GaN cladding layers (see Fig. 2.40).

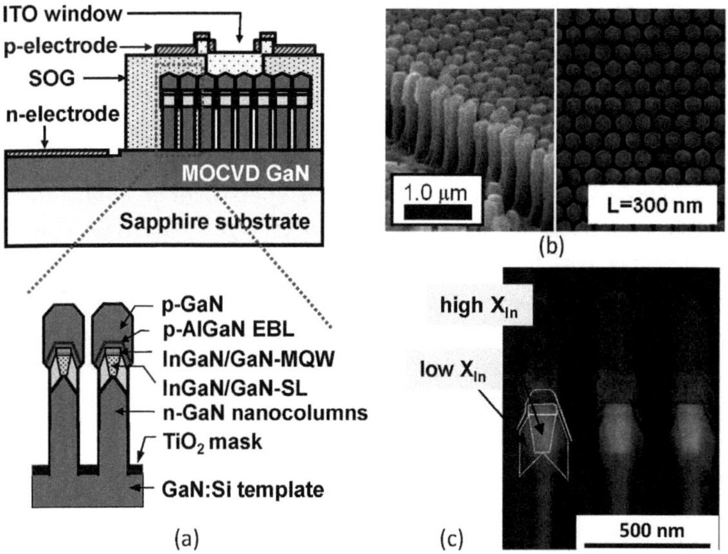

Figure 2.40 (a) Schematic diagrams of nanocolumn LED, (b) bird's-eye-view and top-view SEM images of typical nanocolumn LED crystal with L = 300 nm, and (c) HAADF-STEM images of nanocolumn LED crystal. The crystal boundaries are indicated by dotted lines (after ref. 66).

Figure 2.40c shows a cross-sectional HAADF-STEM image of a sliced nanocolumn LED crystal, thinned to ~100 nm. We unexpectedly observed a bright area sandwiched between the right and left gray areas in the middle region. We intended to grow an InGaN/GaN SL and MQW there, but no periodic structures were observed in the image. EDX analysis showed that both regions corresponding to the SL and MQW had the same In content of ~0.2 and that they were covered by an outer region with a lower In content of ~0.09. The outer diameter of the nanocolumn LED crystal was 240 nm, and the InGaN/GaN MQW and SL regions were confined within the 115 nm-diameter core, embedded in wide-band-gap materials. Thus, the electron–hole pairs were confined within the core region, suppressing surface recombination. As a result, the built-in core–shell structure was self-assembled in the nanocolumn, as discussed in detail in ref. 66. The growth of core-shell InGaN is also discussed in ref. 48.

The device process produced nanocolumn LEDs with ITO emission windows 65 μm in diameter (see ref. 66) and the cross-sectional device structure shown in Fig. 2.40a. The nanocolumn LEDs were evaluated under DC current injection at room temperature using a micro-electroluminescence (micro-EL) system with a 40× objective lens. Pure green emissions were observed through the ITO windows.

Figure 2.41a shows the emission spectra of a nanocolumn LED for injection currents from 2.8 to 15 mA. A spectrum with a single peak was observed over the entire range of the injection current. The peak wavelength was 541.5 nm at 0.2 mA (4.5 A/cm^2) and 536.2 nm at 20 mA (~450 A/cm^2), indicating a blueshift of 5.3 nm. In this experiment, small current-induced spectral blueshifts of 2–11 nm were observed. This range smaller than the blueshift range of conventional c-plane LEDs (20–47 nm) and reasonably close to the range for semi-polar and/or non-polar green LEDs (7–10 nm) [16–19]. The turn-on voltage was approximately 3 V (see the inset of Fig. 2.41a), the applied voltage at 15 mA was 5.4 V, and the reverse current at −5.0 V was 6.4 μA. The current leakage component of this sample was relatively low.

Nanocolumn LED crystals consist of a very large number of nanocolumns in which growth imperfections may be introduced accidentally. These growth imperfections may include areas that lack several nanocolumns and thereby predict current leakage paths. It is essential to suppress current leakage to develop nanocolumn LEDs with high emission efficiency. In this experiment, therefore, passivation of the nanocolumns was performed by atomic layer deposition (ALD) of Al_2O_3 film on the sidewalls to eliminate nanocolumn current leakage paths. Nanocolumns with 35 μm emission windows were fabricated by means of Al_2O_3 passivation. The reduction in the emission window was sufficient to prevent growth imperfection areas.

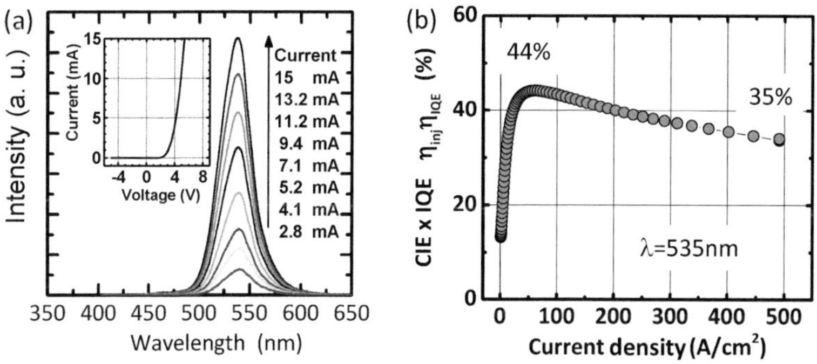

Figure 2.41 (a) Emission spectra of nanocolumn LED (L = 300 nm) with an emission window of 65 μmϕ for injection currents from 2.8 to 15 mA. The inset shows the current-versus-voltage characteristics. (b) $\eta_{nj} \cdot \eta_{IQE}$ of a nanocolumn LED (L = 245 nm) after sidewall passivation with an ALD-Al_2O_3 film, as a function of current density. The light output was measured from the back of the sapphire substrate (after ref. 66).

The light output of the nanocolumn LED was measured from the back of the sapphire substrate. This made it possible to characterize the device efficiency accurately, because the simple layered structure of the GaN template and sapphire enabled us to calculate the light extraction efficiency (LEE: η_{ext}) more simply than in the case of detection from the front of the device, which has a complicated surface structure. Taking into account the contribution of the light escape cone, assuming emission from a point-like source in the semiconductor, η_{ext} for the light incidence from material A (refractive index: n_A) to material B (n_B) ($n_B < n_A$) is expressed as follows [75]:

$$\eta_{ext} = \frac{1}{2}\frac{4n_A n_B}{(n_A + n_B)^2}\left(1 - \sqrt{1-(n_B/n_A)^2}\right), \tag{2.6}$$

where the first term of $4n_An_B/(n_A + n_B)^2$ is Fresnel's light transmission coefficient and the coefficient of 1/2 indicates that half of the light field spreading spherically in A is incident to B.

The refractive indices of GaN and sapphire are 2.42 and 1.77 at 530 nm, respectively; therefore, the η_{ext} from GaN to sapphire is 15.5%. The refracting light in the sapphire is widely diverged and is extracted from the back, but the light extraction efficiency in this case is found by doubling Eq. 2.6 to be 16.2%. Thus a total of 2.5% of the emitted light is extracted from the back of the sapphire substrate [66]. The external quantum efficiency (EQE), η_{EQE}, is given by the formula $\eta_{EQE} = \eta_{ext} \cdot \eta_{inj} \cdot \eta_{IQE}$, where η_{inj} is current injection efficiency (CIE) and η_{IQE} is the internal quantum efficiency (IQE). Therefore, $\eta_{IQE} \cdot \eta_{inj}$ was estimated from η_{EQE} / η_{ext}, using the measured η_{EQE} values for the backside divided by the calculated η_{ext}. Figure 2.41b shows $\eta_{IQE} \cdot \eta_{inj}$ as a function of the injection current density. The $\eta_{inj} \cdot \eta_{IQE}$ had a maximum value of 44% at a current density of ~60 A/cm^2 and then gradually decreased to 33% of the maximum at 450 A/cm^2 (20 mA). Note that the increasing directionality of the radiation increases η_{ext}, resulting in a lower estimated $\eta_{IQE} \cdot \eta_{inj}$, but the nanocolumn LED with a small lattice constant (L = 250 nm) exhibited a broadened Lambertian distribution (see ref. 66), indicating that the above equation for η_{ext} is valid. For a red-light (633 nm) nanocolumn LED, a light output of 26 μW was observed from the back of the sapphire substrate at 20 mA [76], which corresponds to $\eta_{inj} \cdot \eta_{IQE} = 2.7\%$.

2.7.2 Yellow-Light Nanocolumn LEDs with High Directional Radiation Beams

A periodic arrangement of nanocolumns, as shown in Fig. 2.42b, results in a photonic crystal effect. Two-dimensional light confinement in a periodic structure contributes to optically pumped lasing of InGaN/GaN nanocolumn arrays [77, 78], and light diffraction at the photonic band edge at the Γ point induces a highly directional beam radiation from nanocolumn LEDs [66, 79, 80]. In this section, the results of experiments conducted with yellow-emitting (572 nm) InGaN-based nanocolumn LEDs (see Fig. 2.42a), consisting of orderly arranged bottom-up nanocolumns in a triangular lattice, are described. For the LED, a highly directional radiation beam with a radiation angle of ±20° was successfully demonstrated, as shown in Fig. 2.42c [79].

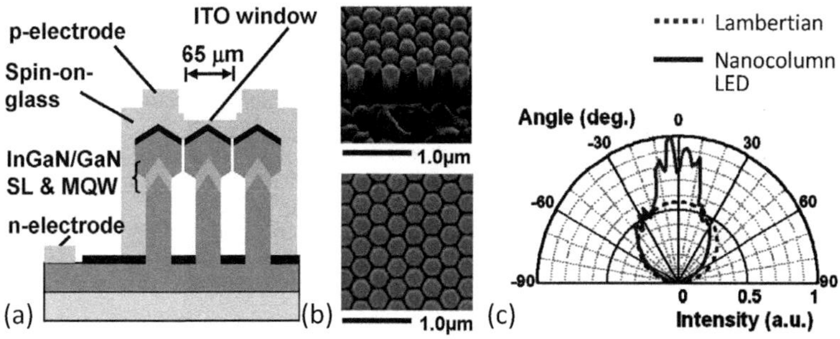

Figure 2.42 (a) Cross-sectional schematic diagram of nanocolumn LED crystals, (b) bird's-eye-view and top-view SEM images of typical nanocolumn LED crystals, in which the nanocolumns were arrayed in an orderly fashion in a triangular lattice with a lattice constant of 300 nm, and (c) angular profile of radiation beam of the yellow-emitting nanocolumn LED. The detection arm was vertically rotated along the m-axis of the nanocolumn system. The dashed line indicates the Lambertian distribution, with light intensity directly proportional to the cosine of the angle from the normal to the emission surface, which was normalized to the same total light power as the nanocolumn LED emission (after ref. 79).

For the fabrication of the nanocolumn LEDs, triangular lattice GaN nanocolumn arrays with a lattice constant of 300 nm were prepared by Ti-mask SAG, followed by growth of a 25-period InGaN/GaN SL buffer, a five-period InGaN (3 nm)/GaN (12 nm) MQW, Mg-doped p-type GaN cladding, and Mg-doped InGaN contact layers. Figure 2.42b shows bird's-eye-view and top-view SEM images of the nanocolumn LED crystals. The standard device process was performed as described in the previous section, and the nanocolumn LEDs were evaluated under DC current injection at room temperature. A uniform yellow emission image was observed in the ITO window of 65 μm. To investigate the directionality of the radiation beam, the emission spectra were measured using a micro-EL system with 40× and 4× objective lenses with NAs of 0.60 and 0.16, respectively, and acceptance angles (2θ max) of 73.7° and 18.4°, respectively. Figure 2.43a shows the emission spectra of the nanocolumn LEDs at injection currents from 1 to 10 mA, measured with the 4× objective lens, which preferentially detects the directional component of a radiation beam. The spectrum obtained using the 4× objective lens exhibited a very narrow linewidth of 8.8 nm; the light diffraction at a photonic band edge causes spectral narrowing. The sharp spectrum had a peak at 571.3 nm, which corresponded to the peak wavelength of the spectrum obtained using the 40× objective lens. The inset in Fig. 2.43a shows the spectrum at 10 mA measured with the 40× objective lens.

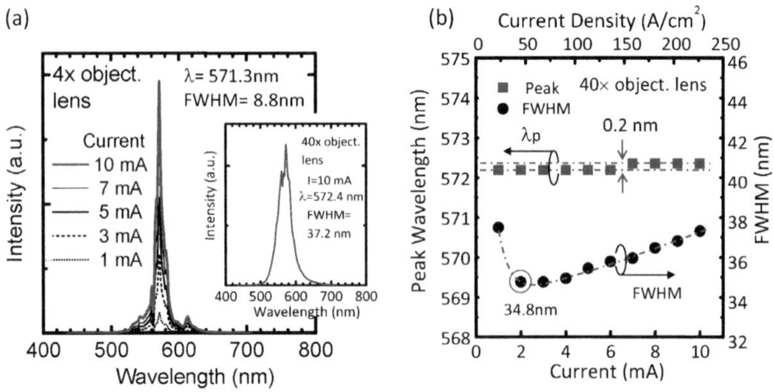

Figure 2.43 (a) Emission spectra of the yellow-emitting nanocolumn LED for injection currents from 1.0 to 10 mA, observed with micro-EL system with 4× objective lens. The inset shows the spectrum at 10 mA observed with a 40× objective lens and (b) peak wavelength and FWHM as a function of injection current and current density measured with a 40× objective lens (after ref. 79).

Figure 2.43b shows the peak wavelength and spectral FWHM of the yellow-emitting nanocolumn LEDs measured with 40 × objective lens as a function of the injection current. A divergent light beam with a wide angle was detected by the 40× objective lens with a large acceptance angle; thus, a widely spread emission component of the LED was included in the spectrum. An increase in the injection current density to 226 A/cm^2 (10 mA) resulted in a current-induced shift in the peak wavelength of less than 0.2 nm. This very small shift was caused by the structural effect of the nanocolumn photonic crystal. The spectral FWHM was minimized at 34.8 nm, and the spectral FWHM increased to 37.3 nm at 226 A/cm^2. The linewidth narrowing that occurred at a low current density resulted from the localized state formed by potential fluctuation, and spectral broadening occurred as a result of state filling with increased current density [15, 66]. The spectral FWHM of InGaN-based film LEDs was reported to be 50 at 594 nm [23]. Thus, the above spectral FWHM is exceptionally narrow for a yellow-emitting InGaN-based LED. In Fig. 2.42c, the dashed line indicates the ordinary Lambertian distribution for film LEDs, exhibiting a radiation angle of ±60°. In contrast, the radiation beam of the nanocolumn LED was directionally extracted in the vertical direction, exhibiting a narrow radiation angle of ±20°.

2.8 Lasing Emission Based on Nanocolumn Photonic Crystal Effect

A periodic arrangement of nanocolumns leads to a photonic crystal effect, contributing to lasing emission based on a two-dimensional distributed feedback (DFB) scheme by optical pumping in the blue [77] and blue-to-green [78] ranges. In the experiment, triangular-lattice GaN nanocolumn arrays were prepared by Ti-mask SAG. The structural parameters were the lattice constant (*L*) and nanocolumn diameter (*D*), which were varied within the ranges of 200–300 nm and 0.6–0.9 L, respectively, on the same substrate. Fifteen pairs of InGaN/GaN multiple quantum well (MQW) active layers with a well layer thickness of ~12 nm and a barrier layer thickness of ~3 nm were grown on top of the nanocolumns. Two wafers of samples A and B with different emission colors, which were controlled by slight differences in the growth temperature of InGaN MQWs, were fabricated. Figure 2.44a shows surface- and bird's-eye-view SEM images of a nanocolumn array with a 275 nm lattice constant and 210 nm diameter. The standard deviation of the nanocolumn diameter was estimated to be 5.3 nm (a coefficient of variation of 2.5%), which is equivalent to the image resolution (5 nm) of the SEM measurement. Thus, a high-quality periodic structure was prepared. The nanocolumn systems were pumped under a high-power excitation density using an yttrium aluminum garnet (YAG) laser (355 nm, 5 ns in pulse width, 20 Hz in repetition rate) at room temperature. Lasing emissions of various nanocolumn systems with different L and D values were obtained, as shown in Fig. 2.44b. Excitation light was focused from the vertical direction onto the nanocolumn surface of the substrate. The excitation areas were approximately 20 μm in diameter, and the emission light was detected in the same vertical direction. The band-edge wavelength of the nanocolumn photonic crystal changed with the nanocolumn parameter, resulting in wavelength shift of the lasing emission. The photonic band diagram of the nanocolumn array was calculated by the two-dimensional (2-D) plane wave expansion method (PWE) [78], proving that the series of lasing emissions illustrated in Fig. 2.44b occurred at the band edges of the Γ point designated as Γ_{11}. At the band edge, a standing wave formed by 2D-DFB effects in the nanocolumn regular array led to the lasing light emission in the vertical direction from the nanocolumn surface, resulting in the operation of surface-emitting nanocolumn lasers.

L = 275 nm, D = 210 nm

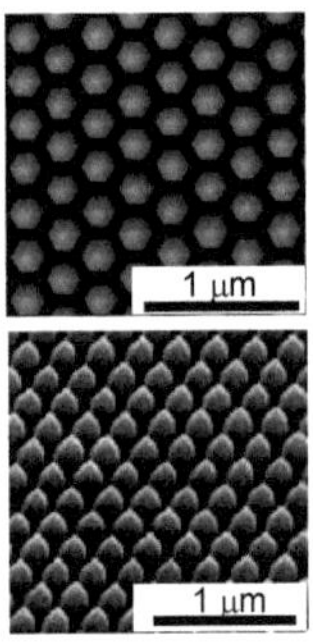

(a)

	L (nm)	D (nm)	λ (nm)
Sample A	230	196.7	477.6
	235	205.9	488.5
	245	209.6	501.1
Sample B	255	198.1	532.1
	260	203.6	535.1
	265	207.6	544.6
	275	210.2	551.7
	280	206.0	556.2
	285	212.9	560.2

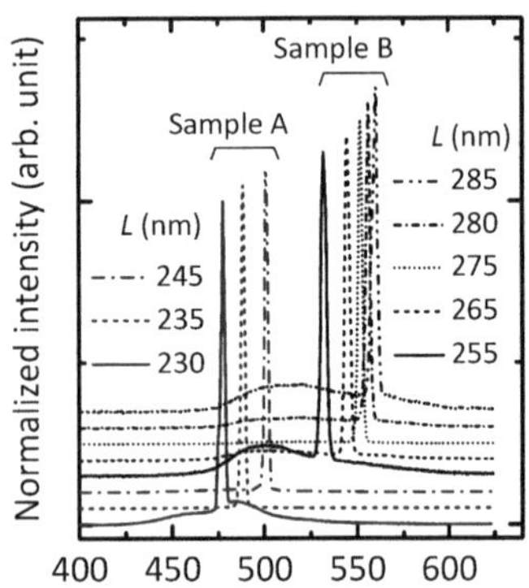

(b)

Figure 2.44 (a) Surface-view and bird's-eye-view SEM images of an InGaN/GaN MQW nanocolumn array with L = 275 nm and D = 210 nm, and (b) lasing emission spectra (λ: peak wavelength) from nanocolumn arrays with the structural parameters shown in the table to the left (after ref. 78).

2.9 Summary and Future Prospects

Since the initial successes achieved in self-assembly of GaN growth to the application of nanocolumns to light emitters, as described above, and has attracted considerable attention among researchers [1].

Nanocolumn technology is expected to contribute to the monolithic integration of InGaN-based three-primary-color (RGB) LEDs. These devices will have a great impact on the full-color application field, as noted in ref. 72. In such devices, red, green, and blue LEDs are integrated in the vicinity of each other, and thus, emissions with a full range of colors are obtained in a small emission spot, such as a 10 × 10 μm^2 area. In principle, this would enable the fabrication of a full high-vision semiconductor video panel with a size of ~2 × 1 cm^2 by arranging micro-LED pixels in an array with a size of 2 k × 1 k. The extended projection of video images on a semiconductor video panel can be used to produce a widescreen TV, which will lead to innovation in LED projection displays, enabling displays with extremely small sizes and low power consumption. A semiconductor video panel can also function as a high-luminous-efficiency white-emitting LED with high color rendering, which would facilitate the super-functional integration of lighting and displays. Light emitted from LEDs is incoherent, but the 10 × 10 μm^2 light emitted by an RGB LED can be controlled in full color, easily enabling imaging on a retina with a small spot by a simple lens system. Such a device could be a promising light source for retina-scanning light beam displays. For these applications, effort will be required to develop RGB-integrated LEDs with highly directional radiation beams [79]. This technology will have a great impact on full-color applications.

Laser projection displays are the next generation of imaging systems that are expected to provide vivid, high-definition, high-brightness images with low power consumption in devices with small sizes, lightweight structures, wide color ranges, and long lifetimes. However, laser projection displays have a serious problem of speckle noise, which is generated through the destructive and constructive interference of laser light scattered from the rough surface of the projection screen. One method for reducing the coherence of the light source, i.e., by increasing the lasing spectral width, decreases the speckle contrast. The use of a random laser with a low spatial coherence suppresses that a transition from photonic band-edge lasing to random lasing speckle noise. Random lasing behavior in GaN-based compounds was first observed in self-organized GaN nanocolumns [81]. It is noteworthy that a transition from photonic band-edge lasing to random lasing can be produced by introducing a certain amount of position and size disorder in nanocolumn photonic crystals. Recently, structurally graded nanocolumn photonic crystals have been produced for broad multi-wavelength optically pumped laser operation and are expected to contribute to speckle contrast reduction in laser-based displays [82].

References

1. Li, S., and Waag, A. (2012). GaN based nanorods for solid state lighting, *J. Appl. Phys.*, **111**, pp. 071101-1–071101-23.
2. Yoshizawa, M., Kikuchi, A., Mori, M., Fujita, N., and Kishino, K. (Sept. 23–27, 1996). Self-organization of GaN nano-structures on C-Al_2O_3 by RF-radical gas source molecular beam epitaxy, *23rd Int. Symp. Compound Semicond.*, **252**, pp. 187–190.
3. Yoshizawa, M., Kikuchi, A., Mori, M., Fujita, N., and Kishino, K. (April 15, 1997). Growth of self-organized GaN nanostructures on Al_2O_3 (0001) by RF-radical source molecular beam epitaxy, *Jpn. J. Appl. Phys.*, **36**, pp. L459–L462.
4. Yoshizawa, M., Kikuchi, A., Fujita, N., Kushi, K., Sasamoto, H., and Kishino, K. (1998). Self-organization of GaN/$Al_{0.18}Ga_{0.82}N$ multi-layer nano-columns on (0001) Al_2O_3 by RF-molecular beam epitaxy for fabricating GaN quantum disks, *J. Cryst. Growth*, **189/190**, pp. 138–141.
5. Han, W., Fan, S., Li, Q., and Hu. Y. (August 29 1997). Synthesis of gallium nitride nanorods through a carbon nanotube-confined reaction, *Science*, **277**, pp. 1287–1289.
6. Sanchez-Garcia, M. A., Calleja, E., Monroy, E., Sanchez, F. J., Calle, F., Muñoz, E., and Beresford, R. (1998). The effect of the III/V ratio and substrate temperature on the morphology and properties of GaN- and AlN-layers grown by molecular beam epitaxy on Si (111), *J. Cryst. Growth*, **183**, pp. 23–30.
7. Sekiguchi, H., Nakazato, T., Kikuchi, A., and Kishino, K. (2007). Structural and optical properties of GaN nanocolumns grown on (0001) sapphire substrates by rf-plasma-assisted molecular beam epitaxy, *J. Cryst. Growth*, **300**, pp. 259–262.

8. Kishino, K., Kikuchi, A., Sekiguchi H., and Ishizawa, S. (2007). InGaN/GaN nanocolumn LEDs emitting from blue to red, *Proc. SPIE,* **6473**, pp. 64730T-1–64730T-12.

9. Kishino, K., and Kikuchi, A. (Aug. 28–Sept. 2, 2005). InGaN/GaN nanocolumn LEDs, *6th Int. Conf. Nitride Semi.*, Mo-PD1-4.

10. Tanaka, J., Kishino, K., Sekiguchi, H., and Kikuchi, A. (September 21–24, 2008). Optical properties of amber-emission InGaN/GaN single quantum disk nanocolumns, *35th Int. Symp. Compound Semicond.*, Mo3.7.

11. Igawa, J., Vadivelu, R., and Kishino, K. (2013). Photoluminescence behaviors of orange-light-emitting InGaN-based nanocolumns exhibiting high internal quantum efficiency (17–22%), *Jpn. J. Appl. Phys.*, **52**, pp. 08JD09-1–08JD09-3.

12. Watanabe, S., Yamada, N., Nagashima, M., Ueki, Y., Sasaki, C., Yamada, Y., Taguchi, T., Tadatomo, K., Okagawa, H., and Kudo, H. (2003). Internal quantum efficiency of highly-efficient $In_xGa_{1-x}N$-based near-ultraviolet light-emitting diodes, *Appl. Phys. Lett.*, **83**, pp. 4906–4908.

13. Kikuchi, A., Kawai, M., Tada, M., and Kishino, K. (2004). InGaN/GaN multiple quantum disk nanocolumn light-emitting diodes grown on (111) Si substrate, *Jpn. J. Appl. Phys.*, **43**, pp. L1524–L1526.

14. Kim, H.-M., Cho, Y.-H., Lee, H., Kim, S., Ryu, S. R., Kim, D. Y., Kang, T. W., and Chung, K. S. (2004). High-brightness light emitting diodes using dislocation-free indium gallium nitride/gallium nitride multiquantum-well nanorod arrays, *Nano Lett.*, **4**, pp. 1059–1062.

15. Funato, M., Ueda, M., Kawakami, Y., Narukawa, Y., Kosugi, T., Takahashi, M., and Mukai, T. (2006). Blue, green, and amber InGaN/GaN light-emitting diodes on semipolar {1122} GaN bulk substrates, *Jpn. J. Appl. Phys.*, **45**, pp. L659–L662.

16. Kojima, K., Funato, M., Kawakami, Y., Nagahama, S., Mukai, T., Braun, H., and Shcwarz, U. T. (2006). Gain suppression phenomena observed in $In_xGa_{1-x}N$ quantum well laser diodes emitting at 470 nm, *Appl. Phys. Lett.*, **89**, pp. 241127-1–241127-3.

17. Yang, Y., Cao, X. A., and Yan, C. H. (2009). Rapid efficiency roll-off in high-quality green light-emitting diodes on freestanding GaN substrates, *Appl. Phys. Lett.*, **94**, pp. 041117-1–041117-3.

18. Lin, Y.-D., Chakraborty, A., Brinkley, S., Kuo, H. C., Melo, T., Fujito, K., Speck, J. S., DenBaars, S. P., and Nakamura, S. (2009). Characterization of blue-green m-plane InGaN light emitting diodes, *Appl. Phys. Lett.*, **94**, pp. 261108-1–261108-3.

19. Kubota, M., Okamoto, K., Tanaka, T., and Ohta, H. (2008). Continuous-wave operation of blue laser diodes based on nonpolar m-plane gallium nitride, *Appl. Phys. Express*, **1**, pp. 011102-1–011102-3.

20. Kawakami, Y., Suzuki, S., Kaneta, A., Funato, M., Kikuchi A., and Kishino, K. (2006). Origin of high oscillator strength in green-emitting InGaN/GaN nanocolumns, *Appl. Phys. Lett.*, **89**, pp. 163124-1–163124-3.

21. Kaganer, V. M., and Belov, A. Yu. (2012). Strain and x-ray diffraction from axial nanowire heterostructures, *Phys. Rev. B*, **85**, pp. 125402.1–125402.9.

22. Kawakami, Y., Kaneta, A., Su, L., Zhu, Y., Okamoto, K., Funato, M., Kikuchi A., and Kishino, K. (2010). Optical properties of InGaN/GaN nanopillars fabricated by postgrowth chemically assisted ion beam etching, *J. Appl. Phys.*, **107**, pp. 023522.1–023522.7.

23. Mukai, T., Yamada, M., and Nakamura, S. (1999). Characteristics of InGaN-based UV/blue/green/amber/red light-emitting diodes, *Jpn. J. Appl. Phys.*, **38**, pp. 3976–3981.

24. Kato, K., Sekiguchi, H., Kishino, K., and Kikuchi, A. (August 2009). Observation of narrow electronluminescence spectrum ($\Delta\lambda$ = 13 nm) from blue-green (507 nm) InGaN/GaN nanocolumn LED with one micron-diameter injection area, *36th Int. Symp. Compound Semicond.*, 2.8.

25. Nguyen, H., Wang, Q., and Mi, Z. (2014). Phosphor-free InGaN/GaN Dot-in-a-wire white light-emitting diodes on copper substrates, *J. Electron. Mater.*, **43**, pp. 868–872.

26. Hayashi, H., Fukushima, D., Tomimatsu, D., Noma, T., Konno, Y., and Kishino, K. (2015). Flip-chip bonding and fabrication of well-ordered nanocolumn arrays on sputter-deposited AlN/Si (111) substrate, *Phys. Status Solidi A*, **212**, pp. 992–996.

27. Hayashi, H., Fukushima, D., Noma, T., Tomimatsu, D., Konno, Y., Mizuno, M., and Kishino, K. (2015). Thermally engineered flip-chip InGaN/GaN well-ordered nanocolumn array LEDs, *IEEE Photonics Tech. Lett.*, **27**, pp. 2343–2346.

28. Sekiguchi, H., Kishino, K., and Kikuchi, A. (2008). Ti-mask selective-area growth of GaN by RF-plasma-assisted molecular-beam epitaxy for fabricating regularly arranged InGaN/GaN nanocolumns, *Appl. Phys. Express*, **1**, pp. 124002-1–124002-3.

29. Kishino, K., Sekiguchi, H., and Kikuchi, A. (2009). Improved Ti-mask selective-area growth (SAG) by RF-plasma-assisted molecular beam epitaxy demonstrating extremely uniform GaN nanocolumn arrays, *J. Cryst. Growth*, **311**, pp. 2063–2068.

30. Sekiguchi, H., Kishino, K., and Kikuchi, A. (2008). GaN/AlGaN nanocolumn ultraviolet light-emitting diodes grown on n-(111) Si by RF-plasma-assisted molecular beam epitaxy, *Electron. Lett.*, **44**, pp. 151–152.

31. Sekiguchi, H., Kato, K., Tanaka, J., Kikuchi, A., and Kishino, K. (2008), Ultraviolet GaN-based nanocolumn light-emitting diodes grown on *n*-(111) Si substrates by RF-plasma-assisted molecular beam epitaxy, *Phys. Status Solidi A*, **205**, pp. 1067–1069.

32. Zhao, S., Connie, A. T., Dastjerdi, M. H. T., Kong, X. H., Wang, Q., Djavid, M., Sadaf, S., and Liu, X. D. (2015). Aluminum nitride nanowire light emitting diodes: Breaking the fundamental bottleneck of deep ultraviolet light sources, *Sci. Rep.*, **5**, pp. 8332-1–8332-5.

33. Zhao, S., Djavid, M., and Mi, Z. (2015). Surface emitting, high efficiency near-vacuum ultraviolet light source with aluminum nitride nanowires monolithically grown on silicon, *Nano Lett.*, **15**, pp. 7006–7009.

34. Yamano, K., Kishino, K., Sekiguchi, H., Oto, T., Wakahara, A., and Kawakami, Y., (2015). Novel selective area growth (SAG) method for regularly arranged AlGaN nanocolumns using nanotemplates, *J. Cryst. Growth*, **425**, pp. 316–321.

35. Hayashi, H., Konno, Y., and Kishino, K. (2016). Self-organization of dislocation-free, high density, vertically aligned GaN nanocolumns involving InGaN quantum wells on graphene/SiO_2 covered with a thin AlN buffer layer, *Nanotechnology*, **27**, pp. 055302-1–055302-7.

36. Shon, J. W., Ohta, J., Ueno, K., Kobayashi, A., and Fujioka, H., (2014). Fabrication of full-color InGaN-based light-emitting diodes on amorphous substrates by pulsed sputtering, *Sci. Rep.*, **4**, pp. 5325-1–5325-4.

37. Hong, Y.-J., Lee, W.-H., Wu, Y., Ruoff, R. S., and Fukui, T. (2012). Van der Waals epitaxy of InAs nanowires vertically aligned on single layer graphene, *Nano Lett.*, **12**, pp. 1431–1436.

38. Munshi, A. M., Dheera, D. L. J., Fauske, V. T., Kim, D.-C., van Helvoort, A. T. J., Fimland, B.-O., and Weman, H. (2012). Vertically aligned GaAs nanowires on graphite and few-layer graphene: Generic model and epitaxial growth, *Nano Lett.*, **12**, pp. 4570–4576.

39. Liang, X., Sperling, B. A., Calizo, I., Cheng, G., Hacker, C. A., Zhang, Q., Obeng, Y., Yan, K., Peng, H., Li, Q., Zhu, X., Yuan, H., Hight Walker, A. R., Liu, Z., Peng, L.-M., and Richter, C. A. (2011). Toward clean and crackless transfer of graphene, *ACS Nano*, **5**, pp. 9144–9153.

40. Sekiguchi, H., Kato, K., Kikuchi A., and Kishino, K. (2008). Effect of Be-doping on InGaN/GaN nanocolumn light-emitting diode

structures by RF-plasma-assisted molecular-beam epitaxy, *Phys. Status Solidi C*, **5**, pp. 3069–3072.

41. Kusakabe, K., Kikuchi, A., and Kishino, K. (2001). Characterization of overgrown GaN layers on nano-columns grown by RF-molecular beam epitaxy, *Jpn. J. Appl. Phys.*, **40**, pp. L192–L194.

42. Kato, K., Kishino, K., Sekiguchi, H., and Kikuchi, A. (2009). Overgrowth of GaN on Be-doped coalesced GaN nanocolumn layer by RF-plasma-assisted molecular-beam epitaxy—Formation of high-quality GaN microcolumns, *J. Cryst. Growth*, **311**, pp. 2956–2961.

43. Sekiguchi, H., Kishino, K., and Kikuchi, A., (2010). Formation of InGaN quantum dots in regularly arranged GaN nanocolumns grown by RF-plasma-assisted molecular beam epitaxy, *Phys. Status Solidi C*, **7**, pp. 2374–2377.

44. Colby, R., Liang, Z., Wildeson, I. H., Ewoldt, D. A., Sands, T. D., Garcia, R. E., and Stach, E. A. (2010). Dislocation filtering in GaN nanostructures, *Nano Lett.*, **10**, p. 1568.

45. Kishino, K., and Ishizawa, S. (2015). Selective-area growth of GaN nanocolumns on Si (111) substrates for application to nanocolumn emitters with systematic analysis of dislocation filtering effect of nanocolumns, *Nanotechnology*, **26**, pp. 225602-1-13.

46. Usui, A., Sunakawa, H., Sakai, A., and Yamaguchi, A. (1997). Thick GaN epitaxial growth with low dislocation density by hydride vapor phase epitaxy, *Jpn. J. Appl. Phys.*, **36**, pp. L899–L902.

47. Sakai, A., Sunakawa, H., and Usui, A. (1997). Defect structure in selectively grown GaN films with low threading dislocation density, *Appl. Phys. Lett.*, **71**, pp. 2259–2261.

48. Oto, T., Mizuno, Y., Yanagihara, A., Miyagawa, R., Kano, T., Yoshida, J., Sakakibara, N., and Kishino K. (2016). Influence of GaN column diameter on structural properties for InGaN nanocolumns grown on top of GaN nanocolumns, *AIP Adv.*, **6**, pp. 115214-1-6.

49. Ertekin, E., Greaney, P. A., Chrzan, D. C., and Sands, T. D. (2005). Equilibrium limits of coherency in strained nanowire heterostructures, *J. Appl. Phys.*, **97**, pp. 114325-1-10.

50. Glas, F. (2206). Critical dimensions for the plastic relaxation of strained axial heterostructures in free-standing nanowires, *Phys. Rev. B*, **74**, pp. 121302R-1-4.

51. Ye, H., Lu, P., Yu, Z., Song, Y., Wang, D., and Wang, S. (2009). Critical thickness and radius for axial heterostructure nanowires using finite-element method, *Nano Lett.* **9**, pp. 1921–1925.

52. Zhang, X., Dubrovskii, V. G., Sibirev, N. V., and Ren, X. (2011). Analytical study of elastic relaxation and plastic deformation in nanostructures on lattice mismatched substrates, *Cryst. Growth Des.*, **11**(12), pp. 5441–5448.

53. Ishizawa, S., Sekiguchi, H., Kikuchi, A., and Kishino, K. (2006). Selective growth of GaN nanocolumns by Al thin layer on substrate, *Proc. Int. WS Nitride Semicond.* (IWN2006), Tu2-1: 91.

54. Ristić, J., Cerutti, L., Calleja, E., Trampert, A., Jahn, U., and Ploog, K. H. (2006). Growth mechanisms and nucleation of GaN nanocolumns on Si(111) and Si(100), *Proc. Int. WS Nitride Semicond.* (IWN2006) Tu-LN9: 109.

55. Hersee, S. D., Sun, X., and Wang, X. (2006). The controlled growth of GaN nanowires, *Nano Lett.*, **6**, pp. 1808–1811.

56. Usui, A., Sunakawa, H., Sakai, A., and Yamaguchi, A. (1997). Thick GaN epitaxial growth with low dislocation density by hydride vapor phase epitaxy, *Jpn. J. Appl. Phys.*, **36**, pp. L899–L902.

57. Nakamura, S., Senoh, M., Nagahama, S., Iwasa, N., Yamada, T., Matsushita, T., Kiyoku, H., Sugimoto, Y., Kozaki, T., Umemoto, H., Sano, M., and Chocho, K. (1997). InGaN/GaN/AlGaN-based laser diodes with modulation-doped strained-layer superlattices, *Jpn. J. Appl. Phys.*, **36**, pp. L1568–L1571.

58. Guha, S., Bojarczuk, N., Johnson, M., and Schetzina, J. (1999). Selective area metalorganic molecular-beam epitaxy of GaN and the growth of luminescent microcolumns on Si/SiO_2, *Appl. Phys. Lett.*, **75**, pp. 463–465.

59. Tang, H., Haffouz, S., and Bardwell, J. A. (2006). Si(111) substrates as highly effective pseudomasks for selective growth of GaN material and devices by ammonia-molecular-beam epitaxy, *Appl. Phys. Lett.*, **88**, pp. 172110-1-3.

60. Ishizawa, S., Sekiguchi, H., Kikuchi, A., and Kishino, K. (2007). Selective growth of GaN nanocolumns by Al thin layer on substrate, *Phys. Status Solidi C*, **244**, pp. 1815–1819.

61. Calleja, E., Ristić, J., Fernández-Garrido, S., Cerutti, L., Sánchez-Garcia, M. A., Grandal, J., Trampert, A., Jahn, U., Sánchez, G., Griol, A., and Sánchez, B. (2007). Growth, morphology, and structural properties of group-III-nitride nanocolumns and nanodisks, *Phys. Status Solidi B*, **244**, pp. 2816–2837.

62. Ishizawa, S., Kishino, K., and Kikuchi, A. (2008). Selective-area growth of GaN nanocolumns on Si(111) substrates using nitrided

Al nanopatterns by RF-plasma-assisted molecular-beam epitaxy, *Appl. Phys. Express*, **1**, pp. 015006-1-3.

63. Kishino, K., Hoshino, T., Ishizawa, S., and Kikuchi, A. (2008). Selective-area growth of GaN nanocolumns on titanium-mask-patterned silicon (111) substrates by RF-plasma-assisted molecular-beam epitaxy, *Electron. Lett.*, **44**, pp. 819–821.
64. Bertness, K. A., Sanders, A. W., Rourke, D. M., Harvey, T. E., Roshko, A., Schlager, J. B., and Sanford, N. A. (2010). Controlled nucleation of GaN nanowires grown with molecular beam epitaxy, *Adv. Funct. Mater.*, **20**, pp. 2911–2915.
65. Schumann, T., Gotschke, T., Limbach, F., Stoica, T., and Calarco, R. (2011). Selective-area catalyst-free MBE growth of GaN nanowires using a patterned oxide layer, *Nanotechnology*, **22**, pp. 095603-1–095603-6.
66. Kishino, K., and Yamano, K. (2014). Green-light nanocolumn light emitting diodes with triangular-lattice uniform arrays of InGaN-based nanocolumns, *IEEE J. Quantum Electron.*, **50**, pp. 538–547.
67. Kishino, K., Sekiguchi, H., and Kikuchi, A. (2008). Improved Ti-mask selective-area-growth (SAG) by RF-MBE demonstrating extremely uniform GaN nanocolumn arrays, *15th Int. Conf. Molecular Beam Epitaxy*, p. 16.
68. Kishino, K., and Kikuchi, A. (2002). Improved molecular beam epitaxy for fabricating AlGaN/GaN heterojunction devices, *Phys. Status Solidi A*, **190**, pp. 23–31.
69. Koblmüller, G., Brown, J., Averbeck, R., Riehert, H., Pongratz, P., and Speck, J. (2005). Ga adlayer governed surface defect evolution of (0001) GaN films grown by plasma-assisted molecular beam epitaxy, *Jpn. J. Appl. Phys.*, **44**, pp. L906–L908.
70. Sekiguchi, H., Kishino K., and Kikuchi, A. (2010). Emission color control from blue to red with nanocolumn diameter of InGaN/GaN nanocolumn arrays grown on same substrate, *Appl. Phys. Lett.*, **96**, pp. 231104-1-3.
71. Takeuchi, T., Amano, H., and Akasaki, I. (2000). Theoretical study of orientation dependence of piezoelectric effects in wurtzite strained GaInN/GaN heterostructures and quantum wells, *Jpn. J. Appl. Phys.*, **39**, pp. 413–416.
72. Kishino, K., Nagashima, K., and Yamano, K. (2013). Monolithic integration of InGaN-based nanocolumn light-emitting diodes with different emission colors, *Appl. Phys. Express*, **6**, pp. 012101-1-3.

73. Kishino, K., Yanagihara, A., Ikeda, K., and Yamano, K. (2015). Monolithic integration of four-color InGaN-based nanocolumn LEDs, *Electron. Lett.*, **51**, pp. 852–854.

74. Yamano, K., Kishino, K., Nagashima, K., Goto, M., and Kikuchi, A. (2010). Green light emission from regularly arranged nanocolumn array LEDs, *2010 MRS Fall Meeting, Boston, Massachusetts*, USA, November 29–December 3.

75. Schubert, E. F. (2003). *Light-Emitting Diodes* (Cambridge University Press).

76. Vadivelu, R., Igawa, Y., and Kishino, K. (2013). 633 nm red emissions from InGaN nanocolumn light-emitting diode by radio frequency plasma assisted molecular beam epitaxy, *Jpn. J. Appl. Phys.*, **52**, pp. 08JE18-1-2.

77. Kouno, T., Kishino, K., Yamano, K., and Kikuchi, A. (2009). Two-dimensional light confinement in periodic InGaN/GaN nanocolumn arrays and optically pumped blue stimulated emission, *Opt. Express*, **17**, pp. 20440–20447.

78. Ishizawa, S., Kishino, K., Araki, R., Kikuchi, A., and Sugimoto, S. (2011). Optically pumped green (530–560 nm) stimulated emissions from InGaN/GaN multiple-quantum-well triangular-lattice nanocolumn arrays, *Appl. Phys. Express*, **4**, pp. 055001-1-3.

79. Yanagihara, A., Ishizawa, S., and Kishino, K. (2014). Directional radiation beam from yellow-emitting InGaN-based nanocolumn LEDs with ordered bottom-up nanocolumn array, *Appl. Phys. Express*, **7**, pp. 112102-1-4.

80. Rangel, E., Matioli, E., Choi, Y.-S., Weisbuch, C., Speck, J. S., and Hu, E. L. (2011). Directionality control through selective excitation of low-order guided modes in thin-film InGaN photonic crystal light-emitting diodes, *Appl. Phys. Lett.*, **98**, pp. 081104-1-3.

81. Sakai, M., Inose, Y., Ema, K., Ohtsuki, T., Sekiguchi, H., Kikuchi, A., and Kishino, K. (2010). Random laser action in GaN nanocolumns, *Appl. Phys. Lett.*, **97**, pp. 151109.1–3.

82. Kishino, K., and Ishizawa, S. (2016). Spectrally-broadened multimode lasing based on structurally graded InGaN nanocolumn photonic crystals suitable for reduction of speckle contrast, *Appl. Phys. Lett.*, **109**, pp. 071106-1-5.

Part II

Materials

Chapter 3

Novel GaNP Nanowires for Advanced Optoelectronics and Photonics

Irina A. Buyanova,[a] Charles W. Tu,[b] and Weimin M. Chen[a]

[a]*Department of Physics, Chemistry and Biology, Linkoping University, Linköping 58183, Sweden*
[b]*Department of Electrical and Computer Engineering, University of California, La Jolla, California, USA*

irb@ifm.liu.se

GaNP alloys are novel III–V semiconductors, which have a great potential for applications in amber-red light-emitting diodes and also as an active material in innovative intermediate band solar cells. These alloys have become available in the one-dimensional geometry only most recently, thanks to the advances in the nanowire growth utilizing molecular beam epitaxy. In this chapter we will discuss structural and optical properties of such novel GaNP nanowires, as well as their potential for future applications in optoelectronics and photonics.

3.1 Introduction

GaNP alloys are obtained by the incorporation of several percentages of nitrogen in GaP. This has a dramatic effect on

Novel Compound Semiconductor Nanowires: Materials, Devices, and Applications
Edited by Fumitaro Ishikawa and Irina Buyanova

ISBN 978-981-4745-76-5 (Hardcover), 978-1-315-36440-7 (eBook)
www.panstanford.com

the band structure of the forming alloy and allows tailoring of its bandgap energy within the green-amber-red spectral ranges, owing to the giant bowing in the bandgap energy [1–3]. It also leads to a substantial increase in radiative efficiency of the material, as GaNP alloys acquire the quasi-direct bandgap even for rather low nitrogen compositions exceeding only 0.4% [4, 5]. As a result, GaNP is currently considered as a promising material for applications in technologically challenging amber-red light-emitting diodes. Moreover, because of the strong perturbation of the host states upon nitrogen incorporation, the conduction band of GaNP splits into two well-separated sub-bands, which makes this material suitable for applications in intermediated band solar cells.

Due to a large miscibility gap between GaP and GaN, the growth of GaNP is known to be rather challenging. In the NW geometry, these alloys with a reasonable structural and high optical quality were first fabricated in 2012, using molecular beam epitaxy (MBE) [6]. In this chapter, we will review our recent efforts devoted to the optimization of GaNP-based NWs for future applications in light-emitting devices and discuss the impacts of structural polymorphism on the radiative efficiency and band structure of this material. We will also show that GaNP NWs can be utilized as a source of linearly polarized light with the polarization direction that is not determined by dielectric mismatch between the NW and its surrounding.

3.2 Optimizing Light Emission Efficiency of GaNP Coaxial Nanowires

The comprehensive optimization of the structural design of coaxial GaNP-based NWs aimed at improving the radiative efficiency of these structures was performed in Ref. 7.

3.2.1 Sample Structure

NWs grown under non-optimized conditions usually suffer from various point and structural defects, which degrade the radiative efficiency. In order to investigate the impacts of nitrogen content, growth mechanisms, and structural design on the defect formation, we have studied several types of NW structures. All of them were grown on (111)-oriented Si substrates by gas-source molecular beam epitaxy (GS MBE) using Ga droplets as a catalyst [6, 8]. The structures of type (A) were 80–100 nm thick Ga(N)P NWs grown via the vapor–liquid–solid (VLS) mechanism. The type (B) structures included GaP/GaN_yP_{1-y} and $GaN_xP_{1-x}/GaN_yP_{1-y}$ ($x < y$) core/shell NWs with the VLS-grown core from type (A) NWs and the 50 nm thick shell layer formed via the step-mediated growth. The type (C) $GaP/GaN_yP_{1-y}/GaN_xP_{1-x}$ ($x < y$) core/shell/shell NWs had an additional passivating outer shell (30–80 nm thick).

To initiate the NW growth for the type-A structures, a seed GaP region was grown prior to igniting a nitrogen plasma [8]. The growth was performed with substrate temperatures T_{sub} of 580–615°C. The V/III incorporation ratio was kept at around 1.5–2.0 during the GaP growth and was increased to 2.5 after a nitrogen plasma was ignited for the growth of GaNP. The same growth conditions were used for the core layers in (B) and (C) structures. The GaNP shells were fabricated via the step-mediated growth with a lower substrate temperature T_{sub} of 450–510°C and a V/III incorporation ratio of 3.5–4.5. The nitrogen composition in the shell layers was varied by changing the power of the rf-plasma and the nitrogen flux. It was estimated by comparing the room-temperature PL data with the band anticrossing (BAC) model [2, 4]. In all types of the studied structures, the NWs displayed a uniform size distribution with an average axial length of approximately 2–3 μm. A detailed description of the growth procedure can be found in Refs. 6 and 8.

Studied NW structures are shown schematically in Fig. 3.1a, where the light (orange online) areas indicate the optically active region of the NWs with the highest nitrogen composition, that is, the outer GaN_yP_{1-y} shell in the type-B structures and the inner GaN_yP_{1-y} shell in the type-C NWs. Scanning electron microscopy (SEM) images of the corresponding NW arrays are presented in Figs 3.1b–d.

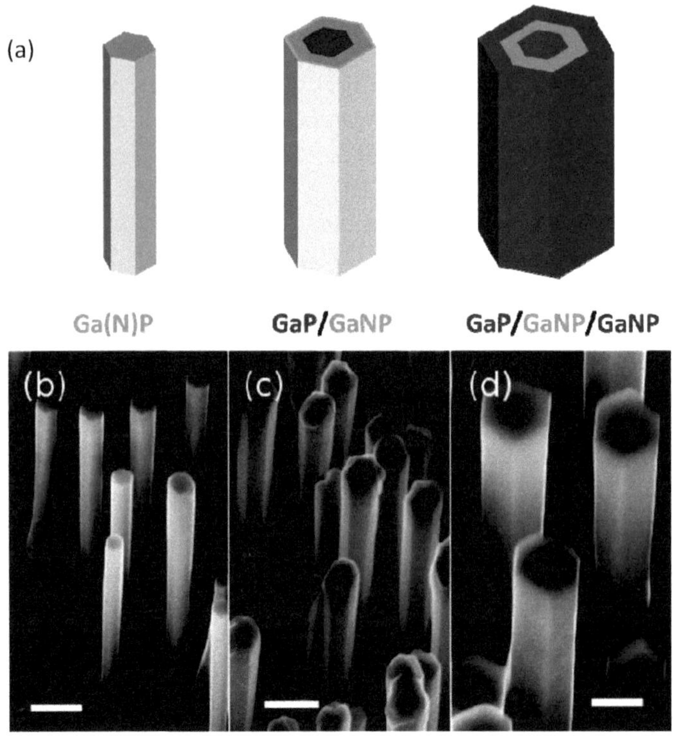

Figure 3.1 (a) Schematic illustrations of the studied NW structures. Light (orange online) areas are optically active regions of the NWs with the highest N. SEM images of arrays of the Ga(N)P NWs (b), GaP/GaNP core/shell NWs (c) and $GaP/GaN_yP_{1-y}/GaN_xP_{1-x}$ core/shell/shell NWs (d) grown on (111) Si substrates. The scale bar displayed in (b)-(d) is 300 nm. Reprinted with permission from Ref. 7 (Stehr et al., *Nano Lett.*, 2015, 15, p. 242). Copyright (2015) American Chemical Society.

3.2.2 Defect Identification

Nitrogen incorporation in bulk dilute nitrides is known to promote the formation of various point defects, which act as centers of efficient non-radiative recombination and suppress radiative efficiency [1]. In order to clarify the role of various point defects in the studied GaNP NWs, we have employed the optically detected magnetic resonance (ODMR) technique, as was first reported in Ref. 7. This technique represents a powerful tool to investigate chemical identity and local structural properties of defects in semiconductors, as it combines chemical sensitivity of magnetic resonance with high sensitivity of the optical method. The ODMR signals were measured by monitoring the near-band-edge emission from the GaNP alloy caused by the recombination of excitons trapped at various N-related centers [1, 10]. The typical photoluminescence (PL) spectrum of the monitored emission is shown in Fig. 3.2a. All investigated structures were found to exhibit the same two-peak ODMR signal, as shown by the open circles in Figs. 3.2c,d. Its chemical identity was revealed from the performed analysis of the ODMR spectra using the Spin-Hamiltonian:

$$\mathcal{H} = \mu_B B g S + \mathrm{SAI} \tag{3.1}$$

Here B denotes an external magnetic field, g and A are the electron g-tensor and the hyperfine interaction tensor, respectively. The splitting between the ODMR peaks was found to remain the same for the two microwave frequencies used in the ODMR experiments (see Figs. 3.2c,d). Based on Eq. 3.1, this provided the evidence that the doublet structure reflects a resolved hyperfine interaction between an unpaired localized electron spin (S = 1/2) and a nuclear spin I = 1/2 with 100% natural abundance, i.e., a ^{31}P nucleus. The spin-Hamiltonian parameters of the center were then determined as $g_{\|}$ = 2.002, $g_{\perp}$ = 2.013, $A_{\|}$ = 330 × 10^{-4} cm^{-1} and $A_{\perp}$ = 130 × 10^{-4} cm^{-1}. The simulated ODMR spectra are shown by the solid line (red online) in Figs. 3.2c,d. The obtained spin-Hamiltonian parameters were found to be identical to those of the so-called DD1 defect [9], which consists of a P_{Ga} antisite (or a P_i interstitial) aligned with a neighboring partner along a <111> crystallographic direction. In planar structures the defect has been shown to reside at the GaNP side of the GaP/GaNP interface [9]. In the case of GaNP-based NWs, the most probable location of DD1 were concluded [7] to include (i) the interfaces between the GaNP shell and the GaP core in the type-B and -C structures and (ii) the surfaces of the NWs. A possible DD1 configuration at the NW surfaces is schematically illustrated in Fig. 3.2b.

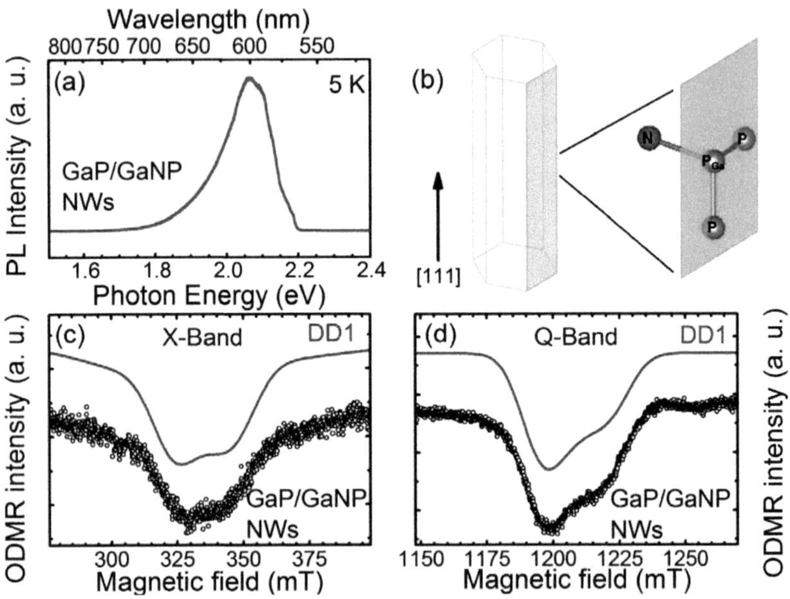

Figure 3.2 (a) A representative PL spectrum measured at 5 K from the GaP/GaNP NW array. (b) A model illustrating a possible configuration of the DD1 defect on the NW surface, taking as an example a P_{Ga} antisite forming a complex with a partnering nitrogen atom. (c) and (d) ODMR spectra of the GaP/GaNP NWs measured at 4 K (the open circles) at X- band (9.4 GHz) and Q-band (34 GHz), respectively. The solid lines are the simulated spectra of DD1 using the spin-Hamiltonian parameters given in the text. Reprinted with permission from Ref. 7 (Stehr et al. *Nano Lett.*, 2015, 15, p. 242). Copyright (2015) American Chemical Society.

The processes responsible for the DD1 formation were single out based on the detailed defect studies of the GaNP NWs as a function of growth conditions and structural design [7]. The formation of DD1 was found to be promoted by the presence of nitrogen and a high growth rate but could be suppressed by reducing the strain between the core and shell layers, as well as by protecting the optically active shell by an outer passivating shell. The experimental results that have led to these conclusions are summarized in Fig. 3.3 and will be briefly discussed below.

First, it was found that the defect could only be detected in N-containing NWs and that its concentration increases with increasing nitrogen content- see the inset in Fig. 3.3. The latter suggested that the nitrogen atom can be involved as a partner atom within the DD1 complex.

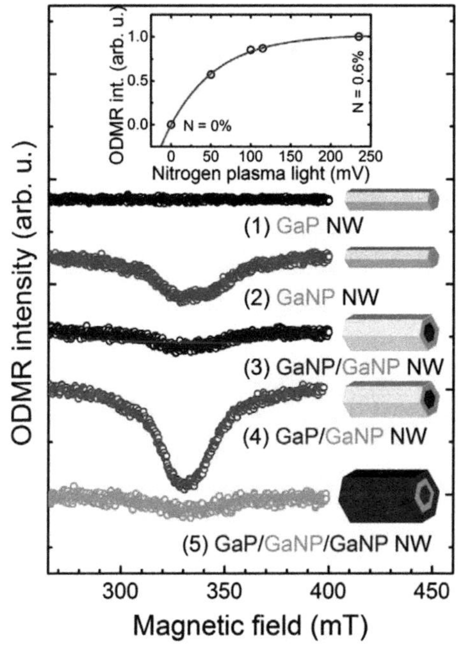

Figure 3.3 ODMR spectra from (1) the GaP NWs, (2) GaNP NWs, (3) $GaN_xP_{1-x}/GaN_yP_{1-y}$ NWs, (4) GaP/GaN_yP_{1-y} NWs and (5) $GaP/GaN_yP_{1-y}/GaN_xP_{1-x}$ NWs, together with illustrations of the corresponding NW structures. The nitrogen composition in the optically active shell was y = 0.006. The inset presents the DD1 ODMR signal intensity as a function of the nitrogen content in the type-A Ga(N)P NWs. The solid line is a guide to the eye. Reprinted with permission from Ref. 7 (Stehr et al., *Nano Lett.*, 2015, 15, p. 242). Copyright (2015) American Chemical Society.

Second, the intensity of the DD1 ODMR signal was found to be substantially reduced in the type-B GaN_xP_x/GaN_yP_{1-y} ($x < y$) core/shell structures (curve 3) as compared with the type-A GaN_yP_{1-y} NWs (curve 2) with the same nitrogen composition y in the optically active layer. The observed suppression of the defect formation was attributed to the reduction of the growth rate from about 150 nm/min during the axial VLS growth of the type-A GaNP NWs [8] down to about 1.8 nm/min during the radial step-mediated growth of the GaNP shell [6] in the type-B structures.

From Fig. 3.3 is also noticeable that a substantial enhancement of the DD1 signal was observed in the GaP/GaN_yP_{1-y} core/shell NWs (curve 4) as compared with that in the GaN_xP_x/GaN_yP_{1-y} structures (curve 3). Since the GaN_yP_{1-y} shell layers in both structures had the same nitrogen composition and were grown under identical conditions, this result suggested that the presence of strain between the GaP core and the GaN_yP_{1-y} shell promotes the formation of DD1.

And finally, based on the comparison of the ODMR spectra (4) and (5) in Fig. 3.3 it is apparent that adding the outer passivating shell of GaN_xP_{1-x} in the type-C GaP/GaN_yP_{1-y}/GaN_xP_{1-x} core/shell/shell NWs, caused a dramatic (by about one order of magnitude) reduction of the DD1 intensity. This provided unambiguous experimental evidence that the majority of the DD1 defects were formed at the NW surfaces.

3.2.3 Impacts of Defects on Carrier Recombination

The ODMR signal was found to be negative, corresponding to a microwave induced decrease of the PL intensity. This implied that the responsible defect acted as a recombination center competing with the monitored PL. Suppression of its formation in the multi-shell NWs, therefore, should lead to a significant enhancement of their radiative efficiency at room temperature (RT). This was indeed demonstrated in Ref. 7 based on μ-PL experiments performed on single GaP/GaN_yP_{1-y} core/shell and single GaP/GaN_yP_{1-y}/GaN_xP_{1-x} core/shell/shell NWs. Although in both structures the PL spectra were governed by the same emission (see Fig. 3.4a) due to the radiative recombination of excitons bound to different N-related centers [10], its intensity was significantly enhanced after the formation of the outer shell, i.e., after suppression of the DD1 formation. The same conclusion also followed from statistics acquired on 50 NWs from each type of the structures. The RT PL intensity was found to follow a Gaussian distribution which peaks at around 20 and 210 counts/sec for the GaP/GaN_yP_{1-y} NWs and GaP/GaN_yP_{1-y}/GaN_xP_{1-x} NWs, respectively—see Fig. 3.4b. The origin of the observed enhancement in the PL intensity was further clarified from transient PL measurements. The transient PL images measured from the arrays of GaP/GaN_yP_{1-y}/GaN_xP_{1-x} NWs and GaP/GaN_yP_{1-y} NWs are shown in Figs. 3.4c,d, respectively. The PL decay was found to be substantially slower in the structures with the outer passivating shell, as is also obvious from Fig. 3.4e, where the PL decays from both NWs arrays are shown. This slower PL

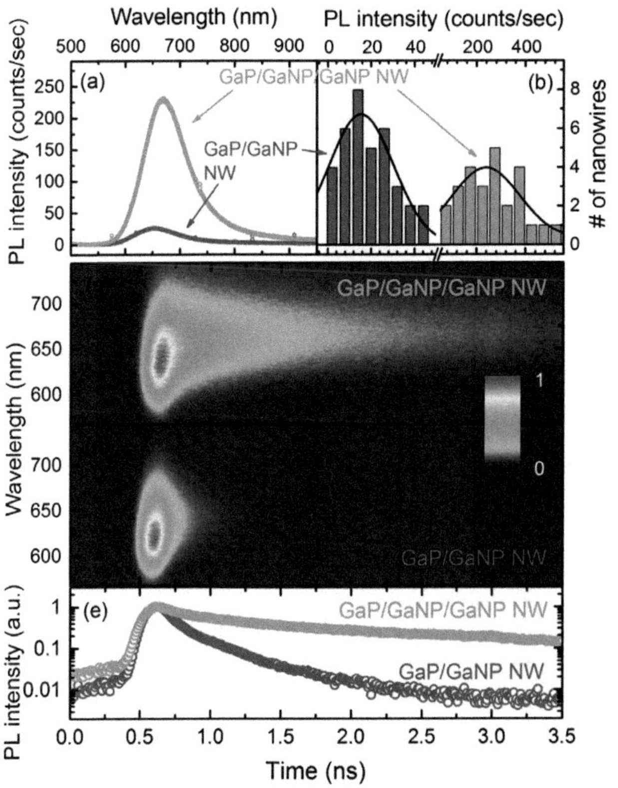

Figure 3.4 (a) Representative μ-PL spectra from a single GaP/GaN$_y$P$_{1-y}$ core/shell NW and a single GaP/GaN$_y$P$_{1-y}$/GaN$_x$P$_{1-x}$ core/shell/shell NW with $y \sim 0.6\%$, as estimated using the band anticrossing model [10, 40]. (b) Statistical distributions of the PL intensity from individual GaP/GaNP core/shell NWs and GaP/GaN$_y$P$_{1-y}$/GaN$_x$P$_{1-x}$ core/shell/shell NWs. (c) and (d) display transient PL images measured from GaP/GaN$_y$P$_{1-y}$/GaN$_x$P$_{1-x}$ and GaP/GaN$_y$P$_{1-y}$ NW arrays, respectively. (e) The corresponding decays of the PL intensity spectrally integrated within the 10 nm range around its peak position. The data in (c)–(e) are normalized to the same peak intensity for an easier comparison. All PL measurements were performed at RT. Reprinted with permission from Ref. 7 (Stehr et al., *Nano Lett.*, 2015, 15, p. 242). Copyright (2015) American Chemical Society.

decay was attributed to a reduction of non-radiative carrier recombination in the multi-shell structure. Indeed, the decay time (τ) of radiative transitions is determined by combined contributions of radiative and non-radiative recombination processes so that $1/\tau = 1/\tau_r + 1/\tau_{nr}$, where τ_r and τ_{nr} denote the radiative and non-radiative lifetimes, respectively. Since the radiative lifetime is intrinsic to the N-related emission, the prolongation of its decay in the multi-shell structure combined with the observed increase in the PL intensity provided the compelling evidence that surface passivation of the active region efficiently suppressed the non-radiative recombination mediated by DD1.

3.2.4 Optimized Design

Based on the results of the ODMR and transient PL studies discussed above, the optimized design of GaNP NWs leading to a high efficiency of light emission was identified [7] as being the Ga(N)P/GaNP/GaNP core/shell/shell structure, where the active inner shell with the highest nitrogen content is grown via the step-mediated mode and has passivated surfaces. We note that the suggested radial core/multi-shell design provides an additional freedom to independently control composition, doping and size of core and shell layers, thereby tuning the band offset across the radial hetero-interfaces to create an efficient radial confinement or to enhance the separation of carriers.

3.3 Effects of Polytypism on Optical Properties and Band Structure

As is extensively documented in the literature, III–V nanowires can be grown with both zinc-blende (ZB) and wurtzite (WZ) lattice structure [11–15]. Since the ZB and WZ phases typically exhibit different bandgap energies, such polytypism can facilitate heterostructure engineering, provided that the crystal structure is well controlled. On the other hand, random formation of different lattice phases, which is usually accompanied by a large number of structural defects, has been reported [15–18] to deteriorate both transport and optical properties of the nanowires. We have analyzed effects of polytypism on the radiative recombination and band structure of Ga(N)P NWs based on correlative TEM and cathodoluminescence (CL) studies. The corresponding results were first reported in Ref. 19 and will be briefly summarized below.

3.3.1 Effects of Polytypism on Radiative Recombination

According to the performed TEM studies, the investigated GaP and GaP/GaNP core/shell NWs predominantly exhibited the ZB structure with a rather high density of randomly nucleated twin planes perpendicular to the growth axis. Close to the top of the NWs, a WZ segment was often formed with the total contribution of the WZ phase being less than 2% of the total volume. On the other hand, in some of the wires, WZ segments could extend to up to 150 nm in length. A characteristic TEM image of such wire is shown in Fig. 3.5b. Effects of the polytypism on optical properties and band structure of Ga(N)P were investigated [19] by performing spatially resolved CL measurements on the NWs, which were first characterized by TEM. The representative results from such measurements are summarized in Fig. 3.5, which shows the CL spectra (c)–(d) collected from different areas of the NW together with the corresponding SEM (a) and TEM (b) images. The spectra contain a broad CL band with a rather asymmetric lineshape. This emission originates from the GaNP shell, as it was not detected from the reference GaP wires. According to the previous PL studies of planar GaNP epilayers [20–22] and GaNP-based NW structures [10, 23], the emission stems from radiative recombination of excitons trapped at various N-related localized cluster states.

From Fig. 3.5b,c, it is noticeable that the CL intensity decreases by about 40% within the central part of the NW, i.e., in the area with the highest density of structural defects. Consequently, we have suggested that these structural defects either directly act as effective non-radiative recombination centers competing with the monitored emission, or they promote formation of various point defects that participate in non-radiative recombination processes. The same trend was also reported for III–V and II–VI NWs [24–26].

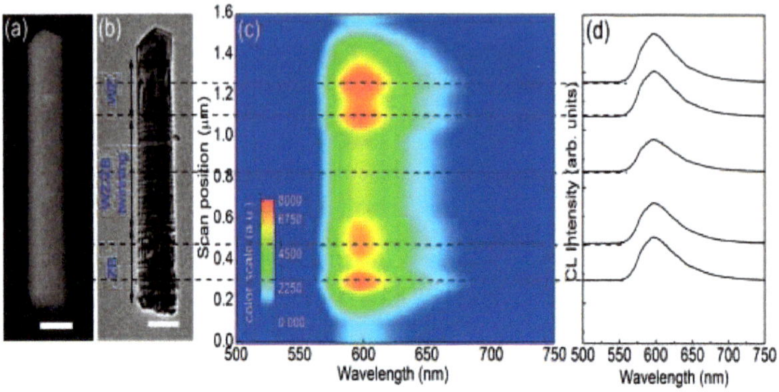

Figure 3.5 (a) SEM and (b) TEM images of a $GaP/GaN_{0.009}P_{0.991}$ NW with a large WZ inclusion. The scale bar is 200 nm. (c) CL line-scan image measured at 50 K along the axis of the same NW as shown in (a) and (b). The CL emission intensity is displayed as a function of the wavelength and emission position along the line-scan. The linear color scale is given in arbitrary units. (d) The corresponding CL spectra collected from the regions of the NW, as indicated by the horizontal dashed lines. Reprinted with permission from Ref. 19 (Dobrovolsky et al., *Nano Lett.*, 2015, 15, p. 4052). Copyright (2015) American Chemical Society.

Most remarkably, the CL spectra recorded from the WZ segment localized within the top region were found to be identical in shape and very similar in intensity to that measured from the ZB region located close to the bottom end of the wire. This implied that the change in the lattice structure from ZB to WZ does not affect the properties of light emission in the GaNP alloy. The same trend was observed for all investigated NWs independent of the N content.

3.3.2 Effects of Nitrogen Content on Optical Properties of Zinc Blende and Wurtzite GaNP Wires

The effect of N incorporation on the optical properties of WZ and ZB Ga(N)P NWs can be analyzed by comparing the low-temperature CL spectra of GaP and GaP/GaNP core/shell nanowires with the N composition varying between 0.9% and 1.9%—see Fig. 3.6. Due to an indirect bandgap, luminescence spectra from ZB GaP usually contain various BE lines that can only be detected at low temperatures. Such optical behavior was indeed observed in the studied GaP NWs where the CL emission was dominated by a series of rather week lines due to BE transitions of unknown origin that could only be detected at $T < 40$ K—see Fig. 3.6. On the other hand, in ZB GaNP alloys the emission spectra are caused by radiative recombination of excitons bound to various N-related cluster states [10, 20–23]. The corresponding optical transitions have a significantly higher oscillator strength due to a strong admixture of the Γ character to the N-related states [3, 5]. This resulted in a significant enhancement of the CL intensity in the GaNP/GaP nanowires as compared with that in the reference GaP structures, as can also be seen from Fig. 3.6. It is also noticeable that the CL emission exhibits a red shift upon N incorporation. The reason for this red shift is as follows. Nitrogen incorporation induces a huge downshift of the CB edge of GaNP, whereas deep states related to the N clusters are pinned in energy to the vacuum level. Consequently, upon increasing N content the defect levels related to different N centers and clusters one-by-one become resonant with the CB states [1, 3], as is schematically shown in the inset in Fig. 3.6. This results in disappearance of the related optical transitions and, consequently, a gradual red shift of the monitored emission. The high-energy cut-off of the emission spectra is determined by the energy position of the CB edge (or the bandgap energy). The latter is indicated by the arrows in Fig. 3.6, and was estimated based on the BAC model [2, 27] using the known parameters for ZB GaNP.

In contrast to ZB Ga(N)P, the band structure of WZ Ga(N)P is yet to be understood and remains under debates even for nitrogen-free GaP [28–32]. Our results from the performed CL mapping of the Ga(N)P NWs (Figs. 3.5 and 3.6) clearly show that both WZ and ZB phases have very similar optical properties, judging from the spectral lineshape, energy positions and also intensities of the detected CL emission. These findings first of all imply that both polytypes have a very similar bandgap energy E_g, as an increase (decrease) of the bandgap should otherwise should lead to a corresponding blue (red) shift of the Ga(N)P emission. Secondly, in spite of the better structural quality of the WZ segments, the emission intensities were found to be comparable from the WZ and ZB Ga(N)P. This suggests that WZ GaP does not acquire a true direct-bandgap character. Indeed, the radiative efficiency of optical transitions in direct bandgap semiconductors should be substantially higher than that in indirect bandgap materials, typically by about two-three orders of magnitude [33]. Because of the band enhancement effect [34], this should remain valid even for BE optical transitions such as isoelectronic excitons bound to different N clusters monitored in the studied GaNP NWs or BE transitions in GaP. Such enhancement, however, was not observed experimentally—see Fig. 3.6.

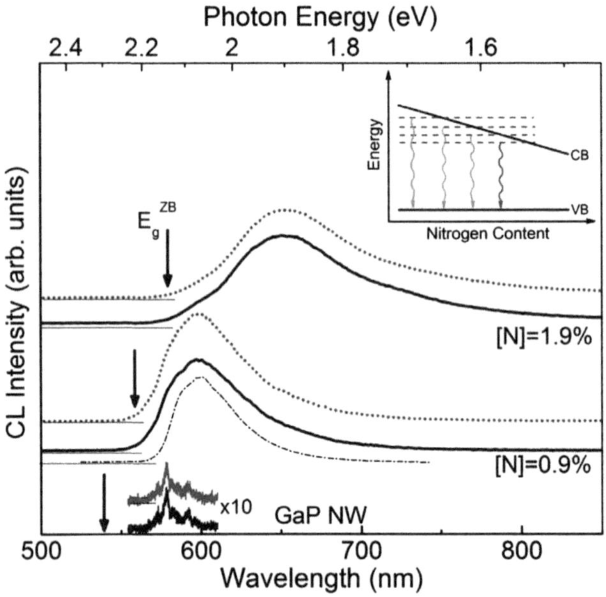

Figure 3.6 Normalized CL spectra collected from the ZB (solid blue lines) and the WZ (red dotted lines) sections of the GaP NW and the GaP/GaNP NWs with [N] = 0.9% and 1.9%. Also shown is a PL spectrum from a ZB $GaN_{0.009}P_{0.991}$ epilayer (the black dash-dotted line). The CL spectra of the N-containing NWs were measured at 50 K, whereas a lower measurement temperature of 30 K was used to record the emission from the GaP wires, due to strong thermal quenching of the monitored emission at higher temperatures. The spectra are vertically offset for clarity, as indicated by horizontal lines. The vertical arrows mark the energy position of the bandgap of ZB GaP with the corresponding nitrogen content. An energy diagram is schematically shown in the insert. Reprinted with permission from Ref. 19 (Dobrovolsky et al., *Nano Lett.*, 15, 2015, p. 4052). Copyright (2015) American Chemical Society.

For the GaP NWs, the aforementioned experimental findings are generally consistent with the results of local density band-structure calculations [30, 32]. It was theoretically derived that

WZ GaP is a pseudo-direct semiconductor with the lowest Γ_{8c} CB state at the zone center that stem from the zone folded L_{6c} states. Consequently, the material should remain optically inactive in the WZ structure, as the direct optical transitions that define the absorption edge are dipole-forbidden [30, 32]. The calculated Γ_{8c} energy in WZ was found to be only slightly below that of the X_{6c} CBM in ZB, i.e., by about 0.06–0.2 eV [30, 31]. Although our results from the WZ GaP did not allow us to accurately determine its bandgap energy E_g^{WZ}, they placed the lowest bound of E_g^{WZ} at 2.22 eV at 30 K, which corresponds to the highest energy of the detected BE emission. We also note that our studies [19] left open the possibility that the lowest CB minimum in WZ GaP remains at the X point of the Brillouin zone.

As to the GaNP alloys, the results of CL mapping proved that the WZ and ZB polytypes have the same bandgap energy, as the high energy cut-off of the CL spectra was found to be identical in both cases. This finding was understood as follows. Phenomenologically, the change of the bandgap energy of GaP upon N incorporation can be described within the band anticrossing model, which suggests that the energy position of the CB edge can be adequately modeled by considering an anticrossing interaction between the extended Γ CB state of the GaP matrix and the localized state (E_N) of a substitutional N_P center [2, 27]. The strength of this interaction is mainly determined by the energy distance between the Γ CB minimum and the E_N position. In ZB, the interacting host states are the Γ_{6c} CB states, whereas their equivalent in WZ are the Γ_{7c} CB states that are theoretically predicted [30, 32] to be located almost at the same energy as the Γ_{6c} states in ZB. To explain the observed independence of the GaNP bandgap energy on the crystal structure we then assumed that the energy position of the N-related states remains the same in the WZ and ZB materials. This assumption took into account the deep-level character of these states and the identical spectral positions of the corresponding BE emissions observed experimentally. Most importantly, our results clearly demonstrated that both WZ and ZB GaNP have a significantly higher efficiency of radiative recombination than that in the parental GaP. This led us to the conclusion that GaNP NWs are promising for potential applications as efficient nanoscale light emitters within the amber-red spectral range.

3.4 Polarized Emitters from GaNP Nanowires

Light absorption and emission in thin NWs is strongly polarized along their axis, due to a large dielectric contrast between a nanowire and its surroundings [35]. Consequently, individual III–V NWs can be used as polarization-sensitive photodetectors or nanoscale sources of linearly polarized light [36, 37]. For device applications, however, it is desirable to control the polarization direction. This, to a certain extent, can be achieved utilizing selection rules of optical transitions. For example, it was recently shown that though polarization of light emission in ZB GaAs and InP nanowires is determined by the antenna effect, their WZ counterparts emit light that is linearly polarized in the direction perpendicular to the NW axis [38, 39]. The WZ phase formation, however, is energetically favorable only in thin NWs [40, 41] with a high surface-to-volume ratio, which imposes size restrictions and also limits maximum achievable polarization in WZ wires due to the counteracting optical antenna effect [35]. We have shown that these limitations can be circumvented in GaNP nanowires and have discussed our findings in Ref. 42.

3.4.1 Polarization of Light Emission

Polarization properties of the light emission were analyzed by employing μ-PL spectroscopy performed on a single NW mechanically transferred onto another Si substrate. The PL intensity (I) was measured as a function of the angle (ϕ) between the polarization axis of the analyzer (E) and the nanowire axis, as shown schematically in Fig. 3.7. The degree of linear PL polarization was calculated by

$$P(\%) = 100 \times \frac{(I_{\parallel} - I_{\perp})}{(I_{\parallel} + I_{\perp})}, \tag{3.2}$$

where $I_{\parallel}$ ($I_{\perp}$) denotes the PL intensity detected in linear polarization parallel (perpendicular) to the direction of the NW axis.

It was found that the PL emission in all wires is preferentially linearly polarized in the direction orthogonal to the NW axis and that this polarization direction remains the same independent on the nanowire diameter (see Fig. 3.8). The measured polarization degree was found to be rather high, falling into the range between –40% and –60%.

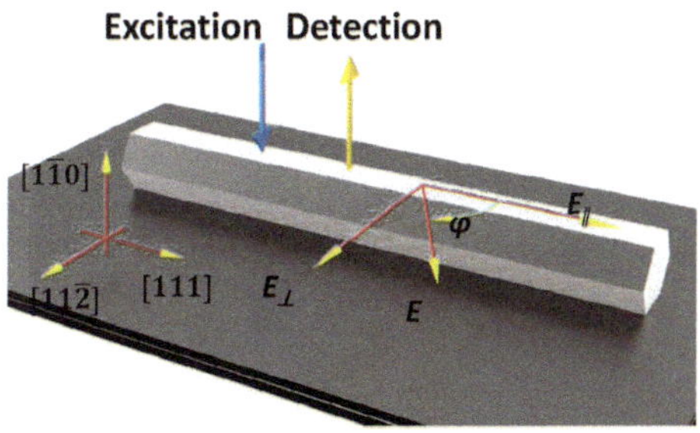

Figure 3.7 Schematic diagram of the measurement geometry utilized for polarization resolved μ-PL measurements. The Cartesian coordinate system is chosen according to the nanowire crystal directions and geometry. $\boldsymbol{E}$ denotes the polarization direction of the analyzer rotated by the angle ϕ relative to the nanowire axis, with $\boldsymbol{E}_{\parallel}$ and $\boldsymbol{E}_{\perp}$ being parallel and perpendicular to the NW axis, respectively. Reprinted with permission from Ref. 42 (Filippov et al., *Nano Lett.*, 14, 2014, p. 5264). Copyright (2014) American Chemical Society.

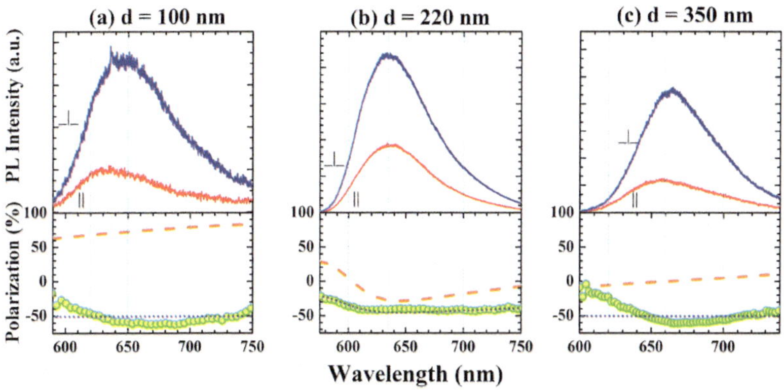

Figure 3.8 Typical PL spectra (solid lines) and linear polarization degree (open circles, green online) measured at 150 K from a GaNP NW (a), GaP/GaNP NW (b) and GaP/GaNP/GaNP NW (c) with an average diameter of 100, 220, and 350 nm, respectively. Labels || and ⊥ denote the PL spectra obtained by detecting linear polarizations parallel and perpendicular to the NW axis, respectively. The dashed and dotted lines are simulated spectral dependences of P expected from the antenna effect for an isotropic and anisotropic emitter, respectively. In the latter case, the degree of linear polarization inside the NW was assumed to be equal to –96% (a), –43% (b) and –82% (c), respectively. Reprinted with permission from Ref. 42 (Filippov et al., *Nano Lett.*, 14, 2014, p. 5264). Copyright (2014) American Chemical Society.

It is known that polarization properties of NWs with diameters smaller than the PL wavelength are modified by the antenna effect to an extent determined by the NW diameter and emission wavelength [35]. In order to estimate the anisotropy of light emission inside the wires, we performed simulations of the expected PL polarization, which can be caused by the dielectric mismatch between the GaNP and the ambient air, for emitters with a varying degree of anisotropy. The best fit to the experimental data yielded the following linear polarization degree of the PL inside the NWs: –96%, –43% and –82% for the NWs with d = 100, 220 and 350 nm, respectively. The internal PL polarization was found to be higher for the thinner NWs grown via the VLS mechanism than for the core/shell structures, where the active GaNP shell was fabricated via the step-mediated growth.

The orthogonal PL polarization was observed at all measurement temperatures over the range of 4–300 K, though the polarization degree somewhat increased at low T. This is demonstrated in Fig. 3.9 which shows temperature variation of the angle θ (Figs. 3.9a–c) and the PL polarization degree (Figs. 3.9d–f) determined at several emission wavelengths.

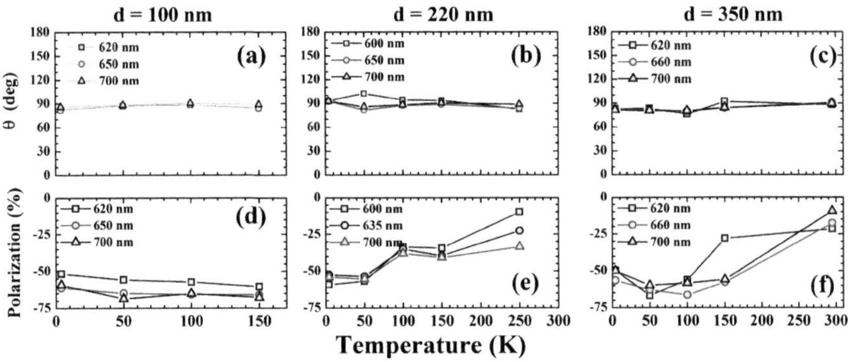

Figure 3.9 Temperature dependences of the angle θ at which the maximum PL intensity was observed (a–c) and the linear polarization degree P (d–f), measured at the specified detection wavelengths. Reprinted with permission from Ref. 42 (Filippov et al., *Nano Lett.*, 14, 2014, p. 5264). Copyright (2014) American Chemical Society.

3.4.2 Origin of the PL Polarization

Let us now discuss the origin of the observed PL polarization that is rather unusual for thin ZB nanowires. We first recall that the

PL emission from GaNP NWs is governed by recombination of excitons trapped at various N-related centers [23]. Electronic structure of isoelectronic excitons bound to such complex centers is determined by combined effects of local field and exchange interaction (or J–J coupling) between a $S_{1/2}$-like electron and a $P_{3/2}$-like hole forming the exciton [43–45]. At sufficiently low temperatures an optical transition from each individual center can be polarized as determined by its local symmetry. As an example, we can consider a polarization response of an axial center with the symmetry of C_{3v} or higher. The axial crystal field, which resembles uniaxial strain, will lift the degeneracy of the $P_{3/2}$-like hole states but would not affect the electron state. The J–J coupling will further split the excitonic states according to the projection m_j of the total angular momentum J along the local symmetry axis [44, 45]. The energy states of bound excitons (BE) for compressive and tensile local strain are shown schematically in Figs. 3.10a,b, respectively. In the case when the strain splitting (2ε) of the hole substates exceeds the electron–hole exchange splitting Δ, the exciton energy levels are composed of two groups of states which are separated in energy comparable with 2ε. From Fig. 3.10 it is obvious that independent of the sign of the local strain, the PL light emitted at low temperatures from a single center should be completely polarized in the plane perpendicular to the center's principal axis. (Lowering down the symmetry will lift the remaining degeneracy and could change the ordering of the exciton states, but may retain polarization of optical transitions from individual excitonic sublevels.)

The polarization anisotropy should be reduced if one takes into account orientational degeneracy of the centers, i.e., the presence of the centers with their principal axes orientated along all possible equivalent crystallographic directions. This affect was analyzed in detail in Ref. 42. It was found that the high experimentally observed values of the polarization degree can only be explained assuming that the N-related centers responsible for the NW emission are not orientation degenerate but possess a particular symmetry axis determined by the growth process. This axis should be oriented so that the emitted light is polarized in the plane perpendicular to the NW axis.

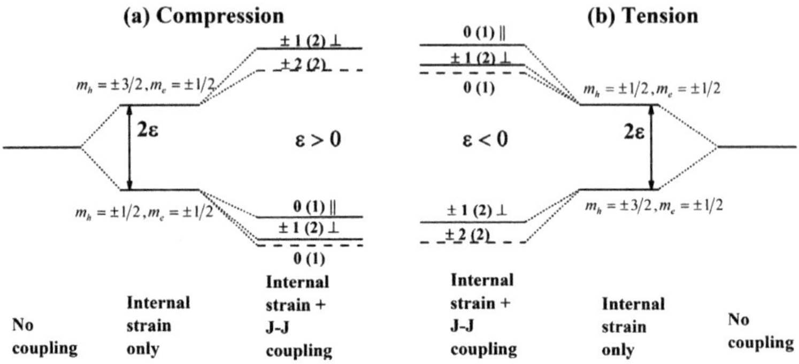

Figure 3.10 (a) Electronic states of an exciton bound to an isoelectronic center with symmetry of C_{3v} or higher in the presence of (a) compressive and (b) tensile local strain. The local strain is assumed to dominate over exchange interaction. The solid (dashed) lines represent the states corresponding to dipole allowed (forbidden) optical transitions. The unbracketed numbers are the m_j values of the exciton sublevels whereas the bracketed numbers indicate their degeneracy. The polarization of the allowed transitions is indicated relative to the principal axis of the center. Reprinted with permission from Ref. 42 (Filippov et al., *Nano Lett.*, 14, 2014, p. 5264). Copyright (2014) American Chemical Society.

The experimental results were consistently explained under the assumption that the principal axis of the N-related centers coincides with the [111] axis of the NWs. The electronic structure of the related BE and its polarization then follow the scheme shown in Fig. 3.10. At cryogenic temperatures, preferential population of the $m_j = \pm 1$ exciton state will lead to complete PL polarization in the plane orthogonal to the NW direction. The sign of the local strain field was singled out based on the measured temperature dependences of the polarization degree. Indeed, in the case of compressive strain (see Fig. 3.10a) the PL polarization should exhibit non-monotonous dependence with increasing T. This is because a temperature increase should first lead to population of the $m_j = 0$ exciton state accompanied by a decrease of polarization degree from −100% down to −33%, when both $m_j = \pm 1$ and $m_j = 0$ states become equally populated. Since the J–J coupling energy Δ for the N-related BE in GaP is of the order of 1 meV [43–45], this decrease in the PL polarization should be observed at rather low measurement temperatures. Polarization anisotropy, however, should be partly restored at higher temperatures due to thermal population of the topmost $m_j = \pm 1$

state. On the other hand, if the N-center experiences tensile local strain, *P* will either remain constant or should gradually decrease with increasing *T*, since the depolarization of PL will appear only when the topmost $m_j = 0$ state becomes populated (see Fig. 3.10b). From Fig. 3.9 it is apparent that the experimental data follows the second scenario, i.e., that the N-related centers experience tensile local strain. This is probably not surprising, considering a considerably smaller covalent radius of an N atom as compared with the replaced P atom. The measured temperature dependences of the PL polarization degree also suggested that the local strain field can be somewhat affected by the growth conditions and is lower for the shallower N-related centers that are responsible for light emission at the high energy side of the PL spectra. This leads to a more pronounced depolarization of the related PL components at elevated temperatures.

We therefore suggested that the strong orthogonal PL polarization in the studied NWs reflected the local structure of the N-related cluster states responsible for the BE emission. Specifically, a majority of these centers experience local tensile strain so that their principal axis is orientated along the NW axis, i.e., the nitrogen clusters are preferentially aligned in one direction. This effect was found to be common for all investigated GaNP NWs irrespective of the growth mechanism responsible for the formation of the GaNP layer. On average, however, the degree of this alignment was higher when the GaNP regions were grown via the VLS mechanism as compared to that when the growth was step mediated. This is based on a higher degree of linear PL polarization inside the GaNP NWs as compared with that in the GaP/GaNP core/shell or GaP/GaNP/GaNP core/shell/shell structures. Although the origin of this alignment is still to be unambiguously established, a possible reason could be a high density of twins orthogonal to the growth direction that affect the local structure of the N-related centers.

3.5 Conclusions and Outlook

In summary, GaNP NWs represent a novel material system with great potential for a variety of optoelectronic and photonic applications. The ability to fabricate GaNP NWs as co-axial multishell heterostructures allows suppression of detrimental surface-related non-radiative recombination, largely improving the radiative efficiency. Although achieving the well-controlled crystalline structure of GaNP in the one dimensional geometry requires further efforts, structural polymorphism does not significantly degrade the optical quality of this material, which is by far superior to that of GaP. In addition to significant improvements in the radiative efficiency, defect engineering via alloying with nitrogen allows realization of polarized nano light sources, which emit light linearly polarized in the direction that is orthogonal to the NW axis even in ZB nanowires of various diameters. The defect engineering can also be used to achieve energy upconversion in GaNP due to a two-step two-photon process via a defect level related to the presence of nitrogen [46]. The current challenges are to optimize these processes and to design strategies of controllably introducing the desired defects during the growth without compromising the material quality.

Acknowledgments

Financial support by the Swedish Energy Agency (grant # P40119-1) is greatly appreciated. The nanowire growth was supported by the U.S. National Science Foundation under Grant # DMR-0907652 and DMR-1106369. The authors also like to acknowledge S. Sukrittanon, Y. Kuang, A. Dobrovolsky, J. E. Stehr, and S. Filippov for their contribution to the results discussed in this chapter.

References

1. Buyanova I. A., Chen W. M. (2004). *Physics and Applications of Dilute Nitrides* (Taylor & Francis Books, Inc., New York).
2. Shan W., Walukiewicz W., Ager J., Haller E., Geisz J., Friedman D., Olson J., Kurtz S. (1999). Band anticrossing in GaInNAs alloys. *Phys. Rev. Lett.*, **82**, pp. 1221–1224.
3. Kent P. R. C., Zunger A. (2004). *Physics and Applications of Dilute Nitrides*, eds., Buyanova I. A., Chen W. M., Chapter 1, an atomistic view of the electronic structure of mixed anion III-V nitrides (Taylor & Francis, London), pp. 23–64.

4. Shan W., Walukiewicz W., Yu K. M., Wu J., Ager J. W., Haller E. E., Xin H. P., Tu C. W. (2000). Nature of the fundamental band gap in GaN_xP_{1-x} alloys. *Appl. Phys. Lett.*, **76**, pp. 3251–3253.

5. Buyanova I. A., Pozina G., Bergman J. P., Chen W. M., Xin H. P., and Tu C. W. (2002). Time-resolved studies of photoluminescence in GaN_xP_{1-x} alloys: Evidence for indirect-direct band gap crossover. *Appl. Phys. Lett.*, **81**, pp. 52–54.

6. Kuang Y. J., Sukrittanon S., Li H., Tu C. W. (2012). Growth and photoluminescence of self-catalyzed GaP/GaNP core/shell nanowires on Si(111) by gas source molecular beam epitaxy. *Appl. Phys. Lett.*, **100**, p. 053108.

7. Stehr J. E., Dobrovolsky A., Sukrittanon S., Kuang Y., Tu C. W., Chen W. M., Buyanova I. A. (2015). Optimizing GaNP coaxial nanowires for efficient light emission by controlling formation of surface and interfacial defects. *Nano Lett.*, **15**, pp. 242–247

8. Sukrittanon S., Dobrovolsky A., Kang W.-M., Jang J.-S., Kim B.-J., Chen W. M., Buyanova I. A., Kuang Y. J., Tu C. W. (2014). Growth and characterization of dilute nitride GaN_xP_{1-x} nanowires and $GaN_xP_{1-x}/GaN_yP_{1-y}$ core/shell nanowires on Si (111) by gas source molecular beam epitaxy. *Appl. Phys. Lett.*, **105**, p. 072107.

9. Dagnelund D., Vorona I. P., Vlasenko L. S., Wang, X. J., Utsumi A., Furukawa Y., Wakahara A., Yonezu H., Buyanova I. A., Chen W. M. (2010). Evidence for a phosphorus-related interfacial defect complex at a GaP/GaNP heterojunction. *Phys. Rev. B*, **81**, p. 115334.

10. Dobrovolsky A., Stehr J. E., Chen S. L., Kuang Y. J., Sukrittanon S., Tu C. W., Chen W. M., Buyanova I. A. (2012). Mechanism for radiative recombination and defect properties of GaP/GaNP core/shell nanowires. *Appl. Phys. Lett.*, **101**, p. 163106.

11. Caroff P., Dick K. A., Johansson J., Messing M. E., Deppert K., Samuelson L. (2009). Controlled polytypic and twin-plane superlattices in III–V nanowires. *Nat. Nanotechnol.*, **4**, pp. 50–55.

12. Algra R. E., Verheijen M. A., Borgström M. T., Feiner L.-F., Immink G., van Enckevort W. J. P., Vlieg E., Bakkers E. P. A. M. (2008). Twinning superlattices in indium phosphide nanowires. *Nature*, **456**, pp. 369–372.

13. Joyce H. J., Wong-Leung J., Gao Q., Hoe T. H., Jagadish C. (2010). Phase perfection in zinc blende and wurtzite III–V nanowires using basic growth parameters. *Nano Lett.*, **10**, pp. 908–915.

14. Spirkoska D., Arbiol J., Gustafsson A., Conesa-Boj S., Glas F., Zardo I., Heigoldt M., Gass M. H., Bleloch A. L., Estrade S., Kaniber M., Rossler J., Peiro F., Morante J. R., Abstreiter G., Samuelson L., Fontcuberta i Morral A. (2009). Structural and optical properties of high quality zinc-blende/wurtzite GaAs nanowire heterostructures. *Phys. Rev. B*, **80**, p. 245325.

15. Bao J., Bell D. C., Capasso F., Wagner J. B., Mårtensson T., Trägårdh J., Samuelson L. (2008). Optical properties of rotationally twinned InP nanowire heterostructures. *Nano Lett.*, **8**, pp. 836–841.

16. Woo R. L., Xiao R., Kobayashi Y., Gao L., Goel N., Hudait M. K., Mallouk T. E., Hicks R. F. (2008). Effect of twinning on the photoluminescence and photoelectrochemical properties of indium phosphide nanowires grown on silicon (111). *Nano Lett.*, **8**, pp. 4664–4669.

17. Perera S., Fickenscher M. A., Jackson H. E., Smith L. M., Yarrison-Rice J. M., Joyce H. J., Gao Q., Tan H. H., Jagadish C., Zhang X., Zou J. (2008). *Appl. Phys. Lett.*, **93**, p. 053110.

18. Parkinson P., Joyce H. J., Gao Q., Tan H. H., Zhang X., Zou J., Jagadish C., Herz L. M., Johnston M. B. (2009). Carrier lifetime and mobility enhancement in nearly defect-free core–shell nanowires measured using time-resolved terahertz spectroscopy. *Nano Lett.*, **9**, pp. 3349–3353.

19. Dobrovolsky A., Persson P. O. Å., Sukrittanon S., Kuang Y., Tu C. W., Chen W. M., Buyanova I. A. (2015). Effects of polytypism on optical properties and band structure of individual Ga(N)P nanowires from correlative spatially resolved structural and optical studies. *Nano Lett.*, **15**, pp. 4052–4058.

20. Buyanova I. A., Rudko G. Y., Chen W. M., Xin H. P., Tu C. W. (2002). Radiative recombination mechanism in GaN_xP_{1-x} alloys. *Appl. Phys. Lett.*, **80**, pp. 1740–1742.

21. Buyanova I. A., Chen W. M., Tu C. W. (2003). Recombination processes in N-containing III–V ternary alloys. *Solid State Electron.*, **47**, pp. 467–475.

22. Liu X., Bishop S. G., Baillargeon J. N., Cheng K. Y. (1993). Band gap bowing in $GaP_{1-x}N_x$ alloys. *Appl. Phys. Lett.*, **63**, pp. 208–210.

23. Dobrovolsky A., Chen S. L., Kuang Y. J., Li H., Sukrittanon S.,Tu C. W., Chen W. M., Buyanova I. A. (2013). Optical properties of GaP/GaNP core/shell nanowires: temperature-dependent study. *Nanoscale Res. Lett.*, **8**, 239.

24. Joyce H. J., Ga, Q, Tan H. H., Jagadish C., Kim Y., Zou J., Smith L. M., Jackson H. E., Yarrison-Rice J. M., Parkinson P., Johnston M. B.

(2011). III–V semiconductor nanowires for optoelectronic device applications. *Prog. Quantum Electron.*, **35**, pp. 23–75.

25. Titova L. V., Hoang T. B., Jackson H. E., Smith L. M., Yarrison-Rice J. M., Kim Y., Joyce H. J., Tan H. H., Jagadish C. (2006). Temperature dependence of photoluminescence from single core-shell GaAs–AlGaAs nanowires. *Appl. Phys. Lett.*, **89**, p. 173126.
26. Svensson C. P. T., Mårtensson T., Trädgårdh J., Larsson C., Rask M., Hessman D., Samuelson L., Ohlsson E. (2008). Monolithic GaAs/InGaP nanowire light emitting diodes on silicon. *J. Nanotechnol.*, **19**, p. 305201.
27. Buyanova I. A., Izadifard M., Kasic A., Arwin H., Chen W. M., Xin H. P., Hong Y. G., Tu C. W. (2004). Analysis of band anticrossing in GaN_xP_{1-x} alloys. *Phys. Rev. B*, **70**, p. 085209.
28. Zhang Z., Senz S., Zhao F., Chen L., Gao X., Liu J.-M. (2012). Phase transition induced vertical alignment of ultrathin gallium phosphide nanowire arrays on silicon by chemical beam epitaxy. *RSC Adv.*, **2**, pp. 8631–8636.
29. Assali S., Zardo I., Plissard S., Kriegner D., Verheijen M. A., Bauer G., Meijerink A., Belabbes A., Bechstedt F., Haverkort J. E. M., Bakkers E. P. A. M. (2013). Direct band gap wurtzite gallium phosphide nanowires. *Nano Lett.*, **13**, pp. 1559–1563.
30. Yeh C.-Y., Wei S.-H., Zunger A. (1994). Relationships between the band gaps of the zinc-blende and wurtzite modifications of semiconductors. *Phys. Rev. B*, 1994, **50**, pp. 2715–2718.
31. De A., Pryor C. E. (2010). Predicted band structures of III-V semiconductors in the wurtzite phase. *Phys. Rev. B*, **81**, p. 155210.
32. Belabbes A., Panse C., Furthmüller J., Bechstedt F. (2012). Electronic bands of III–V semiconductor polytypes and their alignment. *Phys. Rev. B*, **86**, p. 075208.
33. Groves W. O., Herzog A. H., Cradord M. (1971). The effect of nitrogen doping on $GaAs_{1-x}P_x$ electroluminescent diodes. *Appl. Phys. Lett.*, **19**, pp. 184–186.
34. Kash J. A., Collet J. H., Wolford D. J., Thompson J. (1983). Luminescence decays of N-bound excitons in $GaAs_{1-x}P_x$. *Phys. Rev. B*, **27**, pp. 2294–2300.
35. Ruda H. E., Shik A. (2006). Polarization-sensitive optical phenomena in thick semiconducting nanowires. *J. Appl. Phys.*, **100**, p. 024314.
36. Wang J., Gudiksen M. S., Duan X., Cui Y., Lieber C. M. (2001). Highly polarized photoluminescence and photodetection from single indium phosphide nanowires. *Science*, **293**, pp. 1455–1457.

37. Muskens O. L., Diedenhofen S. L., van Weert M. H. M., Borgström M. T., Bakkers E. P. A. M., Gomez J. R. (2008). Epitaxial growth of aligned semiconductor nanowire metamaterials for photonic applications. *Adv. Funct. Mater.*, **18**, pp. 1039–1046.

38. Mishra A., Titova L. V., Hoang T. B., Jackson H. E., Smith L. M., Yarrison-Rice J. M., Kim Y., Joyce H. J., Gao Q., Tan H. H., Jagadish C. (2007). Polarization and temperature dependence of photoluminescence from zincblende and wurtzite InP nanowires. *Appl. Phys. Lett.*, **91**, p. 263104.

39. Hoang B. T., Moses A. F., Ahtapodov L., Zhou H., Dheeraj D. L., van Helvoort A. T. J., Fimland B.-O., Weman H. (2010). Engineering parallel and perpendicular polarized photoluminescence from a single semiconductor nanowire by crystal phase control. *Nano Lett.*, **10**, pp. 2927–2933.

40. Caroff P., Dick K. A., Johansson J., Messing M. E., Deppert K., Samuelson L. (2008). Controlled polytypic and twin-plane superlattices in III–V nanowires. *Nat. Nanotech.*, **4**, pp. 50–55.

41. Joyce H. J., Gao Q., Tan H., Jagadish C., Kim Y., Zou J., Smith L. M., Jackson H. E., Yarrison-Rice J. M., Parkinson P. (2011). III–V semiconductor nanowires for optoelectronic device applications. *Prog. Quantum Eelectron.*, **35**, pp. 23–75.

42. Filippov S., Sukrittanon S., Kuang Y., Tu C. W., Persson P. O. Å., Chen W. M., Buyanova I. A. (2014). Origin of strong photoluminescence polarization in GaNP nanowires. *Nano Lett.*, **14**, pp. 5264–5269.

43. Thomas D. G., Hopfield J. J. (1966). Isoelectronic traps due to nitrogen in gallium phosphide. *Phys. Rev. Lett.*, **313**, pp. 680–683.

44. Morgan J. W., Morgan T. N. (1970). Stress effects on excitons bound to axially symmetric defects in semiconductors. *Phys. Rev. B.*, **1**, pp. 739–749.

45. Gil B., Camassel J., Merle P., Mathieu H. (1982). J–J coupling and local-field effects on N-N pair spectra in GaP. *Phys. Rev. B*, **25**, pp. 3987–3996.

46. Dobrovolsky A., Sukrittanon S., Kuang Y. J., Tu C. W., Chen W. M., Buyanova I. A. (2014). Energy upconversion in GaP/GaNP core/shell nanowires for enhanced near-infrared light harvesting. *Small*, **10**, pp. 4403–4408.

Chapter 4

GaNAs-Based Nanowires for Near-Infrared Optoelectronics

Irina A. Buyanova,[a] Fumitaro Ishikawa,[b] and Weimin M. Chen[a]

[a]*Department of Physics, Chemistry and Biology, Linkoping University, Linköping 58183, Sweden*
[b]*Graduate School of Science and Engineering, Ehime University, Matsuyama 790-8577, Japan*

irb@ifm.liu.se

Recent progress in fabrication techniques and extensive investigations of physical properties of III–V semiconductor nanowires (NWs), such as GaAs NWs, have demonstrated their potential for a multitude of advanced electronic and photonics applications. Alloying of GaAs with nitrogen can further enhance the performance and extend the device functionality via intentional defect and heterostructure engineering in GaNAs and GaAs/GaNAs coaxial NWs. In this chapter, we will analyze the impacts of alloying with nitrogen on structural properties and recombination processes in GaNAs NWs. We will also discuss possible innovative applications of these structures in advanced nano-emitters, where the incorporation of nitrogen induces the formation of self-assembled quantum dot-like states embedded in the NWs.

Novel Compound Semiconductor Nanowires: Materials, Devices, and Applications
Edited by Fumitaro Ishikawa and Irina Buyanova

ISBN 978-981-4745-76-5 (Hardcover), 978-1-315-36440-7 (eBook)
www.panstanford.com

4.1 Introduction

Similar to the GaNP alloys discussed in Chapter 3, Ga(In)NAs alloys are obtained by the substitution of a small amount of group V As atoms in Ga(In)As by N atoms. Due to the giant bowing effect, the bandgap energy of the forming alloy could be significantly smaller than that of the parental Ga(In)As, which makes Ga(In)NAs materials attractive for applications in semiconductor light-emitting diodes (LEDs) and lasers emitting within the spectral window of 1.3–1.55 μm, important for fiber-optic communications [1–3]. Moreover, Ga(In)NAs exhibits extraordinary spin functionalities at room temperature [4–6] and, therefore, can be utilized as a spin filter or spin amplifier that operates at room temperature and does not require external magnetic fields. Additionally, the ability to grow high-quality GaNAs nanowires on Si substrates paves the way for integration of III-V NW-based photonic devices (including those made from dilute nitrides) with the mature microelectronic technology based on Si.

Currently, GaNAs alloy in the NW geometry can be fabricated utilizing molecular beam epitaxy (MBE). In this chapter, we will discuss recombination processes in such GaNAs-based nanowires and also new opportunities for heterostructure engineering provided due to nitrogen incorporation.

4.2 Growth and Structural Properties of GaNAs Nanowire Structures

As described in details in Refs. 7–9, GaAs and GaAs/GaNAs core/shell NWs can be grown by plasma-assisted MBE on (111) Si substrates using Ga droplets as a catalyst. The nitrogen composition in the shell layer of the NWs to be discussed in this chapter was varied between 0.1% and 0.5%. The growth was performed as follows. The substrate was preheated to 570°C under an As_4 beam at an equivalent pressure of 1.3×10^{-5} Torr. The growth of NWs was initiated by the formation of Ga droplets. The GaAs nanowire core was then formed via Ga-induced vapor-liquid-solid growth. The Ga supply was set to match a planar growth rate of 1.0 ML/s on (001) GaAs. The GaAs core was grown for 15 min at the requisite temperature to initiate longitudinal wire growth. Next, the growth was interrupted for 15 min and the equivalent pressure of the As_4 beam was increased to 5×10^{-5} Torr to induce the crystallization of the Ga catalyst on the tip of the NWs. After the crystallization of the catalyst, the lateral growth became dominant, which allowed the formation of core–shell structures with precisely controlled shell diameters [7, 8]. The Ga flux was further supplied for 15 min to continue radial growth of the GaAs core to the total thickness of 100 nm. During this time the substrate temperature was lowered down to 430°C, to improve efficiency of the N incorporation during the subsequent growth of a GaNAs shell. The latter was performed for 30 min with a N plasma ignited. The nitrogen composition in the shell layer was varied by changing parameters of the rf-plasma. The so-produced GaNAs shell had a thickness of 100 nm. As a reference, we also grew GaAs/GaAs NW structures that were produced under the identical conditions as the GaAs/GaNAs NWs but without N flux irradiation. The schematic illustration of the utilized growth process is shown in Fig. 4.1.

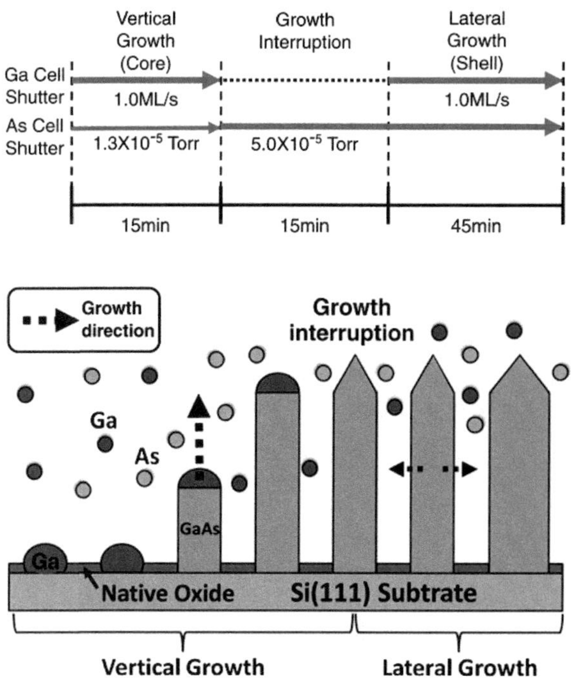

Figure 4.1 Growth sequence and mechanism of the NWs growth. Reprinted from Ref. 8, copyright (2014) of the Japan Society of Applied Physics.

Based on the performed scanning electron microcopy (SEM) measurements, the so-obtained nanowires have a hexagonal cross section with a total diameter of 300–400 nm, and are 3–4.5 μm long. A representative SEM image of the NW array is shown in Fig. 4.2.

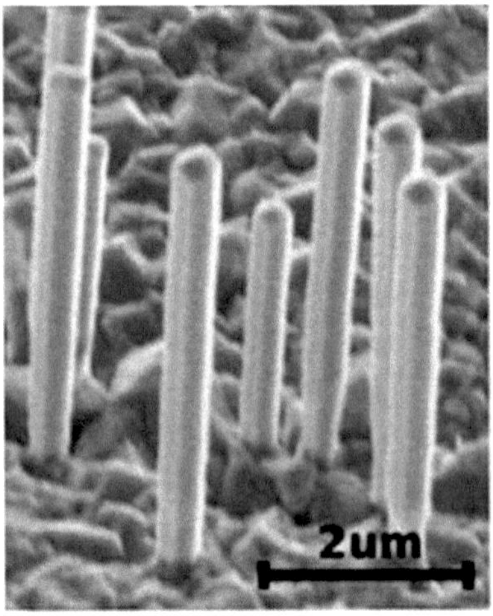

Figure 4.2 A representative SEM image of the GaNAs/GaAs NW array. Reprinted from Ref. 7 with the permission from IOP Publishing.

The structural properties of these NWs were investigated by transmission electron microcopy (TEM) [9]. Typical TEM images are shown in Fig. 4.3 with overview TEM images of a GaAs NW and a GaAs/GaNAs core/shell NW being presented in Figs. 4.3a,e, respectively. It was found that the NWs predominantly crystallize in the zinc blende (ZB) structure and exhibit only few (less than 5 twins/μm) structural defects within the first 1–2 μm. The corresponding areas are labeled as ZB in Fig. 4.3a,e with the twin positions being marked by the arrows. The twin density progressively increases towards the top part of the NWs (labeled as the twinned ZB (tZB) in Fig. 4.3). These polymorphic regions (referred to as the "mixed" regions, for simplicity) may also contain short (below 200 nm) wurtzite (WZ) segments. The total contribution of the WZ phase is found to be below 5% of the NW volume, independent of the N content in the NW. High resolution TEM (HR-TEM) images of the pure ZB, WZ and twinned ZB areas are shown in Fig. 4.3b–d and Fig. 4.3f–h for the GaAs/GaAs and GaAs/GaNAs core/shell NWs, respectively.

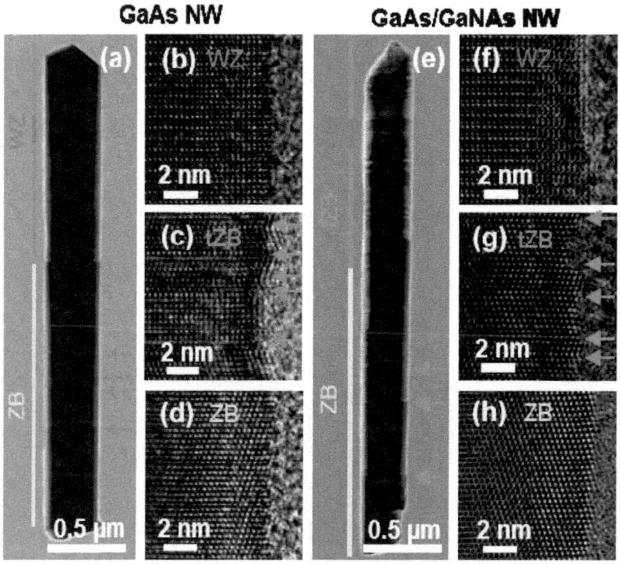

Figure 4.3 TEM images of the representative NWs. (a) and (e) are the overview images of a GaAs NW and a GaAs/$GaN_{0.005}As_{0.995}$ NW, respectively. The yellow, green and blue bars mark the nanowire segments with pure ZB, tZB and WZ lattice structure, respectively. (b-d) and (f-h) show HR-TEM images of the specified areas of the GaAs NW and GaAs/GaNAs NW, respectively. The horizontal arrows indicate the positions of the rotational twin planes within the ZB regions. The twin positions are not marked in (a) and (e) within the tZB segments due to a high density, but are indicated in the HR-TEM images shown in (c) and (g). Reproduced from Ref. 9 with the permission from the Royal Chemical Society.

4.3 Effects of Nitrogen Incorporation on Recombination Processes

4.3.1 Mechanism for Radiative Recombination and Long-Range Localization Effects

The effects of nitrogen incorporation on radiative recombination in GaNAs-based NWs were first studied in Ref. 10. It was shown that the PL spectra of GaAs NWs contain several overlapping peaks which stem from free-to-bound transitions at residual impurities and also indirect recombination of carriers localized at the interfaces between ZB and WZ regions of the nanowire—see Fig. 4.4. This emission is substantially quenched in the GaAs/GaNAs core/shell structures, where the PL spectra are instead dominated by a rather broad band which shifts towards lower energies within increasing [N]. This indicates that the PL emission in the core/shell structures originates from the GaNAs shell, where the bandgap is reduced upon N incorporation. In addition, the PL band acquired an asymmetric line shape with a sharp high energy cut-off and a long low-energy tail. Such lineshape is commonly observed in GaNAs alloys [1, 11] where the PL emission at low temperatures is due to radiative recombination of excitons trapped within the long-range fluctuations of the band edges due to alloy disorder.

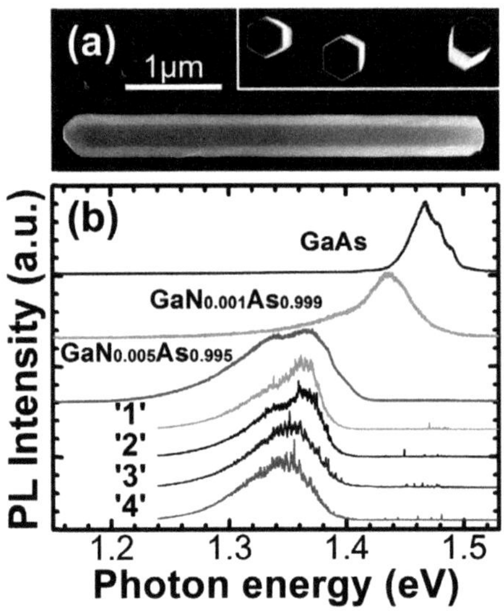

Figure 4.4 (a) An SEM image of the studied GaAs/GaNAs core/shell NWs with [N] = 0.5%. (b) Representative PL spectra of the investigated structures obtained at 5 K. The three top spectra were measured from ensembles of GaAs NWs, GaAs/$GaN_{0.001}As_{0.999}$ and GaAs/$GaN_{0.005}$ $As_{0.995}$ core/shell NWs. The four lower spectra labeled as "1"–"4" are results from the μ-PL measurements performed on four single GaAs/ $GaN_{0.005}As_{0.995}$ core/shell NWs. Reproduced from *Appl. Phys. Lett.*, **105**, 253106 (2014), with the permission of AIP Publishing.

The distribution of localized states in individual NW was analyzed by using μ-PL measurements performed on single NWs, which were mechanically transferred onto another Si substrate. A typical SEM image of such transferred nanowire is shown in Fig. 4.4a. The so-obtained μ-PL spectra from four NWs are shown by the curves "1"–"4" in Fig. 4.4b. The observed shift of the PL maximum position between different wires indicated variations of [N] between the wires, as was further collaborated from temperature-dependent μ-PL and μ-PLE measurements described below. It is also noticeable that the PL spectra of individual nanowires contain a large number of sharp lines superimposed on the broad background due to the localized exciton (LE) emission. These PL lines can be attributed to recombination of excitons strongly localized within certain spatial regions of the NWs by short-range compositional disorder, as will be discussed in detail in Section 4.4.

The suggested origin of the broad PL emission as being due to the LE recombination was independently confirmed from temperature dependent μ-PL measurements. It was found that the peak position of the LE band shifts to lower energies with increasing measurement temperature *T*—see Fig. 4.5. This was accompanied by an appearance of an additional band (labeled as "FE" in Fig. 4.5a) which become dominant when $T > 130$ K. The FE band also shifts to lower energies with increasing *T* but at a substantially lower rate as compared with that for the LE emission. This thermal behavior results in the so-called "S-shape" dependence of the PL maximum position, as can also be seen from Fig. 4.5b.

Such "S-shape" temperature dependence is typical for dilute nitrides and is characteristic for the LE emission [11, 12]. The red shift of the LE peak at low *T* reflects thermal depopulation of excitons starting from the shallower localized states (and, therefore, emitting at the high energy side of the LE band). At higher temperatures, complete thermal depopulation of the localized states will lead to the activation of radiative recombination from free excitons/carriers giving rise to the FE PL component. The weaker *T*-induced shift of this emission will follow the *T*-induced variation of the GaNAs bandgap energy E_g^{GaNAs}.

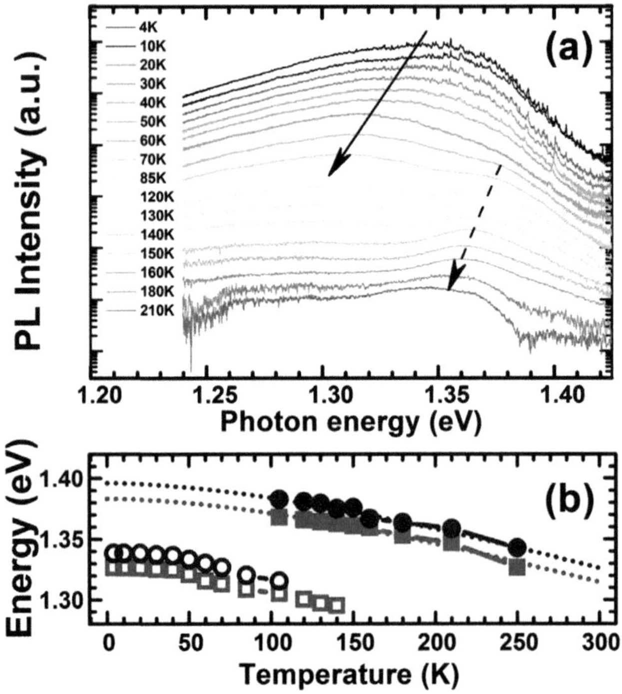

Figure 4.5 (a) Temperature dependent PL spectra measured from the single NW. The solid and dashed arrows mark the peak positions of the LE and FE PL components, respectively. (b) The peak position of the LE (the open symbols) and FE (the solid symbols) PL bands measured as a function of temperature from the NW '2' (the circles) and NW '4' (the squares). The dotted lines are expected T-dependences of the GaNAs bandgap energy based on the BAC model for the alloys with [N] = 0.5% (the upper curve) and 0.58% (the lower curve). Reproduced from *Appl. Phys. Lett.*, **105**, 253106 (2014), with the permission of AIP Publishing.

The results of the T-dependent μ-PL measurements were then used to estimate variations in the nitrogen composition between different wires. This was done by fitting the T-dependence of the FE position measured from different wires using the band anticrossing model (BAC) [2, 13]. According to this model, the $E_g(T)$ dependence in dilute nitrides can be modeled using the following equation:

$$E_{\mathrm{g}}^{\mathrm{GaAs}}(T)=\frac{1}{2}\left[(E_{\mathrm{g}}^{\mathrm{GaAs}}(T)+E_{\mathrm{N}})-\sqrt{(E_{\mathrm{g}}^{\mathrm{GaAs}}(T)-E_{\mathrm{N}})^{2}+4xC_{\mathrm{NM}}^{2}}\right] \quad (4.1)$$

Here $E_{\mathrm{g}}^{\mathrm{GaNAs}}(T)$ is the temperature-dependent bandgap of GaAs, E_{N} = 1.65 eV is the energy of the localized N level relative to the valence band edge that is assumed to be temperature independent, and x is the N fraction. C_{NM} is the coupling parameter determined by the strength of coupling between the localized and extended states that is equal to 2.5 eV for GaAs [13]. By fitting measured temperature dependences of the FE PL maximum position using Eq. 4.1, the N composition in the individual NWs was estimated to vary between 0.50% and 0.58%, i.e., within 0.08%.

Homogeneity of the GaNAs alloy within the individual NW was further evaluated utilizing μ-PLE measurements. It was found that the cut-off of the PLE spectra depends on the detection energy E_{det}. This is demonstrated in Fig. 4.6 were μ-PLE spectra of the NW "2" detected at E_{det} = 1.355 eV (the solid line, blue online) and 1.302 eV (the dashed line, red online), are shown. For easy reference, we also show the PL spectrum (the dotted line, green online) of this NW where the detection energies used for the PLE measurements are indicated by the vertical lines. Since PLE spectra are mainly determined by optical absorption transitions between extended states, these changes in the PLE cut-off energy should be related to variations of the N composition *within the NW* either in radial or axial directions. By linearly extrapolating the dependence of the squared PL intensity (I_{PL}) as a function of E_{exc}, as shown in the inset in Fig. 4.6, we can estimate that the local bandgap energy within the NW varies between 1.388 and 1.394 eV. From Eq. 4.1, this corresponds to the variation of N composition within 0.04%. This indicates that the broadening of the GaNAs-related PL is caused not only by alloy disorder responsible for the formation of band tail states but also by long-range variations in the N composition affecting energies of the extended states. We note that according to the PLE measurements performed on a reference GaNAs epilayer (not shown here), this effect is not specific for the NW growth but is equally present in planar GaNAs.

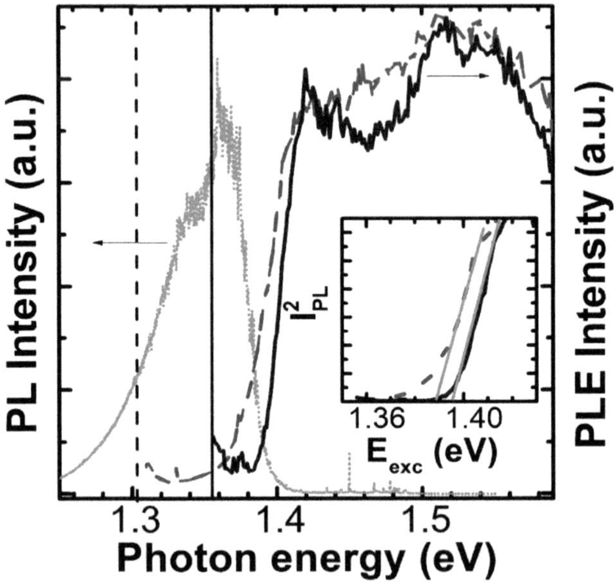

Figure 4.6 μ-PLE spectra measured from the NW "4" at the detection energies of 1.352 eV (the solid line, blue online) and 1.313 eV (the dashed line, red online), respectively. The inset shows the dependence of the squared PL intensity (I_{PL}) as a function of the excitation energy E_{exc} used to extrapolate the GaNAs bandgap energy (solid green lines). The PL spectrum of this NW is shown by the dotted line (green online), where the detection energies used for the PLE measurements are indicated by the vertical lines. Reproduced from *Appl. Phys. Lett.*, **105**, 253106 (2014), with the permission of AIP Publishing.

4.3.2 Recombination Dynamics

An important material parameter that affects performance of the NW-based devices is carrier lifetime. In solar cells, it limits a carrier diffusion length whereas in light-emitting devices it governs radiative efficiency and threshold current in laser structures. The carrier lifetime is known to be controlled by combined contributions of radiative and non-radiative recombination (NRR) processes characterized by their corresponding lifetimes. The non-radiative lifetime is largely affected by the material quality both in bulk and within near-surface regions. The contribution of the surface-related recombination is known to be especially severe in GaAs-based NW structures due to a large surface-to-volume ratio and the presence of surface states participating in the NRR processes [14–16]. Physical processes that limit carrier lifetime in GaAs/GaNAs core/shell NWs were first discussed in Ref. 17 from comprehensive temperature-dependent studies of exciton and carrier dynamics based on time-resolved PL spectroscopy. Below we will summarize the main results from this study.

Time-integrated PL spectra of the GaAs/GaNAs core/shell NWs with [N] = 0.1 and 0.5%, respectively, measured at 5 K with the excitation power (W_{exc}) of 4 mW are shown as the solid curves in Fig. 4.7a,b. For comparison, Fig. 4.7c also shows the PL spectrum from the $GaN_{0.005}As_{0.995}$ epilayer measured under the identical excitation conditions (the solid curve). As in cw-PL measurements discussed in Section 4.3.1, the spectra are dominated by the LE emission within the GaNAs region. Increasing excitation power, e.g., to 55 mW, leads to a saturation of the LE states. Under these conditions, the PL spectra (shown by the dotted curves in Fig. 4.7a,b) contain an additional PL component that is located above the high energy cut-off of the LE band and is caused by the FE emission in GaNAs (marked by the vertical dotted lines). The same tendency is also observed in the reference GaNAs epilayer (see the dotted curve in Fig. 4.7c). Recombination dynamics of these transitions was evaluated by analyzing their evolution as a function of a time delay (Δt_d) after an excitation laser pulse, as shown in Fig. 4.7d–f. First of all, it was found that within very short time delays $\Delta t_d < 30$ ps, the PL spectrum from each structure peaks at an energy close to the FE position. They are exponentially broadened at the high energy side with the PL tail extending above the FE position. This indicates a significant contribution of the FE component in the overall PL spectra. The slope of the high energy tail of the FE reflects a population distribution among the FE states determined by the exciton temperature. At later times, the PL emission is dominated by the LE transitions and undergoes a spectral red shift as marked by the arrows in Fig. 4.7d–f.

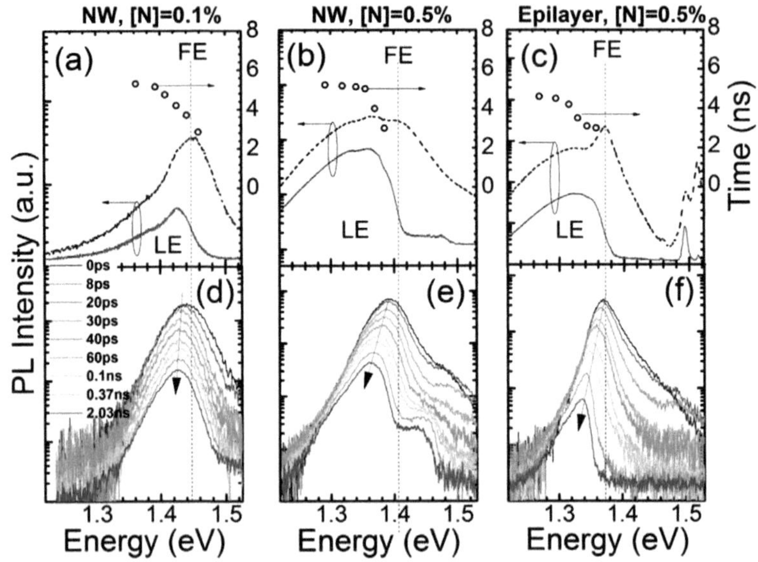

Figure 4.7 (a–c) PL spectra of the investigated structures measured at 5 K with the excitation power levels of 4 mW (the solid curves, red online) and 55 mW (the dashed curves, blue online), respectively. The open circles represent the measured PL lifetimes as a function of emission energy, with the excitation power of 55 mW. (d–f) PL spectra of the investigated structures detected at different time delays after an excitation pulse. The arrows indicate the shifts of the peak positions of the LE emission. The vertical dashed lines mark the FE spectral positions. Reproduced from Ref. 17 with the permission from the *Nature* Publishing Group.

4.3.2.1 Localized exciton dynamics

We will now analyze the temporal behavior of the LE and FE transitions starting from the LE emission. Consistent with the previous studies of planar dilute nitrides [11, 12], the LE decays are mostly single-exponential and speed-up for higher detection energies (E_{det}). Consequently, the LE maximum position shifts to lower energies with increasing Δt_{d}. The observed behavior reflects energy relaxation from the shallow localized states to the deeper ones. The deduced decay times (τ_{LE}) of the LE transitions are plotted as the open circles in Fig. 4.7a–c. In all structures, the energy transfer between the localized states become inefficient for the deep LE as τ_{LE} saturates at the low energy side of the

LE band. The saturation values were found to be very similar among all NWs and also epilayer structures and were comparable to the radiative lifetime in bulk GaAs [18]. The latter indicated that recombination of the deep LE is predominantly radiative, owing to localization of the exciton wavefunction within the limited volume of the GaNAs alloy, which reduces probability of exciton trapping by NRR centers.

It was also noted that the saturation value of τ_{LE} (and, therefore, the radiative lifetime τ_{r}) is constant within the temperature range of T = 4–100 K, i.e., at temperatures when the localization effects dominate. Since such behavior is known to be typical for excitonic transitions within zero-dimensional (0D) structures, the observed independence of τ_{LE} from T suggested that the deeply localized states likely experience three-dimensional quantum confinement. This conclusion is consistent with our recent results of micro-PL measurements performed on individual GaAs/GaNAs nanowires to be discussed in Section 4.4.

4.3.2.2 Free exciton/carrier dynamics

As was shown from temperature-dependent cw-PL measurements discussed in Section 4.3.1, localization effects become unimportant for T > 130 K. Under these conditions, the PL spectra are due to the FE/carrier recombination and no longer exhibit a red shift with increasing Δt_{d}, as can be seen from Fig. 4.8, which shows time-resolved PL spectra (the solid curves) and decay times (the symbols) from the investigated structures at 140 K. The FE decays can be fitted by a bi-exponential function of the form

$$I_{\mathrm{PL}} = A_{\mathrm{f}}\exp\left(-\frac{t}{\tau_{\mathrm{f}}}\right) + A_{\mathrm{s}}\exp\left(-\frac{t}{\tau_{\mathrm{s}}}\right) \tag{4.2}$$

where $A_{\mathrm{f}}(A_{\mathrm{s}})$ and $\tau_{\mathrm{f}}(\tau_{\mathrm{s}})$ are the amplitude and the decay time constant of the fast (slow) decay component, respectively. The bi-exponential decay of the excitonic emission usually implies that the monitored transitions occur in different spatial regions of the NWs with distinctly different lifetimes, e.g., in bulk and near-surface regions of the NWs. Because of a large amount of surface-related electronic states acting as efficient NRR centers, the

PL lifetime (τ) in the near-surface regions is typically rather fast, as it is determined by combined contributions from the radiative and nonradiative (τ_{nr}) lifetimes as described by $\frac{1}{\tau}=\frac{1}{\tau_r}+\frac{1}{\tau_{nr}}$. Therefore, the fast (slow) decay component was attributed to the FE recombination within the near-surface (volume) region of the GaNAs shell. This assignment was corroborated by the transient behavior of the FE emission detected in the reference $GaN_{0.005}As_{0.995}$ epilayer where the fast decay component is largely suppressed—see also Fig. 4.9. The determined FE lifetimes even of the slow component (τ_f = 50 ps at T = 100 K for the GaAs/$GaN_{0.005}As_{0.995}$ core/shell NW) are substantially shorter than that for the LE emission, which indicates an enhanced contribution of NRR in the FE decay. This is probably not surprising taking into account that the free excitons are mobile in the lattice and, therefore, have a higher chance to encounter non-radiative recombination centers.

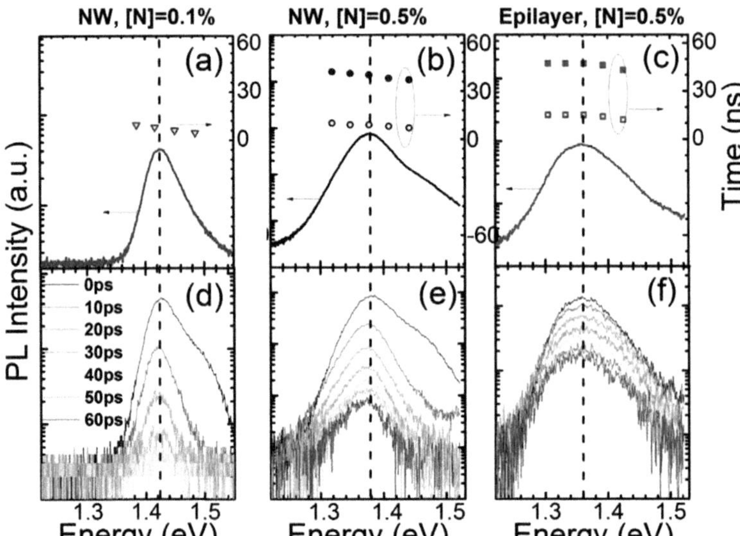

Figure 4.8 (a–c) Time-resolved PL spectra of the investigated structures measured at 140 K. The open (filled) symbols represent the deduced fast (slow) PL lifetimes as a function of emission energy. (d–f) PL spectra of the investigated structures detected at different time delays after the excitation pulse. The vertical dashed lines mark the FE spectral positions. Reproduced from Ref. 17 with the permission from the Nature Publishing Group.

The FE decays accelerate with increasing T, see also Fig. 4.9a, where representative PL decays of the FE/free carrier emission at 100 K, 200 K and 300 K are shown. The observed reduction of carrier lifetimes with rising T reflects thermal activation of competing NRR processes degrading the PL efficiency. Simultaneously, the fast PL decay component gains its intensity and becomes dominant at room temperature not only in the core/shell NWs but also in the epilayer structure. Temperature dependences of the deduced lifetimes and the A_s/A_f ratio are shown in Fig. 4.9b,c, respectively.

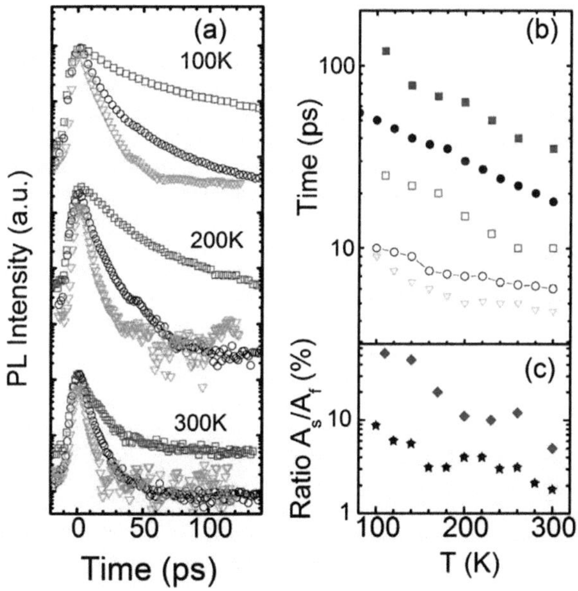

Figure 4.9 (a) Transient profiles of the FE/free carrier emission measured from the investigated structures at the specified temperatures. The measured decays were fitted by the bi-exponential function given by equation (4.2) and the deduced fast and slow time constants are shown in (b) by the open and filled symbols, respectively. In (a) and (b), the triangles (green online), the circles (blue online) and the squares (red online) are the results from the $GaAs/GaN_{0.001}As_{0.999}$ core/shell NWs, $GaAs/GaN_{0.005}As_{0.995}$ core/shell NWs and the $GaN_{0.005}As_{0.995}$ epilayer, respectively. The determined A_s/A_f ratio is displayed in (c) for the NW (stars, blue online) and the epilayer (diamonds, red online) structures with [N] = 0.5%. For the NW structure with [N] = 0.1%, A_s was set to zero. Reproduced from Ref. 17, with the permission from the *Nature* Publishing Group.

These results of the FE/free carrier dynamics provided further information regarding recombination processes and material quality of the studied GaAs/GaNAs core/shell nanowires. First of all, both τ_s (the filled symbols, Fig. 4.9b) and τ_f (the open symbols, Fig. 4.9b) were found to be shorter in the GaNAs shell layer (the circles) than that in the reference GaNAs epilayer with the same nitrogen content (the squares). This indicated a stronger contribution of both volume and surface-related NRR processes, possibly due to a larger number of structural defects in the NWs caused by structural polytypism, i.e., inclusions of zinc blende and wurtzite phases, as revealed by the performed TEM measurements. We should note, however, that the shorter τ_s value in the NWs may not be solely determined by the volume quality of the GaNAs alloy. Indeed, the thickness of the GaNAs shell layer is about 150 nm, which is significantly lower than the thickness of the reference epilayer (~1 μm). As a result, volume recombination in the NW shell occurs relatively closely to the surface (and also to the interface with the GaAs core) that may also shorten the measured τ_s.

Second, the temperature dependent plots of the A_s/A_f ratio shown in Fig. 4.8b, implied that the contribution of the surface recombination become increasing important with raising *T*. This effect was ascribed to the *T*-induced decrease in the carrier diffusion length caused by shortening of the carrier lifetime at elevated temperatures. Indeed, taking into account that the alloying with nitrogen does not substantially affect an absorption coefficient (α) for energies exceeding 2.2 eV [19] and using the tabulated values of α [20], the penetration depth of the excitation light in our experiments can be estimated as being 15 nm. Therefore, excitation of the GaNAs volume regions occurred as a result of carrier diffusion governed initially by τ_f and then by τ_s when the photo-generated carriers reach the volume region. Since both of these time constants decrease at elevated temperatures, the photo-excited region of the structure became restricted to the near-surface region where the surface recombination dominated.

And finally, a comparison of the PL dynamics in the core/shell NWs with different N compositions led to the conclusion that effects of surface recombination become less severe with increasing N composition in the GaNAs shell. Indeed, the FE/free carrier decay time in the $GaN_{0.001}As_{0.999}$ shell is very fast, i.e., of the order of 9 ps at 100 K and further accelerates with increasing *T*. As a result, the majority of the photo-generated carriers recombined within the near-surface region, which degraded the overall emission efficiency. On the other hand, the decay times τ_f are longer in the GaAs/$GaN_{0.005}As_{0.995}$ core/shell NWs (see Fig. 4.9). Consequently, the contribution of the volume-related recombination was concluded to be more pronounced in this structure, as the photo-excited excitons/free carriers no longer solely reside within the near-surface region. This was obvious from the appearance of the slow decay component in the PL transients—see Fig. 4.9. The observed slow-down of the surface recombination rate led to a substantial (by about 20 times) increase in the PL intensity when the N composition in the shell was changed from 0.1% to 0.5%.

The observed suppression of the surface recombination was attributed to partial nitridation of the GaNAs surface during the growth. Previous studies of GaAs surfaces [21] and NWs [22] have established that chemical nitridation of the GaAs surface substantially reduces surface defect density due to the formation of stable Ga-N bonds at the GaAs surface. This causes passivation of Ga dangling bonds and inhibits surface oxidation. We speculated that the same process may in principle occur during the growth of the GaNAs shell as a result of the supply of N atoms from a N plasma. Indeed, the nanowire growth was carried out under As and N overpressure conditions. At the end of the nanowire growth, the flux of group III Ga was steeply terminated by the shutter control. On the other hand, the atmospheric As and residual N were kept supplied to the nanowires, as these processes cannot be completely controlled by a shutter [23, 24]. In contrast to the desorption of As, the active nitrogen adsorbs on the surface with unit sticking efficiency [24]. Consequently, the residual N would substitute the topmost As, providing an increased number of Ga-N bonds at the nanowire surface and, therefore, its supplementary nitridation. This is also consistent with our results from micro-Raman measurements which shows that the total number of Ga-N bonds in the NW structures exceeds what is expected for a given alloy composition. Such a reduction in surface recombination could, therefore, be of importance for improving efficiency of radiative recombination in GaNAs-based nanostructures.

4.4 Quantum Dot-Like Emitters Embedded in GaNAs NWs

Possibilities provided by heterostructure engineering can be further extended utilizing quantum confinement effects. In the case of one-dimensional NW structures, these effects become significant only in very thin wires, which unfortunately suffer from severe surface recombination. Alternatively, the NWs can be fabricated with zero-dimensional (0D) quantum dots (QDs), which can be formed as a result of lateral confinement [25–28], deliberate QD polymorphism in the axial direction [29], or alloy fluctuations [30].

So far, efforts devoted to fabrication and characterization such hybrid 0D-1D structures were mainly focused on conventional III-V material systems such GaAs/AlGaAs, GaAs/GaAsP, InAsP/InP, and In(Ga)As/GaAs. Most recently, we have shown that alloying with nitrogen provides an alternative approach for fabrication of self-assembled QD-like emitters embedded in the GaNAs-based nanowires [9], as will be briefly discussed below.

As was already mentioned previously, the low temperature μ-PL spectra of GaNAs/GaAs NWs contain numerous sharp lines superimposed onto the LE background. In order to understand their origin, we performed μ-PL measurements on the same wires as those characterized by TEM (shown in Fig. 4.3). For each NW, the measurements were conducted by selectively exciting either the pure ZB or the mixed regions. The corresponding results are shown in Fig. 4.10. It is obvious that the presence of the sharp lines is induced by alloying with nitrogen as none of them could be detected from the pure GaAs NWs. Moreover, the appearance of the sharp lines changes among the wires and is strongly spot dependent. They also exhibit extremely narrow linewidth with a FWHM of 50–130 μeV—see the inset in Fig. 4.10b, which points to the suppression of exciton motion. Consequently, these sharp lines were attributed to optical transitions associated with strongly localized, e.g., QD-like, states that are randomly distributed within the wire. The observed large number (more than 100) of the sharp lines indicated a large number of independent light emitters within the excitation volume, setting the upper bound of the size of the confinement potentials to be much less than 60 nm. Their formation seems to be more probable within the mixed regions, as the number of the sharp lines in these areas was about 50% higher than that in the pure ZB segments of the nanowires. The localization cannot simply be caused by carrier confinement in narrow WZ/ZB segments within the polymorphic region of the GaNAs shell, however, as intense narrow lines can also be detected from the pure ZB regions.

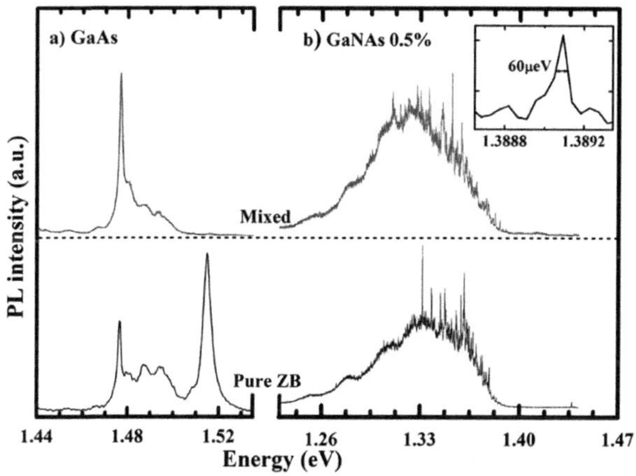

Figure 4.10 Representative PL spectra measured from the pure ZB and the mixed regions of the GaAs (a) and GaAs/$GaN_{0.005}As_{0.995}$ NWs (b) shown in Fig. 4.3. The GaAs/GaNAs NW PL spectra contain numerous sharp lines superimposed with a broad background due to the LE emission. The inset in (b) shows a magnified part of the spectrum from the GaAs/$GaN_{0.005}As_{0.995}$ NW. Reproduced from Ref. 9 with permission from the Royal Society of Chemistry.

There are two possible sources of the confining potential. First of all, it can be caused by various N-related clusters, which contain several nitrogen atoms and introduce a range of deep levels within the GaNAs bandgap (determined by the number and mutual arrangement of the involved N atoms). Second, it can be introduced by random local fluctuations in nitrogen composition. Due to the very strong bowing in the bandgap energy, such short-range fluctuations can create a 3D confinement potential within some regions of the nanowires and, therefore, act as quantum dots. Based on the previous studies of planar GaNAs epilayers [31–37], strong exciton localization induced by either of the two aforementioned mechanisms can lead to the appearance of sharp PL peaks with a linewidth of 70–300 μeV, i.e., of the same order of magnitude as the linewidth of the sharp lines detected in the studied GaAs/GaNAs core/shell NWs. However, the ultra-narrow PL lines from the N cluster bound excitons in bulk GaAs were so far detected only at substantially higher energies within the range of 1.443–1.5 eV [31–36]. This makes the first scenario somewhat less likely. We therefore attributed the 3D confining potential to short-range fluctuations in the N content.

Further information regarding the electronic structure of these QD-like states was gained based on detailed polarization-resolved μ-PL studies of individual lines. Since the broad PL background from the LE may mask the actual polarization degrees of the sharp lines, it was first subtracted from the PL spectra. Representative dependences of the PL intensity as a function of the detection energy and the angle, and the corresponding angular plots are shown on Fig. 4.11a–c and d–f, respectively. The angular variation of the PL intensity from each individual line can be fitted with the following expression:

$$I = I_{min}\sin^2(\varphi-\theta) + I_{max}\cos^2(\varphi-\theta), \qquad (4.3)$$

where $I_{max}(I_{min})$ are the maximum (minimum) PL intensity and φ is the angular position of the linear polarization analyzer—see Fig. 4.11g. θ is the angle at which the PL intensity reaches its maximum value and corresponds to the angle between the optical polarization axis and the [111] growth axis of the nanowire. Overall, three types of lines were observed: (1) the lines polarized orthogonally to the NW growth axis (Fig. 4.11a,d), (2) the lines with arbitrary polarization angles (Fig. 4.11b,e), and (3) the lines that are preferentially polarized along the NW (Fig. 4.11c,f). Their polarization degree was defined as follows:

$$P = (I_{max} - I_{min})/(I_{max} + I_{min})\cdot 100\% \qquad (4.4)$$

The determined values of P were found to vary between 10% and 100%. The same polarization behavior was also observed for the NWs with [N] = 0.1%.

As the diameter of the wire are around 300 nm, the polarization of the emitted light is not expected to be affected by the antenna effect due to mismatch between the dielectric constants of the NW and the surrounding media. Therefore, the observed polarization was attributed to preferential quantization directions of the corresponding QD-like localization potential and the electronic structure of the trapped excitons. The most probable direction of the quantization axis was determined from a statistical analysis of the angle θ—see Fig. 4.12a, where the corresponding results from both the pure ZB and the mixed part of the nanowires are shown. (θ is the angle at which the PL intensity reaches its maximum value and corresponds to the angle between the optical polarization axis and the [111] growth axis of the nanowire.) The statistical distributions of the polarization degree P is shown in Fig. 4.12b, whereas the distributions of the measured polarization degree P and the polarization angle θ within the ZB and the mixed regions are given in Fig. 4.12c,d) respectively. It is notable that the lines with $\theta \sim 90°$ are strongly polarized with $P \sim 55\%$. On the other hand, randomly oriented lines seemed to have a smaller polarization degree (around 30–40%).

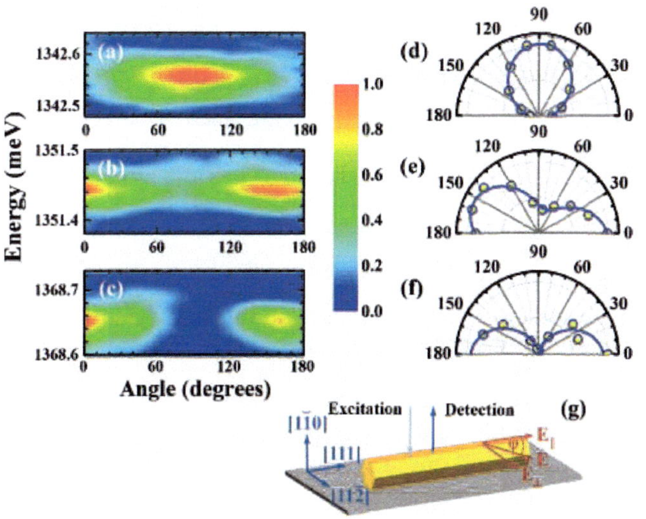

Figure 4.11 (a–c) 2D color plots of the PL intensity as a function of the detection energy and the angle ϕ for several representative PL lines. (d–f) The corresponding polar plots of the PL intensity. The solid lines are the simulated curves by Equation (4.3). The angles are counted relatively to the NW axis. (g) Schematic illustrating the measurement geometry used for polarization resolved μ-PL measurements. E is the polarization direction of the analyzer rotated by the angle ϕ relative to the nanowire axis. Its components, which are parallel and perpendicular to the NW axis, are given by $E_{\parallel}$ and $E_{\perp}$, respectively. Reproduced from Ref. 9 with the permission from the Royal Chemical Society.

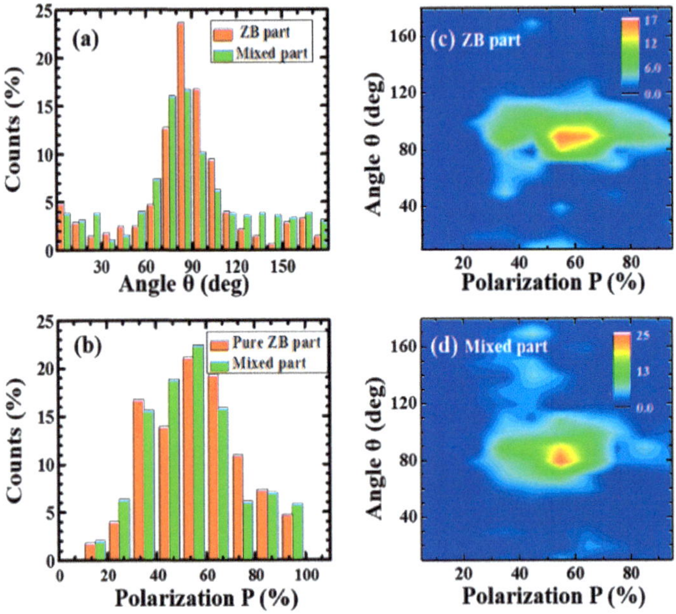

Figure 4.12 Statistical distributions of the angle θ of the polarization axis (a) and the polarization degree P (b) measured from the ZB (red) and the mixed (green) regions of the studied GaAs/GaNAs NW structures. (c) and (d) show the number of the PL lines as a function of both angle θ and polarization degree P measured from the ZB and mixed regions, respectively. The total numbers of the QD-like emission lines used in the statistical distributions are 274 and 442 for the ZB and the mixed regions, respectively. Reproduced from Ref. 9 with the permission from the Royal Chemical Society.

The observed orthogonal polarization of light emission (i.e., θ around 90°) in our NWs implied that the predominant quantization axis of the QDs coincided with the NW axis. This preferential alignment can be caused by shear strain in the GaNAs shell created due to the lattice mismatch with the GaAs core layer that has a larger lattice constant. Indeed, as it was demonstrated in previous studies [38], overall uniaxial anisotropy of nanowires has significant effects on strain in coherent core/shell structures from lattice mismatched semiconductors. In addition to the hydrostatic component, the predominant strain contribution was found to arise from the axial ε_{zz} component. This will elongate the exciton confinement potential along the NW axis and will cause splitting between heavy hole (hh) and light hole (lh) valence band states at the Γ point with hh states being the ground state, due to a predominantly uniaxial tensile strain field in the GaNAs shell. This splitting will be further increased due to quantum confinement effects (and the crystal field in the WZ regions) along the same direction. The expected values of the PL polarization would then range between 100% (if the monitored emission is due to recombination of hh excitons alone and the hh states retain a pure character) and 33% (if both hh and lh states are equally populated or the hh-lh splitting is smaller than the PL linewidth). As shown in Fig. 4.12a, the majority of the monitored sharp lines have $P > 33\%$, which means that the hh-states are preferentially populated. Fluctuations in the nitrogen composition will likely affect the strain distribution causing some tilting of the local strain field away from the NW axis. This will cause mixing between the hh and lh states, to a degree determined by the electronic structure of a particular dot, which will decrease the measured values of P from 100%. If the splitting between the hh and lh states exceeds the PL linewidth, population of lh states will lead to the appearance of the emission lines that are polarized along the nanowire (i.e., with θ close to 0°), as is indeed observed experimentally. From Fig. 4.12a, some of the QDs have their principal axis also oriented in random directions. Similar behavior was also observed in self-assembled QDs embedded in GaAs/AlGaAs core/shell NWs [29] and was attributed to surface effects. We note that the suggested exciton localization within the QD states is also consistent with the results of the performed power-dependent PL measurements, which imply formation of biexcitons at a high excitation power level that is typical for QD systems.

4.5 Conclusions

In summary, dilute nitride GaAsN/GaAs heterostructure NWs were successfully grown on Si(111) substrates using plasma-assisted MBE. The NWs so far exhibit polymorphic crystal structure, which calls for further optimization of the growth process. Our optical studies have shown that alloy disorder in the optically active GaNAs shell leads to the short- and long-range localization effects, which affect recombination processes in GaNAs/GaAs nanowires. Some of the states created due to short-range fluctuations in the nitrogen composition form optically active states with a quantum-dot like electronic structure. The light emission from these self-embedded nano-emitters is found to be spectrally narrow (~50–130 μeV) and is highly polarized (up to 100%) with the preferable polarization direction orthogonal to the NW axis, suggesting a preferential orientation of the localization potential. From detailed temperature-dependent time-resolved PL measurements, we also showed that GaNAs NWs have a lower rate of the detrimental surface recombination and tentatively attributed this effect to the N-induced nitridation of the NW surface. The revealed optical properties of the GaNAs-based nanowire structures may be attractive for future optoelectronic applications in advanced nano-sized light emitters which could be integrated with Si technology.

Acknowledgments

Financial support by the Swedish Energy Agency (grant # P40119-1) and the Swedish Research Council (grant # 2015-05532) is greatly appreciated. The authors also like to acknowledge M. Jansson, J. E. Stehr and S. Filippov, P. O. Å. Persson, and J. Palisaitis for their contribution to the results discussed in this chapter.

References

1. Buyanova I. A., Chen W. M. (2004). *Physics and Applications of Dilute Nitrides* (Taylor & Francis Books, Inc., New York).
2. Shan W., Walukiewicz W., Ager J., Haller E., Geisz J., Friedman D., Olson J., Kurtz S. (1999). Band anticrossing in GaInNAs alloys. *Phys. Rev. Lett.*, **82**, pp. 1221–1224.
3. Kent P. R. C., Zunger A. (2004) *Physics and Applications of Dilute Nitrides*, eds., Buyanova I. A., Chen, W. M., Chapter 1, An atomistic view of the electronic structure of mixed anion III–V nitrides (Taylor & Francis, London), pp. 23–64.
4. Wang X. J., Buyanova I. A., Zhao F., Lagarde D., Balocchi A., Marie X., Tu C. W., Harmand J. C., Chen W. M. (2009). Room-temperature defect-engineered spin filter based on a non-magnetic semiconductor. *Nat. Mater.*, **8**, pp. 198–202.
5. Puttisong, Y., Wang X. J., Buyanova I. A., Geelhaar L., Riechert H., Ptak A. J., Tu C. W., Chen W. M. (2013). Efficient room-temperature nuclear spin hyperpolarization of a defect atom in a semiconductor. *Nat. Commun.*, **4**, p. 1751.

6. Puttisong Y., Buyanova I. A., Ptak A. J., Tu C. W., Geelhaar L., Riechert H., Chen W. M. (2013). Room-temperature electron spin amplifier based on Ga(In)NAs alloys. *Adv. Mater.*, **25**, pp. 738–742.

7. Araki Y., Yamaguchi M., Ishikawa F. (2013). Growth of dilute nitride GaAsN/GaAs heterostructure nanowires on Si substrates. *Nanotechnology*, **24**, p. 065601.

8. Ahn N., Araki Y., Kondow M., Yamaguchi M., Ishikawa F. (2014). Effects of growth interruption, As and Ga fluxes, and nitrogen plasma irradiation on the molecular beam epitaxial growth of GaAs/GaAsN core–shell nanowires on Si(111). *Jpn. J. Appl. Phys.*, **53**, p. 065001.

9. Filippov S., Jansson M., Stehr J. E., Palisaitis J., Persson P. O. Å., Ishikawa F., Chen W. M., and Buyanova I. A. (2016). Strongly polarized quantum-dot-like light emitters embedded in GaAs/GaNAs core/shell nanowires. *Nanoscale*, DOI: 10.1039/C6NR05168E.

10. Chen S. L., Filippov S., Ishikawa F., Chen W. M., Buyanova I. A. (2014). Origin of radiative recombination and manifestations of localization effects in GaAs/GaNAs Core/shell nanowires. *Appl. Phys. Lett.*, **105**, p. 253106.

11. Buyanova I. A., Chen W. M., Pozina G., Bergman J. P., Monemar B., Xin H. P., Tu C. W. (1999). Mechanism for low-temperature photoluminescence in GaNAs/GaAs structures grown by molecular-beam epitaxy. *Appl. Phys. Lett.*, **75**, pp. 501–503.

12. Grenouillet A. L., Bru-Chevallier C., Guillot G., Gilet P., Duvaut P., Vannuffel S., Millon A., Chenevas-Paule A. (2000). Evidence of strong carrier localization below 100 K in a GaInNAs/GaAs single quantum well. *Appl. Phys. Lett.*, **76**, pp. 2241–2243.

13. Suemune I., Uesugi K., Walukiewicz W. (2000). Role of nitrogen in the reduced temperature dependence of band-gap energy in GaNAs. *Appl. Phys. Lett.*, **77**, pp. 3021–3023.

14. Lee H. H., Racicot R. J., Lee S. H. (1989). Surface passivation of GaAs. *Appl. Phys. Lett.*, **54**, pp. 724–726.

15. Parkinson P. Joyce H. J., Gao Q., Tan H. H., Zhang X. Zou J., Jagadish C., Herz L. M., Johnston M. B. (2009). Carrier lifetime and mobility enhancement in nearly defect-free core–shell nanowires measured using time-resolved terahertz spectroscopy. *Nano Lett.*, **9**, pp. 3349–3353.

16. Demichel O., Heiss M., Bleuse J., Mariette H., Fontcuberta i Morral A. (2010). Impact of surfaces on the optical properties of GaAs nanowires. *Appl. Phys. Lett.*, **97**, p. 201907.

17. Chen S. L., Chen W. M., Ishikawa F., Buyanova I. A. (2015). *Scientific Reports*, **5**, Article number: 11653.

18. P't Hooft G. W., Van der Poel W. A. J. A., Molenkamp L. W., Foxon C. T. (1987). Giant oscillator strength of free excitons in GaAs. *Phys. Rev. B*, **35,** pp. R8281–R8284.

19. Walukiewicz W., Shan W., Wu J., Yu K. M. (2004) *Physics and Applications of Dilute Nitrides*, eds., Buyanova, I. A., Chen, W. M., Chapter 2, Band anticrossing in III-N-V alloys: Theory and Experiments (Taylor & Francis, London), pp. 23–64.

20. Blakemore J. S. (1982). Semiconducting and other major properties of gallium arsenide. *J. Appl. Phys.*, **53,** pp. R123–R181.

21. Berkovits V. L., Ulin V. P., Losurdo M., Capezzuto P., Bruno G., Perna G., Capozzi V. (2002). Wet chemical nitridation of GaAs (100) by hydrazine solution for surface passivation. *Appl. Phys. Lett.*, **80,** pp. 3739–3741.

22. Alekseev P. A., Dunaevskiu M. S., Ulin V. P., Lvova T. V., Filatov D. O., Nezhdanov A. V., Mashin A. I., Berkovits V. L. (2015). Nitride surface passivation of GaAs nanowires: Impact on surface state density. *Nano Lett.*, **15**, pp. 63–68.

23. Ishikawa F., Wu S., Kato M., Uchiyama M., Higashi K., Kondow M. (2009). Unintentional source incorporation in plasma-assisted molecular beam epitaxy. *Jap. J. Appl. Phys.*, **48**, p. 125501.

24. Harris Jr., J. S., Kudrawiec R., Yuen H. B., Bank S. R., Bae H. P., Wistey M. A., Jackrel D., Pickett E. R., Sarmiento T., Goddard L. L., Lordi V., Gugov T. (2007). Nitrogen development of GaInNAsSb alloys: Growth, band structure, optical properties and applications. *Phys. Stat. Solidi* (b), **244**, pp. 2707–2729.

25. Borgström M. T., Zwiller V., Müller E., Imamoglu A. (2005). Optically bright quantum dots in single nanowires. *Nano Lett.*, **5**, pp. 1439–1443.

26. Versteegh M. A. M., Reimer M. E., Jöns K. D., Dalacu D., Poole P. J., Gulinatti A., Giudice A., Zwiller V. (2014). Observation of strongly entangled photon pairs from a nanowire quantum dot. *Nat. Commun.*, **5**, p. 5298.

27. Van Weert M. H. M., Akopian N., Perinetti U., van Kouwen M. P., Algra R. E., Verheijen M. A., Bakkers E. P. A. M., Kouwenhoven L. P., Zwiller V. (2009). Selective excitation and detection of spin states in a single nanowire quantum dot. *Nano Lett.*, **9**, pp. 1989–1993

28. Heiss M., Fontana Y., Gustafsson A., Wüst G., Magen C., O'Regan D. D., Luo J. W., Ketterer B., Conesa-Boj S., Kuhlmann A. V., Houel J., Russo-Averchi E., Morante J. R., Cantoni M., Marzari N., Arbiol J., Zunger A., Warbutron R. J., Fontcuberta i Morral A. (2013). Self-assembled

quantum dots in a nanowire system for quantum photonics. *Nat. Mater.*, **12**, pp. 439–444.

29. Spirkoska D., Arbiol J., Gustafsson A., Conesa-Boj, S., Glas, F., Zardo, I., Heigoldt, M., Gass M. H., Bleloch A. L., Estrade S., Kaniber M., Rossler J. Peiro F., Morante J. R., Abstreiter G., Samulson L., Fontcuberta I Morral A. (2009). Structural and optical properties of high quality zinc-blende/wurtzite GaAs nanowire heterostructures. *Phys. Rev. B*, **80**, p. 245325.

30. Jeon N., Loitsch B., Morkoetter S., Abstreiter G., Finley J., Krenner H. J., Koblmueller G., Lauhon L. J. (2015). Alloy fluctuations act as quantum dot-like emitters in GaAs-AlGaAs core–shell nanowires. *ACS Nano*, **9**, pp. 8335–8343.

31. Ikezawa M., Sakuma Y., Zhang L., Sone Y., Mori T., Hamano T., Watanabe M., Sakoda K., Masumoto Y. (2012). Single-photon generation from a nitrogen impurity center in GaAs. *Appl. Phys. Lett.*, **100**, p. 042106.

32. Éthier-Majcher G., St-Jean P., Boso G., Tosi A., Klem J. F., Francoeur S. (2014). Complete quantum control of exciton qubits bound to isoelectronic centres. *Nat. Commun.*, **5**, p. 3980.

33. Kita T., Harada Y., Wada O. (2008). Fine structure splitting of isoelectronic bound excitons in nitrogen-doped GaAs. *Phys. Rev. B*, **77**, p. 193102.

34. Karaiskaj D., Mascarenhas A., Klem J. F., Volz K., Stolz W., Adamcyk M., Tiedje T. (2007). Excitons bound to nitrogen pairs in GaAs as seen by photoluminescence of high spectral and spatial resolution. *Phys. Rev. B*, **76**, p. 125209.

35. Karaiskaj D., Mascarenhas A., Adamcyk M., Young E. C., Tiedje T. (2006). Ultranarrow photoluminescence transitions of nitrogen cluster bound excitons in dilute GaAsN. *Phys. Rev. B*, **74**, p. 035208.

36. Francoeur S., Klem J. F., Mascarenhas A. (2004). Optical spectroscopy of single impurity centers in semiconductors. *Phys. Rev. Lett.*, **93**, p. 067403.

37. Mintairov, A. M., Kosel, T. H., Merz, J. L., Blagnov, P. A., Vlasov, A. S., Ustinov, V. M., Cook, R. E. (2001). Near-field magnetophotoluminescence spectroscopy of composition fluctuations in InGaAsN. *Phys. Rev. Lett.*, **87**, P. 277401.

38. Wojnar P., Zielinski M., Janik E., Zaleszczyk W., Wojciechowski T., Wojnar R., Szymura, M., Kłopotowski Ł., Baczewski L. T., Pietruchik A., Wiater M., Kret S., Karczewski G., Wojtowicz, Kossut J. (2014). Strain-induced energy gap variation in ZnTe/ZnMgTe core/shell nanowires. *Appl. Phys. Lett.*, **104**, p. 163111.

Chapter 5

Dilute Bismide Nanowires

Wojciech M. Linhart,[a] Szymon J. Zelewski,[a] Fumitaro Ishikawa,[b] Satoshi Shimomura,[b] and Robert Kudrawiec[a]

[a]*Faculty of Fundamental Problems of Technology, Wrocław University of Science and Technology, Wybrzeże Wyspiańskiego 27,50–370 Wrocław, Poland*
[b]*Graduate School of Science and Engineering, Ehime University*, 3 Bunkyo-cho, Matsuyama, Ehime 790-8577, Japan

wojciech.linhart@pwr.edu.pl, robert.kudrawiec@pwr.edu.pl

This chapter provides an overview of the existing research on dilute III-V-Bi nanowires, particularly dilute GaAsBi nanowires. The development of dilute III–V bismides nanowires is in the first stages because the growth of these nanostructures is a major challenge to be overcome. So far, in the family of III-V-Bi nanowires, only dilute GaAsBi nanowires have been grown by molecular beam epitaxy or metalorganic vapor phase epitaxy. Here we present the recent progress in the growth and structural and optical properties of these structures. Additionally, we will show that photoacoustic spectroscopy is a powerful tool to study the energy gap of semiconductor nanowires.

Novel Compound Semiconductor Nanowires: Materials, Devices, and Applications
Edited by Fumitaro Ishikawa and Irina Buyanova

ISBN 978-981-4745-76-5 (Hardcover), 978-1-315-36440-7 (eBook)
www.panstanford.com

5.1 Introduction

Bismuth in group-V is the heaviest non-radioactive element, with metal atoms (Ga and In) creates compounds (GaBi and InBi), which have a negative band gap and a very large spin-orbit splitting [1–3]. Dilute bismides are formed from regular zinc blende III–V semiconductors such as GaAs, GaSb, InSb, or InP by incorporating small amounts of bismuth into the group V sublattice [3–20]. Small amount of Bi into III–V host (GaAs, InP, etc.) very strongly modifies the electronic band structure of III–V host (i.e., the band gap and the spin-orbit splitting) and leads to very interesting and unusual electronic and optical properties [3–17]. This is due to major effect of bismuth states on the valence bands. The localized Bi 6*p*-like states interact with the host valence bands and result in the formation of valence subbands and reducing the band gap of the material, which can be described within the band anticrossing (BAC) model [5]. Recent DFT calculations clearly show that the incorporation of Bi into III–V host influences both the valence and conduction band [17, 18], but chemical trends are in accordance with the BAC model: Bi-related changes in the valence band are the strongest for mixing atoms of very different electronegativity (e.g., P and Bi) and weaker when the electronegativity difference between atoms is small (e.g., Sb and Bi) [18].

The band gap vs. the lattice constant for dilute bismides is shown in Fig. 5.1 [20]. In this plot the Bi-related reduction of band gap is taken after Ref. [18]. It is clearly visible that the bang gap of dilute bismides decreases very strongly with the increase in the lattice constant (i.e., increase in Bi concentration), such strong change is not observed for other III–V alloys (GaInAs or InAsSb). Moreover, it is visible that alloying GaBi or InBi with regular III–V compounds allows to close the band gap in III–V alloys. This feature of dilute bismides opens perspectives for their applications in mid-infrared detectors and other devices [19, 20]. In the case of laser applications, the incorporation of Bi atoms into GaSb-based quantum wells structures (i.e., GaInAsSb/GaSb quantum wells) can enhance the quantum confinement in the valence band due to Bi-related change in the valence band [20].

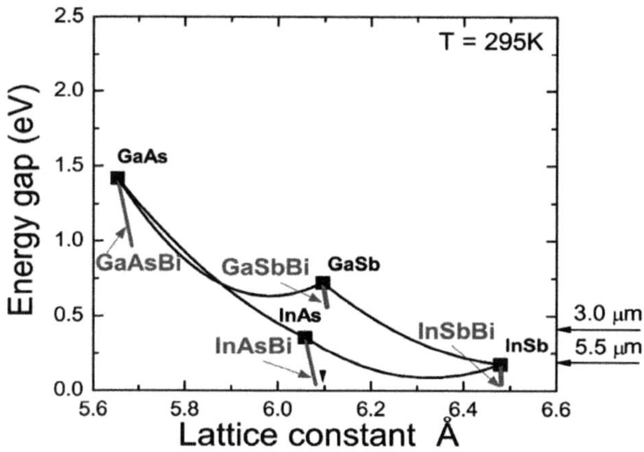

Figure 5.1 Energy gap versus lattice constants for binary III–V materials (solid squares), ternary III–V alloys (black lines), and ternary III-V-Bi alloys up to 5% Bi (red lines). Reprinted figure from [20]. Reprinted from [20], with the permission of AIP Publishing.

It is well known that electronic devices based on conventional III–V alloys suffer from significant intrinsic losses, especially above room temperature. The degradation in device performance with increasing temperatures is mainly caused by the presence of Auger recombination process [21, 22]. Therefore, the perspective of suppression of Auger processes in GaAs-based lasers is very interesting. In general, an improvement can be attributed to the spin orbit splitting, exceeding the band gap, which leads to the suppression of the Auger recombination mechanism [23, 24]. Achieving the band gap smaller than spin orbit splitting in GaAs-based heterostructures can be realized by incorporation of Bi into Ga(In)As host [10]. Therefore, in recent years, GaAsBi alloy was very intensively studied, but many complications in the growth process have limited the impact of this alloy [25, 26]. One of them is defect formation or weak incorporation of Bi atoms into Ga(In)As host.

Defect formation may be limited during the growth of GaAsBi nanowires due to termination mechanism at the growth front close to a side facet of wires, limiting growth-induced defects resulting from the large mismatch and miscibility gap of the constituents [27]. Therefore, dilute bismides can be a very interesting component in novel electronic devices containing nanowires.

5.2 GaAsBi Nanowires Grown by Molecular Beam Epitaxy

Epitaxial III-V-Bi alloys have been widely investigated in the last years, however, detailed studies of GaAsBi alloying nanowires are very scarce. Only three papers have been published yet on III-V-Bi nanowires [28–30]. Furthermore, those two articles are about GaAsBi nanowires. F. Ishikawa et al. have grown the multi-shell GaAs/GaAsBi nanowires by molecular beam epitaxy on phosphorus-doped n-type Si(111) substrate [28]. Figure 5.2a shows secondary electron (SE) images of the 45°-tilted GaAs/GaAsBi nanowire sample taken at 2 kV. Figure 5.2b presents 45°-tilted reference GaAs nanowires, grown under identical conditions with the GaAs/GaAsBi nanowire but without Bi-flux throughout growth. The GaAs/GaAsBi nanowires have exhibited rather corrugated surface morphologies, in contrast to GaAs nanowires which show sharp facetted sidewall, as it has been commonly observed for GaAs-related NWs. The density of GaAs/GaAsBi nanowires has been found to be 1.4×10^7 cm^{-2}.

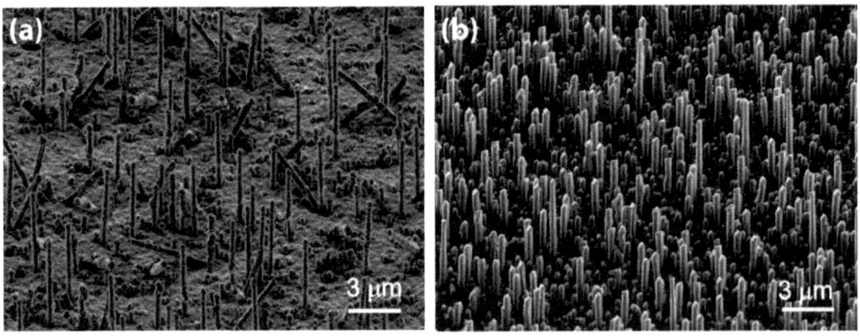

Figure 5.2 (a) SE image of the 45°-tilted GaAs/GaAsBi nanowires, (b) SE image of the 45°-tilted reference GaAs nanowire sample grown under identical conditions as for GaAs/GaAsBi nanowire, without Bi-flux. Reprinted with permission from [28]. Copyright 2015 American Chemical Society.

Bright field transmission electron microscopy (BF-TEM), Scanning transmission electron spectroscopy (STEM) and energy-dispersive X-ray spectroscopy (EDS) have shown the formation of GaAsBi shell with inhomogeneous Bi distribution within the layer surrounded by the outermost GaAs. This is presented in Fig 5.3. EDS elemental mapping of the wire clearly showed strong intensities of Ga and As with intensities of approximately 50% each, indicating the formation of the nanowire predominantly consisting of GaAs. A week intensity of Bi was observed throughout the entire nanowire, with its concentration of ~0.5%. However, an accumulation of Bi with concentration ~2% was observed at the tip of the hemispherical top crystal and near the edge of the middle part, which corresponds to the part of GaAsBi shell. The clear existence of the GaAsBi shell is presented in the profile scan in Fig. 5.4, which was created by the integration of the imaging area along the indicated positions by the red arrows.

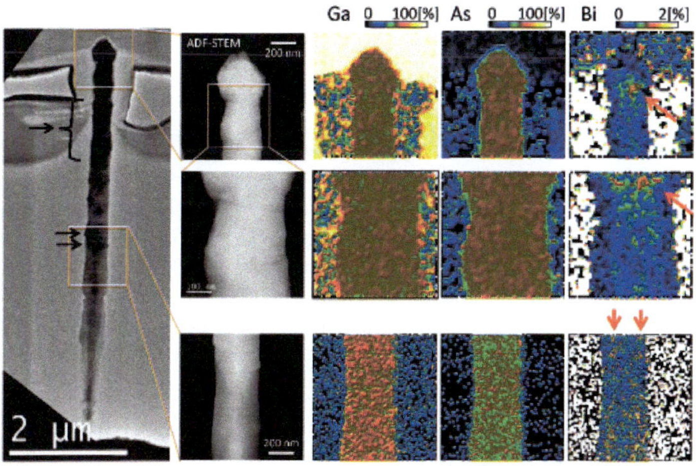

Figure 5.3 BF-TEM image of a single GaAs/GaAsBi nanowire (left most). (Right series) ADF-STEM and EDS elemental mapping images for areas delimited by the squares in the BF-TEM or ADF-STEM images. The black arrows label the twin defects. The red arrows in the Bi EDS-mapping images indicate high Bi concentrations. Reprinted with permission from [28]. Copyright 2015 American Chemical Society.

More analysis has revealed also a strong structural disorder. The nanowires exhibit a predominantly zinc-blend structure from the bottom to the middle part. In the upper parts, nanowires have polytypic wurtzite structure, which creates twin defects which are indicted by the black arrows in the BF-TEM image shown in Fig. 5.3. The inhomogeneity of the GaAsBi nanowires has also manifested in cathodoluminescence (CL) spectra. Figure 5.5 shows SE images and CL spectra of the GaAs/GaAsBi nanowires taken at room temperature from the spots indicated in the SE image. A broad emission between 1.1 and 1.5 eV has been observed, accompanied by the appearance of multiple peaks. A distinct peak at 1.24 eV agrees well with the energy of the reduced band gap of GaAsBi alloy with Bi content of 2%. The low-temperature (20 K) CL measurements revealed the existence of localized states energetically and spatially dispersed throughout the nanowire.

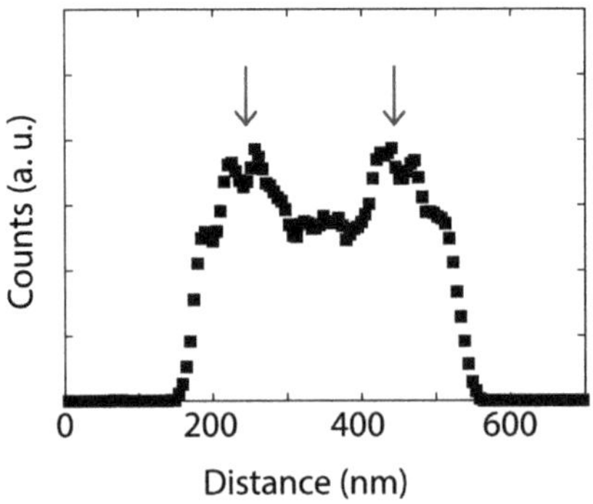

Figure 5.4 Profile scan across the GaAs/GaAsBi nanowire over the area indicated on the bottom-right corner of Fig 5.3: EDS Bi elemental mapping along the middle part of nanowire. The red arrows indicate positions with high Bi counts. Reprinted with permission from [28]. Copyright 2015 American Chemical Society.

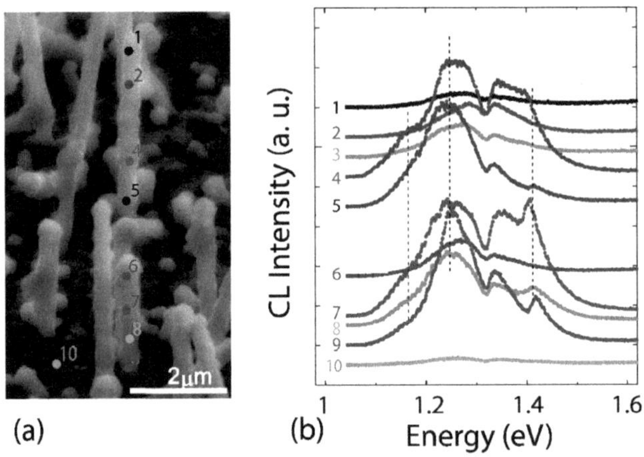

Figure 5.5 (a) SE image of the 45°-tilted GaAs/GaAsBi nanowires for CL measurements. (b) Series of CL spectra taken from the spots indicated in (a). Vertical lines in CL spectra are eye-guides for observed peak positions. Reprinted with permission from [28]. Copyright 2015 American Chemical Society.

The summarized CL spectra of GaAs/GaAsBi nanowire samples taken at 20 K and the corresponding monochromatic CL images are shown in Fig. 5.6. The top spectrum in Fig. 5.6a, labeled as Ensemble, is an average CL spectrum of hundreds of nanowires and exhibits a peak at 1.23 eV with a full width at half maximum (FWHM) of 0.17 eV, while the CL spectra labeled as Wire 1 and Wire 2, were collected from a single nanowires. Even for individual nanowires a broad CL spectra were observed with peak positions at 1.23 and 1.20 eV and FWHM of 0.18 and 0.21 eV for Wire 1 and Wire 2, respectively. The monochromatic CL images of Wire 1 and Wire 2 for three different detection energies are presented in Fig. 5.6b,c. It can be seen that for the chosen detection energies the CL emission originates from the lower part of the NWs, which exhibits a lower defect density and consists mainly a zinc blende structure. The CL obtained at 1.30 eV for Wire 1 and at 1.29 eV for Wire 2 were found at well-resolved NW segments. While for lower detection energies (1.24 and 1.15 eV for Wire 1 and 1.20 and 1.12 eV for Wire 2), the segment structure of the CL image was smeared out due to a carrier transfer toward lower energy states.

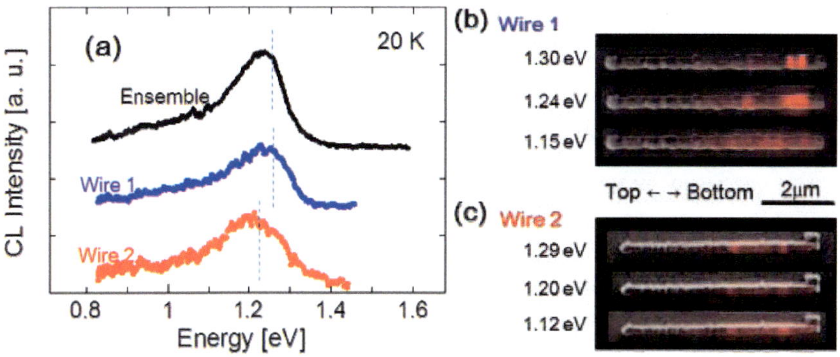

Figure 5.6 (a) CL spectra of GaAs/GaAsBi NWs. The top spectrum is dedicated to an ensemble of NWs. CL spectra labeled as Wire 1 and Wire 2 are collected from single NWs. (b) and (c) Monochromatic CL maps superimposed on the respective SEM images for Wire 1 and Wire 2, respectively, using the indicated CL detection energies. All images and spectra were taken at 20 K. the middle part of nanowire. The red arrows indicate positions with high Bi counts. Reprinted with permission from [28]. Copyright 2015 American Chemical Society.

5.3 GaAsBi Nanowires Grown by Metalorganic Vapor Phase Epitaxy

Essouda et al. synthesized GaAsBi nanowires by metalorganic vapor phase epitaxy (MOVPE) on Zn-doped GaAs(001) [29]. They reported that initially grown bismuth nanoislands act as local points of the nucleus for the growth of the GaAsBi nanowires. SEM image of Bi droplets deposited on GaAs substrate are shown in Fig. 5.7.

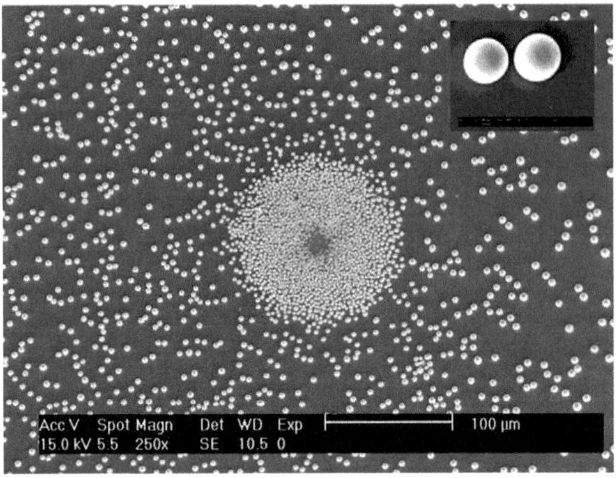

Figure 5.7 SEM image of Bi droplets deposited on GaAs substrate. The inset shows a zoom of Bi islands. Reprinted with permission from [29]. Copyright 2013 Elsevier.

The low eutectic temperature of Bi-Ga binary system enables the nanowire growth at low temperatures. Such behavior of Bi-Ga binary system has attracted much attention, due to its importance for technological applications. The growth of nanowires at low temperatures resulted in columnar GaAsBi structures with the same orientation and uniform diameter (~150 nm) and length (~1 μm) with Bi particles at the top of the structures. The substrate temperature and the flow rate of Bi precursor are two the most important parameters governing the structural properties of GaAsBi nanowires. The doubling of Bi precursor flow rate leads to a decrease of nanowires density and a change in their geometrical shapes. The high-resolution X-ray diffraction scan (shown in Fig. 5.8) indicated that GaAsBi nanowires grown on GaAs are epitaxially grown. The shape, density, and orientation of nanowires are dictated by the initial formation of Bi droplets and growth parameters.

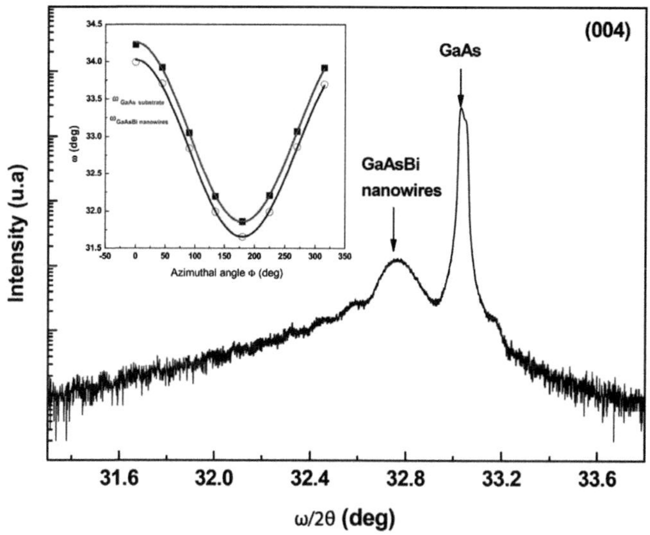

Figure 5.8 HRXRD (004) $w/2\theta$ curve of GaAsBi nanowires grown on GaAs substrate. Inset shows the diffraction angles variation of GaAs substrate and GaAsBi nanowires versus azimuthal angles φ. Reprinted with permission from [29]. Copyright 2013 Elsevier.

5.4 Electronic and Optical Properties of GaAsBi Nanowires

Investigations of the electronic and optical properties of GaAsBi nanowires have been performer by first principle calculations by Ding et al. [30]. Authors proposed a typical model of $Ga_{31}As_{31}$ nanowire in which the concentration of Bi was ranging from 3.2% to 12.8% and taking into account the substitutional method and keeping the supercell at constant volume. They observed that the band gap is shrinking from 1.0773 to 0.9049 eV mainly due to a shift of the valence band to the higher energy range and a shift of the conduction band to the lower energy range. Because Bi is a heavy atom the spin orbit coupling (SOC) effect may play a key role in the electronic and optical properties of GaAsBi nanowires.

Indeed, introducing the SOC effect into the calculations results in more severe band gap shrinkage with increasing Bi content. Additionally, the Bi 6*p* states showed a dominant contribution for the electronic band structure around the Fermi level.

The optical properties of GaAsBi nanowires can be represented by the dielectric function $\varepsilon(\omega) = \varepsilon_1(\omega) + i\varepsilon_2(\omega)$, where $\varepsilon_1(\omega)$ and $\varepsilon_2(\omega)$ are the real and imaginary parts of the dielectric function, respectively. Figure 5.9 shows the imaginary part of the dielectric function. The sharp peaks between 2.0 and 4.0 eV correspond to the optical transitions from Ga-*p*, As-*p* and Bi-*p* states to Ga-s/*p*, As-*p* and Bi-*s*/*p*/*d* states that create the first peak, while the second peak corresponds to the transitions from Ga-*s*/*p* and Bi-*p* states to Ga-*p*/*d* and Bi-*s*/*p*/*d* states. With increasing the Bi content, the imaginary part of the dielectric function moves toward lower energies, which confirms the band gap reduction.

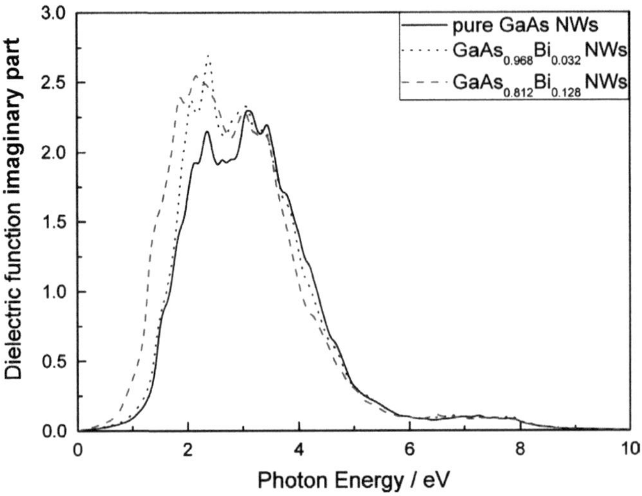

Figure 5.9 The imaginary part of the dielectric function $\varepsilon_2(\omega)$ of pure GaAs and GaAsBi nanowires. Reprinted with permission from [30]. Copyright 2013 Elsevier.

5.5 Photoacoustic Spectroscopy of GaAsBi Nanowires

In the case of the semiconductor nanowires for which the band gap energy is expected to be higher than the band gap of substrate (i.e., GaAs NWs grown on Si, InP grown on Si, etc.) the use of standard transmission method for the determination of the optical absorption spectrum would indicate only the edge associated with Si substrate. In addition, the unwanted scattering would potentially affect the results, causing the overestimation of the overall light extinction.

Photoacoustic spectroscopy (PAS) has been used for obtaining optical absorption spectra of GaAs/GaAsBi nanowires grown by MBE [31]. Periodically modulated monochromatic light beam was focused on a sample enclosed inside the measurement cell. Due to nonradiative processes some of the absorbed light energy was converted into a heat, causing the periodic pressure changes in the surrounding gas, known as acoustic waves. PAS provides a unique ability to control the depth from which the signal is generated by tuning the light modulation frequency (f) according to $d \propto f^{-0.5}$ relation [32]. In addition, there is no need to analyze the transmitted light. These led to a conclusion that PAS is an effective method in optical characterization of semiconductor nanowires [31].

At first, both reference GaAs and core–shell GaAsBi samples grown on Si substrate were measured at the modulation frequency of 14 Hz, typical value for photoacoustic experiments due to maximization of the signal amplitude. In this case the absorption edge was dominated by a signal generated from Si substrate. The measurements were repeated at frequencies of 120 and 240 Hz keeping the same experimental configuration. Photoacoustic spectra taken at different frequencies are shown in Fig. 5.10. For GaAs sample the substrate contribution was significantly reduced, revealing the absorption edge in the 1.4–1.5 eV range associated with GaAs band gap. A similar transition was observed also for GaAsBi sample. The band gap values for GaAs and GaAsBi were determined to be 1.42 and 1.26 eV, respectively in good agreement with the values obtained from CL measurements for these samples [28]. It is worth to note that the standard absorption measurements cannot be applied for these samples due to Si substrate which is not transparent at 1.26 eV. On the other hand, reflectance (or modulated reflectance) does not work for these samples due to light scattering by nanowires. In this context, photoacoustic spectroscopy is a powerful technique to study the band edge of nanowires.

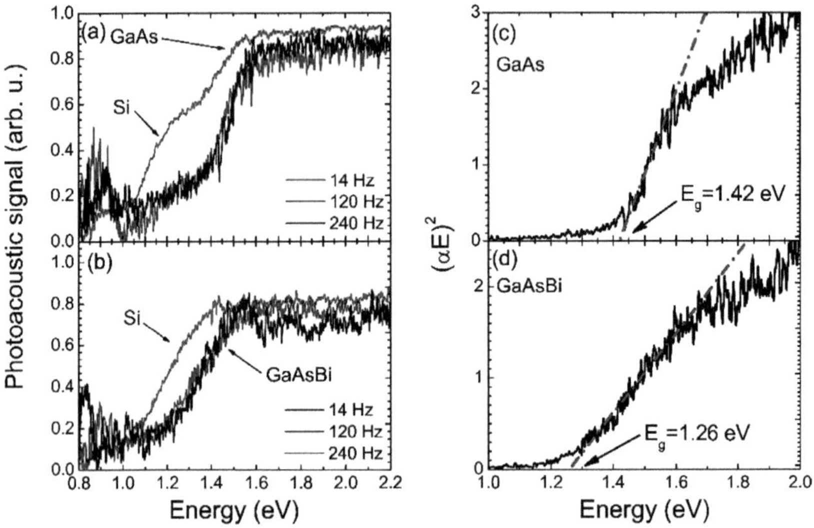

Figure 5.10 Photoacoustic signal taken at different frequencies for (a) reference GaAs nanowires and (b) GaAsBi nanowires. Photoacoustic absorption edge for (c) reference GaAs nanowires and (d) GaAsBi nanowires. Reprinted from [31], with the permission of AIP Publishing.

5.6 Summary

In this chapter, we have presented the recent progress in the growth and structural and optical properties of III-V-Bi nanowires grown by MBE and MOVPE methods. The GaAs/GaAsBi core–shell nanowires grown by MBE have revealed a strong roughening, a strong structural disorder at the wire surface, and an inhomogeneous Bi distribution within the layer surrounded by the outermost GaAs. A room-temperature cathodoluminescence of these structures has exhibited a broad emission between 1.1 and 1.5 eV, accompanied by multiple peaks. A distinct peak at 1.24 eV agrees well with the energy of the reduced GaAsBi alloy band gap by the incorporation of 2% Bi. The GaAsBi nanowires developed by the MOVPE method were grown on the nanoislands of bismuth, which act as a local point of the nucleus for the growth. The growth conditions affect the initial nucleation of GaAsBi structures, resulting in different numbers and shapes of nanowires. Unfortunately, no optical characterization has been performed on MOVPE-grown structures so far.

A first-principles study has shown that the band gap of GaAsBi nanowires shrinks with the increasing Bi concentration and the Bi 6*p* states dominate contribution for the electronic structure around the Fermi level. The photoacoustic spectroscopy has been shown to be very powerful in nanowire materials, where measurements of absorption are difficult due to light scattering and/or non-transparent substrate in the spectral range of band gap of nanowires. The band gap values for GaAs and GaAs/GaAsBi nanowires were determined by photoacoustic spectroscopy to be 1.42 and 1.26 eV, respectively. These values are in good agreement with cathodoluminescence results.

The development of III–V bismides nanowires is still in the first stages because the growth of these nanostructures is a major challenge to be overcome. A rigorous fundamental research is required to understand the III-V-Bi nanowires growth mechanisms and optimize material properties in order to enable potential device applications to be evaluated. Photoacoustic spectroscopy is a powerful tool to study the energy gap of semiconductor nanowires.

Acknowledgments

F. Ishikawa greatly acknowledges the TEM/EDS study by F. Uesugi, Y. Nakayama, and M. Takeguchi of the National Institute for Materials Science under the Nanotechnology Platform Project operated by the Ministry of Education, Culture, Sports, Science and Technology (MEXT), Japan. W. M. Linhart acknowledges support from the National Science Center (NCN) grant no 2014/13/D/ST3/01947.

References

1. Janotti, A., Wei, S.-H., and Zhang, S. B. (2002). Theoretical study of the effects of isovalent coalloying of Bi and N in GaAs, *Phys. Rev. B* **65**, 115203
2. Ferhat, M., and Zaoui, A. (2006). Structural and electronic properties of III–V bismuth compounds, *Phys. Rev. B* **73**, 115107.
3. Fluegel, B., Francoeur, S., Mascarenhas, A., Tixier, S., Young, E. C., and Tiedje, T. (2006). Giant spin-orbit bowing in $GaAs_{1-x}Bi_x$, *Phys. Rev. Lett.* **97**, 067205.

4. Zhang, Y., Mascarenhas, A., and Wang, L.-W. (2005). Similar and dissimilar aspects of III–V semiconductors containing Bi versus N, *Phys. Rev. B* **71**, 155201.

5. Alberi, K., Wu, J., Walukiewicz, W., Yu, K. M., Dubon, O. D., Watkins, S. P., Wang, C. X., Liu, X., Cho, Y.-J., and Furdyna, J. (2007). Valence-band anticrossing in mismatched III–V semiconductor alloys, *Phys. Rev. B* **75**, 045203.

6. Tiedje, T., Young, E., and Mascarenhas, A. (2008). Growth and properties of the dilute bismide semiconductor $GaAs_{1-x}Bi_x$ a complementary alloy to the dilute nitrides, *Int. J. Nanotechnol.* **5**, 963.

7. Kudrawiec, R., Kopaczek, J., Misiewicz, J., Petropoulos, J. P., Zhong, Y., and Zide, J. M. O. (2011). Contactless electroreflectance study of E_0 and $E_0 + \Delta_{SO}$ transitions in $In_{0.53}Ga_{0.47}Bi_xAs_{1-x}$ alloys, *Appl. Phys. Lett.* **99**, 251906.

8. Kudrawiec, R., Kopaczek, J., Sitarek, P., Misiewicz, J., Henini, M., and Novikov, S. V. (2012). Unusual broadening of E_0 and $E_0 + \Delta_{SO}$ transitions in GaAsBi studied by electromodulation spectroscopy, *J. Appl. Phys.* **111**, 066103.

9. Rajpalke, M. K., Linhart, W. M., Birkett, M., Yu, K. M., Scanlon, D. O., Buckeridge, J., Jones, T. S., Ashwin, M. J., and Veal, T. D. (2013). Growth and properties of GaSbBi alloys, *Appl. Phys. Lett.* **103**, 142106.

10. Sweeney, S. J., and Jin, S. R. (2013). Bismide-nitride alloys: Promising for efficient light emitting devices in the near- and mid-infrared, *J. Appl. Phys.* **113**, 043110.

11. Rajpalke, M. K., Linhart, W. M., Birkett, M., Yu, K. M., Alaria, J., Kopaczek, J., Kudrawiec, R., Jones, T. S., Ashwin, M. J., and Veal, T. D. (2014). High Bi content GaSbBi alloys, *J. Appl. Phys.* **116**, 043511.

12. Kopaczek, J., Kudrawiec, R., Linhart, W. M., Rajpalke, M. K., Jones, T. S., Ashwin, M. J., and Veal, T. D. (2014). Low- and high-energy photoluminescence from $GaSb_{1-x}Bi_x$ with $0 < x \leq 0.042$, *Appl. Phys. Express* **7**, 111202.

13. Rajpalke, M. K., Linhart, W. M., Yu, K. M., Birkett, M., Alaria, J., Bomphrey, J. J., Sallis, S., Piper, L. F. J, Jones, T. S., Ashwin, M. J., and Veal, T. D. (2014). Bi-induced band gap reduction in epitaxial InSbBi alloys, *Appl. Phys. Lett.* **105**, 212101.

14. Kopaczek, J., Kudrawiec, R., Polak, M. P., Scharoch, P., Birkett, M., Veal, T. D., Wang, K., Gu, Y., Gong, Q., and Wang, S. (2014). Contactless electroreflectance and theoretical studies of band gap and spin-orbit splitting in $InP_{1-x}Bi_x$ dilute bismide with $x \leq 0.034$, *Appl. Phys. Lett.* **105**, 222104.

15. Kopaczek, J., Rajpalke, M. K., Linhart, W. M., Jones, T. S., Ashwin, M. J., Kudrawiec, R., and Veal, T. D. (2014). Photoreflectance spectroscopy of GaInSbBi and AlGaSbBi quaternary alloys, *Appl. Phys. Lett.* **105**, 112102.

16. Kopaczek, J., Linhart, W. M., Baranowski, M., Richards, R. D., Bastiman, F., David, J. P. R., and Kudrawiec, R. (2015). Optical properties of GaAsBi/GaAs quantum wells: Photoreflectance, photoluminescence and time-resolved photoluminescence study, *J. Phys. D* **30**, 094005.

17. Kudrawiec, R., Kopaczek, J., Polak, M. P., Scharoch, P., Gladysiewicz, M., Misiewicz, J., Richards, R. D., Bastiman, F., and David, J. P. R (2014). Experimental and theoretical studies of band gap alignment in $GaAs_{1-x}Bi_x$/GaAs quantum wells, *J. Appl. Phys.* **116**, 233508.

18. Polak, M., Scharoch, P., and Kudrawiec, R. (2015). First-principles calculations of bismuth induced changes in the band structure of dilute Ga–V–Bi and In–V–Bi alloys: Chemical trends versus experimental data, *Semicond. Sci. Technol.* **30**, 094001.

19. Gladysiewicz, M., Kudrawiec, R., and Wartak, M. S. (2015). 8-band and 14-band kp modeling of electronic band structure and material gain in Ga(In)AsBi quantum wells grown on GaAs and InP substrates, *J. Appl. Phys.* **118**, 055702.

20. Gladysiewicz, M., Kudrawiec, R., and Wartak, M. S. (2016). Electronic band structure and material gain of III-V-Bi quantum wells grown on GaSb substrate and dedicated for mid-infrared spectral range, *J. Appl. Phys.* **119**, 075071.

21. Higashi, T., Sweeney, S. J., Phillips, A. F., Adams, A. R., O'Reilly, E. P., Uchida, T., and Fujii, T. (1999). Experimental analysis of temperature dependence in 1.3 μm AlGaInAs-InP strained MQW lasers, *IEEE J. Selected Topics Quantum Electron.* **5**, 413.

22. Silver, M., O'Reilly, E. P., and Adams, A. R. (1997). Determination of the wavelength dependence of Auger recombination in long-wavelength quantum-well semiconductor lasers using hydrostatic pressure, *IEEE J. Quantum Electron.* **33**, 1557.

23. Sweeney, S. J., Batool, Z., Hild, K., Jin, S. R., and Hosea, T. J. C. (2011). The potential role of bismide alloys in future photonic devices *in Proceedings of the 13th International Conference on Transparent Optical Networks*, Stockholm, Sweden.

24. Cheetham, K. J., Krier, A., Marko, I. P., Aldukhayel, A., and Sweeney, S. J. (2011). Direct evidence for suppression of Auger recombination in GaInAsSbP/InAs mid-infrared emitting diodes, *Appl. Phys. Lett.* **99**, 141110.

25. Bastiman, F., Mohmada, A. R. B., Ng, J. S., David, J. P. R., and Sweeny, S. J. (2012). Non-stoichiometric GaAsBi/GaAs (100) molecular beam epitaxy growth, *J. Crystal Growth* **338**, 57.

26. Ptak, A. J., France, R., Beaton, D. A., Alberi, K., Mascarenhas, J. S., and Jiang, C.-S. (2012). Kinetically limited growth of GaAsBi by molecular-beam epitaxy, *J. Crystal Growth* **338**, 107.

27. Kishino, K., Kikuchi, A., Sekiguchi, H., and Ishizawa, S. (2007). InGaN/GaN nanocolumn LEDs emitting from blue to red, *Proc. SPIE* **6473**, 64730T.

28. Ishikawa, F., Akamatsu, Y., Watanabe, K., Uesugi, F., Asahina, S., Jahn, U., and Shimomura, S. (2015). Metamorphic GaAs/GaAsBi heterostructured nanowires, *Nano Lett.* **15**, 7265.

29. Essouda, Y., Fitouri, H., Boussafa, R., Elayech, N., Rebey, A., and Jani, B. E. (2015). Bismuth catalyzed growth of GaAsBi nanowires by metalorganic vapor phase epitaxy, *Mater. Lett.* **152**, 298.

30. Ding, L., Lu, P., Cao, H., Cai, N., Yu, Z., Gao, T., and Wang, S. (2013). Bismuth alloying properties in GaAs nanowires, *J. Solid State Chem.* **205**, 44.

31. Zelewski, S. J., Kopaczek, J., Linhart, W. M., Ishikawa, F., Shimomura, S., and Kudrawiec, R. (2016). Photoacoustic spectroscopy of absorption edge for GaAsBi/GaAs nanowires grown on Si substrate, *Appl. Phys. Lett.* **109**, 182106.

32. Rosencwaig, A., and Gersho, A. (1976). Theory of the photoacoustic effect with solids, *J. Appl. Phys.* **47**, 64.

Chapter 6

Ferromagnetic MnAs/III–V Hybrid Nanowires for Spintronics

Shinjiro Hara

Research Center for Integrated Quantum Electronics,
Hokkaido University, North 13 West 8, Sapporo 060-8628, Japan

hara@rciqe.hokudai.ac.jp

This chapter introduces some of the experimental results of hybrid nanowires between ferromagnetic MnAs and III–V compound semiconductors, which have been demonstrated for possible magneto-nanoelectronic or spintronic device applications of the future by the author's group. The purpose of this study is to synthesize vertical ferromagnetic/semiconducting heterojunction nanowires by combing the catalyst-free selective-area growth of III–V compound semiconductor nanowires and the endotaxial nanoclustering of MnAs. The first part of this chapter demonstrates the formation of MnAs/GaAs hybrid nanowires, and then, MnAs/InAs heterojunction nanowires are demonstrated in the second part. As a result of endotaxy, MnAs nanoclusters are typically embedded in the six ridges of hexagonal GaAs nanowires. However, the MnAs nanoclusters are formed not only at the six ridges, but on the surfaces of six {0–11} crystal facets of hexagonal GaAs nanowires, when we decrease the growth temperature from

Novel Compound Semiconductor Nanowires: Materials, Devices, and Applications
Edited by Fumitaro Ishikawa and Irina Buyanova

ISBN 978-981-4745-76-5 (Hardcover), 978-1-315-36440-7 (eBook)
www.panstanford.com

600 to 400°C. Current and voltage characteristics of MnAs/GaAs hybrid nanowires are investigated using two-terminal device structures of free-standing hybrid nanowires. The MnAs/GaAs hybrid nanowires formed on Zn-doped p-type GaAs (111)B substrates show ohmic characteristics, while those formed on Si-doped n-type GaAs (111)B substrates show clear rectifying characteristics. The hybrid nanowires show p-type conductivity possibly owing to the formation of p-type GaAs layers near the nanowire surfaces caused by the thermal diffusion of manganese atoms into GaAs nanowires during the endotaxial nanoclustering of MnAs. In the case of InAs nanowire templates, MnAs penetrates the InAs nanowires to form nanoclusters. The surface migration length of manganese adatoms on the nanowires, which is estimated to be 600 nm at 580°C, is a key to the successful fabrication of vertical MnAs/InAs heterojunction nanowires with atomically abrupt heterointerfaces. MnAs nanoclusters grown at 580°C in and on InAs nanowires have marked single magnetic domains, and the observed results for magnetization directions are consistent with the structural characterization results, i.e., the magnetic easy axes, *a*-axes, are perpendicular to the <111>B direction of host InAs nanowires.

6.1 Introduction

Vertical free-standing semiconducting nanowires (NWs) have recently been demonstrating extraordinary versatility and extremely high possibility for use in potential applications to next-generation electronic, photonic, and bio-chemical sensing devices [1–12]. For future electronic industry, for example, practical ways to install vertical surrounding gate field-effect transistors (FET) using NWs to current integrated circuits based on the CMOS technologies were intensively demonstrated [13]. Most of these NWs reported elsewhere have thus far been synthesized and demonstrated by using some of the most popular bottom-up fabrication techniques in the world, i.e., vapor–liquid–solid (VLS) methods that typically use gold nanoparticles as catalysts. Conventional metal catalysts used in the VLS method, however, possibly lead to the deterioration of performance of NW devices mainly due to deep level formation by incorporated metal impurities. In addition, the VLS-grown NWs are randomly distributed on a semiconducting substrate in most cases, which might hinder the practical applications of NWs in devices in the future. We, on the other hand, have demonstrated the catalyst-free formation of vertical free-standing semiconducting NW arrays, e.g., GaAs, InGaAs, InP, InAs, GaAs/AlGaAs core–shell, and GaAs/GaAsP core–shell NWs, by selective-area metal-organic vapor phase epitaxy (SA-MOVPE) using partially SiO_2-masked {111}A and B substrates, which has enabled us to control the size, aspect ratio, position, and density of NWs on various semiconducting substrate materials [14, 15]. Nanoelectronic devices such as vertical surrounding gate FETs have been fabricated, and their performance has been demonstrated using our semiconducting NWs grown by SA-MOVPE for future device applications [16, 17].

The perspective and outlook for possible and attractive magneto-nanoelectronic or nanospintronic device applications in semiconducting NWs have not been fully discussed in one of the most important and latest papers that reviewed future applications of semiconducting NWs [18]. (There was no section dealing with "NW spintronics.") Activities to achieve diluted magnetic semiconducting layers (DMS) [19] and the heteroepitaxy of ferromagnetic layers and III–V compound semiconductors (FM III–V hybrids) have attracted a great deal of attention in recent semiconductor nanospintronics research owing to possible additional functionalities that have been added to current semiconducting devices and integrated circuits. Of these approaches, granular hybrid structures in which ferromagnetic nanoclusters (NC) are embedded into semiconductor layers are one of the most attractive candidate materials for future nanospintronic devices because they have been reported to have huge magnetoresistance (MR) effects and a relatively long spin-relaxation time [20–22]. These devices have been fabricated on semiconducting substrates mostly by low-temperature molecular beam epitaxy in combination with conventional top-down fabrication techniques. Various kinds of NWs using FM III–V hybrids or DMS [23] have recently been synthesized to produce NW spintronic devices using several techniques based on conventional VLS methods as bottom-up fabrication approaches, e.g., GaMnAs NWs [24, 25], GaAs/GaMnAs core–shell NWs [26, 27], MnAs nanoparticles on InAs [28] and GaAs [29] NWs,

GaAs/MnAs core–shell NWs [30, 31], GaN-related DMS NWs [23, 32, 33], heterojunction NWs between MnGe alloys and Ge [34, 35] and between MnSi alloys and Si [36], and vertical multisegment AuGe/Ge NWs [37]. However, it is also crucial to avoid and overcome the possible problems arising in the current nanoelectronic and photonic NW devices fabricated by these conventional VLS methods particularly when fabricating practical NW spintronic devices of the future. We have, therefore, developed hybridization techniques of ferromagnetic MnAs NCs [38] on vertical free-standing semiconducting NW templates grown by our catalyst-free SA-MOVPE technique [14, 15] in the research fields of semiconductor nanospintronics, and reported on the formation and characterizations of MnAs NC/GaAs hybrid NWs [39, 40]. The author's group developed an SA-MOVPE technique in our previous studies to directly synthesize single-crystal ferromagnetic MnAs NCs on a semiconducting substrate, and we demonstrated that the NCs could be used as promising building blocks for fabricating nanospintronic devices of the future since they demonstrated large angle-dependent MR effects. However, we realized that it is still difficult to form vertical heterojunction NWs between ferromagnetic MnAs and semiconducting layers, which would enable us to inject spin-polarized carriers and currents into semiconducting NW channels. It is also difficult to control the crystal orientations of MnAs NCs in a <111>B-oriented GaAs NW template [39, 40] and to obtain atomically abrupt heterointerfaces between MnAs and GaAs even when using our hybridization techniques. It is likely that the difficulty in forming vertical heterojunction NWs was mainly due to the relatively stable crystal facet of the c-plane, i.e., the {0001}-oriented surface, at the surface of hexagonal NiAs-type MnAs NCs grown by SA-MOVPE [39]. We believe that the combination of a magnetic tunnel junction (e.g., CoFeB/MgO [41] and CoFe/MgO [42]) electrode and vertical ferromagnetic/semiconducting heterojunction NWs with atomically abrupt heterointerfaces [35] could overcome major obstacles of conductance mismatch and poor controllability in the heterointerface formation in the current spin MOSFETs [43]. In addition, the {111}B-oriented semiconducting NW channels possibly have a great advantage for enhancing spin lifetime in the channels of spin MOSFETs [44].

Therefore, the author of this book chapter introduces the experimental results of the synthesis of high-quality vertical heterojunction NWs with atomically abrupt heterointerfaces between ferromagnetic MnAs NC and III–V compound semiconductors, in particular, InAs, on semiconducting GaAs (111)B substrates utilizing the so-called endotaxy of MnAs NCs after the SA-MOVPE of semiconducting NW templates and by adjusting the synthesis conditions. In the first part of this chapter, in addition to the detailed cross-sectional observation results, the author presents the results of the electrical characterization of the hybrid structures, since there is no information on the conductivity of the MnAs/GaAs hybrid NWs fabricated by combining SA-MOVPE of GaAs NWs and endotaxial nanoclustering of MnAs. Fundamental MOVPE growth condition dependence of the formation of MnAs NCs on the GaAs NW array templates is also discussed. In the second part, to obtain atomically abrupt heterointerfaces between MnAs and semiconducting layers in NWs, a possible mechanism of the formation of single-crystal MnAs NCs in and on the InAs NW templates is discussed on the basis of the detailed dependences of the formation of MnAs NCs on synthesis conditions and structural characterizations obtained from scanning and transmission electron microscopies (SEM and TEM, respectively). The author believes that our vertical ferromagnetic MnAs NC/semiconducting InAs heterojunction NWs in the current work are definitely exhibiting new possibilities and versatility for creating novel magneto-nanoelectronic or nanospintronic devices using vertical free-standing semiconducting NWs, e.g., spin-NW-MOSFETs and spin-NW-light-emitting diodes, as was reported and discussed in our review paper [45].

The contents of this book chapter are mostly based on those of the author's papers published in *Japanese Journal of Applied Physics* [39, 40, 46]. (IOP Publishing, http://iopscience.iop.org/journal/1347-4065, The Japan Society of Applied Physics.) Two articles listed in the References section of this chapter, Refs. [40, 46], were published as an "open access" paper in *Japanese Journal of Applied Physics*.

6.2 Experimental

6.2.1 Selective-Area Growth of Semiconducting Nanowire Templates

The templates of the typical GaAs NW arrays on GaAs (111)B substrates were fabricated by our typical SA-MOVPE process, which has been described in detail elsewhere [47, 48]. We first fabricated the initial circular openings, which were defined in a SiO_2 thin-film mask by electron beam (EB) lithography (JBX-6300FS system, JEOL), with a diameter, d_0, of approximately 200 nm and a distance between them, or a period, a, in the range from 0.5 to 3.0 μm. The SiO_2 thin-film mask, whose typical thickness was estimated to be approximately 20 to 30 nm, was deposited on GaAs (111)B wafers by plasma-enhanced chemical vapor deposition (PE-CVD). The growth temperature T_g and the V/III ratio for the GaAs NW growth were 750°C and approximately 180, respectively. Conventional organometallic and hydride sources, such as $(CH_3)_3Ga$ and 20% AsH_3 diluted in H_2, were used as the group III and group V sources for all the growth experiments, respectively.

Next, the author used InAs NW arrays as a template, which were fabricated heteroepitaxially on semi-insulating (S.I.) GaAs (111)B substrates by utilizing the SA-MOVPE process, for creating high-quality vertical ferromagnetic MnAs NC/semiconducting InAs heterojunction NWs. First, we again prepared the initial circular openings, which were arranged and defined in SiO_2 thin films by EB lithography. There were typically two types of the observed diameter, d_0, of the initial circular openings: one was approximately 80 to 90 nm and the other was approximately 110 to 140 nm. The distances between the initial circular openings or the periods, a, were 0.5, 1.0, and 3.0 μm on the substrates. The SiO_2 thin films, whose typical thicknesses were estimated to be approximately 20 to 30 nm, were deposited on GaAs (111)B wafers by plasma sputtering in this case. The growth temperature T_g and the growth time t for InAs NWs were 580°C and 30 min, respectively. Conventional organometallic and hydride sources, such as $(CH_3)_3In$ and 20% AsH_3 diluted in H_2, were used as the group III sources for the former and group V sources for the latter in all the growth experiments. The estimated partial pressures of $(CH_3)_3In$ [49] and 20% AsH_3 diluted in H_2 were 4.9×10^{-7} and 1.3×10^{-4} atm, respectively, for the SA-MOVPE of undoped InAs NW templates.

6.2.2 Endotaxy of Ferromagnetic MnAs Nanoclusters

For the MnAs NC growth after the growth of undoped semiconducting NW templates, we utilized the phenomenon of "endotaxy" of MnAs in GaAs and InAs. We have observed that single-crystal MnAs NCs were grown in GaInAs (111)A layers even without any supply of an AsH_3 source as a result of endotaxy [50]. This is the key technique for forming ferromagnetic MnAs NCs "into" undoped semiconducting, GaAs and InAs, NWs grown by SA-MOVPE [39, 40, 46]. Endotaxy is associated with a diffusion process that leads to the redistribution of substances, i.e., MnAs NCs in host crystals, GaAs and InAs NWs in the current work, and the formation of new stable phases. The detailed explanation for the phenomenon of endotaxy was reported elsewhere [51].

During the endotaxial growth of MnAs NCs into the undoped host GaAs and InAs NW templates, we only supplied the organometallic source of $(CH_3C_5H_4)_2Mn$ diluted in H_2, i.e., no AsH_3 supply. The growth temperature T_g was changed from 400 to 600°C in the case of GaAs NW templates, and the growth time t was set to 1 min for the MnAs NC growth. The estimated partial pressure of $(CH_3C_5H_4)_2Mn$ was 1.1×10^{-6} atm [38]. Zn-doped p-type and Si-doped n-type GaAs (111)B substrates were used for the selective-area growth and the electrical characterizations of hybrid structures between MnAs NCs and undoped GaAs NWs. Carrier concentrations of the p-type and n-type wafers were estimated to be approximately 1.0×10^{19} and 2.5×10^{18} cm^{-3}, respectively, in the current experiments. In the case of the endotaxial growth of MnAs NCs into the undoped host InAs NW templates, next, the growth temperatures T_g for the MnAs NCs were changed from 400 to 580°C, and their growth times t were all 1 min. The estimated partial pressure of $(CH_3C_5H_4)_2Mn$ was 3.0×10^{-6} atm. During the decrease in temperature during the purging process after MnAs NC growth, 20% AsH_3 diluted in H_2 was supplied.

6.2.3 Characterization Methods

Structural characterizations in terms of the size, density, and position of MnAs NCs, GaAs NWs, and InAs NWs were carried out by SEM. TEM was also used for obtaining lattice images of NCs and NWs, and we conducted detailed structural characterizations, such as the analyses of crystal structures and solid compositions of the NCs and NWs, by electron-beam diffraction (ED) and energy dispersive X-ray (EDX) spectroscopy in combination with TEM, using an electron beam with a spot diameter of approximately 1 nm. (All the TEM characterizations in the current work, including the ED measurements and EDX spectroscopy, were carried out by a professional characterization company, Toray Research Center, Inc.) All of the current and voltage curves, i.e., *I–V* characteristics, for hybrid structures between MnAs NCs and undoped GaAs NWs were obtained at room temperature. For the magnetic characterization of MnAs NCs in and on undoped InAs NWs, we used magnetic force microscopy (MFM), Nanoscope IIIa system (Digital Instruments), first in the conventional phase detection (PD) mode of the system at room temperature and under zero magnetic field condition after applying the external magnetic fields, **B**, of approximately 5,700 Gauss. For MFM measurements, vertical ferromagnetic MnAs NC/semiconducting InAs heterojunction NWs were separated from GaAs (111)B substrates mechanically by ultrasonic vibration in isopropanol solution and deposited on SiO_2/Si substrates. To ensure that magnetic responses of MnAs NCs were observed, we also conducted amplitude detection (AD) mode measurements in the MFM system. In the AD mode measurements, the resonance curve of the high-resolution-type MFM cantilever coated with Co/Cr alloy materials is shifted when it is affected by stray magnetic fields. At the same time, the amplitude change of the MFM cantilever is detected at the point of drive frequency, which is set to be lower or higher than the resonance frequency of a free-vibrating high-resolution-type MFM cantilever. The change in drive frequency gives the reversals in contrasts of the magnetized region in MFM images.

6.3 Results and Discussion

6.3.1 MnAs/GaAs Hybrid Nanowires

6.3.1.1 Endotaxial formation of MnAs NCs in GaAs NWs

To realize the hybrid structures between MnAs NCs and GaAs NWs, we investigated the endotaxial formation of MnAs NCs in GaAs NWs grown by SA-MOVPE. Figure 6.1a shows an SEM image of a typical template structure of the GaAs NW array before the MnAs growth. In Fig. 6.1a, the distance between the GaAs NWs, or the period, *a*, was 1.0 μm. The diameter and height of the GaAs NWs were typically estimated to be approximately 130 nm and 1.5 μm, respectively. As shown in the highly magnified SEM images in Fig. 6.1b, hexagonal GaAs NWs surrounded with six {0–11} facets were successfully formed. For the observation shown in Fig. 6.1b, the distance between the GaAs NWs, *a*, was 3.0 μm, and the typical height of the NWs was approximately 1.0 μm. The height of the GaAs NWs decreased with increasing *a*. The results obtained in the current work were consistent with those discussed and reported in detail elsewhere [47]. Subsequently, to investigate the detailed growth condition dependence of the endotaxial MnAs nanoclustering in the GaAs NWs, we conducted growth experiments under various T_g conditions. Figure 6.2 shows typical SEM images of the hybrid structures between MnAs NCs and GaAs NWs, in which the MnAs NCs were grown at T_g values from 400 to 600°C. We have found that it is crucial to choose the atmospheric conditions, i.e., only H_2 carrier gas or the mixture of 20%-AsH_3 and H_2, during the decrease in T_g after the endotaxy of MnAs NCs because the diameter and density of the NCs were strongly influenced by the gas supply conditions in this purging process [39]. Therefore, we introduced, in the current work, only H_2 carrier gas in the MOVPE reactor during the decrease in T_g after the endotaxy of the MnAs NCs. In Fig. 6.2, the distance between the NWs, *a*, was set to 3.0 μm for all the observed samples. The size of the MnAs NCs was markedly increased, as shown in Figs. 6.2a,b, when we decreased T_g for the endotaxy of the MnAs NCs from 600 to 500°C. With decreasing T_g for the MnAs NC formation to 400°C, on the other hand, the size of the MnAs NCs markedly decreased, and, in addition, the NCs were formed not only on the six ridges, which are the areas between two of the {0–11} facets, of hexagonal GaAs NWs from the top to the bottom part of the NWs, but also on the {0–11} facet surfaces, as shown in Fig. 6.2c.

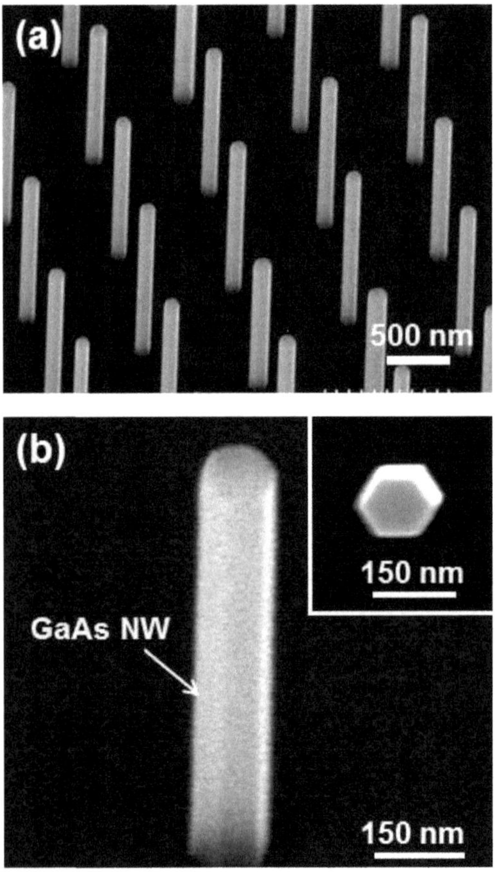

Figure 6.1 Bird's-eye view obtained by SEM for (a) the typical template structure of GaAs NW arrays, in which the distance, or the period, between the NWs, *a*, was 1.0 μm. Highly-magnified bird's-eye views obtained by SEM for a GaAs NW (*a* = 3.0 μm) (b) before the MnAs growth. The inset in Fig. 6.1b is the top view of the NW without the NCs (Ref. [40], open access). Copyright (2012) The Japan Society of Applied Physics.

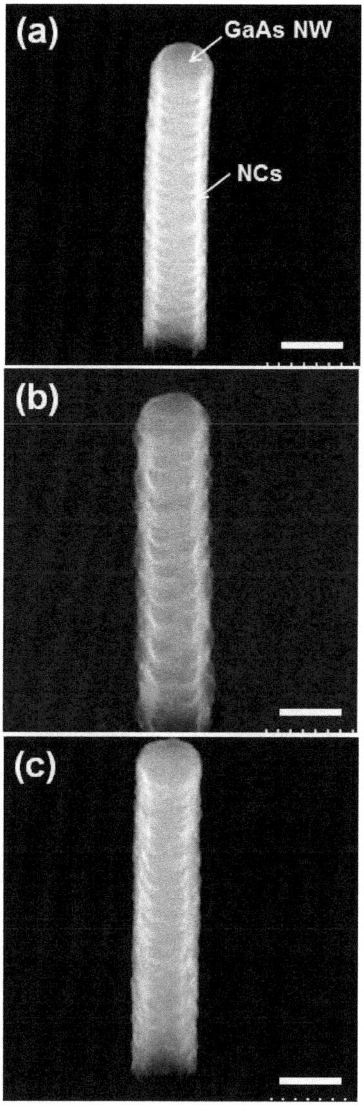

Figure 6.2 Bird's-eye views of typical GaAs NWs obtained by SEM with MnAs NCs grown at various growth temperatures, T_g, of MnAs NCs: (a) 600, (b) 500, and (c) 400°C. Only H_2 was supplied during the decrease in temperature after the growth of MnAs NCs. The distance, a, and the estimated diameters (width) of the GaAs NWs were 3.0 μm and approximately 130 nm, respectively. White scale bars in all the figures represent 150 nm (Ref. [40], open access). Copyright (2012) The Japan Society of Applied Physics.

Next, we observed the cross-sectional images of the hybrid structures between MnAs NCs and GaAs NWs from the top to the bottom part of the NWs by TEM. The growth conditions for the NCs and the atmospheric conditions after the NC growth for the observed structure were the same as those for the sample shown in Fig. 6.2a, but the distance between the NWs, *a*, was 0.5 μm. Figures 6.3a,b show a typically observed image of the whole structure of a hybrid NW and the schematic illustrations and the crystal orientations for the cross-sectional observations, respectively. All the cross-sectional images were taken in the [0–11] direction of the zinc-blende-type GaAs crystal structure, as schematically illustrated in Fig. 6.3b. The height and diameter of the observed structure in Fig. 6.3a were estimated to be 1.7 μm and 210 nm, respectively. We confirmed from Fig. 6.3a that the

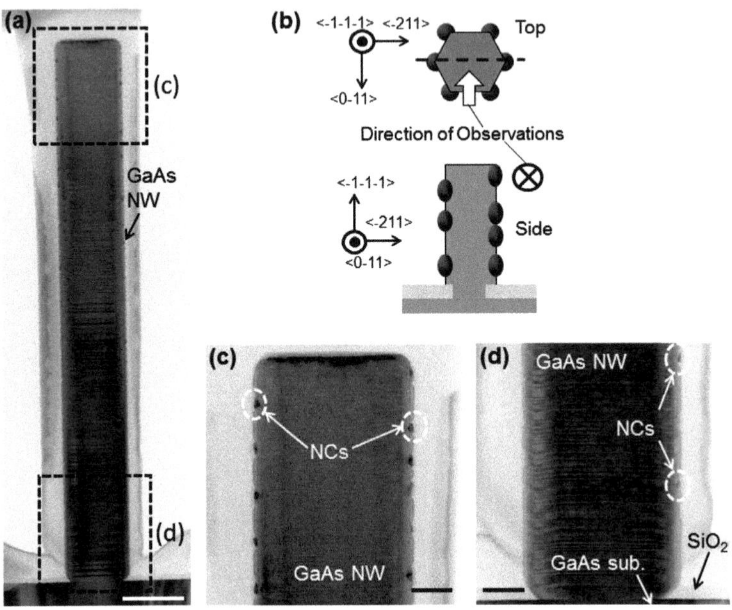

Figure 6.3 (a) Cross-sectional overview of the observed GaAs NW with MnAs NCs at the ridges of the NW and (b) schematic illustrations and crystal directions of the NW with the NCs for all the cross-sectional TEM observations in the current work. White scale bar in Fig. 6.3a represents 200 nm. Highly magnified cross-sectional images of MnAs NCs and GaAs NW at around (c) the top and (d) the bottom parts of the NW. Black scale bars in Figs. 6.3c,d represent 50 nm (Ref. [40], open access). Copyright (2012) The Japan Society of Applied Physics.

hybrid structures had the same diameter, 210 nm, over the whole structure from the top to the bottom, by which we can nearly ensure that the {0–11} facets were formed as side walls of hexagonal GaAs NWs from the top to the bottom. From the cross-sectional TEM images at the top and bottom parts of the NWs in Figs. 6.3c,d, respectively, we estimated the average width and depth of the NCs in the NWs to be approximately 14.8 and 6.0 nm at the top part of the NWs, and approximately 10.7 and 4.0 nm at the bottom one, respectively. The average density of the NCs, which was defined as the number of NCs per ridge of the NWs, i.e., /μm, was also estimated to be approximately 19.6 at the top part of the NWs and approximately 15.5 at the bottom one. The lattice image in Fig. 6.4 and the ED pattern in the inset of the figure show that the NCs had the hexagonal NiAs-type crystal structure, whereas GaAs NWs had the zinc-blende-type crystal structure (not shown here). From the observed lattice images of the samples including those shown in Fig. 6.4, it appeared that the *a*-axes of the NCs were tilted by approximately 17° to 34° from the <111>B direction of the GaAs NWs, although we have not observed the marked dependences of the growth directions of the MnAs NCs on the crystal direction of the host GaAs NWs. On the observed top surfaces of the GaAs NWs, i.e., the {111}B crystal facets, we found no clear evidence, so far, showing that MnAs NCs were formed on the {111}B surfaces of the NWs (not shown here). Therefore, in the case of the MnAs NC formation on the GaAs NWs, the NCs presumably grew at the areas, such as the ridge areas of the NWs between two of the {0–11} crystal facets, in which many atomic steps were formed on the surfaces. It also appeared that the MnAs NCs tended to form near the twin dislocations formed in the {111}B planes of the host GaAs NWs, although further growth experiments and structural characterizations have to be conducted in detail to clarify the relationship between the twin dislocations of the host NWs and NC formation. To estimate solid compositions in the NCs and NWs, in addition, analysis by EDX spectroscopy was conducted for the three elements manganese, arsenic, and gallium in the NCs and NWs. Figure 6.5 shows the typical results of the EDX analyses of the line across the NWs and NCs. The line scan started at one of the NCs on one side of the NW and finished at another NC on the opposite side of the NW. Here, for the EDX analyses, the atomic percentages of the three elements were estimated after eliminating the possible external contaminations by carbon, oxygen, and silicon atoms presumably during the sample preparation processes and originating from the material of the sample holders for the TEM observations. The solid compositions (atomic compositions in %) of manganese, arsenic, and gallium in one of the NCs were estimated to be approximately 44%, 50%, and 6.0%, and, in another NC on the same NW, to be approximately 33%, 56%, and 11%, respectively. In the middle of the NW, we detected no manganese element, but approximately 48% of arsenic and 52% of gallium elements. Therefore, a small amount of gallium atoms from 6.0% to 11% was possibly included in the MnAs NCs presumably owing to the redistribution of the host materials as a result of the endotaxy of MnAs NCs in the host GaAs NWs, but we confirmed in our previous study [39] that all the crystal structures of the MnAs or Mn(Ga)As NCs and the GaAs NWs observed were of the hexagonal NiAs- and cubic zinc-blende types, as determined from ED measurements, respectively.

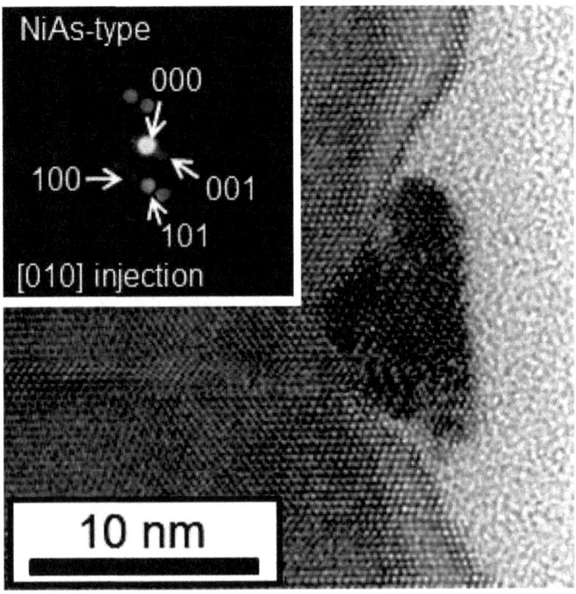

Figure 6.4 Cross-sectional lattice image of MnAs NC and GaAs NW. The inset in the figure shows an ED pattern for the MnAs NC (Ref. [39]). Copyright (2012) The Japan Society of Applied Physics.

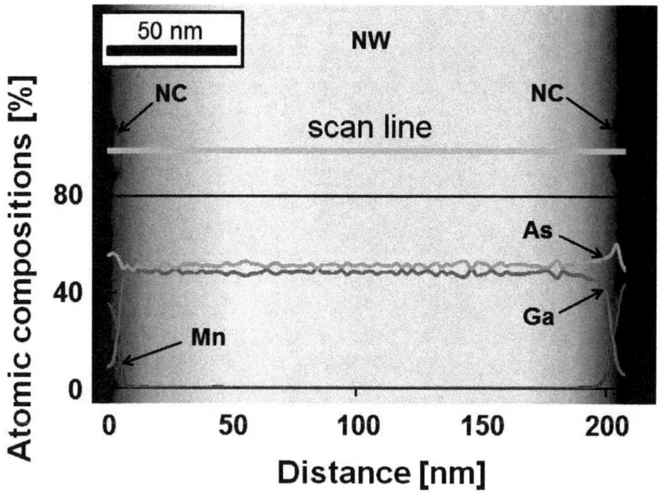

Figure 6.5 One-dimensional line profile of solid compositions of the GaAs NW and MnAs NCs analyzed by EDX spectroscopy. The percentages of gallium, arsenic, and manganese atoms were estimated by eliminating the possible external contaminations by chemicals or atoms, such as carbon, oxygen, and silicon, during the sample preparation processes and originating from the material of the sample holders (Ref. [39]). Copyright (2012) The Japan Society of Applied Physics.

Taking together all the findings obtained in these growth experiments and structural characterizations, we discuss the possible explanation for the growth behaviors of MnAs NCs on GaAs NWs. In the SA-MOVPE growth process, there possibly are three major contributions in the diffusion of source molecules (or species): the first is the diffusion from the vapor phase, the second is the surface diffusion of adatoms on SiO_2-mask surfaces, and the final is the diffusion between the surfaces of crystal facets. In the case of the MnAs NCs grown at 400°C, as shown in Fig. 6.2, we observed NC formation not only on the six ridges of hexagonal GaAs NWs, but also on the six {0–11} crystal facet surfaces. This experimental result presumably shows the decrease in the diffusion rate of manganese adatoms between the {0–11} facets. In addition, the diffusion of manganese atoms from the vapor phase was much more predominant in our experiments than the surface diffusion of manganese adatoms on the SiO_2-mask surfaces, because the size and density of the NCs at the top part of the NWs were larger than those at the bottom part of the

NWs, as observed in Fig. 6.3. It appears that these results are consistent with the growth behaviors of the NCs in terms of the different growth parameters discussed in our previous study [39].

6.3.1.2 Electrical characterization of MnAs/GaAs hybrid NWs

Next, to evaluate electrical properties of hybrid structures between MnAs NCs and GaAs NWs, we first developed a device process. Figure 6.6 shows schematic illustrations for the developed fabrication procedure for the two-terminal device structures. Corresponding images taken by SEM for each of the steps of the device process are also shown in Fig. 6.6. We have modified the developed procedure for the fabrication of nanowire light-emitting diodes reported in our previous study [52]. After the growth of hybrid structures between MnAs NCs and undoped GaAs NWs (Fig. 6.6a), the space between the hybrid structures was filled with polymer resin, i.e., benzocyclobutene (BCB) in this experiment, by spin coating, to put an electrode on the top part of hybrid structures and ensure the insulation between two electrodes for the electrical measurements. After the spin coating of BCB, the samples were annealed at 300°C for 15 min for the solidification of BCB (Fig. 6.6b). Subsequently, the top portion of the hybrid structures overlaid with an excessive amount of BCB was exposed by reactive ion etching (RIE) using the gases of CF_4 and O_2. The height of the exposed hybrid structures after the RIE was estimated to be approximately 100 nm (Fig. 6.6c). Next, after the photolithography process to form top electrode patterns for each of the hybrid structure arrays on the substrates, 30 nm-thick Cr and 50 nm-thick Au film electrodes were deposited onto the exposed hybrid structures by evaporation in vacuum, and then, lift-off was carried out to remove the electrodes outside the areas with hybrid structure arrays. Finally, the back-side electrodes were formed on the substrates. For Zn-doped p-type GaAs (111)B substrates, 30 nm-thick Cr and 50 nm-thick Au film electrodes were deposited, and, for Si-doped n-type ones, composite film electrodes of Au/Ge/Ni/Au, whose thicknesses were 25, 20, 15, and 30 nm, respectively, were deposited by evaporation in vacuum (Fig. 6.6d).

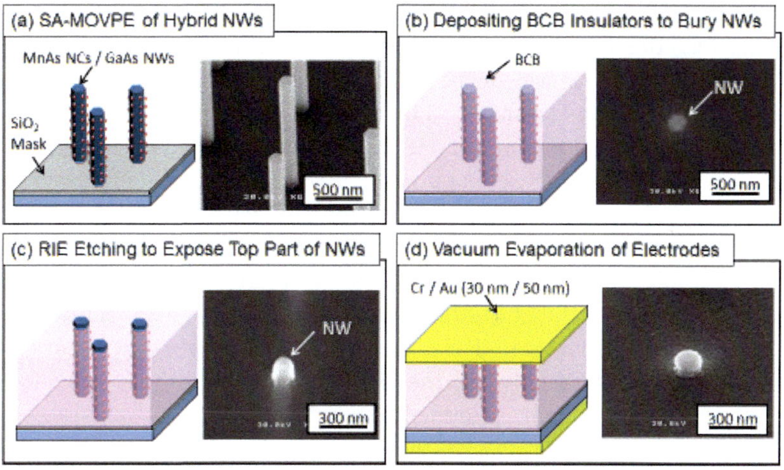

Figure 6.6 Process for the electrical characterization and fabrication of two-terminal devices with corresponding images taken by SEM and schematic illustrations at each of the steps (Ref. [40], open access). Copyright (2012) The Japan Society of Applied Physics.

Using the developed device process in Fig. 6.6, we fabricated two-terminal device structures for evaluating electrical properties of hybrid structures between MnAs NCs and undoped GaAs NWs. The growth temperature T_g for the endotaxy of MnAs NCs was 600°C for all the samples used in the electrical characterizations. Figure 6.7a shows a summary of the typical current and voltage (I–V) characteristics of the hybrid structures, in which the distances between the NWs, a, were 0.5, 1.0, and 3.0 μm, fabricated on Zn-doped p-type GaAs (111)B substrates. For the sample with a = 1.0 μm, the diameter of the GaAs NWs was estimated to be approximately 150 nm, and approximately 8,500 NWs were included under one electrode of the device. On the p-type substrates, *I–V* curves obtained in the current work showed approximately linear characteristics, i.e., ohmic behavior, for all the samples. The resistance of the samples with a = 0.5 and 1.0 μm was estimated to be approximately 1.0 and 6.4 MΩ at room temperature, respectively. In the case of the sample with a = 3.0 μm, as shown in the highly magnified *I–V* characteristic in Fig. 6.7b, the resistance was much higher than that of the other cases, but the sample also showed approximately an ohmic behavior. Although we have to optimize the device structures and processes to improve the resistance of the hybrid structures in the current work, the resistance decreased with decreasing the distance between the NWs from a = 3.0 to 0.5 μm, as shown in Fig. 6.7. From the rough estimation, it appeared, as we expected, that the resistance decreased approximately linearly with increasing number of the hybrid structures under the electrodes. In the case of the hybrid structures, in which MnAs NCs were formed at a T_g of 500°C, I–V curves also clearly showed an ohmic behavior on the p-type substrates (not shown here). Figure 6.8a shows a summary of the typical I–V characteristics for the hybrid structures with the distances between the NW templates, a, of 0.5, 1.0, and 3.0 μm fabricated on Si-doped n-type GaAs (111)B substrates. On the n-type substrates, on the other hand, I–V curves obtained in the current work showed clear rectifying properties for all the samples. As shown in the highly magnified I–V characteristic of the sample with a = 3.0 μm in Fig. 6.8b, the sample also showed a rectifying diode property within the current range, i.e., nanoampere, nA, similar to that shown in Fig. 6.7b. These results showed that the hybrid nanowires had p-type conductivity.

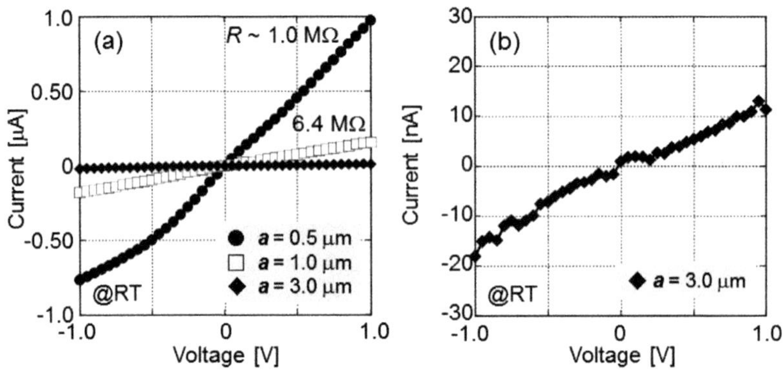

Figure 6.7 I–V characteristics for hybrid structures of ferromagnetic MnAs NCs and semiconducting GaAs NWs fabricated on Zn-doped p-type GaAs (111)B substrates. The distances between the NW templates, a, were 0.5 (solid circles), 1.0 (open squares), and 3.0 μm (solid diamonds). All the measurements were carried out at room temperature (Ref. [40], open access). Copyright (2012) The Japan Society of Applied Physics.

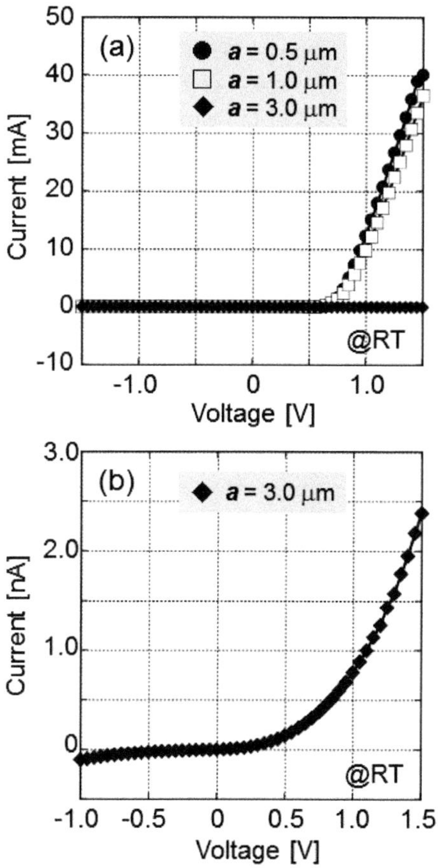

Figure 6.8 I–V characteristics for hybrid structures of ferromagnetic MnAs NCs and semiconducting GaAs NWs fabricated on Si-doped n-type GaAs (111)B substrates. The distances between the NW templates, *a*, were 0.5 (solid circles), 1.0 (open squares), and 3.0 μm (solid diamonds). All the measurements were carried out at room temperature (Ref. [40], open access). Copyright (2012) The Japan Society of Applied Physics.

In our previous study [53], we found the electrical and magnetotransport characteristics of the samples with MnAs NCs formed on GaInAs/InP (111)B layers by conventional MOVPE. The MnAs NCs were self-assembled on GaInAs (111)B surfaces at the growth temperature T_g of 650°C, and all the samples showed p-type conductivity with nearly the same carrier concentrations of approximately 1.7 × 10^{18} cm^{-3} estimated at 280 K [53]. Therefore, as in the case of the self-assembled formation of MnAs NCs on GaInAs (111)B surfaces, the manganese atoms provided during the endotaxy of MnAs NCs at a T_g of 600°C thermally diffuse into the undoped GaAs NWs in the current work. Then, they are unintentionally incorporated as an acceptor yielding p-type GaAs layers, or possibly p-type paramagnetic dilute magnetic semiconductor layers of GaMnAs, near the NW surfaces. We observed relatively large differences in the electrical currents measured for the present hybrid structures between the p-type and n-type substrates, i.e., microampere, μA, in Fig. 6.7a and milliampere, mA, in Fig. 6.8a, possibly owing to the poor reproducibility in the fabrication processes of the electrodes and/or the differences in carrier concentrations near the NW surfaces. However, we believe that the electrical characteristics of hybrid structures between ferromagnetic NCs and semiconducting NWs are much improved by introducing InAs NWs, which have been demonstrated in our vertical surrounding gate transistors [16, 17], as an electrical channel of core NWs, in combination with the optimization of the device processes. Although further detailed growth experiments are required for controlling the endotaxial nanoclustering of MnAs in InAs NWs, the author introduces the latest experimental results for MnAs/InAs hybrid NWs in the next subsection.

6.3.2 MnAs/InAs Hybrid Nanowires

6.3.2.1 Endotaxial formation of MnAs/InAs heterojunction NWs

We first grew a template structure of InAs NW arrays by SA-MOVPE to fabricate vertical ferromagnetic MnAs NC/ semiconducting InAs heterojunction NWs. Figure 6.9a shows a typical bird's-eye view obtained by SEM of the template structure of InAs NW arrays before the endotaxy of MnAs NCs. The period of InAs NWs, a, in Fig. 6.9a was 1.0 μm. The diameter of the InAs NWs was estimated to be approximately 150 nm and their height was estimated to be approximately 1.5 μm, when we used the initial circular openings with the diameters, d_0, of 110 to 140 nm in the SiO_2 thin films. We observed InAs NWs typically with a diameter of approximately 120 nm for the initial circular openings with the d_0 values of 80 to 90 nm from rough estimates. The inset in Fig. 6.9a is a highly magnified top view of one of the InAs NWs. We observed that InAs NWs with a hexagonal prismatic shape were surrounded by six {0–11} crystal facets and grew in the <111>B direction on GaAs (111)B substrates. The tops of the NWs revealed that flat {111}B crystal facets surrounded by tilted {–1–10} ones were formed. Figure 6.9b shows typical ferromagnetic MnAs NC/semiconducting InAs hybrid NWs grown at 580°C for 1 min. The NCs were formed on the top {111}B crystal facets of the InAs NWs, and we also observed that some additional NCs were formed around the middle parts of the NWs.

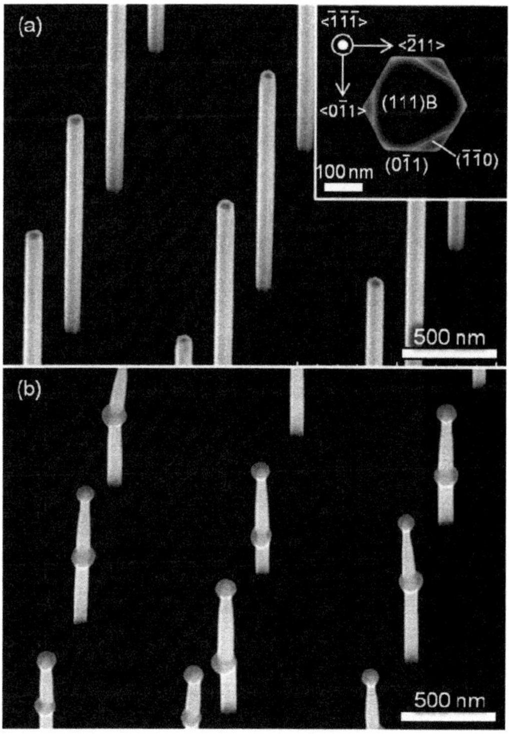

Figure 6.9 Bird's-eye views of SEM images of typical (a) host InAs NW array template before the synthesis of MnAs NCs and (b) MnAs/InAs heterojunction NWs. Inset of (a) is a top view of SEM image of crystal facets typically formed on top of host InAs NWs (Ref. [46], open access). Copyright (2016) The Japan Society of Applied Physics.

Next, we conducted structural characterizations by TEM for comparably similar hybrid NWs to carefully investigate structural characteristics of the ferromagnetic MnAs NC/semiconducting InAs hybrid NWs shown in Fig. 6.9b. Figures 6.10a,b show highly magnified cross-sectional bright-field TEM images of MnAs NCs that were formed in the middle and on the top {111}B crystal facets of the host InAs NWs, respectively. The width and height of the NC in Fig. 6.10a were estimated to be approximately 85 nm for the former and 55 nm for the latter, and those in Fig. 6.10b were estimated to be approximately 72 nm for the former and 54 nm for the latter. We confirmed from TEM measurement results that rotational twin defects were randomly formed in the InAs NWs, and that no dislocations or defects, on the other hand, were observed in the MnAs NCs. We conducted EDX spectroscopy of the MnAs NCs formed in the middle and on the top {111}B crystal facets of NWs in addition to the cross-sectional TEM observations, as shown in Figs. 6.10c,d, respectively. The atomic compositions of NCs and NWs in terms of three elements, i.e., arsenic, indium, and manganese, were estimated from the line profiles obtained by EDX spectroscopy. We eliminated possible external contamination by chemicals (or atoms), such as carbon, oxygen, and silicon, which were possibly introduced during the sample preparation processes and from the materials of the sample holders used for the TEM observations to precisely examine the solid compositions of NCs and NWs. The solid compositions (atomic compositions in %) of arsenic, indium, and manganese elements were estimated to correspond to approximately 52%, 1.0%, and 47% in the regions of the NC in Fig. 6.10c. We also concluded that the MnAs NCs that were formed on the top {111}B crystal facets of the host InAs NWs had similar solid compositions, i.e., they corresponded to approximately 51%, 0%, and 49% of arsenic, indium, and manganese elements, as shown in Fig. 6.10d. Figures 6.10e,f show the ED pattern of the MnAs NC and a lattice image around the interface between the NW and the NC that were formed in the middle of NW, respectively. The ED pattern in Fig. 6.10e revealed that the MnAs NC had a hexagonal NiAs-type crystal structure. The host InAs NWs, on the other hand, had a zinc-blende-type crystal structure, as shown in the ED patterns for the whole structure of a heterojunction NW in Fig. 6.11.

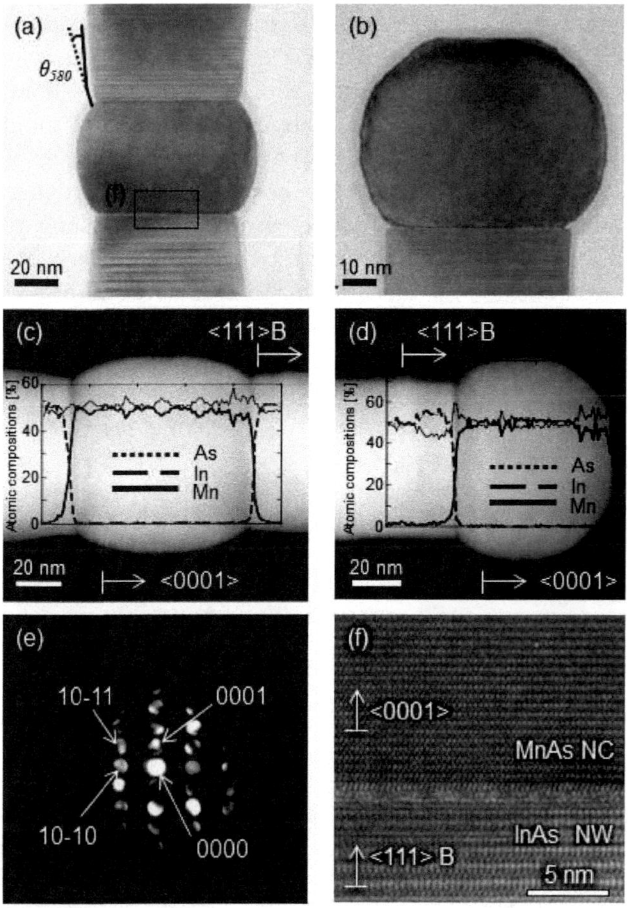

Figure 6.10 Cross-sectional TEM images of ferromagnetic MnAs NCs formed (a) in the middle and (b) on top {111}B crystal facet of semiconducting InAs NWs. MnAs NCs were grown at 580°C for 1 min. Cross-sectional TEM images with corresponding line profile obtained by EDX spectroscopy for (c) middle and (d) top of NW. The percentages of arsenic, indium, and manganese atoms were estimated by eliminating possible external contamination by chemicals or atoms, such as carbon, oxygen, and silicon, which were possibly introduced during sample preparation processes and from materials of sample holders used for TEM observations. (e) ED pattern for MnAs NCs showing the NCs with hexagonal NiAs-type crystal structure. (f) Lattice image of atomically abrupt heterointerface between InAs and MnAs that was formed in the middle of NW (Ref. [46], open access). Copyright (2016) The Japan Society of Applied Physics.

The *c*-axes, i.e., the <0001> directions, of the NiAs-type MnAs NCs were approximately parallel to the <111>B directions of the host zinc-blende-type InAs NWs. We confirmed that atomically abrupt heterointerfaces between MnAs NCs and InAs NWs were formed, as shown in Fig. 6.10f. It appeared that a small number of the observed MnAs NCs possibly rotated along the c-axes parallel to the <111>B directions of the host InAs NWs, judging from other cross-sectional lattice images of the NCs (not shown here). Similar results in terms of the heterointerfaces were obtained for the NCs formed on the top {111}B crystal facets of NWs. We concluded from these results of structural characterization that the vertical ferromagnetic MnAs NC/semiconducting InAs heterojunction NWs with the atomically abrupt heterointerfaces were successfully formed by utilizing the endotaxy of MnAs NCs at the T_g of 580°C.

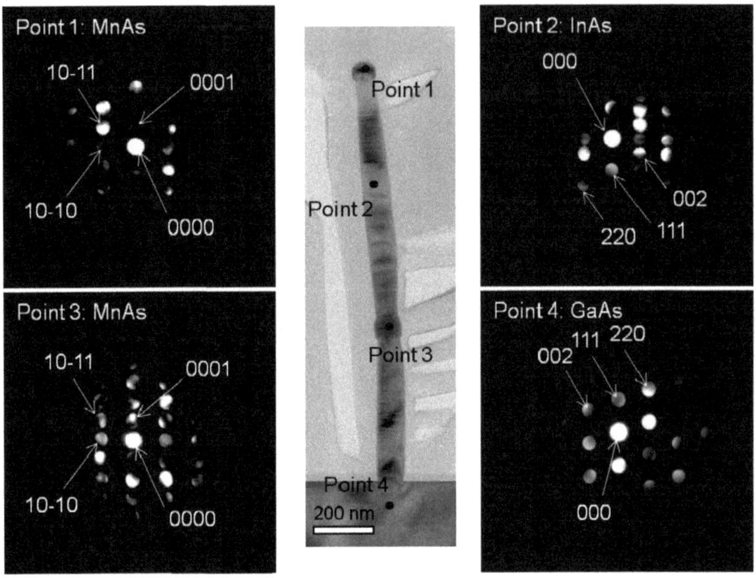

Figure 6.11 Cross-sectional TEM image and ED patterns of vertical ferromagnetic MnAs NC/semiconducting InAs heterojunction NWs on GaAs (111)B substrates: Points 1 and 3 for MnAs NCs, Point 2 for InAs NWs, and Point 4 for GaAs (111)B substrate. MnAs NCs were grown at 580°C for 1 min (Ref. [46], open access). Copyright (2016) The Japan Society of Applied Physics.

6.3.2.2 Magnetic characterization of MnAs/InAs heterojunction NWs

Subsequently, we conducted MFM observations of the vertical ferromagnetic MnAs NC/semiconducting InAs heterojunction NWs. Figures 6.12a,b show SEM and corresponding MFM images of one of the comparable heterojunction NWs, respectively. It was quite difficult to examine the same heterojunction NW observed by TEM in Fig. 6.10 by MFM since a large number of NWs were formed at the same time on the same GaAs (111)B substrate with various types of SiO_2-mask openings, i.e., different d_0 and a values. However, it was highly possible, as determined from the SEM image in Fig. 6.12a and the TEM observation results shown in Fig. 6.10, that most of the NCs penetrated the host InAs NWs. It was likely in this case, therefore, that the c-axes of the NCs were parallel to the <111>B direction of the NW (i.e., the magnetic easy axes, i.e., a-axes, of the NCs were perpendicular to the <111>B direction of the NW). Most of MnAs NCs, e.g., NCs "I" and "II," had a marked single magnetic domain, although some of the NCs like NC "III" had two magnetic domains. For the NCs I and II, a dark area was observed at the center of the NC, and the bright and dark areas were aligned perpendicular to the <111>B direction of the NW. It was revealed that the magnetized direction for the NCs I and II was possibly along one of the a-axes, which were presumably parallel to the applied **B** direction. To ensure that the detected dark and bright contrasts in the images were due to magnetic responses, we also conducted MFM measurements of the NCs I to III in the AD mode, as shown in Figs. 6.12c to 6.12f. The drive frequencies were set higher in Figs. 6.12c,e, and lower in Figs. 6.12d,f than the resonance frequency of a free-vibrating MFM cantilever. It was revealed that changing the drive frequency resulted in reversal contrasts in the regions of NCs I, II, and III, which showed magnetic responses from the NCs. We have not determined the Curie temperature T_c of the MnAs NCs in InAs NWs. However, the hysteresis curves [54] and the increased T_c of 340 K [53] were observed for the comparable samples of NiAs-type MnAs NCs on InGaAs layers in our previous studies. Some researchers have reported on a single magnetic domain of MnAs that was epitaxially grown on GaAs NWs and observed the hysteresis curve of the MnAs and temperature-dependent curves, which revealed the T_c of MnAs on GaAs NWs to be 313 K [55].

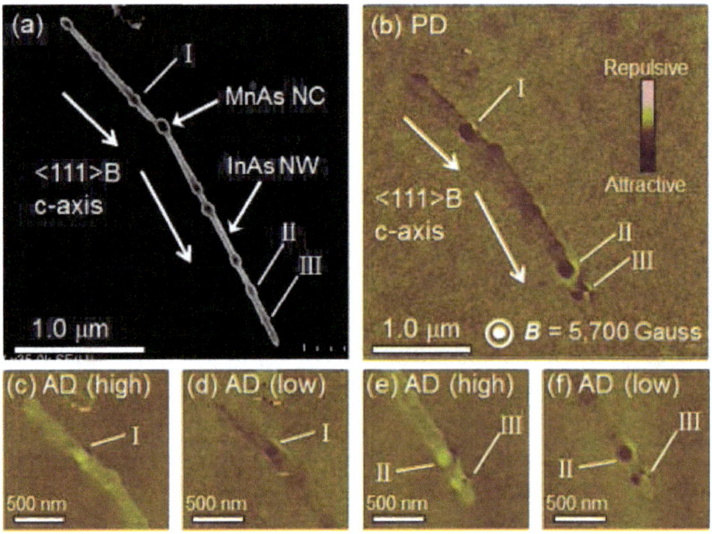

Figure 6.12 (a) Top views of SEM image and (b) corresponding MFM image of one of vertical ferromagnetic MnAs NC/semiconducting InAs heterojunction NWs. The MFM image was observed in the conventional PD mode of the MFM system. It was quite difficult to examine the same heterojunction NW observed by TEM in Fig. 6.10 that was investigated by MFM since a large number of the NWs were formed at the same time on the same GaAs (111)B substrate with various types of SiO_2-mask openings, i.e., different d_0 and a values. (c)–(f) Highly-magnified MFM images of NCs I, II, and III, which were observed in the AD mode of the MFM system. (c, e) High and (d, f) low drive frequencies were set in the AD mode measurements. The reversals in contrasts of NCs I, II, and III by changing the drive frequency clearly showed that magnetic responses were obtained from them (Ref. [46], open access). Copyright (2016) The Japan Society of Applied Physics.

6.3.2.3 Endotaxial growth mechanism of MnAs NCs in InAs NWs

We observed the endotaxial phenomenon of MnAs nanoclustering in the case of the MnAs NCs grown at 490°C for 1 min, which was first reported in our previous paper [56]. The detailed TEM observations, ED measurements, and EDX spectroscopy in the current study revealed that the MnAs NCs were not only formed similarly on the top {111}B crystal facets of the host InAs NWs, but also grew partially into the NWs from the {0–11} sidewall crystal facets and/or six ridges between them (i.e., the MnAs NCs did not completely penetrate the NWs to form heterointerfaces within the cross-sectional NW areas), as shown in Figs. 6.13a,b. In the case of the NCs grown at 490°C, some of the *c*-axes of MnAs NCs were not parallel to the <111>B directions of InAs NWs, as shown in Fig. 6.13b. We observed that the hexagonal truncated pyramidal shapes of the NCs formed on the top of host InAs NWs were rotated by 30° against the host NW hexagonal prisms, as shown in a top view of SEM image of the

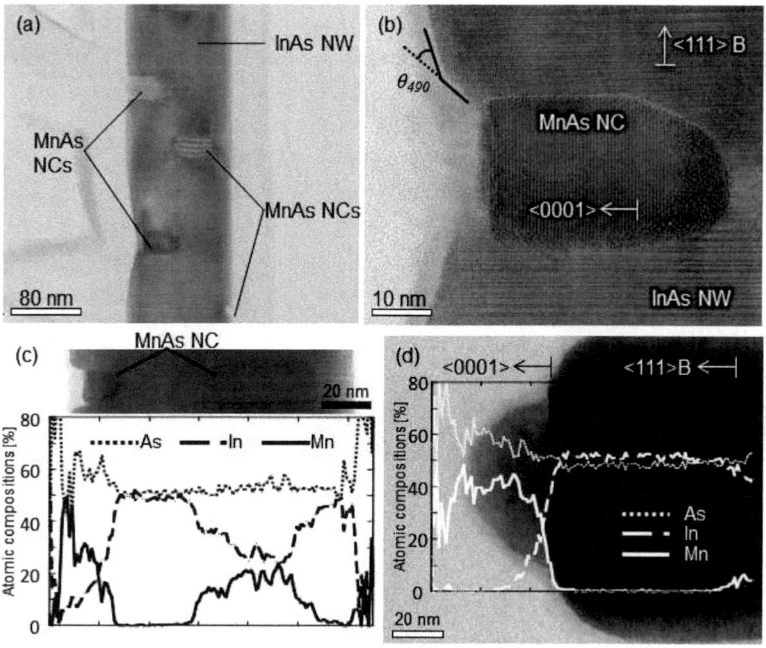

Figure 6.13 (a) Highly magnified cross-sectional TEM image of middle of vertical ferromagnetic MnAs NC/semiconducting InAs hybrid NWs in which MnAs NCs were grown at 490°C for 1 min. (b) Lattice image for one of MnAs NCs formed in the middle of host InAs NW. Line profiles obtained by EDX spectroscopy for (c) middle and (d) top parts of MnAs/InAs hybrid NW. The percentages of arsenic, indium, and manganese elements were estimated by eliminating possible external contamination by chemicals (or atoms), such as carbon, oxygen, and silicon (Ref. [46], open access). Copyright (2016) The Japan Society of Applied Physics.

MnAs NC in Fig. 6.14. This was consistent with the results of our previous study [56]. We found that the MnAs NCs grown at 490°C for 1 min, which were mainly composed of arsenic and manganese elements, had a hexagonal NiAs-type crystal structure, similarly to the NCs in Fig. 6.10. However, in the MnAs NCs grown at 490°C, small amounts of the indium element were observed even in the MnAs NC regions, as shown in Figs. 6.13c,d. This observation suggests two possibilities. One is that the indium elements of the InAs NWs behind the MnAs NCs were possibly detected in the NCs formed in the middle because the MnAs NCs did not completely penetrate the InAs NWs. The other is that InMnAs was grown during the synthesis of MnAs NCs. Neither of the possibilities can still be excluded in the current work. Some researchers have reported on the endotaxial nanoclustering of MnP in GaP (001) substrates [57]. The large diffusion coefficients of the interfacial diffusion of manganese atoms during the endotaxy of MnP in the GaP substrates are discussed in this article. The endotaxial depth of MnP NCs in the GaP substrates increased with increasing T_g for the MnP growth. We roughly estimated the depths of endotaxial MnAs nanoclustering, d, and the diffusion coefficients of manganese atoms, $D = d^2/4t$, from the TEM images of two types of NWs in which MnAs NCs were grown at 490 and 580°C for 1 min. For the NWs with

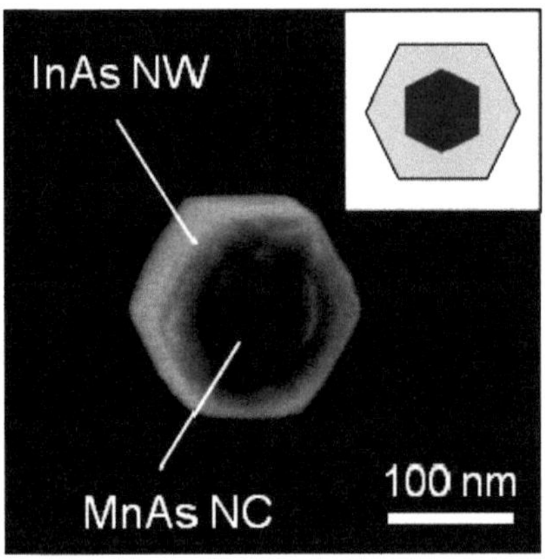

Figure 6.14 A top view of SEM image of MnAs NCs grown at 490°C. The inset is a schematic in which MnAs NC is drawn as a dark gray hexagon on a bright gray one of InAs NW (Ref. [46], open access). Copyright (2016) The Japan Society of Applied Physics.

MnAs NCs grown at 490°C, d and D were estimated to be 46 nm and 9 $nm^2\ s^{-1}$, respectively. For the NWs with MnAs NCs grown at 580°C, on the other hand, d and D were 83 nm and 28 $nm^2\ s^{-1}$, respectively. The d and D increased with increasing T_g for MnAs NCs. Therefore, it was consistent with the experimental results reported previously [57]. In addition, the results suggest that the growth mechanism of the MnP/GaP system was similar to that proposed by W. Braun et al. for the growth of a MnSb film on GaSb [58]. The gallium atoms at the MnP/GaP interface were replaced by manganese atoms and diffused to the surface, and then such out-diffused gallium atoms combined with phosphorus atoms, which are present at the surface during the growth of MnP NCs. The substitution reaction from gallium to manganese atoms at the MnP/GaP interface is reasonable, because the standard enthalpies for the formation reactions of the elements, ΔH_f, at 900 K (approximately 630°C) for MnP and GaP binary compounds are −115.564 and −109.086 kJ mol^{-1}, respectively [59], although N. Nateghi et al. did not discuss ΔH_f in their report [57]. However, it is somewhat difficult to expect the substitution reaction from indium to manganese atoms in our system, because ΔH_f at 800 K (approximately 530°C) for MnAs and InAs are −45.766 and −64.622 kJ mol^{-1}, respectively [59]. (In the case of our MnAs NC/GaAs NW system, ΔH_f at 800 K for GaAs is −82.011 kJ mol^{-1} [59]). This suggests that other growth mechanisms worked in our MnAs NC/InAs NW system (and, in addition, MnAs NC/GaAs NW system). We observed that the total volume of MnAs NCs in one InAs NW increased with increasing T_g. The increase in the total volume of MnAs NCs was roughly estimated from the obtained TEM images to be approximately 140% by changing the T_g from 490 to 580°C. In addition, the TEM and SEM measurement results revealed that the host InAs NW diameters in the vicinities of MnAs NCs tended to slightly and gradually decrease after the MnAs NC growth, compared with those around the middle of the NWs away from the MnAs NCs, as shown in Figs. 6.10a and 6.13b. Bending of NW sidewalls was observed in the vicinities of NCs. Here, we defined the bending angles of NW sidewalls, θ, and the θ values for the NCs grown at 580 and 490°C were defined as θ_{580} and θ_{490}, respectively, as shown in Figs. 6.10a and 6.13b. We confirmed that θ_{580} was much shallower than θ_{490}, as estimated in Figs. 6.10a and 6.13b. It is reasonable because the desorption rate of indium and arsenic atoms should increase with increasing T_g. The indium and arsenic atoms in the host InAs NWs are easily desorbed during the endotaxy of MnAs NCs because only $(CH_3C_5H_4)_2Mn$ and H_2 are supplied. When we supplied only H_2 at 490°C for 1 min immediately after the growth of the InAs NW templates grown at 580°C for 30 min, we actually observed the decrease in NW height after the H_2 treatment. The decrease was estimated to be approximately 70%. A slight and negligible decrease in NW diameter was also observed. Therefore, we clearly observed that the indium and arsenic atoms were markedly desorbed from the host InAs NWs under the current synthesis conditions for MnAs NCs. Figure 6.15a shows the dependences of MnAs NC formation on the period, *a*, of the host InAs NWs. The MnAs NCs were grown at 580°C for 1 min in all the experiments. The average height of the host InAs NWs and vertical ferromagnetic MnAs NC/semiconducting InAs heterojunction NWs were estimated using 30 randomly chosen NWs observed in the SEM images. The decreases in NW height after the endotaxy of MnAs NCs were estimated to be approximately 30% at least. In addition, the differences in NW height between before and after the endotaxy of MnAs NCs were almost constant (approximately 30% at least) for all the periods, *a*, of 3.0, 1.0, and 0.5 μm, i.e., no significant dependence of the decreases on the periods, *a*, was observed. These results clearly indicated that the incorporation of manganese atoms, which reacted with arsenic atoms in the host InAs NWs, suppressed the desorption of arsenic and/or indium atoms from the host NWs. Therefore, it was highly possible that the decreases in the NW height were mainly caused by the desorption of indium and arsenic atoms from the host InAs NWs during the endotaxy of MnAs NCs. The results suggest that the desorption of indium and arsenic atoms from the host InAs NWs was a major trigger of the synthesis of MnAs NCs in the InAs NWs. Some of the desorbed arsenic atoms were incorporated into the solid phase after the chemical reactions with the supplied manganese atoms from the vapor phase. That resulted in the nucleation of MnAs on the NW surface. The synthesis of MnAs NCs proceeded "into" the host InAs NWs from the nuclei because of the following reasons. One is that the arsenic atoms in the host InAs NWs were consumed for the MnAs NC formation. Another reason is that the *D* of manganese atoms in the {111}B plane was possibly larger than those in other planes since we observed that the crystal facets of the NCs and the MnAs/InAs heterointerfaces parallel to the InAs {111}B plane were quite flat and abrupt compared with the other facets. Moreover, the penetration depth from the NW sides for almost all of the NCs was larger than these height in the InAs <111>B direction.

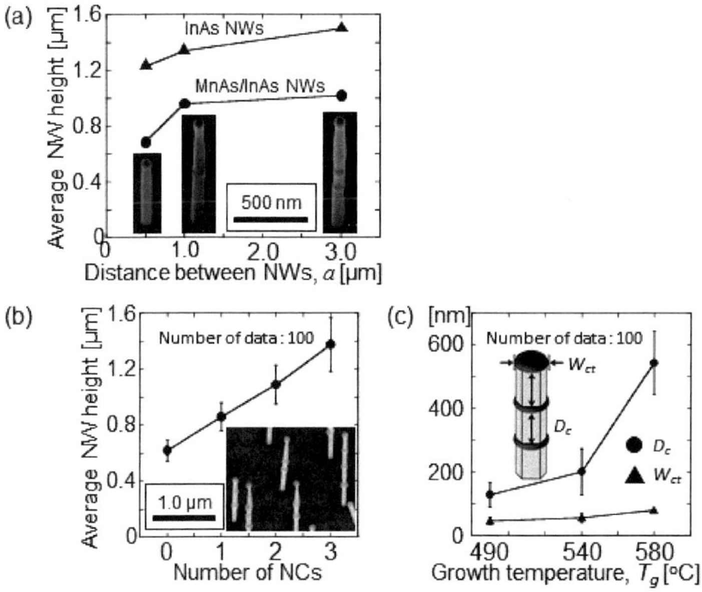

Figure 6.15 (a) Average height of host InAs NWs and vertical ferromagnetic MnAs NC/semiconducting InAs heterojunction NWs were estimated using 30 randomly chosen NWs observed in SEM images. They are plotted as a function of the host InAs NW periods a. Insets show bird's-eye views of SEM images of vertical ferromagnetic MnAs NC/semiconducting InAs heterojunction NWs with a values of 0.5, 1.0, and 3.0 μm, in order from left to right. MnAs NCs were grown at 580°C for 1 min by utilizing endotaxy. (b) The average height of NWs was estimated using 100 randomly chosen NWs observed in SEM images. They are plotted as a function of the number of MnAs NCs formed in the middle of NWs. Inset shows a bird's-eye view of SEM image of vertical ferromagnetic MnAs NC/semiconducting InAs heterojunction NWs, which had MnAs NCs grown at 580°C for 1 min. (c) The average size of MnAs NCs formed on top {111}B crystal facets of host InAs NWs, W_{ct}, and the distance between MnAs NCs, D_c, were estimated using 100 randomly chosen NWs observed in SEM images. They are plotted as a function of growth temperatures of NCs, T_g, and are defined in the inset (Ref. [46], open access). Copyright (2016) The Japan Society of Applied Physics.

The insets in Fig. 6.15a, SEM images, show that the host InAs NW height itself and the number of MnAs NCs formed in one NW decreased with decreasing the period, *a*, of the NWs. The decrease in NW height was possibly caused by the change in the amount of the indium source supplied per NW. The amount of indium source supplied per NW decreased with decreasing *a* because the surface migration length of indium atoms on the substrates was sufficiently larger than *a*. This resulted in the decrease in host InAs NW height at a relatively small *a*. This decrease in NW height led to the decrease in the number of NCs. The supplied manganese atoms were first adsorbed physically on the InAs NW surfaces after the diffusion from the vapor phase and then migrated on the surfaces. The manganese adatoms reached one of the possible chemical adsorption sites, where the indium and arsenic atoms presumably desorbed, and were then adsorbed chemically at vacant sites after indium and arsenic atoms had desorbed from the host InAs NWs. In the case of a small *a*, more of the manganese adatoms possibly reached one certain site because the surface migration length was sufficiently larger than the NW height. Therefore, the smaller the height of NWs, the fewer NCs were formed. In the case of a large *a*, on the other hand, the height of the host InAs NWs increased. A smaller number of manganese adatoms possibly reached one certain site because the surface migration length was not sufficiently larger than the NW height. This resulted in the growth of a large number of MnAs NCs in relatively long NWs. The inset in Fig. 6.15b shows a typical bird's-eye view of a SEM image of the vertical MnAs NC/InAs heterojunction NWs, in which several NCs were formed in the middle of NWs depending on the NW height. The MnAs NCs were grown at a T_g of 580°C for 1 min. The NWs had a diameter of approximately 80 nm. We estimated the average heights of NWs as a function of the number of NCs formed in the middle of NWs from SEM images. As shown in Fig. 6.15b, the average height of NWs increased with increasing number of NCs. This suggests that the number of NCs was strongly affected by NW height and the surface migration length of manganese adatoms on the NW surface.

Finally, we investigated the T_g dependences for the endotaxy of MnAs NCs. We measured the size of NCs formed on the top

{111}B crystal facets, W_{ct}, and the vertical distance between the NCs, D_c, to carefully examine NC formation in and on the NWs for the NWs in which MnAs NCs were grown at 490, 540, and 580°C. Figure 6.15c summarizes the T_g dependences of average W_{ct} and D_c, which were estimated from 100 randomly chosen NWs observed in the SEM images shown in Figs. 6.16a to 6.16c. The inset in Fig. 6.15c is a schematic that illustrates the definitions of W_{ct} and D_c. Both W_{ct} and D_c markedly increased with increasing T_g, as plotted in Fig. 6.15c. It was difficult to estimate D_c for the MnAs NCs grown at 400°C for 1 min in the host InAs NWs because the NCs formed too closely to one another to identify. However, we observed a tendency consistent with that observed in Fig. 6.15c, as shown in Fig. 6.16d. These experimental results suggest that the surface migration length of manganese adatoms on the sidewalls of the host InAs NWs is one of the key factors for the endotaxial formation of MnAs NCs. The larger the surface migration length became at higher T_g values, the larger the number of manganese adatoms that possibly reached one certain site. In addition, the desorption rates of indium and arsenic atoms and the D of manganese atoms increased at higher T_g values. These phenomena led to the increase in the size of NCs at a certain site and the decrease in the number of NCs. However, when the surface migration length decreased at lower T_g values, a small number of manganese adatoms possibly reached one certain site. The manganese adatoms in this case were possibly incorporated into other different sites within the range of their surface migration lengths. The desorption became less active and D became smaller at lower T_g values. As a result, it led to the decrease in the size of NCs and the increase in their number. Thus, the size and number of NCs were changed by T_g which affected the surface migration length of manganese adatoms, the desorption rate of indium and arsenic atoms, and the D of manganese atoms. The average height of NWs, at which no NC was formed in the middle, and the D_c obtained for the NCs grown at 580°C were both approximately 600 nm, as shown in Figs. 6.15b,c. It was therefore possible for the surface migration length of manganese adatoms at a T_g of 580°C to be roughly estimated from these results to be at least 600 nm.

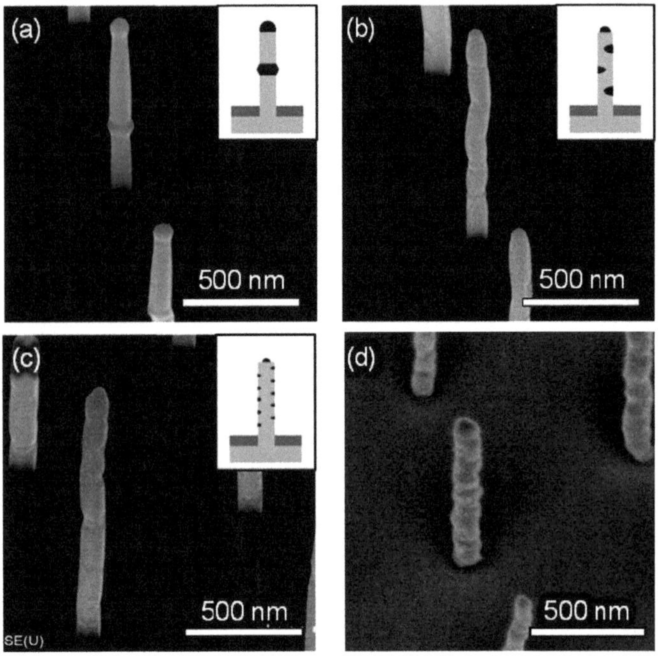

Figure 6.16 Bird's-eye view of SEM images of vertical ferromagnetic MnAs NC/semiconducting InAs hybrid NWs in which MnAs NCs were grown at (a) 580, (b) 540, (c) 490, and (d) 400°C for 1 min. Insets in (a)–(c) are schematics in which MnAs NCs are indicated by dark gray dots in and on bright gray InAs NWs. It was highly possible that the host InAs NWs in (d) was already lower than the other NWs in (a)–(c) before the NC formation. It was difficult to estimate D_c for the MnAs NCs grown at 400°C for 1 min in the host InAs NWs because the NCs formed too closely to one another to identify. However, we observed a tendency consistent with that obtained in Fig. 6.15c (Ref. [46], open access). Copyright (2016) The Japan Society of Applied Physics.

We concluded from these results that the period, a, of the host InAs NWs and T_g of MnAs NCs are the main factors for MnAs NC formation. At higher T_g values such as 580°C, MnAs NCs penetrated the host InAs NWs with atomically abrupt heterointerfaces, as shown in Fig. 6.10. Although the possibilities of the surface segregation of MnAs NCs and/or the merging among several NCs, which were reported elsewhere for a different-materials system [60], are not excluded, they might be low in the current study because no dislocations nor defects were observed in the MnAs NCs and because atomically abrupt heterointerfaces were formed between MnAs and InAs. Since rotational twin defects were randomly and densely formed in the host InAs NWs, there are supposed to be a large number of relatively unstable InAs bonds at the edges of atomic steps owing to the twin defects. These unstable InAs bonds at the edges can be the first sites for the nucleation of MnAs because the indium and arsenic atoms are easily desorbed there. Therefore, the difference between the densities of the twin defects might determine where the NCs are formed in the host NWs, although further experiments are required before we can conclude.

6.4 Conclusions and Outlook

In the first part of this chapter, the author fabricated MnAs/GaAs hybrid NWs by combining SA-MOVPE of undoped GaAs NWs and endotaxial nanoclustering of MnAs. As a result of endotaxy, MnAs NCs were typically embedded in the six ridges of hexagonal GaAs NWs. However, the MnAs NCs were formed not only at the six ridges, but also on the surfaces of six {0–11} crystal facets of hexagonal GaAs NWs, when we decreased the growth temperature T_g from 600 to 400°C. From the cross-sectional characterizations by TEM, the size and density of the MnAs NCs formed at the top part of the GaAs NWs were much larger than those formed at the bottom part of the NWs. To characterize electrical properties of MnAs/GaAs hybrid NWs, two-terminal device processes for free-standing hybrid NWs were developed. I–V characteristics of MnAs/GaAs hybrid NWs showed that the hybrid NWs formed on Zn-doped p-type GaAs (111)B substrates had ohmic characteristics, while those formed on Si-doped n-type GaAs (111)B substrates had clear rectifying characteristics. The hybrid NWs showed p-type conductivity possibly owing to the formation of p-type GaAs layers near the NW surfaces caused by the thermal diffusion of manganese atoms into GaAs NWs during the endotaxial nanoclustering of MnAs in GaAs.

The author introduced in the second part of this chapter some of the latest results on the synthesis and structural characterizations of vertical ferromagnetic MnAs NC/semiconducting InAs heterojunction NWs by combining the SA-MOVPE of InAs NWs and the endotaxial nanoclustering of

MnAs NCs. The *c*-axes, i.e., the <0001> directions, of the hexagonal NiAs-type MnAs NCs were approximately parallel to the <111>B directions of the host zinc-blende-type InAs NWs in the heterojunction NWs with the atomically abrupt complete heterointerfaces. Some of the MnAs NCs had marked single magnetic domains, and the observed results for magnetization directions were consistent with the structural characterization results. The detailed growth condition dependences revealed that the parameters T_g and a are the key factors for the formation of MnAs NCs into InAs NWs. The diffusion coefficients of manganese atoms, D, the surface migration length of manganese adatoms on the host InAs NWs, and the desorption rates of indium and arsenic atoms in the NWs strongly depended on the growth temperature T_g, and NW height changed according to the period of the host InAs NWs, a. At a relatively low T_g of 490°C, the D, surface migration length, and desorption rate decreased. This led to a relatively large number of small MnAs NCs, which were shallowly formed into the NWs from the {0–11} sidewalls. When T_g increased, on the other hand, the D, surface migration length, and desorption rate increased. This resulted in relatively large NCs that formed deeply into the host NWs and the decrease in the number of NCs. At a T_g of 580°C, the MnAs NCs, which formed from the {0–11} sidewalls of the NWs, penetrated the host InAs NWs. This phenomenon successfully led to the formation of vertical ferromagnetic MnAs NC/semiconducting InAs heterojunction NWs with atomically abrupt complete heterointerfaces between MnAs and InAs layers.

The author has recently been investigating magnetotransport properties of single ferromagnetic MnAs NC/semiconducting InAs heterojunction NW as a function of temperature and applied external magnetic fields, **B**, up to 10 T [61]. Linear negative magnetoresistance (MR) effects of approximately 10% have been observed at low temperatures up to 10 K for a MnAs/InAs heterojunction NW, in which two MnAs NCs are formed in an InAs NW between two electrodes, while an InAs NW without any MnAs NC has shown universal conductance fluctuations at low temperatures up to 20 K accompanied with large positive MR effects up to approximately 160%. Further experiments are required for fully discussing magnetotransport characteristics and MR effects of the MnAs/InAs heterojunction NWs before we can conclude, through our international collaborative work between Japan and Germany [61].

Selective-area growth of vertical free-standing III–V semiconducting NWs has shown extremely high applicability in next-generation electronic and photonic devices in our previous studies [62, 63]. (The author has also recently reported on preliminary results of MOVPE-grown GaAs NWs on amorphous glass substrates using the crystallization techniques of amorphous Al_2O_3 layers as interlayers on amorphous glass substrates to promote the nucleation before the AlGaAs buffer layer growth for GaAs NWs [64].) In addition, our selectively-grown MnAs NCs on semiconducting GaAs (111)B and Si (111) substrates have shown interesting magnetic and magnetotransport characteristics [65–71]. The author, therefore, believes that our vertical ferromagnetic MnAs NC/III–V semiconducting heterojunction NWs in this book chapter have demonstrated new possibilities and versatility for creating future novel magneto-nanoelectronic or nanospintronic NW devices [45]. The combination of vertical ferromagnetic/semiconducting heterojunction NWs and magnetic tunnel junction electrodes with atomically abrupt heterointerfaces can be a key to overcoming major obstacles of conductance mismatch and poor controllability in the heterointerface formation in the current nanospintronic devices, in particular, e.g., in spin MOSFETs.

Acknowledgments

The author sincerely thanks Professor P. J. Klar, Dr. M. T. Elm, Dr. M. Fischer, Professor T. Fukui, and Professor J. Motohisa for the stimulating and productive discussions the author had with them, and R. Kodaira, K. Kabamoto, R. Horiguchi, H. Fujimagari, M. Yatago, H. Kato, Y. Kobayashi, Dr. S. Sakita, and Dr. Y. Kohashi for supporting the MOVPE growth experiments. This work was financially supported in part by Grants-in-Aid for Scientific Research (B) (KAKENHI Grant Number 23360129) and Challenging Exploratory Research (KAKENHI Grant Numbers 25600034 and 16K13671) from the Japan Society for the Promotion of Science (JSPS), and the Japan–Germany Research Cooperative Program from the JSPS and German Academic Exchange Service (DAAD).

References

1. Duan, X., Huang, Y., Cui, Y., Wang, J., and Lieber, C. M. (2001). Indium phosphide nanowires as building blocks for nanoscale electronic and optoelectronic devices, *Nature*, **409**, pp. 66–69.
2. Wang, J., Gudiksen, M. S., Duan, X., Cui, Y., and Lieber, C. M. (2001). Highly polarized photoluminescence and photodetection from single indium phosphide nanowires, *Science*, **293**, pp. 1455–1457.
3. Cui, Y., and Lieber, C. M. (2001). Functional nanoscale electronic devices assembled using silicon nanowire building blocks, *Science*, **291**, pp. 851–853.
4. Thelander, C., Matensson, T., Björk, M. T., Ohlsson, B. J., Larsson, M. W., Wallenberg, L. R., and Samuelson, L. (2003). Single-electron transistors in heterostructure nanowires, *Appl. Phys. Lett.*, **83**, pp. 2052–2054.
5. Duan, X., Huang, Y., Agarwal, R., and Lieber, C. M. (2003). Single-nanowire electrically driven lasers, *Nature*, **421**, pp. 241–245.
6. Zhang, D., Li, C., Liu, X., Han, S., Tang, T., and Zhou, C. (2003). Doping dependent NH_3 sensing of indium oxide nanowires, *Appl. Phys. Lett.*, **83**, pp. 1845–1847.
7. Offermans, P., Crego-Calama, M., and Brongersma, S. H. (2010). Gas detection with vertical InAs nanowire arrays, *Nano Lett.*, **10**, pp. 2412–2415.
8. Tian, B., Zheng, X., Kempa, T. J., Fang, Y., Yu, N., Yu, G., Huang, J., and Lieber, C. M. (2007). Coaxial silicon nanowires as solar cells and nanoelectronic power sources, *Nature*, **449**, pp. 885–890.
9. Wallentin, J., Anttu, N., Asoli, D., Huffman, M., Åberg, I., Magnusson, M. H., Siefer, G., Fuss-Kailuweit, P., Dimroth, F., Witzigmann, B., Xu, H. Q., Samuelson, L., Deppert, K., and Borgström, M. T. (2013). InP nanowire array solar cells achieving 13.8% efficiency by exceeding the ray optics limit, *Science*, **339**, pp. 1057–1060.
10. Han, N., Wang, F., Hou, J. J., Yip, S. P., Lin, H., Xiu, F., Fang, M., Yang, Z., Shi, X., Dong, G., Hung, T. F., and Ho, J. C. (2013). Tunable electronic transport properties of metal-cluster-decorated III–V nanowire transistors, *Adv. Mater.*, **25**, pp. 4445–4451.
11. Hong, Y. J., Yang, J. W., Lee, W. H., Ruoff, R. S., Kim, K. S., and Fukui, T. (2013). Van der Waals epitaxial double heterostructure: InAs/single-layer graphene/InAs, *Adv. Mater.*, **25**, pp. 6847–6853.
12. Ren, P., Hu, W., Zhang, Q., Zhu, X., Zhuang, X., Ma, L., Fan, X., Zhou, H., Liao, L., Duan, X., and Pan, A. (2014). Band-selective infrared

photodetectors with complete-composition-range $InAs_xP_{1-x}$ alloy nanowires, *Adv. Mater.*, **26**, pp. 7444–7449.

13. Bao, T. H., Yakimets, D., Ryckaert, J., Ciofi, I., Baert, R., Veloso, A., Boemmels, J., Collaert, N., Roussel, P., Demuynck, S., Raghavan, P., Mercha, A., Tokei, Z., Verkest, D., Thean, A. V.-Y., and Wambacq, P. (2014). Circuit and process co-design with vertical gate-all-around nanowire FET technology to extend CMOS scaling for 5 nm and beyond technologies, *Proc. 44th European Solid-State Device Research Conf. (ESSDERC 2014)*, IEEE, pp. 102–105.

14. Sato, T., Kobayashi, Y., Motohisa, J., Hara, S., and Fukui, T. (2008). SA-MOVPE of InGaAs nanowires and their compositions studied by micro-PL measurement, *J. Cryst. Growth*, **310**, pp. 5111–5113.

15. Tomioka, K., Motohisa, J., Hara, S., and Fukui, T. (2008). Control of InAs nanowire growth directions on Si, *Nano Lett.*, **8**, pp. 3475–3480.

16. Tanaka, T., Tomioka, K., Hara, S., Motohisa, J., Sano, E., and Fukui, T. (2010). Vertical surrounding gate transistors using single InAs nanowires grown on Si substrates, *Appl. Phys. Express*, **3**, p. 025003.

17. Kobayashi, Y., Kohashi, Y., Hara, S., and Motohisa, J. (2013). Selective-area growth of InAs nanowires with metal/dielectric composite mask and their application to vertical surrounding-gate field-effect transistors, *Appl. Phys. Express*, **6**, p. 045001.

18. Dasgupta, N. P., Sun, J., Liu, C., Brittman, S., Andrews, S. C., Lim, J., Gao, H., Yan, R., and Yang, P. (2014). Semiconductor nanowires–synthesis, characterization, and applications, *Adv. Mater.*, **26**, pp. 2137–2184.

19. Gould, C., Pappert, K., Schmidt, G., and Molenkamp, L. W. (2007). Magnetic anisotropies and (Ga,Mn)As-based spintronic devices, *Adv. Mater.*, **19**, pp. 323–340.

20. Hai, P. N., Ohya, S., Tanaka, M., Barnes, S. E., and Maekawa, S. (2009). Electromotive force and huge magnetoresistance in magnetic tunnel junctions, *Nature*, **458**, pp. 489–493.

21. Hai, P. N., Ohya, S., and Tanaka, M. (2010). Long spin-relaxation time in a single metal nanoparticle, *Nat. Nanotechnol.*, **5**, pp. 593–596.

22. Tanaka, M., Ohya, S., and Hai, P. N. (2014). Recent progress in III-V based ferromagnetic semiconductors: Band structure, Fermi level, and tunneling transport, *Appl. Phys. Rev.*, **1**, p. 011102.

23. Radovanovic, P. V., Barrelet, C. J., Gradečak, S., Qian, F., and Lieber, C. M. (2005). General synthesis of manganese-doped II–VI and III–V semiconductor nanowires, *Nano Lett.*, **5**, pp. 1407–1411.

24. Kim, H. S., Cho, Y. J., Kong, K. J., Kim, C. H., Chung, G. B., Park, J., Kim, J. Y., Yoon, J., Jung, M. H., Jo, Y., Kim, B., and Ahn, J. P. (2009). Room-temperature ferromagnetic $Ga_{1-x}Mn_xAs$ ($x \leq 0.05$) nanowires: Dependence of electronic structures and magnetic properties on Mn content, *Chem. Mater.*, **21**, pp. 1137–1143.

25. Borschel, C., Messing, M. E., Borgström, M. T., Paschoal Jr., W., Wallentin, J., Kumar, S., Mergenthaler, K., Deppert, K., Canali, C. M., Pettersson, H., Samuelson, L., and Ronning, C. (2011). A new route toward semiconductor nanospintronics: Highly Mn-doped GaAs nanowires realized by ion-implantation under dynamic annealing conditions, *Nano Lett.*, **11**, pp. 3935–3940.

26. Rudolph, A., Soda, M., Kiessling, M., Wojtowicz, T., Schuh, D., Wegscheider, W., Zweck, J., Back, C., and Reiger, E. (2009). Ferromagnetic GaAs/GaMnAs core–shell nanowires grown by molecular beam epitaxy, *Nano Lett.*, **9**, pp. 3860–3866.

27. Yu, X., Wang, H., Pan, D., Zhao, J., Misuraca, J., von Molnár, S., and Xiong, P. (2013). All zinc-blende GaAs/(Ga,Mn)As core–shell nanowires with ferromagnetic ordering, *Nano Lett.*, **13**, pp. 1572–1577.

28. Ramlan, D. G., May, S. J., Zheng, J. G., Allen, J. E., Wessels, B. W., and Lauhon, L. J. (2006). Ferromagnetic self-assembled quantum dots on semiconductor nanowires, *Nano Lett.*, **6**, pp. 50–54.

29. Wolff, M. F. H., Görlitz, D., Nielsch, K., Messing, M. E., and Deppert, K. (2011). Synthesis and magnetic characterization of MnAs nanoparticles via nanoparticle conversion, *Nanotechnology*, **22**, p. 055602.

30. Hilse, M., Takagaki, Y., Herfort, J., Ramsteiner, M., Herrmann, C., Breuer, S., Geelhaar, L., and Riechert, H. (2009). Ferromagnet-semiconductor nanowire coaxial heterostructures grown by molecular-beam epitaxy, *Appl. Phys. Lett.*, **95**, p. 133126.

31. Liang, J., Wang, J., Paul, A., Cooley, B. J., Rench, D. W., Dellas, N. S., Mohney, S. E., Engel-Herbert, R., and Samarth, N. (2012). Measurement and simulation of anisotropic magnetoresistance in single GaAs/MnAs core/shell nanowires, *Appl. Phys. Lett.*, **100**, p. 182402.

32. Choi, H.-J., Seong, H.-K., Chang, J., Lee, K.-I., Park, Y.-J., Kim, J.-J., Lee, S.-K., He, R., Kuykendall, T., and Yang, P. (2005). Single-crystalline diluted magnetic semiconductor GaN: Mn nanowires, *Adv. Mater.*, **17**, pp. 1351–1356.

33. Seong, H.-K., Kim, J.-Y., Kim, J.-J., Lee, S.-C., Kim, S.-R., Kim, U., Park, T.-E., and Choi, H.-J. (2007). Room-temperature ferromagnetism in Cu doped GaN nanowires, *Nano Lett.*, **7**, pp. 3366–3371.

34. Lensch-Falk, J. L., Hemesath, E. R., and Lauhon, L. J. (2008). Syntaxial growth of Ge/Mn-germanide nanowire heterostructures, *Nano Lett.*, **8**, pp. 2669–2673.

35. Tang, J., Wang, C.-Y., Hung, M.-H., Jiang, X., Chang, L.-T., He, L., Liu, P.-H., Yang, H.-J., Tuan, H.-Y., Chen, L.-J., and Wang, K.-L. (2012). Ferromagnetic germanide in Ge nanowire transistors for spintronics application, *ACS Nano*, **6**, pp. 5710–5717.

36. Lin, Y.-C., Chen, Y., Shailos, A., and Huang, Y. (2010). Detection of spin polarized carrier in silicon nanowire with single crystal MnSi as magnetic contacts, *Nano Lett.*, **10**, pp. 2281–2287.

37. Li, X., Meng, G., Qin, S., Xu, Q., Chu, Z., Zhu, X., Kong, M., and Li, A.-P. (2012). Nanochannel-directed growth of multi-segment nanowire heterojunctions of metallic $Au_{1-x}Ge_x$ and semiconducting Ge, *ACS Nano*, **6**, pp. 831–836.

38. Hara, S., Kawamura, D., Iguchi, H., Motohisa, J., and Fukui, T. (2008). Self-assembly and selective-area formation of ferromagnetic MnAs nanoclusters on lattice-mismatched semiconductor surfaces by MOVPE, *J. Cryst. Growth*, **310**, pp. 2390–2394.

39. Yatago, M., Iguchi, H., Sakita, S., and Hara, S. (2012). Growth and characterization of MnAs nanoclusters embedded in GaAs nanowires by metal–organic vapor phase epitaxy, *Jpn. J. Appl. Phys.*, **51**, p. 02BH01.

40. Hara, S., Sakita, S., and Yatago, M. (2012). Selective-area growth and electrical characterization of hybrid structures between semiconducting GaAs nanowires and ferromagnetic MnAs nanoclusters, *Jpn. J. Appl. Phys.*, **51**, 11PE01 (Open Access).

41. Inokuchi, T., Marukame, T., Ishikawa, M., Sugiyama, H., and Saito, Y. (2009). Electrical spin injection into n-GaAs channels and detection through MgO/CoFeB electrodes, *Appl. Phys. Express*, **2**, p. 023006.

42. Jiang, X., Wang, R., Shelby, R. M., Macfarlane, R. M., Bank, S. R., Harris, J. S., and Parkin, S. S. P. (2005). Highly spin-polarized room-temperature tunnel injector for semiconductor spintronics using MgO(100), *Phys. Rev. Lett.*, **94**, p. 056601.

43. Sugahara, S., and Nitta, J. (2010). Spin-transistor electronics: An overview and outlook, *Proc. IEEE*, **98**, pp. 2124–2154.

44. Biermann, K., Hernandez-Minguez, A., Hey, R., and Santos, P. V. (2012). Electrically tunable electron spin lifetimes in GaAs(111)B quantum wells, *J. Appl. Phys.*, **112**, p. 083913.

45. Elm, M. T., and Hara, S. (2014). Transport properties of hybrids with ferromagnetic MnAs nanoclusters and their potential for new magnetoelectronic devices, *Adv. Mater.*, **26**, pp. 8079–8095.

46. Kodaira, R., Hara, S., Kabamoto, K., and Fujimagari, H. (2016). Synthesis and structural characterization of vertical ferromagnetic MnAs/semiconducting InAs heterojunction nanowires, *Jpn. J. Appl. Phys.*, **55**, p. 075503 (Open Access).

47. Noborisaka, J., Motohisa, J., and Fukui, T. (2005). Catalyst-free growth of GaAs nanowires by selective-area metalorganic vapor-phase epitaxy, *Appl. Phys. Lett.*, **86**, p. 213102.

48. Noborisaka, J., Motohisa, J., Hara, S., and Fukui, T. (2005). Fabrication and characterization of freestanding GaAs/AlGaAs core–shell nanowires and AlGaAs nanotubes by using selective-area metalorganic vapor phase epitaxy, *Appl. Phys. Lett.*, **87**, p. 093109.

49. Shenai-Khatkhate, D. V., DiCarlo Jr., R. L., and Ware, R. A. (2008). Accurate vapor pressure equation for trimethylindium in OMVPE, *J. Cryst. Growth*, **310**, pp. 2395–2398; Shenai-Khatkhate, D. V. et al. reported on an update on the vapor pressure (P) equation of $(CH_3)_3In$, i.e., log P (Torr) = 10.98–3204/T (K). In the current work, however, we used the estimated partial pressure for $(CH_3)_3In$ obtained using a conventionally used old equation, i.e., log P (Torr) = 10.52–3014/T (K), for comparison with our previous papers.

50. Iguchi, H., Hara, S., Motohisa, J., and Fukui, T. (2008). Metal–organic vapor phase epitaxial growth condition dependences of MnAs nanocluster formation on GaInAs (111)A surfaces, *Jpn. J. Appl. Phys.*, **47**, pp. 3253–3256.

51. Bonev, I. (1972). On the terminology of the phenomena of mutual crystal orientation, *Acta Crystallogr. A*, **28**, pp. 508–512.

52. Maeda, S., Tomioka, K., Hara, S., and Motohisa, J. (2012). Fabrication and characterization of InP nanowire light-emitting diodes, *Jpn. J. Appl. Phys.*, **51**, p. 02BN03.

53. Elm, M. T., Michel, C., Stehr, J., Hofmann, D. M., Klar, P. J., Ito, S., Hara, S., and Krug von Nidda, H.-A. (2010). Comparison of the magnetic properties of GaInAs/MnAs and GaAs/MnAs hybrids with random and ordered arrangements of MnAs nanoclusters, *J. Appl. Phys.*, **107**, p. 013701.

54. Hara, S., and Fukui, T. (2006). Hexagonal ferromagnetic MnAs nanocluster formation on GaInAs/InP (111)B layers by metal-organic vapor phase epitaxy, *Appl. Phys. Lett.*, **89**, p. 113111.

55. Hubmann, J., Bauer, B., Körner, H. S., Furthmeier, S., Buchner, M., Bayreuther, G., Dirnberger, F., Schuh, D., Back, C. H., Zweck, J., Reiger, E., and Bougeard, D. (2016). Epitaxial growth of room-temperature ferromagnetic MnAs segments on GaAs nanowires via sequential crystallization, *Nano Lett.*, **16**, pp. 900–905.

56. Hara, S., Fujimagari, H., Sakita, S., and Yatago, M. (2013). Difference in formation of ferromagnetic MnAs nanoclusters on III–V semiconducting nanowire templates, *Proc. SPIE*, **8820**, p. 88200V.

57. Nateghi, N., Ménard, D., and Masut, R. A. (2014). Large interface diffusion in endotaxial growth of MnP films on GaP substrates, *J. Appl. Phys.*, **116**, p. 133512.

58. Braun, W., Trampert, A, Kaganaer, V. M., Jenichen, B., Satapathy, D. K., and Ploog, K. H. (2007). Endotaxy of MnSb into GaSb, *J. Cryst. Growth*, **301–302**, pp. 50–53.

59. Barin, I. (1989). *Thermochemical Data of Pure Substances* (Wiley-VCH, Weinheim).

60. Hara, S., and Kuramata, A. (2005). Ferromagnetic nanoclusters hybridized in Mn-incorporated GaInAs layers during metal–organic vapour phase epitaxial growth on InP layers under low growth temperature conditions, *Nanotechnology*, **16**, pp. 957–965.

61. Hara, S., Elm, M. T., and Klar, P. J. (2016). *Unpublished*, as an international collaborative work between Hokkaido University, Japan, and Justus-Liebig University of Giessen, Germany.

62. Tomioka, K., Ikejiri, K., Tanaka, T., Motohisa, J., Hara, S., Hiruma, K., and Fukui, T. (2011). Selective-area growth of III–V nanowires and their applications, *J. Mater. Res.*, **26**, pp. 2127–2141.

63. Tomioka, K., Tanaka, T., Hara, S., Hiruma, K., and Fukui, T. (2011). III–V nanowires on Si substrate: Selective-area growth and device applications, *IEEE J. Sel. Top. Quantum Electron.*, **17**, pp. 1112–1129.

64. Sakita, S., and Hara, S. (2015). Growth of AlGaAs nanostructures on crystallized Al_2O_3 interlayers for semiconducting nanowire growth on glass substrate, *Jpn. J. Appl. Phys.*, **54**, p. 075504.

65. Sakita, S., Hara, S., Elm, M. T., and Klar, P. J. (2016). Selective-area growth and magnetic characterization of MnAs/AlGaAs nanoclusters on insulating Al_2O_3 layers crystallized on Si(111) substrates, *Appl. Phys. Lett.*, **108**, p. 043108.

66. Fischer, M., Elm, M. T., Kato, H., Sakita, S., Hara, S., and Klar, P. J. (2015). Analysis of magnetic random telegraph noise in individual arrangements of a small number of coupled MnAs nanoclusters, *Phys. Rev. B*, **92**, p. 165306.

67. Hara, S., and Komagata, K. (2015). Selective-area growth and magnetic reversals of ferromagnetic nanoclusters on semiconducting substrate for magnetic logic applications, *Phys. Status Solidi B*, **252**, pp. 1925–1933.

68. Kato, H., Sakita, S., and Hara, S. (2015). Selective-area growth and magnetic characterization of lateral MnAs nanowires, *J. Cryst. Growth*, **414**, pp. 151–155.

69. Fischer, M., Elm, M. T., Sakita, S., Hara, S., and Klar, P. J. (2015). Magnetoresistance effects and spin-valve like behavior of an arrangement of two MnAs nanoclusters, *Appl. Phys. Lett.*, **106**, p. 032401.

70. Elm, M. T., Klar, P. J., Ito, S., Hara, S., and Krug von Nidda, H.-A. (2011). Effect of the cluster magnetization on the magnetotransport at low temperatures in ordered arrays of MnAs nanoclusters on (111)B GaAs, *Phys. Rev. B*, **84**, p. 035309.

71. Elm, M. T., Klar, P. J., Ito, S., and Hara, S. (2011). Influence of ordered arrangements of cluster chains on the hopping transport in GaAs: Mn/MnAs hybrids at low temperatures, *Phys. Rev. B*, **83**, p. 235305.

Chapter 7

GaAs-Fe_3Si Semiconductor–Ferromagnet Core–Shell Nanowires for Spintronics

Maria Hilse, Bernd Jenichen, and Jens Herfort

Paul-Drude-Institut für Festkörperelektronik,
Hausvogteiplatz 5-7, 10117 Berlin, Germany

dr.maria.hilse@t-online.de

Semiconductor–ferromagnet core–shell nanowires (SFCSNWs) with a high-quality ferromagnet like the binary Heusler alloy Fe_3Si are regarded as promising candidates to enable the fabrication of new forward-looking devices in the field of spintronics and magnetic recording. This chapter presents an overview of the fabrication and properties of such GaAs-Fe_3Si SFCSNWs. SFCSNW hybrid structures were grown by molecular beam epitaxy and analyzed by scanning and transmission electron microscopy, X-ray diffraction, Mössbauer spectroscopy, superconducting quantum interference device magnetometry, and magnetic force microscopy. The analysis showed that the SFCSNWs indeed possess all the prerequisites to enable new high-quality devices for spintronics.

Novel Compound Semiconductor Nanowires: Materials, Devices, and Applications
Edited by Fumitaro Ishikawa and Irina Buyanova

ISBN 978-981-4745-76-5 (Hardcover), 978-1-315-36440-7 (eBook)
www.panstanford.com

7.1 Introduction

Spintronics is based on the generation, manipulation, and detection of spin-polarized currents. Among other techniques for generation and detection, all-electrical spin-injection and detection is regarded as the only technique that is suitable for large-scale device production. This technique benefits from the fact that electrical transport in ferromagnetic metals is always spin-polarized. Thus, ferromagnetic contacts are used to inject or detect a certain spin orientation into or from a semiconductor device. These contacts are built by a thin ferromagnetic metal layer on top of the semiconducting part.

Due to the magnetic shape anisotropy energy, the magnetization of such thin ferromagnetic layers underlies geometrical constraints. In general, these lead to an equilibrium magnetization that lies in the layer plane and a magnetic hard axis in the direction along the layer normal. Ways to overcome this equilibrium condition are first to turn the magnetization along the hard axis by applying high external magnetic fields—in the order of a few Tesla [16, 53, 59]; second, to make extensive efforts in designing advanced, and well-defined stacks of layers consisting of specific materials to reach conditions where the shape anisotropy energy is dominated by the inter-layer dipole-interaction and the direction normal to the thin film becomes a magnetic easy axis [5, 34, 35, 80, 82]; or third, to use very limited and highly specific ferromagnetic materials that by nature possess a very weak shape anisotropy energy [7, 66]. Due to the high technical requirements that are needed to generate high magnetic fields, the extensive efforts to realize complex superlattices, and the strict limitation in the choice of materials, applications of spintronic hybrid structures are limited to spin-polarized currents with a spin-direction that lies in the ferromagnetic contact layer plane.

Anyhow, magnetizations of thin ferromagnetic layers, i.e., contact films perpendicular to the layer plane or the substrate are a crucial prerequisite for the fabrication of several forward-looking spintronic device applications. Such applications are, for instance, ready-to-use circular light emitting diodes (spin-LEDs) operating in the absence of any external magnetic field, that enable optical spin information transfer over macroscopic distances without heat development associated with moving charges [25, 65, 76]. This has been demonstrated using optically communicating spin-LEDs [15]. As further applications with great technological importance, spin valves or magnetic tunnel junctions based on spin transfer torque (STT) switching are to be mentioned. STT devices with perpendicular magnetization, i.e., perpendicular magnetic anisotropy, were demonstrated to have great advantages over the in-plane ones. These include strong thermal stability, low critical switching current density, and no restriction of shape aspect ratio [86]. Moreover, also three-dimensional recording concepts with unsurpassed data storage capacities, such as the magnetic domain wall racetrack memory, which are considered as an alternative to the conventionally used two-dimensional magnetic recording technology that nowadays attains its maximum storage density, rely on magnetizations and device architectures perpendicular to the substrate [55, 62].

As a fourth way to overcome the geometrical constraints of the magnetization orientation in thin ferromagnetic layers, semiconductor–ferromagnet core–shell nanowires (SFCSNWs) finally come into play. By this way, in principle, there is no limitation regarding the choice of materials, no high external fields are needed, and no elaborate superlattice must be designed. It is based on the unique geometry of SFCSNWs and their geometrical constraints to the magnetization of the ferromagnetic shells. For long and very thin magnetic nanotubes, i.e., magnetic shells of nanowires (NWs), the magnetization can either form vortices—vortex configuration —or align along the NW, e.g., in bar magnets—parallel configuration [8, 44, 45, 68, 78]. In the latter case, the orientation of the core NWs perpendicular to the substrate in principle aligns the magnetization in the required direction with respect to the substrate. However, the vortex or even onion state—that is supported by ferromagnetic nanotubes but not by ferromagnetic NWs—is a stray-free state that has a clear advantage with regard to possible memory applications [2, 79].

Whether the magnetization of the NW shells aligns along the wire or adopts the vortex state depends on the wire geometry and the anisotropy constants of the used ferromagnetic material [14, 13]. For a given geometry, the parallel configuration dominates for materials with high saturation magnetizations

(M_S) and large anisotropy constants. Thus, in theory, with this method it is possible for any used material to tailor the core–shell NW geometry in a way that the magnetization adopts a given configuration. Furthermore, NWs of various lengths and diameters can be fabricated in large arrays comprising thousands of elements each on the nanometer scale [83]. Actually, spin-LED devices can in principle be fabricated within one single SFCSNW—consisting of a semiconductor NW LED structure coated with a ferromagnetic shell—arrays of such SFCSNW-spin-LEDs would enable on-chip optical communication of spin information [36, 43, 47, 57, 58, 69, 73]. Thus, SFCSNWs are promising candidates for the realization of forward-looking spintronic device applications like the ones mentioned above.

Epitaxial SFCSNWs were first presented in 2009 independently by two research groups using materials combinations of GaAs with the ferromagnet MnAs, and the dilute ferromagnet Ga(Mn)As respectively [32, 60]. Up to now, work was published using the same materials' combination [4, 9, 33, 46, 70, 85] as well as GaAs-Fe [56, 72], GaAs-Ni [2, 61, 79], and ZnTe-Co [48]. Also, work on oxide-based SFCSNWs was published [24]. Finally, GaAs-Fe_3Si SFCSNWs grown by molecular beam epitaxy (MBE) were presented [31, 39, 40, 41]. This combination of materials has some essential advantages compared to the previously studied SFCSNWs using ferromagnetic materials that cannot reach the high qualities of a binary Heusler alloy like Fe_3Si. First, the high Curie temperature of about 840 K of Fe_3Si is a prerequisite for the realization of spintronic applications as mentioned above and second, the perfect lattice matching to GaAs allows for MBE of high-quality hybrid structures [27, 30, 42]. Third, the cubic Fe_3Si phase shows a robust stability against stoichiometric variations—that covers a range from 10% to 26% at. Si [12, 26] with only slightly modified magnetic properties [28]—and fourth, its thermal stability against chemical reactions at the ferromagnet/semiconductor interface is considerably higher than that for conventional ferromagnets such as Fe, Co, Ni, and Fe_xCo_{1-x} [29].

This chapter gives an overview of our research published about GaAs-Fe_3Si SFCSNWs [31, 39–41]. It covers a description of the growth process, morphology, crystal quality, and magnetic properties of these wires.

7.2 Epitaxial Growth

GaAs-Fe_3Si core–shell NWs are grown by a two-step growth in a MBE system equipped with two separate growth chambers: one for group III arsenide and one for metal growth. Thus, the NWs are entirely processed under ultra-high vacuum conditions avoiding any oxidation of the semiconductor–ferromagnet interface.

In the first growth step, GaAs NWs are fabricated by the Ga-assisted vapor–liquid–solid (VLS) growth method [6, 37]. Here, Si(111) substrates covered with a thin native Si-oxide layer are used. During the growth process at temperatures of about 580°C, the sticking coefficient for Ga is higher at the exposed Si substrate in tiny pin-holes of the Si-oxide than at the surrounding Si-oxide. Thus, the supplied Ga forms liquid droplets in these pin-holes of the Si-oxide. In the course of growth, the Ga droplets serve as a reservoir for further supplied Ga and As and as a growth promoter for the NWs as GaAs growth occurs at the triple-phase line between the surrounding vapor, the liquid droplet, and the solid GaAs. Thus, the droplet is driven upward and a NW develops. The growth parameters used for this first step are a Ga flux equivalent to a growth rate of 100 nm/h of a two-dimensional GaAs layer and a V/III flux ratio of 1.0. Under these conditions, the NWs typically grew 1 μm per hour in length at diameters of about 80 nm. After the NW growth, the Ga-droplets were consumed by the arsenic atmosphere and the samples were then cooled down and transferred to the growth chamber for metals under ultra-high vacuum conditions.

In the second growth step, the GaAs NWs are overgrown with a thin Fe_3Si layer. The growth temperature for this step is systematically varied in the range of 100–350°C—the typical growth temperatures in Fe_3Si epitaxy [29]. Within this growth step, the supplied Fe and Si fluxes were equivalent to a growth rate of 23 nm/h of a planar Fe_3Si layer and the flux ratio was calibrated by X-ray diffraction on previously grown layers in order to produce stoichiometric films consisting of 25% Si and 75% Fe. The growth was terminated after 3 h. After the growth, the planar Fe_3Si layer between the NWs was about 70 nm thick. Due to geometrical reasons and the low growth temperatures, the NW shells had a thickness of about 14 nm [32].

7.3 Morphology

Scanning electron microscopy (SEM) images of as-grown SFCSNWs and close-up images of the middle part of single NWs are shown in Fig. 7.1. The rotund structure visible at the top of several NWs most probably results from the Ga droplet which initiated the GaAs NW growth and which was not fully solidified after the GaAs growth finished.

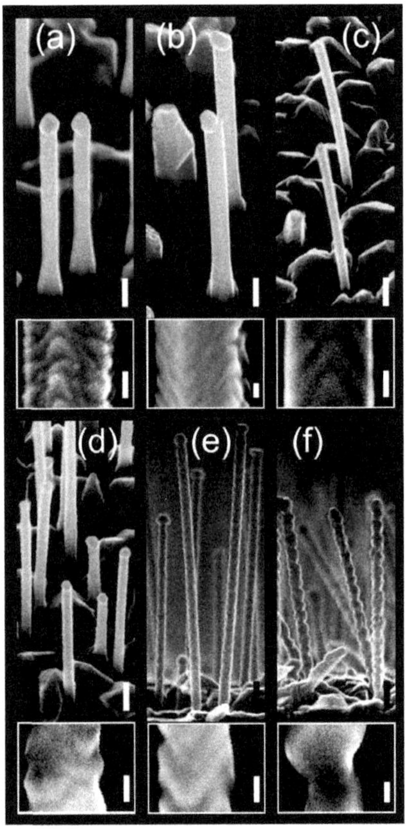

Figure 7.1 Scanning electron microscopy images of GaAs-Fe_3Si core shell NWs with shells grown (a) 100°C, (b) 150°C, (c) 200°C, (d) 250°C, (e) 300°C, and (f) 350°C. Each micrograph combines an image of typical as-grown NWs (scale bar of 200 nm) and a close-up image of the middle part of a single NW (scale bar of 20 nm). Reprinted with permission from Ref. [31]. Copyright (2016) American Chemical Society.

At low growth temperatures, the NW side walls exhibit a short-range surface roughness, which is most pronounced for NWs with a Fe_3Si growth temperature of 100°C (Fig. 7.1a). This short-range roughness flattens when the Fe_3Si growth temperature is increased to 150°C (Fig. 7.1b) and almost vanishes at a Fe_3Si growth temperature of 200°C, Fig. 7.1c. Most probably, the highly restricted adatom diffusion at 100°C leads to the formation of Fe_3Si shells consisting of various grains of different crystal orientation and stoichiometry [27, 29]. When the shell growth process is governed by higher adatom diffusion, the Fe_3Si shells grow more uniform in crystal orientation and stoichiometry and therefore the NW side wall surface flattens when the Fe_3Si growth temperature is increased.

For Fe_3Si growth temperatures above 200°C, a long-range surface roughness of the NW side walls that is most distinct at 350°C can be observed in Fig. 7.1(d–f). As it is known from planar Fe_3Si layers on GaAs substrates, Fe very intensely tends to react with Ga and As in the high-temperature regime. In Fe_3Si epitaxy on GaAs, it has been shown that for Fe_3Si growth temperatures of about 400°C these reactions lead to grains of different orientation and/or composition and thus to rough surface and interface structures [27, 29, 30]. Thus, the observed large-scale roughening of the NW side walls in the high-temperature regime in Fig. 7.1(d–f) indicates an ongoing reaction and interdiffusion of the involved elements Fe, Si, Ga, and As. These intermixing reactions starting at the core/shell interface set in at much lower growth temperatures compared to planar Fe_3Si layers [30]. This is most probably due to the high aspect ratio and the small diameter of the NWs that in general promote instabilities at surfaces and interfaces [11, 20, 33, 54, 64].

7.4 Structural Properties and Crystal Quality

Figures 7.2a,b display bright- and dark-field transmission electron microscopy (TEM) images of GaAs-Fe_3Si core–shell NWs with a shell growth temperature (T_g) of 100°C taken under two-beam conditions using g = 111 of Si. Core and shell regions can be well distinguished. In contrast to Fe epitaxy on GaN NWs where the Fe forms superparamagnetic corncob-like clusters that do not completely coat the nanowire side walls [19], the Fe_3Si completely wets the GaAs NW side walls forming a uniform cover layer with smooth core/shell interfaces, which are prerequisites for the application in high-quality electronic devices. In the bright-field image in Fig. 7.2a, the contrast of the GaAs core is homogeneously dark, which is typical for single crystal material. On the other hand, the Fe_3Si shell consists of inhomogeneous dark and bright regions in correspondence to various crystallites existing in different orientations. This points to a polycrystalline nature of the shell. In the dark-field image in Fig. 7.2b, the single-crystalline GaAs core shows the same homogeneous contrast like the Si substrate (except a strain-related contour close to the interface). Thus, it consists of high-quality and pure zincblende material without planar defects. The Fe_3Si is hardly visible due to the large misalignment. In addition, in the corresponding high-resolution image displayed in Fig. 7.2c, parallel lattice fringes are visible inside the shell, whereas the NW core again is a single crystal. Despite the presence of different crystallites within the Fe_3Si, in the present region the (220) planes of the GaAs core and the Fe_3Si shell are parallel. The direction of the incident beam was $[11\bar{2}]$.

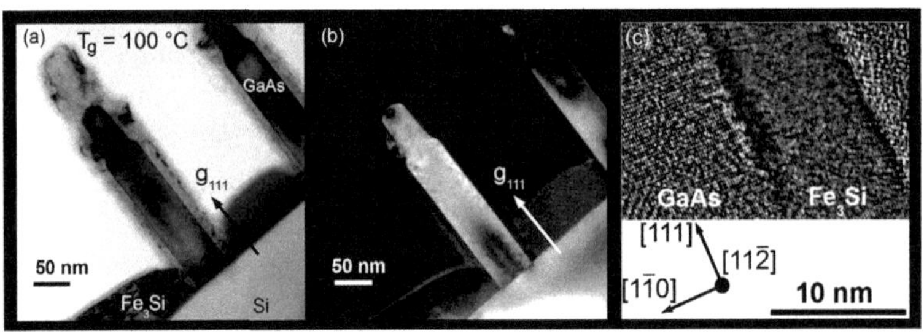

Figure 7.2 Transmission electron microscopy (TEM) images of as-grown GaAs-Fe_3Si core–shell nanowires at 100°C. (a) Bright-field TEM image near the Si (111) reflection of nanowires. (b) Corresponding dark-field TEM image. (c) Corresponding high-resolution TEM image of the GaAs core and the Fe_3Si shell (Fourier filtered image). Reprinted with permission from Ref. [31]. Copyright (2016) American Chemical Society.

Figure 7.3a shows a TEM bright-field image of a plan-view along the axis of a GaAs-Fe_3Si core–shell NW with T_g = 100°C [41]. The single-crystalline GaAs core exhibits an almost homogenous contrast. The shell is 26 ± 7 nm thick. The thickest regions are found at the corners. As the variation within one NW is large compared to the differences between them, typical NWs were chosen to determine the thickness of the Fe_3Si shell. Figure 7.3b shows a TEM dark-field image of a similar GaAs-Fe_3Si core–shell NW in the vicinity of (a). Here only some parts of the Fe_3Si shell are fulfilling the diffraction condition, i.e., the shell is polycrystalline. However, the ferromagnetic Fe_3Si shells cover the GaAs core NWs completely and continuously without any holes. From that point of view, almost perfect SFCSNWs are fabricated. At T_g = 100°C, the hexagonal shape of the GaAs core is reproduced by the Fe_3Si shell despite the textured polycrystal structure of the shell.

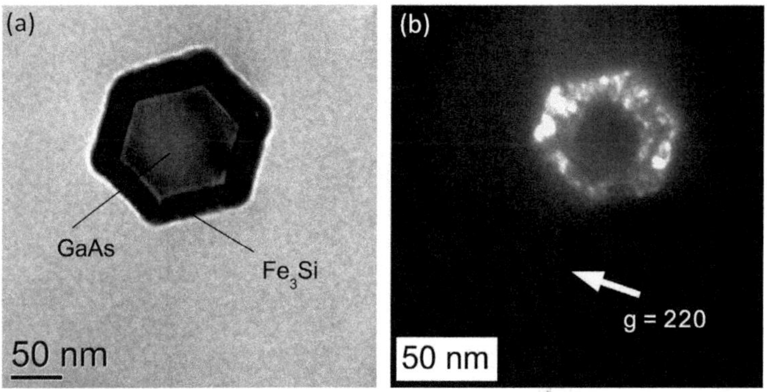

Figure 7.3 (a) TEM bright-field image of a cross section perpendicular to the axis of a GaAs-Fe_3Si core–shell NW (T_g = 100°C). (b) TEM dark-field image of a similar core–shell NW in the vicinity of (a). Reprinted from *Journal of Crystal Growth*, 410, B. Jenichen, M. Hilse, J. Herfort, and A. Trampert, Real structure of lattice matched GaAs-Fe_3Si core-shell nanowires, 1, Copyright (2016), with permission from Elsevier.

Figure 7.4 shows a multi-beam bright-field TEM micrograph and the corresponding selected area diffraction (SAD) pattern illustrating the orientational relationship of a GaAs-Fe_3Si core–shell NW on Si(111) with T_g = 200°C [40]. At that shell growth temperature, the orientations of core and shell coincide, hence the Fe_3Si growth is predominantly epitaxial on the GaAs. In addition, the crystallographic orientation of the Fe_3Si shell was confirmed by high-resolution TEM. The gray line drawn on the SAD pattern near the $11\overline{1}$ reflection is oriented perpendicular to the characteristic Fe_3Si shell nanofacets visible in the corresponding micrograph. The separation of the diffraction spots along this line corresponds to the (111) net plane distance of Fe_3Si and indicates that the nanofacets are mainly (111)-oriented. In the SAD pattern from a single core–shell NW the Fe_3Si maxima are stronger probably due to the larger volume fraction of the shell. In order to measure the SAD, the substrate was first oriented near the [011] zone axis. Then the sample had to be tilted a little further in order to reach the [011] zone axis of the GaAs NW, since the axis of the NW was not exactly perpendicular to the Si surface. In the NW SAD pattern shown in Fig. 7.4, the fundamental reflections of the Fe_3Si are more intense than the superlattice maxima [42]. Nevertheless, the $2\bar{2}2$, $3\bar{3}3$, and $4\bar{4}4$ maxima can be distinguished indicating that the NW is properly oriented and the crystallographic orientations of core and shell basically coincide. This illustrates that a growth temperature of 200°C results in a highly perfect Fe_3Si shell structure.

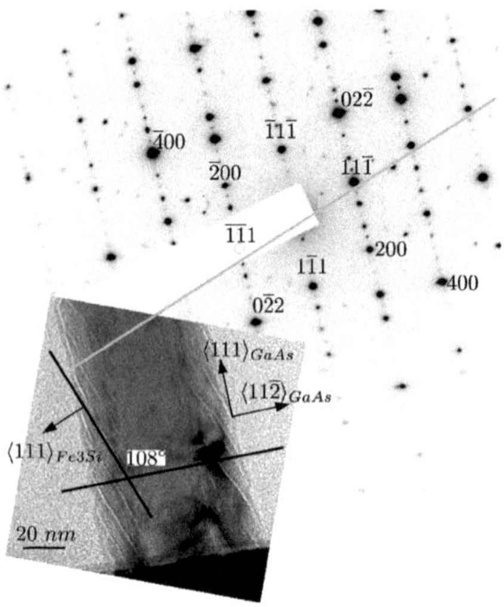

Figure 7.4 Multi-beam bright-field TEM micrograph and the corresponding SAD pattern illustrating the orientational relationship of a GaAs-Fe_3Si core–shell NW with a shell growth temperature of 200°C. The straight gray line near the $11\bar{1}$ reflection is oriented perpendicular to the nanofacets of Fe_3Si visible in the corresponding micrograph. Reprinted from *Journal of Crystal Growth*, 427, B. Jenichen, M. Hilse, J. Herfort, and A. Trampert, Facetted growth of Fe_3Si shells around GaAs nanowires on Si(111), 21, Copyright (2016), with permission from Elsevier.

Figure 7.5 demonstrates in-plane X-ray diffraction (XRD) reciprocal space maps recorded in grazing incidence geometry with zero incidence angle. The experiments were performed at the ESRF (beamline BM02). Here the detected signal comes only from the NWs. Diffraction by the substrate and the parasitic Fe_3Si film between the NWs could be successfully avoided. Figure 7.5 displays reciprocal space maps of the ($\bar{2}$20)-reflection of GaAs NWs (left) and GaAs-Fe_3Si core–shell NWs (center and right). The growth temperatures of the Fe_3Si shells were 200°C (center) and 300°C (right).

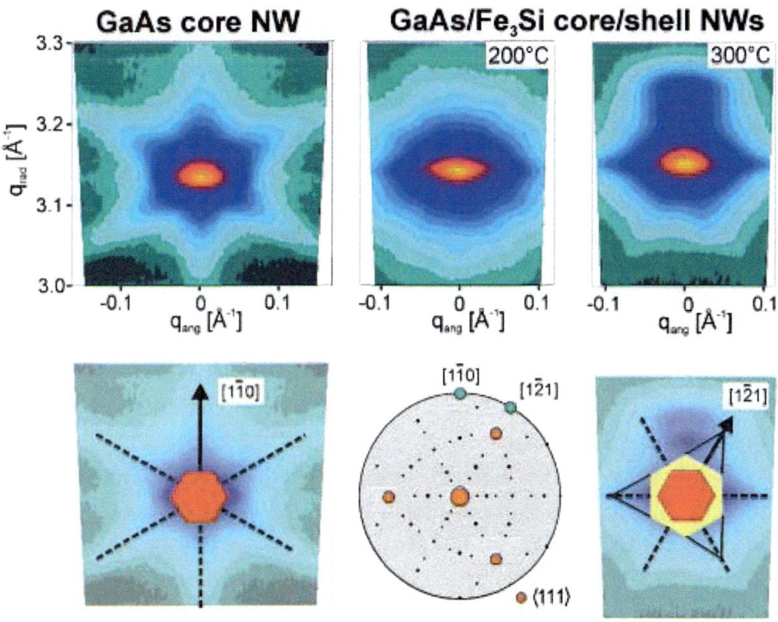

Figure 7.5 (220) in-plane reciprocal space maps of [111]-oriented GaAs NWs (upper left) and GaAs-Fe_3Si core–shell NWs (upper center and upper right). The growth temperatures of the Fe_3Si shells are given above. The crystallographic directions together with the corresponding sidewalls are sketched below. In the symbolic stereogram (middle) the [1$\bar{1}$0] and [1$\bar{2}$1] directions are marked by blue circles and the different {111} directions by red circles. The radial direction of the scans is drawn vertically and the angular direction is directed horizontally in the figure. Copyright authors of Ref. [39] licensed under a Creative Commons Attribution license (http://creativecommons.org/licenses/by/4.0/).

The crystallographic directions are sketched below. The radial direction (ω-2Θ) of the scans is vertical and the angular direction (ω) horizontal. The facets of the pillar-shaped cores with hexagonal cross section are clearly distinguished by streaks in the diffuse scattering pointing along the $\langle 110 \rangle$ directions. Remarkably, in the central map core streaks along the $\langle 110 \rangle$ directions are still visible although they are superimposed by additional streaks along the $\langle 1\bar{2}1 \rangle$ directions originating from the facets of the Fe_3Si shells. Thus, at T_g = 200°C, the hexagonal cross section of the GaAs core NWs is still conserved in shape by the Fe_3Si cover layer but the hexagonal cross section of the SFCSNWs is rotated by 30° with respect to the GaAs core cross section leading to the superimposed streaks along the $\langle 1\bar{2}1 \rangle$ directions visible in the top center image of Fig. 7.5. A sketch of the two twisted hexagons of the GaAs core in red and the Fe_3Si shell in yellow is given in the bottom right image of Fig. 7.5. This is in contrast to the GaAs-Fe_3Si core shell NWs grown at T_g = 100°C, where the SFCSNW cross section conserves also the orientation of the GaAs core wires cross section as shown in Fig. 7.3. In the map shown on the right side of Fig. 7.5, only streaks of the shell facets remain. This and additional maxima above the main peak indicate chemical reactions between Fe_3Si and GaAs occurring at T_g = 300°C near the interface similar as those observed for planar films on GaAs [22, 29].

Figure 7.6 shows a multi-beam bright-field TEM micrograph, further illustrating the orientations of the Fe_3Si facets of GaAs-Fe_3Si core–shell NWs with a Fe_3Si shell growth temperature of 200°C. In Fig. 7.6(a,b), the incident electron beam is oriented along the $[1\bar{1}0]$, and the $[11\bar{2}]$ direction, respectively. In Fig. 7.6a, nanofacets of the Fe_3Si shell can be observed as tilted black linear contrast steps similar to Fig. 7.4. However, in Fig. 7.6b smooth interfaces and surfaces are visible. The roughening due to the facetting is not observed along this direction. The Fe_3Si nanofacets consist of $\{111\}$ planes tilted around the $\langle 011 \rangle$ axes towards the $\langle 211 \rangle$ direction, i.e., for [111] oriented NWs the $(\bar{1}11)$ facets are most pronounced, forming a regular pattern along the GaAs NWs.

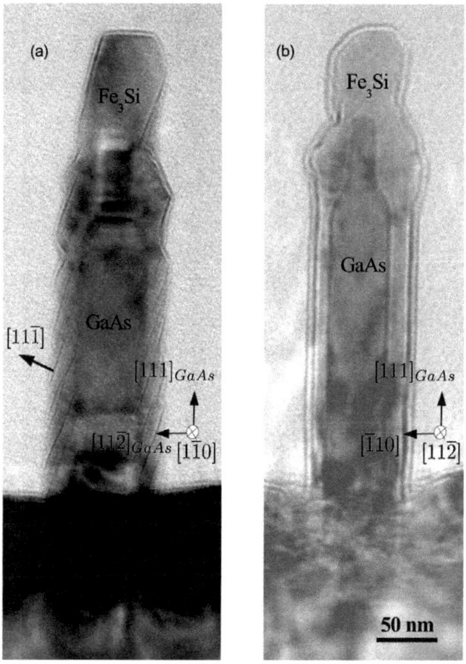

Figure 7.6 Multi-beam bright-field TEM micrographs illustrating the orientations of the Fe_3Si shell facets with respect to the GaAs cores for core–shell NWs with T_g = 200°C. The cores and shells are distinguished clearly. For the $[1\bar{1}0]$ direction of the electron beam (a) facets corresponding to tilted (111) planes can be seen. When the electron beam is running along the $[11\bar{2}]$ direction, no facets are observed (b). The images are slightly defocused in order to increase the contrast of the surfaces and interfaces. Copyright authors of Ref. [39] licensed under a Creative Commons Attribution license (http://creativecommons.org/licenses/by/4.0/).

The surface nanofacets of the GaAs-Fe_3Si core–shell NWs grown at T_g = 200°C are inclined to the (111) net planes parallel to the Si surface by an angle of approximately 108° and lead to an increase of the surface area. However, the formation of facets reduces the overall surface energy and evidences nonnegligible material transport over distances small compared to the NW lengths [10, 81]. Figure 7.7 shows the results of density functional theory (DFT) calculations of surface energies of Fe_3Si

for a wide range of chemical potentials and several possible surface terminations. Fe_3Si surfaces were found to be Si-rich [23, 67]. In that case, the Si-terminated $Fe_3Si(111)$ planes are most favorable energetically, even if geometrical factors, i.e., the inclination angle between GaAs (110) and Fe_3Si (111) planes of $\Psi = 35.3°$, are taken into account. The (110) surface has an energy of $\gamma_{110} = 140$ meV·Å^{-2}. Hence the criterion for (111) facet formation is that the (111) surfaces have an energy less than $\gamma_{110} \cdot \cos(\Psi) = 114.25$ meV·Å^{-2}. As a result, the DFT calculations confirmed that the formation of facets reduces the overall surface energy. On a larger length scale, the Fe_3Si shell is reproducing the hexagonal shape of the GaAs core NWs twisted by 30° for $T \geq 200°C$ as described earlier. Fe_3Si shells grown at $T_g = 200°C$ are fully epitaxial with a nanofacetted surface forming a regular pattern around the GaAs NWs. This facetting is probably the result of the Vollmer–Weber island growth mode of Fe_3Si on GaAs.

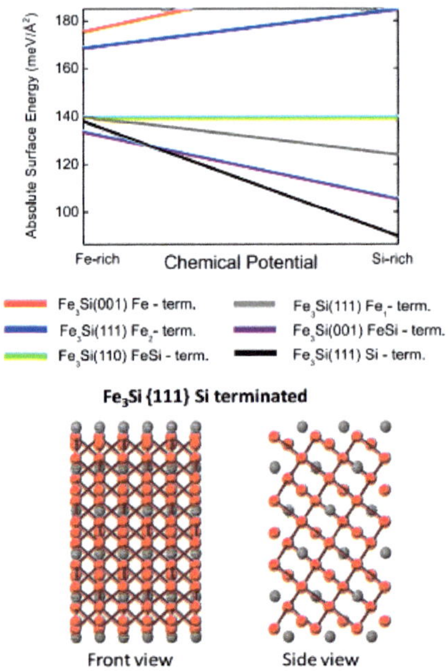

Figure 7.7 Surface energy of differently terminated (term.) surfaces calculated by DFT in generalized gradient approximation. The structure of the Si-terminated Fe_3Si {111} surface is illustrated below, Fe atoms are symbolized as red balls, whereas Si atoms are shown as gray balls. Copyright authors of Ref. [39] licensed under a Creative Commons Attribution license (http://creativecommons.org/licenses/by/4.0/).

Despite the very pronounced long-range NW side wall surface roughness observed in SEM, TEM reveals closed shells also for NWs with a Fe_3Si shell growth temperature of 350°C, which can be seen in the multi-beam TEM image in Fig. 7.8. However, the interface between the core and the shell is not smooth anymore. In addition, a distinct surface roughness of the Fe_3Si can be observed and an extended crater-shaped defect region below the NW foot surround GaAs-Fe_3Si core–shell NWs grown at 350°C. The depth of the craters ranges between 50 and 100 nm. This points to reactions at the core/shell interface and indicates, that also the Si/Fe_3Si, and Si/GaAs interfaces are not stable any more [84]. Melt back etching of the Si surface near Ga droplets at high temperatures was observed earlier [63, 71, 75, 87]. Probably a similar kind of reaction could be the reason for the formation of the craters. Precipitates were formed on the substrate side of the interface as well.

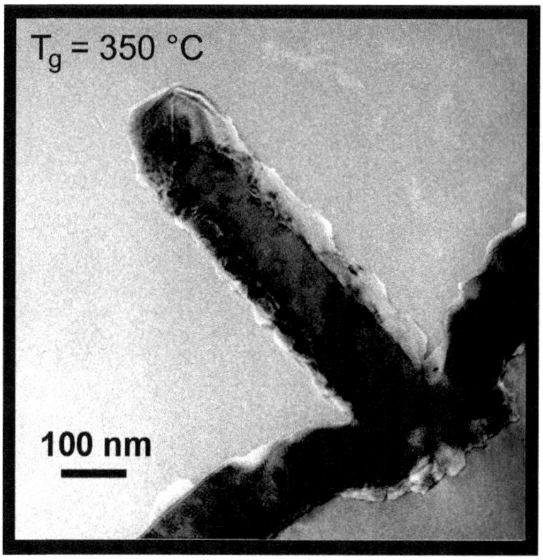

Figure 7.8 Multi-beam TEM image of a SFCSNW with T_g = 350°C. Reprinted with permission from Ref. [31]. Copyright (2016) American Chemical Society.

The orientation of the crystallites in the Fe_3Si shell and the formation of reaction products at the core–shell interface were further analyzed by high-resolution XRD. Here, all GaAs-Fe_3Si core–shell samples shown in Fig. 7.1 and additionally a planar

Fe_3Si layer grown at 100°C on a Si(111) substrate covered with a thin native oxide and a GaAs nanowire template sample were measured. The obtained ω-2Θ scans are shown in Fig. 7.9.

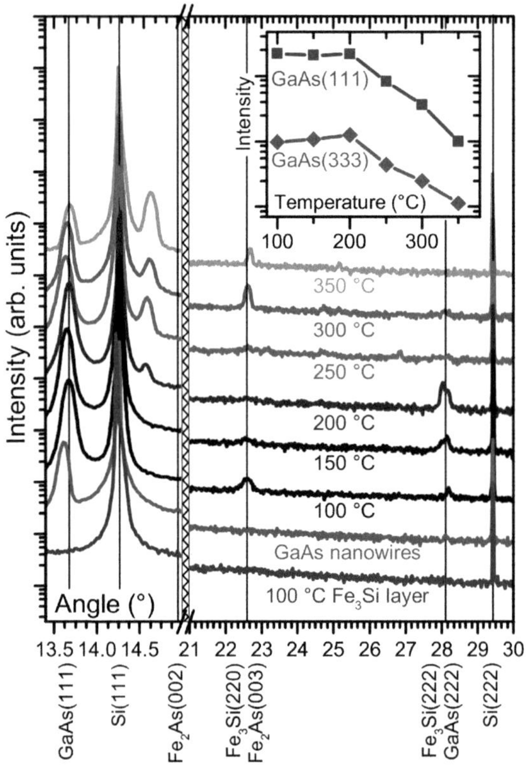

Figure 7.9 X-ray diffraction ω-2Θ scans measured at Cu-$K_{\alpha 1}$ of the six GaAs-Fe_3Si core–shell NW samples with different shell growth temperatures, of a GaAs NW template sample and a planar Fe_3Si layer grown at 100°C on a Si(111) substrate covered with a thin native oxide. The inset shows the evolution of the integrated intensities of the GaAs-related diffraction peaks with Fe_3Si growth temperature. Reprinted with permission from Ref. [31]. Copyright (2016) American Chemical Society.

For the planar Fe_3Si layer grown at 100°C on a Si(111) substrate covered with a thin native oxide (green curve), only the Si-related diffraction peaks are visible. This shows that the Fe_3Si film is either amorphous like the underlying silicon oxide or of a polycrystalline nature with very small grains.

The ω-2Θ scan of the GaAs NW template (gray curve) shows the Si and GaAs-related (111) diffraction peaks, proving the good epitaxial alignment of the GaAs nanowires [1, 18].

Additional diffraction peaks arise in the ω-2Θ scans of the six GaAs-Fe_3Si core–shell NW samples. SFCSNWs T_g = 100°C (curve in dark blue) show peaks at 22.61° And 28.09° that are attributed to Fe_3Si diffraction of net-planes (220) and (222), respectively [77]. As the planar Fe_3Si between the NWs corresponding to the green curve does not lead to any Fe_3Si-related contribution, the ordered Fe_3Si crystal phases can be attributed to the Fe_3Si NW shells. Thus, at 100°C the polycrystalline ferromagnetic NW shells show a certain amount of grains in (110) and (111) orientations. Therefore, the amount of Fe_3Si in the shells contributing to these diffraction intensities is ordered at least in the B2 phase. In this phase one of the three Fe occupied sublattices intermixes with the Si occupied sublattice.

The Fe_3Si(222) diffraction peak at 28.09° is present in the core–shell NW samples up to Fe_3Si growth temperatures of about 200°C, whereas the Fe_3Si(220) diffraction peak at 22.61° is barely visible for a Fe_3Si growth temperature of 150°C and completely absent for the sample grown at 200°C. Thus, with increasing shell growth temperature up to 200°C more Fe_3Si grains are oriented in the epitaxial (111) phase.

At T_g = 200°C and above, a diffraction peak at 14.51° arises in the ω-2Θ scans that shifts with increasing shell growth temperature to higher angles. Such diffraction is unknown for Fe_3Si [77]. In addition, at Fe_3Si shell growth temperatures of 300 and 350°C, again a peak is visible slightly above the Fe_3Si(220) diffraction peak at 22.61° that shows a similar shift with Fe_3Si shell growth temperature like the peak at 14.51°. These continuously shifting diffraction peaks lie slightly below the (002) and (003) diffraction peaks of Fe_2As at 14.91° and 22.71°, respectively [52, 74]. In planar Fe_3Si epitaxy on GaAs at elevated temperatures, precipitates of the antiferromagnetic Fe_2As phase at the interface were detected [21, 30]. Thus, the already discussed reaction and interdiffusion at the GaAs/Fe_3Si interface lead most probably to the formation of Fe_2As in the NWs already for Fe_3Si growth temperatures of 200°C and above. The deviation of the observed Fe_2As-related peaks to the exact reported positions may be due to strain in the Fe_2As and/or different stoichiometric composition.

Also the observed small shift of the GaAs (111) peak might be related to slight variations in the strain state of the NW cores. Necessarily, interdiffusion at the core/shell interface also alters the GaAs core which is further supported by a consistent decrease of the GaAs-related (111) and (333) diffraction peak intensities with shell growth temperatures above 200°C, shown in the inset of Fig. 7.9. Because of a certain amount of Fe_2As and possibly also other phases consisting of Ga, As, Fe, and Si at the core/shell interface, the magnetic properties for GaAs-Fe_3Si core–shell NWs might alter with rising shell growth temperature.

7.5 Magnetic Properties

Most important for a potential device fabrication is an appropriate magnetic characterization of such SFCSNWs. This can be done either by ensemble measurement techniques or single-NW investigation methods. Ensemble measurements have the advantage that they give quite good statistics by averaging over a large number of NWs on a sample but as a disadvantage they contain also contributions from magnetic material on the sample that does not belong to the NW shells, such as the parasitic ferromagnetic layer between the NWs. Due to the two-step epitaxial growth method where semiconductor NWs are coated by ferromagnetic material, the covering layer does not necessarily stick to the NW shells exclusively. It rather grows where the material flux reaches the sample. This is due to the quite low growth temperatures that are used and the therefore high sticking coefficient almost every where on the sample, i.e., also between the NWs. Thus, as-grown samples always comprise SFCSNWs as well as a parasitic ferromagnetic layer between the NWs. The insight of ensemble measurements on the SFCSNWs themselves that were reported so far by using superconducting quantum interference device magnetometry (SQUID), small angle neutron scattering and ferromagnetic resonance techniques are therefore quite equivocal [31, 32, 56, 60, 70, 72, 85]. Single-NW magnetic investigation, on the other hand, gives precise information about only the selected structure—one NW shell—but thus it reflects only the information obtained from one single event of measurement—from one NW. To draw conclusions from such measurements over all other NWs is hence quite limited. To investigate single NWs, electrical measurements of the magnetoresistance, magnetic force microscopy (MFM) and nano-SQUID techniques were used so far in literature [2, 3, 31, 46, 61, 79]. For GaAs-Fe_3Si core–shell NW samples the capability of the commonly used magnetic ensemble and single-NW magnetic characterization were directly compared.

7.5.1 Properties of SFCSNW Ensembles

To investigate ensembles of GaAs-Fe_3Si core–shell NWs, i.e., as-grown samples, Mössbauer spectroscopy (MB) and SQUID were used.

The obtained MB spectra for the GaAs-Fe_3Si core–shell NW samples with a Fe_3Si shell growth temperature of 100 and 350°C are shown in Fig. 7.10(a,b), respectively. Only the sample with T_g = 100°C shows several magnetically split Zeeman sextets typical for magnetic materials, like Fe_3Si. Here, the ratio of the relative intensities of the lines L_1, L_2, and L_3 of the fitted sextet sites 2, 3, and 4 gives 3:4:1. This ratio directly refers to an orientation of the spins perpendicular to the γ-beam, i.e., the substrate normal. Thus, the magnetization of the Fe_3Si in this sample is oriented parallel to the substrate surface. The parameters of the fitted Fe sites are scattered around the literature values for Fe_3Si [31]. This means that either only a small fraction of the Fe_3Si is present in the completely ordered $D0_3$ phase or that the sample contains a relatively large stoichiometric variety.

In contrast, the sample with T_g = 350°C contains only paramagnetic phases—doublet splitting—and no signs of Fe_3Si anymore. The parameters of the fitted Fe sites correspond to various compounds of Fe and Si with a high Si content such as β-$FeSi_2$, FeSi, or Fe_2Si_2, that may form due to the intermixing processes at the interfaces and in the crater-shaped defect region below the NW feet [31]. However, the non-magnetic doublets might also be ascribed to Fe_2As. Although this phase is antiferromagnetic with a Neel-temperature of 356 K [21], the small particle size or the content of Ga or Si might reduce this Neel-temperature below room temperature so that the phase appears as a paramagnetic doublet structure.

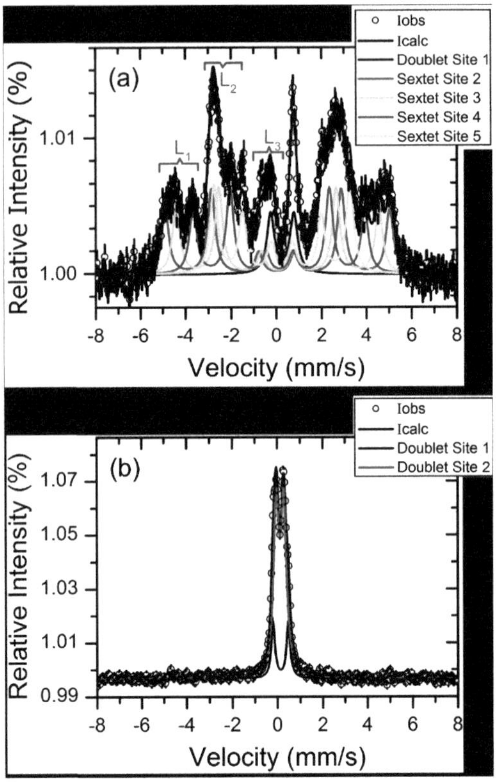

Figure 7.10 Conversion electron Mössbauer spectra of two GaAs-Fe_3Si core–shell NW samples with shell growth temperatures of (a) 100°C and (b) 350°C. Reprinted with permission from Ref. [31]. Copyright (2016) American Chemical Society.

By SQUID, three different sample configurations were measured at room temperature. The first was an as-grown GaAs-Fe_3Si core–shell NW sample that contains GaAs-Fe_3Si core–shell NWs as well as the parasitic planar Fe_3Si layer between the NWs. After the measurement of this sample, the SFCSNWs were removed. This was done in an ultrasonic bath and checked with SEM. The obtained sample now only consists of the parasitic Fe_3Si film and with Fe_3Si covered GaAs islands that typically grow as well during the self-assisted VLS GaAs NW growth mechanism. Finally an equally thick planar Fe_3Si layer was grown under identical conditions that were used for the core–shell NW growth on the same substrate, i.e., a Si(111) substrate covered with a thin native oxide layer. All three sample configurations were prepared with Fe_3Si growth temperatures of 100, 200, 300, and 350°C. After the SQUID measurements along the out-of-plane [111] direction and the two major in-plane directions $[1\bar{1}0]$ and $[11\bar{2}]$, the paramagnetic contribution of the Si substrate and the GaAs NWs and islands was subtracted from the field sweeps. From the obtained M–H curves, the anisotropy fields as well as M_S was extracted. This data is not shown here. All three sample configurations gave coinciding results regarding the anisotropy fields as well as M_S. To calculate the volume of the ferromagnetic material for the evaluation of M_S, a mathematical estimation of the NW shell volume was used. For the estimation, the NWs and NW shells were described by simple cylinders and typical values found by TEM and SEM investigations were used. These are 1–9 × 10^7 cm^{-2}, and 1 μm for the NW density and length, 25–50 nm for the GaAs NW diameter, and 14–21 nm and 70 nm for the shell and the layer thickness, respectively. The so-found overall NW shell volume amounts only to (0.5–12)% of the parasitic layer volume. It is, hence, possible that the magnetic signal of the parasitic layer may dominate the comparably small contribution of the NW shells in the ensemble measurements. At a Fe_3Si growth temperature of 350°C, the as-grown GaAs-Fe_3Si core–shell NW sample as well as the sample with removed NWs showed an almost zero magnetic moment. At this temperature, a defect region with diameters in the range of several 100 nm and a depth of about 100 nm forms below each NW in the Si substrate, which can be seen in the TEM image displayed in Fig. 7.8. Furthermore, also the interface between GaAs and Fe_3Si is not stable anymore. These interface reactions finally lead to an almost zero magnetic moment of GaAs-Fe_3Si core–shell NW samples with a Fe_3Si growth temperature of 350°C in SQUID that was earlier as well observed by MB.

However, it can be deduced that it is neither by MB nor by a careful SQUID ensemble investigation possible to elucidate a discrete magnetic contribution from the Fe_3Si NW shells. In these measurements GaAs-Fe_3Si core–shell NWs as well as the parasitic layer between the SFCSNWs seem to possess the same magnetic configuration or the parasitic layer between the NWs dominates all measurements, whereas the latter is the more likely as the parasitic layer is by far larger in volume than the NW coating. Thus, ensemble measurements are not suited to investigate the magnetic texture of ferromagnetic NW shells. For this purpose, an investigation of single NWs is indispensable.

7.5.2 Properties of Single SFCSNWs

To investigate the magnetic properties of single GaAs-Fe_3Si core–shell NWs, MFM was carried out on NWs that were first removed from the as-grown samples via an ultrasonic bath in a solvent and then transferred to a second clean substrate. The obtained MFM phase maps of four different GaAs-Fe_3Si core–shell NWs with a Fe_3Si growth temperature of 100°C are shown in Fig. 7.11a. Every NW exhibits a dark contrast corresponding to an attractive force acting on the ferromagnetic probing tip at one end and a bright contrast corresponding to a repellent force acting on the probing tip at the other end. This coincides with a magnetic stray field $\vec{H}$ around each NW that points out of the substrate plane at the bright NW end and points into the substrate plane at the dark end which is indicated in the left-hand image of Fig. 7.11a.

In magnetic nanotubes, there are only two stable equilibrium magnetic configurations when no external field is present. These are first the magnetization along the wire—parallel configuration—and second the magnetization perpendicular to the wire axis—vortex state [44, 68]. Whether the equilibrium orientation of the spontaneous magnetization will form the vortex or the magnetization along the wire depends on nanotube geometry, crystal anisotropy energies, and saturation magnetization of the magnetic material [13, 14]. As long as anisotropy energies are neglected, relaxation simulation runs using the numerical Phython-based finite element micromagnetic simulation tool Nmag [17] showed that a magnetization along the wire is energetically favored for long and thin magnetic tubes.

To convert the observed MFM phase contrast into a specific magnetization in the Fe_3Si NW shells, the magnetic field of the sample as well as the magnetic moment of the needle have to be calculated precisely [38, 49, 50, 51]. In general, this is not possible but in a first and simple approximation of the tip's moment in the direction perpendicular to the sample, the observed phase shift is proportional to the first or second derivative of the sample's magnetic field in this direction [51]. Thus, the magnetic field of one GaAs-Fe_3Si core–shell NW with parallel and vortex magnetization configuration needs to be calculated. This has been done with Nmag [17]. The results are not shown here. Whereas the vortex state gives an opposing maximum and minimum of magnetic field along the axis of the wire that is uniformly spread along the whole length of the tube, the parallel configuration leads to a significant magnetic field contribution at the ends of the NW only. As illustrated in Fig. 7.11b, such magnetic stray field distributions correspond exactly to that of a magnetic bar with a magnetic north (N) and a magnetic south pole (S). Because the magnetic field in the middle of the bar lies completely in the substrate plane, no stray field can be observed in this region by MFM. Thus, the magnetization $\vec{M}$ of the Fe_3Si shells is oriented along the NWs, which is indicated by the dark blue arrow in the left-hand image of Fig. 7.11a.

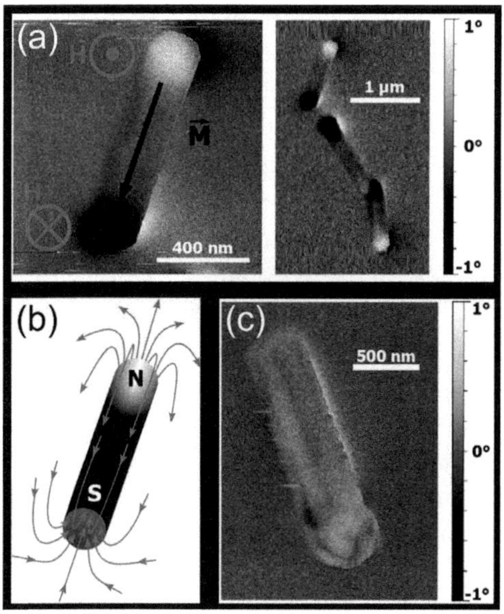

Figure 7.11 Color-coded face shift maps of GaAs-Fe_3Si core–shell NWs with shell growth temperatures of (a) 100°C and (c) 150°C measured by a magnetic force microscope. The phase shift directly corresponds to magnetic stray fields ($\vec{H}$) coming out of and pointing into the scanning plane. (b) Sketch of the magnetic stray field of a magnetic bar. Reprinted with permission from Ref. [31]. Copyright (2016) American Chemical Society.

Single-NW investigation, hence, revealed a spontaneous magnetic moment in the Fe_3Si NW shells parallel to the wire, i.e., along the out-of-plane direction of the as-grown samples and hence to be completely different from the moment's orientation obtained by the above-discussed ensemble MB and SQUID measurements. Thus, the parasitic planar layer between the GaAs-Fe_3Si core–shell NWs dominates in magnetic ensemble investigation.

In contrast, no magnetic stray fields were found in MFM for core–shell NWs grown above 100°C, which is exemplarily shown in Fig. 7.11c for the sample grown at 150°C. Here, interdiffusion of Fe, Si, As, and Ga at the interfaces already alters the ferromagnetic behavior of Fe_3Si.

7.6 Summary

GaAs-Fe_3Si core–shell NWs prepared under specific growth conditions by MBE unite the following morphological and crystal properties. The Fe_3Si completely wets the GaAs NW side walls, which leads to closed shells and smooth NW morphologies when grown at low temperatures. The hexagonal shape of the original GaAs core NW is conserved in shape and orientation during the overgrowth with Fe_3Si at 100°C and the shells show polycrystalline characteristics. The SFCSNWs can be fabricated with smooth interfaces in a highly perfect epitaxial crystalline structure for T_g = 200°C. Here, the Fe_3Si shells form characteristic (111) oriented Si-terminated nanofacets that reduce the overall surface energy. On a larger length scale, the Fe_3Si shell reproduces the hexagonal shape of the GaAs core NWs twisted by 30° for this and higher Fe_3Si growth temperatures. For GaAs-Fe_3Si core–shell NWs with Fe_3Si growth temperatures above 200°C, interdiffusion processes at the core/shell and the NW/substrate interface lead to severe reactions and intermixing altering the composition of the NW shells as well as the NW cores and form large precipitates in the shape of extended craters in the Si substrate.

Ensemble magnetic measurements by MB and SQUID of GaAs-Fe_3Si core–shell NWs were completely dominated by the extensive parasitic Fe_3Si layer between the NWs. Thus, with such ensemble techniques it is not possible to draw any conclusions about the magnetic properties of the NW shells of SFCSNWs. To gain insights about these properties, single-NW characterization is absolutely essential. This is of general interest, as it applies to all the presented SFCSNWs so far where a significant amount of magnetic material grows as well between the NWs and should be taken into account when analyzing ensemble magnetic measurements. Nevertheless, as a side note, ensemble magnetic measurements are a powerful tool to investigate the magnetic properties of SFCSNWs in the case of a suppressed parasitic layer growth. In principle, this can be realized, for example, by increasing the NW density and length such that the sample space between the NWs is shadowed by the NWs themselves and the incoming material fluxes cannot reach the sample surface any more during the growth. However, such high NW densities have not been demonstrated for SFCSNW growth so far.

The single measurement technique carried out by MFM measurements and micromagnetic computational modeling showed for the GaAs-Fe_3Si core–shell NWs in contrast to the ensemble investigation a magnetization along the wire and a maximum magnetic moment when the Fe_3Si shell is grown at 100°C. Thus, magnetization of the SFCSNWs oriented perpendicular to the substrate forming nano-bar magnets can indeed be achieved. Such properties and magnetizations are prerequisites to enable the fabrication of high-quality spintronic devices such as spin-LEDs operable at zero external magnetic field and three-dimensional magnetic recording devices. Thus, GaAs-Fe_3Si core–shell NW nano-bar magnets are indeed promising candidates for new forward-looking device applications in the field of spintronics and magnetic recording.

References

1. Biermanns, A., Breuer, S., Davydok, A., Geelhaar, L., and Pietsch, U. (2011). Structural evolution of self-assisted GaAs nanowires on Si(111), *Phys. Status Solidi RRL* 5, p. 156.

2. Buchter, A., Nagel, J., Rüffer, D., Xue, F., Weber, D. P., Kieler, O. F., Weimann, T., Kohlmann, J., Zorin, A. B., Russo-Averchi, E., Huber,

R., Berberich, P., Fontcuberta i Morral, A., Kemmler, M., Kleiner, R., Koelle, D., Grundler, D., and Poggio, M. (2013). Reversal mechanism of an individual ni nanotube simultaneously studied by torque and SQUID magnetometry, *Phys. Rev. Lett.* 111(6), p. 067202.

3. Butschkow, C., Reiger, E., Rudolph, A., Geißler, S., Neumaier, D., Soda, M., Schuh, D., Woltersdorf, G., Wegscheider, W., and Weiss, D. (2013). Origin of negative magnetoresistance of GaAs/(Ga,Mn)As core-shell nanowires, *Phys. Rev. B* 87(24), p. 245303.
4. Butschkow, C. H., Reiger, E., Geißler, S., Rudolph, A., Soda, M., Schuh, D., Woltersdorf, G., Wegscheider, W., and Weiss, D. (2011). Magnetoresistance of individual ferromagnetic GaAs/(Ga,Mn)As core-shell nanowires, arXiv:1110.5507 [cond-mat.mes-hall], p. 18.
5. Carcia, P. F., Meinhaldt, A. D., and Suna, A. (1985). Perpendicular magnetic anisotropy in Pd/Co thin film layered structures, *Appl. Phys. Lett.* 47, p. 178.
6. Colombo, C., Spirkoska, D., Frimmer, M., Abstreiter, G., and Fontcubertai Morral, A. (2008). Ga-assisted catalyst-free growth mechanism of GaAs nanowires by molecular beam epitaxy, *Phys. Rev. B* 77(15), p. 155326.
7. Dau, M. T., Jenichen, B., and Herfort, J. (2015). Perpendicular magnetic anisotropy in the Heusler alloy Co_2 TiSi/GaAs(001) hybrid structure, *AIP Adv.* 5, p. 057130.
8. Daub, M., Knez, M., Goesele, U., and Nielsch, K. (2007). Ferromagnetic nanotubes by atomic layer deposition in anodic alumina membranes, *J. Appl. Phys.* 101(9), p. 09J111.
9. Dellas, N. S., Liang, J., Cooley, B. J., Samarth, N., and Mohney, S. E. (2010). Electron microscopy of GaAs/MnAs core/shell nanowires, *Appl. Phys. Lett.* 97(072505), p. 072505.
10. Dhanaraj, G., Byrappa, K., and Prasad, V. (2010). *Handbook of Crystal Growth* (Springer, Berlin, Heidelberg).
11. Duan, H., Weissmüller, J., and Wang, Y. (2008). Instabilities of core-shell heterostructured cylinders due to diffusions and epitaxy: Spheroidization and blossom of nanowires, *J. Mech. Phys. Solids* 56(5), p. 1831.
12. Elliot, R. P. (1965). *Constitution of Binary Alloys, Suppl 1*(McGraw-Hill, New York).
13. Escrig, J., Landeros, P., Altbir, D., and Vogel, E. (2007a). Effect of anisotropy in magnetic nanotubes, *J. Magn. Magn. Mater.* 310(2), p. 2448.

14. Escrig, J., Landeros, P., Altbir, D., Vogel, E., and Vargas, P. (2007b). Phase diagrams of magnetic nanotubes, *J. Magn. Magn. Mater.* 308(2), p. 233.

15. Farshchi, R., Ramsteiner, M., Herfort, J., Tahraoui, A., and Grahn, H. T. (2011). Optical communication of spin information between light emitting diodes, *Appl. Phys. Lett.* 98(16), p. 162508.

16. Fiederling, R., Keim, M., Reuscher, G., Ossau, W., Schmidt, G., Waag, A., and Molenkamp, L. W. (1999). Injection and detection of a spin-polarized current in a light-emitting diode, *Nature* 402(6763), p. 787.

17. Fischbacher, T., Franchin, M., Bordignon, G., and Fangohr, H. (2007). A Systematic approach to multiphysics extensions of finite-element-based micromagnetic simulations: Nmag, *IEEE Trans. Magn.* 43(6), pp. 2896–2898.

18. Fontcuberta i Morral, A., Colombo, C., Abstreiter, G., Arbiol, J., and Morante, J. R. (2008). Nucleation mechanism of gallium-assisted molecular beam epitaxy growth of gallium arsenide nanowires, *Appl. Phys. Lett.* 92(6), p. 63112.

19. Gao, C., Farshchi, R., Roder, C., Dogan, P., and Brandt, O. (2011). GaN/Fe core/shell nanowires for nonvolatile spintronics on Si, *Phys. Rev. B* 83(24), p. 245323.

20. Givargizov, E. (1973). Periodic instability in whisker growth, *J. Cryst. Growth* 20(3), p. 217.

21. Grandjean, F., Gerard, A., and Sobry, R. (1973). Study by Mössbauer spectroscopy of the antiferro-paramagnetic transition of Fe_2As, *Int. J. Magn.* 4, p. 1.

22. Gusenbauer, C., Ashraf, T., Stangl, J., Hesser, G., Plach, T., Meingast, A., Kothleitner, G., and Koch, R. (2011). Interdiffusion in heusler film epitaxy on gaas(001), *Phys. Rev. B* 83, p. 035319.

23. Hafner, J., and Spišák, D. (2007). Structure and stability of the low-index surfaces of Fe_3Si: Ab initio density functional investigations, *Phys. Rev. B* 75(19), p. 195411.

24. Han, S., Li, C., Liu, Z., Lei, B., Zhang, D., Jin, W., Liu, X., Tang, T., and Zhou, C. (2004). Transition metal oxide coreshell nanowires: Generic synthesis and transport studies, *Nano Lett.* 4(7), pp. 1241–1246.

25. Hanna, T., Yoshida, D., and Munekata, H. (2011). Preparation characterization of MnSb-GaAs spin LED, *J. Cryst. Growth* 323 (1), p.383.

26. Hansen, M. (1958). *Constitution of Binary Alloys* (McGraw-Hill, New York).

27. Herfort, J., Jenichen, B., Kaganer, V., Trampert, A., Schönherr, H.-P., and Ploog, K. (2006a). Epitaxial Heusler alloy Fe_3Si films on GaAs(001) substrates, *Phys. E* 32(1–2), p. 371.

28. Herfort, J., Schönherr, H.-P., Friedland, K.-J., and Ploog, K. H. (2004). Structural and magnetic properties of epitaxial Fe[sub 3] Si/GaAs(001) hybrid structures, *J. Vac. Sci. Technol. B* 22(4), p. 2073.

29. Herfort, J., Schönherr, H.-P., and Ploog, K. H. (2003). Epitaxial growth of Fe_3Si/GaAs(001) hybrid structures, *Appl. Phys. Lett.* 83(19), p. 3912.

30. Herfort, J., Trampert, A., and Ploog, K. (2006b). Growth, interface structure, and magnetic properties of Fe/GaAs and Fe_3Si/GaAs hybrid systems, *Int. J. Mat. Res.* 97(7), p. 1026.

31. Hilse, M., Herfort, J., Jenichen, B., Trampert, A., Hanke, M., Schaaf, P., Geelhaar, L., and Riechert, H. (2013). GaAs-Fe_3Si core-shell nanowires: nanobar magnets, *Nano Lett.* 13(12), pp. 6203–6209.

32. Hilse, M., Takagaki, Y., Herfort, J., Ramsteiner, M., Herrmann, C., Breuer, S., Geelhaar, L., and Riechert, H. (2009). Ferromagnet-semiconductor nanowire coaxial heterostructures grown by molecular-beam epitaxy, *Appl. Phys. Lett.* 95(13), p. 133126.

33. Hilse, M., Takagaki, Y., Ramsteiner, M., Herfort, J., Breuer, S., Geelhaar, L., and Riechert, H. (2011). Strain in GaAs-MnAs core-shell nanowires grown by molecular beam epitaxy, *J. Cryst. Growth,* 323(1), p. 307.

34. Hövel, S., Gerhardt, N. C., Hofmann, M. R., Lo, F.-Y., Ludwig, A., Reuter, A., Dand Wieck, Schuster, E., Wende, H., Keune, W., Petracic, O., and Westerholt, K. (2008a). Room temperature electrical spin injection in remanence, *Appl. Phys. Lett.* 93, p. 021117.

35. Hövel, S., Gerhardt, N. C., Hofmann, M. R., Lo, F.-Y., Reuter, D., Wieck, A. D., Schuster, E., Keune, W., Wende, H., Petracic, O., and Westerholt, K. (2008b). Electrical detection of photoinduced spins both at room temperature and in remanence, *Appl. Phys. Lett.* 92(24), p. 242102.

36. Hua, B., Motohisa, J., Kobayashi, Y., Hara, S., and Fukui, T. (2009). Single GaAs/GaAsP coaxial core-shell nanowire lasers, *Nano Lett.* 9(1), p. 112.

37. Jabeen, F., Grillo, V., Rubini, S., and Martelli, F. (2008). Self-catalyzed growth of GaAs nanowires on cleaved Si by molecular beam epitaxy, *Nanotechnology* 19(27), p. 275711.

38. Jackson, J. D. (1975). *Classical Electrodynamics*, 2nd ed. (John Wiley & Sons), ISBN 0-471-43132-X.

39. Jenichen, B., Hanke, M., Hilse, M., Herfort, J., Trampert, A., and Erwin, S. C. (2016). Diffraction at GaAs/Fe_3Si core/shell nanowires: The formation of nanofacets, *AIP Adv.* 6, p. 055108.

40. Jenichen, B., Hilse, M., Herfort, J., and Trampert, A. (2015a). Facetted growth of Fe_3Si shells around GaAs nanowires on Si(111), *J. Crystal Growth* 427, p. 21.

41. Jenichen, B., Hilse, M., Herfort, J., and Trampert, A. (2015b). Real structure of lattice matched GaAs-Fe_3Si core-shell nanowires, *J. Cryst. Growth* 410, p. 1.

42. Jenichen, B., Kaganer, V., Herfort, J., Satapathy, D., Schönherr, H., Braun, W., and Ploog, K. (2005). Long-range order in thin epitaxial Fe_3Si films grown on GaAs(001), *Phys. Rev. B* 72(7), p. 075329.

43. Kikuchi, A., Kawai, M., Tada, M., and Kishino, K. (2004). In-GaN/GaN multiple quantum disk nanocolumn light-emitting diodes grown on (111) Si Substrate, *Jpn. J. Appl. Phys.* 43(12A), p. L1524.

44. Landeros, P., Escrig, J., Altbir, D., Laroze, D., dAlbuquerque e Castro, J., and Vargas, P. (2005). Scaling relations for magnetic nanoparticles, *Phys. Rev. B* 71(9), p. 094435.

45. Lee, J., Suess, D., Schrefl, T., Oh, K. H., and Fidler, J. (2007). Magnetic characteristics of ferromagnetic nanotube, *J. Magn. Magn. Mater.* 310(2), p. 2445.

46. Liang, J., Wang, J., Paul, A., Cooley, B. J., Rench, D. W., Dellas, N. S., Mohney, S. E., Engel-Herbert, R., and Samarth, N. (2012). Measurement and simulation of anisotropic magnetoresistance in single GaAs/MnAs core/shell nanowires, *Appl. Phys. Lett.* 100(18), p. 82402.

47. Limbach, F., Hauswald, C., Lähnemann, J., Wölz, M., Brandt, O., Trampert, A., Hanke, M., Jahn, U., Calarco, R., Geelhaar, L., and Riechert, H. (2012). Current path in light emitting diodes based on nanowire ensembles, *Nanotechnology* 23(46), p. 465301.

48. Misiuna, P., Duewski, P., Wojciechowski, T., Miliska, E., Kurowska, B., Wiater, M., Wawro, A., Wojtowicz, T., and Baczewski, L. (2015). Structural and magnetic properties of hybrid ferromagnetic metal/ semiconductor (ZnTe)/Co core-shell nanowires, *J. Crystal Growth* 412, pp. 80–86.

49. Nolting, W. (2004). *Grundkurs Theoretische Physik 3 Elektrody-Namik*, 7th ed. (Springer-Verlag), ISBN 3-540-20509-8.

50. NT-MDT (2014a). 2.3 Cantilever linear oscillations, https://www.ntmdt-si.com/data/media/files/spm_basics/linear_oscillations_of_cantilever.pdf.

51. NT-MDT (2014b). Magnetic Force Microscopy: Quantitative Results Treatment, https://www.ntmdt-si.com/data/media/files/spm_basics/magnetic_force_microscopy_quantitative_results_treatment.pdf.

52. Nuss, J., Wedig, U., and Jansen, M. (2006). Geometric variations and electron localizations in intermetallics: PbFCl type compounds, *Z. Kristallogr.* 221, p. 554.

53. Ohno, Y., Young, D., Beschoten, B., and Matsukura, F. (1999). Electrical spin injection in a ferromagnetic semiconductor heterostructure, *Nature* 402, p. 790.

54. Oliveira, D. S., Galvão Tizei, L. H., Ugarte, D. M., and Cotta, M. A. (2013). Spontaneous periodic diameter oscillations in InP nanowires: The role of interface instabilities, *Nano Lett.* 13, pp. 9–13.

55. Parkin, S. S. P., Hayashi, M., and Thomas, L. (2008). Magnetic domain-wall racetrack memory, *Science* 320(5873), p. 190.

56. Pimpinella, R. E., Zhang, D., McCartney, M. R., Smith, D. J., Krycka, K. L., Kirby, B. J., O'Dowd, B. J., Sonderhouse, L., Leiner, J., Liu, X., Dobrowolska, M., and Furdyna, J. K. (2013). Magnetic properties of GaAs/Fe core/shell nanowires, *J. Appl. Phys.* 113(17), p. 17B520.

57. Qian, F., Gradečak, S., Li, Y., Wen, C.-Y., and Lieber, C. M. (2005). Core/multishell nanowire heterostructures as multicolor, high-efficiency light-emitting diodes, *Nano Lett.* 5(11), p. 2287.

58. Qian, F., Li, Y., Gradečak, S., Wang, D., Barrelet, C. J., and Lieber, C. M. (2004). Gallium nitride-based nanowire radial heterostructures for nanophotonics, *Nano Lett.* 4(10), p. 1975.

59. Ramsteiner, M., Brandt, O., Flissikowski, T., Grahn, H. T., Hashimoto, M., Herfort, J., and Kostial, H. (2008). Co_2FeSi/GaAs/(Al,Ga)As spin light-emitting diodes: Competition between spin injection and ultrafast spin alignment, *Phys. Rev. B* 78(12), p. 121303(R).

60. Rudolph, A., Soda, M., Kiessling, M., Wojtowicz, T., Schuh, D., Wegscheider, W., Zweck, J., Back, C., and Reiger, E. (2009). Ferromagnetic GaAs/GaMnAs core-shell nanowires grown by molecular beam epitaxy, *Nano Lett.* 9(11), p. 3860.

61. Rüffer, D., Huber, R., and Berberich, P. (2012). Magnetic states of an individual Ni nanotube probed by anisotropic magnetoresistance, *Nanoscale* 4, p. 4989.

62. Ryu, K.-S., Thomas, L., Yang, S.-H., and Parkin, S. S. P. (2012). Current induced tilting of domain walls in high velocity motion along

perpendicularly magnetized micron-sized Co/Ni/Co racetracks, *Appl. Phys. Exp.* 5, p. 093006.

63. Sakai, S., Matyi, R. J., and Shichijo, H. (1988). Growth of GaAs on GaAs-coated Si by liquid-phase epitaxy, *J. Appl. Phys.* 63(4), p. 1075.
64. Schmidt, V., McIntyre, P., and Gösele, U. (2008). Morphological instability of misfit-strained core-shell nanowires, *Phys. Rev. B* 77, p. 235302.
65. Schuster, E., Keune, W., Lo, F.-Y., Reuter, D., Wieck, A., and Westerholt, K. (2005). Preparation and characterization of epitaxial Fe(001) thin films on GaAs(001)-based LED for spin injection, *Superlattices Microstruct.* 37(5), p. 313.
66. Sellmyer, D. J., Kirby, R. D., and Jaswal, S. S. (1995). *The Magnetism of Amorphous Metals and Alloys* (Singapore: World Scientific).
67. Starke, U., Schardt, J., Weiss, W., Meier, W., Polop, C., de Andres, P. L., and Heinz, K. (2001). Structural and compositional reversible phase transitions on low-index Fe_3Si surfaces, *Europhys. Lett.* 56(6), p. 822.
68. Sui, Y. C., Skomski, R., Sorge, K. D., and Sellmyer, D. J. (2004). Nanotube magnetism, *Appl. Phys. Lett.* 84(9), p. 1525.
69. Svensson, C. P. T., Mårtensson, T., Trägårdh, J., Larsson, C., Rask, M., Hessman, D., Samuelson, L., and Ohlsson, J. (2008). Monolithic GaAs/InGaP nanowire light emitting diodes on silicon, *Nanotech. Nol.* 19(30), p. 305201.
70. Takagaki, Y., Herfort, J., Hilse, M., Geelhaar, L., and Riechert, H. (2011). Swingback in magnetization reversal in MnAs-GaAs coaxial nanowire heterostructures, *J. Phys.: Condensed Matter* 23(126002), pp. 1–9.
71. Takemoto, K., Murakami, H., Iwamoto, T., Matsuo, Y., Kangawa, Y., Kumagai, Y., and Koukitu, A. (2006). Growth of GaN directly on Si(111) substrate by controlling atomic configuration of Si surface by metalorganic vapor phase epitaxy, *Japanese J. Appl. Phys.* 45(18), pp. L478–L481.
72. Tivakornsasithorn, K., Pimpinella, R. E., Nguyen, V., Liu, X., Dobrowolska, M., and Furdyna, J. K. (2012). Magnetic anisotropy of GaAs/Fe/Au core-shell nanowires grown by MBE, *J. Vac. Sci. Technol. B* 30(2), p. 02B115.
73. Tomioka, K., Motohisa, J., Hara, S., Hiruma, K., and Fukui, T. (2010). GaAs/AlGaAs core multishell nanowire-based light-emitting diodes on Si, *Nano Lett.* 10(5), p. 1639.

74. Van der Kraan, A. M., and Buschow, K. H. J. (1986). The 57Fe Mössbauer isomer shift in intermetallic compound of iron, *Phys. B* 138, p. 55.

75. Van der Ziel, J. P., Dupuis, R. D., Logan, R. A., and Pinzone, C. J. (1987). Degradation of gaas lasers grown by metalorganic chemical vapor deposition on Si substrates, *Appl. Phys. Lett.* 51(2), pp. 89–91.

76. Van Roy, W., Van Dorpe, P., De Boeck, J., and Borghs, G. (2006). Spin injection in LED's and in unipolar devices, *Mater. Sci. Eng. B* 126(2–3), p. 155.

77. Waliszewski, J., Dobrzynski, L., Malinowski, A., Satula, D., Szymanski, K., Prandl, W., Bruckel, T., and Scharpf, O. (1994). Magnetic moment distribution in $Fe_{3-x}Cr_x$ Si alloys, *J. Magn. Magn. Mater.* 132, p. 349.

78. Wang, Z., Lim, H., Liu, H., Ng, S., Kuok, M., Tay, L., Lockwood, D., Cottam, M., Hobbs, K., Larson, P., Keay, J., Lian, G., and Johnson, M. (2005). Spin waves in nickel nanorings of large aspect ratio, *Phys. Rev. Lett.* 94(13), p. 137208.

79. Weber, D. P., Rüffer, D., Buchter, A., Xue, F., Russo-Averchi, E., Huber, R., Berberich, P., Arbiol, J., Fontcuberta, I., Morral, A., Grundler, D., and Poggio, M. (2012). Cantilever magnetometry of individual Ni nanotubes, *Nano Lett.* 12, p. 6139.

80. Weller, D., Farrow, R. F. C., Hurst, J. E., Notarys, H., Brändle, H., Rühr ig, M., and Hubert, A. (1994). CoPt multilayers and alloys for short wavelength magneto-optical recording, *Opt. Memory Neural Networks* 3, pp. 353–368.

81. Wulff, G. V. (1901). Zur frage der geschwindigkeit des wachstums und der auflsung von krystallflchen, *Z. Kristallogr.* 34, pp. 449–530.

82. Yamamoto, K., Matsuda, T., Nishibayashi, K., Kitamoto, Y., and Munekata, H. (2013). Low-power photo-induced precession of magnetization in ultra-thin Co/Pd multilayer films, *IEEE Trans. Magn.* 49, p. 3155.

83. Yan, R., Gargas, D., and Yang, P. (2009). Nanowire photonics, *Nat. Photonics* 3(10), p. 569.

84. Yoshitake, T., Nakagauchi, D., and Nagayama, K. (2003). Ferromagnetic iron silicide thin films prepared by pulsed-laser deposition, *Jpn. J. Appl. Phys.* 42, Part 2(7B), p. L849.

85. Yu, X., Wang, H., Pan, D., Zhao, J., Misuraca, J., von Molnár, S., and Xiong, P. (2013). All zinc-blende GaAs/(Ga,Mn) As core-shell nanowires with ferromagnetic ordering, *Nano Lett.* 13(4), p. 1572.

86. Zhang, X., Wang, C., Liu, Y., Zhang, Z., Jin, Q. Y., and Duan, C.-G. (2016). Magnetization switching by combining electric field and spin-transfer torque effects in a perpendicular magnetic tunnel junction, *Sci. Rep.* 6, p. 18719.

87. Zhu, D., Wallis, D. J., and Humphreys, C. J. (2013). Prospects of III-nitride optoelectronics grown on Si, *Rep. Prog. Phys.* 76(10), p. 106501.

Chapter 8

$GaAs/AlGaO_x$ Heterostructured Nanowires

Fumitaro Ishikawa[a] **and Naoki Yamamoto**[b]

[a]*Graduate School of Science and Engineering, Ehime University, 3 Bunkyo-cho, Matsuyama, Ehime 790-8577, Japan*
[b]*Department of Physics, Tokyo Institute of Technology, Ohokayama, Meguro-ku, Tokyo 152-8551, Japan*

ishikawa@ee.ehime-u.ac.jp

We synthesize semiconductor/oxide composite nanowires (NWs) via molecular beam epitaxial growth and subsequent wet oxidation. Selective oxidation applied to the GaAs/AlGaAs core–shell NWs grown on silicon substrates was used to fabricate $GaAs/AlGaO_x$ NWs. Nonselective oxidation conditions resulted in $GaO_x/AlGaO_x$ NWs. The oxidized amorphous $AlGaO_x$ shell produced luminescence related to oxygen over the entire visible spectrum. The wire core was buried in the shells when the diameter of the shell increased, forming a planar structure. At low temperatures, the luminescence from the ensembles of vertically aligned $(Al,Ga)O_x$ microwires is found to be spatially homogeneous and extend over the entire visible range. As the temperature increases from 10 to 300 K, the photoluminescence intensity is observed to increase. Together with the energy of the luminescence bands,

Novel Compound Semiconductor Nanowires: Materials, Devices, and Applications
Edited by Fumitaro Ishikawa and Irina Buyanova

ISBN 978-981-4745-76-5 (Hardcover), 978-1-315-36440-7 (eBook)
www.panstanford.com

these opposing dependencies demonstrate that the emission stems from F centers in the (Al,Ga)O_x shells. These μ-wires are promising structures for obtaining white light emission on a Si platform. To realize efficient white light emitters, it is essential to prevent the transfer of carriers to the GaAs core and the shell surface. The system can provide extended functions for both semiconductor and oxide-material systems, possibly utilized for light emitters, as well as solar cells in which we can utilize their absorbance.

8.1 Introduction

One-dimensional composite nanowires (NWs), including semiconductors, metal oxides, and organic materials, have attracted extensive research efforts because of their numerous potential applications in not only nanoelectronics but also catalysis, chemical sensing, and energy storage [1–8]. Composite systems comprising various materials have been used to produce high-performance electronic devices, such as metal-oxide–semiconductor transistors [9]. The combination of materials exhibiting contrasting properties is expected to provide superior functions that are not achievable from the individual materials.

The introduction of III–V compound semiconductors in NWs provides dynamic control over the electronic band structure of these systems [1, 10]. These heterostructures can serve as charge carriers and light waveguides, making NWs attractive building blocks in nanoelectronics and nanophotonics [11–14]. Moreover, advanced epitaxial techniques used to fabricate NWs can overcome the large mismatches between the lattice constants and thermal coefficients of III–V semiconductors and Si, enabling the heteroepitaxy of these materials [15, 16]. The integration of semiconductor NWs with various Si substrates opens the door for large-scale integrated systems with enhanced electronic and optical properties [17, 18]. In addition to their usefulness and ease of fabrication, core–shell NWs present tunable properties and provide passivation for core material surfaces, resulting in an extensive range of high-performance device applications [19]. For example, NWs can be capped with a larger bandgap material for either passivation or modulation doping. Strain induced by a lattice mismatch between the core and shell may be used to further alter the energy structure within the NW. Conversely, oxides displaying advantageous dielectric, thermal, and resistive properties, which cannot be achieved in semiconductors, make their combination with semiconductors appealing. Further, even in the amorphous state, these materials present useful properties for developing advanced devices [20–22]. For example, Ga_2O_3 [23–25], In_2O_3 [26], and Al_2O_3 [27] NWs have been studied to control various functions, such as plasmonic and excitonic behaviors as well as for field emitter array applications.

The AlGaAs/GaAs system was subjected to wet oxidation to integrate oxide functions into the semiconductor. The well-established selective oxidation technique is used for vertical-cavity surface emitting lasers, [28–30] while the recent nonselective oxidation technique, which transforms entire semiconductor systems into oxides, can be exploited for optical waveguides [31, 32]. In addition to structural precision at the nanoscale, this technique enables selective reaction with the initial AlGaAs/GaAs semiconductor. Wet oxidation transforms the AlGaAs compound semiconductor into the amorphous oxide $AlGaO_x$, resulting in high refractive index contrast between the materials before oxidation as well as current confinement because of the insulating nature of the oxides. Varying the original semiconductor compositions precisely tunes the properties of the semiconductors such as the refractive index [33]. Controlling the extent of oxidation through selective and nonselective processes further increases the versatility of oxidized AlP/InP NWs including native oxides [34] and the reversible conversion of GaN/Ga_2O_3 NWs [35]. AlGaAs/GaAs NWs promise to play a major role in advanced electrical and optical applications [36–41]. A method for fabricating $(Al,Ga)O_x$ oxide shells around GaAs NWs grown by molecular beam epitaxy on Si(111) substrates was recently reported by our group in Ref. 42. This fabrication technique, based on the wet oxidation of the shell of GaAs/(Al,Ga)As core/shell NWs, takes advantage of the fact that the growth of III–V semiconductors in the form of NWs is going to mature and paves the way for the realization of oxide NWs integrated on the Si platform [42]. Under electron irradiation, the oxide shells of $GaAs/(Al,Ga)O_x$ give rise to light emission ranging from the near-infrared to the ultraviolet [42–46]. This observation, together with the facts that NWs with a high crystalline quality can be grown on Si substrates [13, 15, 47] and that the NW geometry facilitates efficient coupling between light and matter [48–50], suggests that core/shell GaAs/$(Al,Ga)O_x$ μ-wires can be used as white-light sources.

An entirely buried semiconductor structure is synthesized, in which the GaAs/AlGaAs NW is surrounded by the oxides. This system is expected to benefit from passivation by the surrounding oxides and lead to future device applications because it easily forms top and bottom contact electrodes [51]. When sufficiently buried by the low-refractive-index oxides, the structure itself may exhibit specific optical functions, such as the photonic crystal effect [52]. Therefore, the combination of semiconductors and oxides within monolithically synthesized NWs and the burying of individual NWs in the coalesced oxide would have strong potential for future functional devices. In this study, semiconductor/oxide heterostructure NWs were fabricated by molecular beam epitaxy followed by wet oxidation of the compound semiconductor heterostructure system. Moreover, GaAs/AlGaAs heterostructure NWs were buried in the oxide shells by expanding the shell diameter to form planar structures that can be applied to future devices.

8.2 Nanowire Growth

GaAs/AlGaAs core–shell NWs were grown by molecular beam epitaxy on Si substrates. By extension and coalescence of the AlGaAs shell, the system forms a fully buried structure. We studied the growth mechanism and structural characteristics of these samples at various growth stages. At first, the NWs' AlGaAs shells grew individually, preserving the single crystals until they coalesced with adjacent wires, forming the polycrystalline structure. The lateral growth of the AlGaAs shell promoted the formation of energetically stable facets, which then became a large hexagonal rod containing a certain number of GaAs cores. The polycrystalline AlGaAs shell continued to extend, eventually forming the fully buried structure.

8.2.1 Experimental Details

GaAs/AlGaAs NWs were grown on phosphorous-doped n-type Si(111) substrates by molecular beam epitaxy using Ga as a catalyst [42, 53–56]. Epi-ready substrate surfaces were not treated prior to NW growth, but were covered with a native oxide layer with a typical thickness of several nanometers. The Al and Ga supplies were set to match a planar growth rate of 1.0 ML/s (monolayer per second) on a GaAs (001) substrate calibrated by the reflection of high-energy electron diffraction oscillation [57]. This planar growth rate corresponds to an F_{As}/F_{Ga} atomic flux ratio of approximately 10 estimated from the stoichiometric conditions identified by the surface reconstruction transition between the As-rich (2 × 4) and Ga-rich (4 × 2) areas for the GaAs growth plane on GaAs (00l) [58]. The samples were heated to 570°C and subjected to a constant As_4 beam at an equivalent pressure of 1.7×10^{-5} mbar. The GaAs core was grown for 15 min at these conditions into a longitudinal wire. Next, the growth was interrupted for 15 min to crystallize the Ga catalyst, and the As_4 pressure was increased to 3.3×10^{-5} mbar during this time. Finally, Al and Ga were added, and the AlGaAs shell containing 90% Al was allowed to grow. During this period, the lateral growth became dominant, generating core–shell NWs. GaAs NWs were buried in the AlGaAs shells as the shell's diameters expanded. $Al_{0.9}Ga_{0.1}As$ shells with various thicknesses were grown by varying the deposition time to be 30 min, 2 h, or 6 h [42].

Structural characteristics of the NWs were investigated by scanning electron microscopy (SEM) using an S-5200 instrument (Hitachi, Japan) and transmission electron microscopy (TEM). The cross-sectional analysis was conducted using a scanning transmission electron microscope (STEM, JEM-ARM200F, JEOL, Japan) operating at 200 kV equipped with an energy dispersive X-ray (EDX) spectroscopy imaging system (JED-2300T, JEOL, Japan). For the TEM measurements, finely sliced samples displaying thicknesses smaller than 200 nm were fabricated using a focused ion beam system (FEI Strata DB 235). STEM images were obtained in bright field (BF) and high-angle annular dark-field (HAADF) modes. The crystallinity was investigated by electron diffraction using a TEM (Hitachi H-9000UHR-II) operating at

300 kV in the <112> azimuth direction. The optical characteristics of the NWs were investigated via cathodoluminescence (CL) measurements using a STEM (JEM-2100F, JEOL, Japan) combined with a light detection system, operating at an acceleration voltage of 200 kV. To characterize their optical properties, individual NWs were transferred to a carbon-coated supporting film. The cathodoluminescence spectra were obtained by subtracting the background spectrum of the surrounding area from the raw data. The diameter and beam current of the electron beam were 1 nm and 1 nA, respectively [42, 59–61]. All measurements were performed at room temperature.

8.2.2 Growth Mechanism of Core–Shell GaAs/AlGaAs Nanowires

The formation mechanism of the core–shell GaAs/AlGaAs NW is proposed in Fig. 8.1a [54, 55]. First, NW growth is initiated

by Ga droplet formation on the native oxide-coated substrate surface. This oxide layer has pinholes that create nanocraters that act as nucleation sites for the NWs [53, 56]. The density of NWs in an area is therefore very sensitive to the substrate surface conditions. The wire predominantly starts to grow longitudinally for 15 min. During the subsequent growth interruption, the increase in As flux promotes Ga catalyst consumption and Ga crystallization, allowing the 100 nm-diameter NWs to grow longer. After this interruption, Al and Ga addition mostly resulted in the lateral growth of the AlGaAs shell layer, producing the core–shell NWs. Buried GaAs/AlGaAs heterostructure NWs were obtained by enlarging the shell diameters.

SEM images (Figs. 8.1b,c) of NWs comprising AlGaAs shells containing 80% Al after growth provided clear evidence for NW formation. The NWs exhibited a preferable vertical alignment, which is suitable for device applications, and a density of approximately 4 μm^{-1}. SEM images (Figs. 8.1d,e) showed that a typical single GaAs/AlGaAs NW had a diameter and length of approximately 300 nm and 3.0 μm, respectively, and possessed a straight sidewall with a hexagonal cross-section. The NW morphologies provided information about their growth mechanism [62–64].

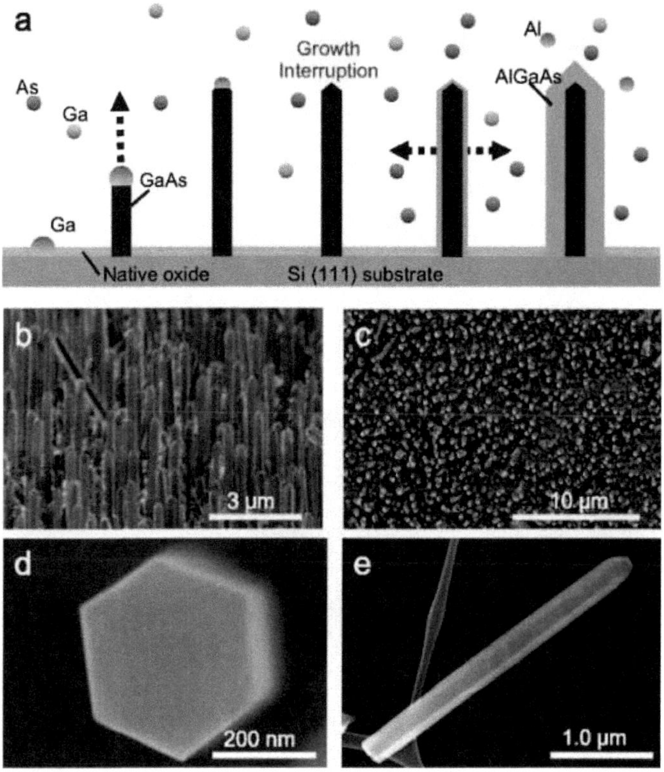

Figure 8.1 (a) Schematic of the NW growth mechanism. (b) Bird's-eye view and (c) planar SEM images of NWs comprising $Al_{0.8}Ga_{0.2}As$ shells. (d) Enlarged cross-sectional and (e) side-view SEM images of a typical NW. The background mesh corresponds to the porous carbon film used as a support for the NWs. Reprinted with permission from [42]. Copyright 2014 American Chemical Society.

8.2.3 Growth of Buried Nanowire Structure

Figures 8.2a–c show planar and cross-sectional SEM images of the NWs whose shells were grown for 30 min, 2 h, and 6 h, respectively, revealing that the NW diameter expanded with increasing shell growth time at ~500 nm/h [42, 55]. Moreover, neighboring wires coalesced over time, burying the entire structure further. Specifically, the wire diameter increased monotonically, and the wire density decreased with increasing growth time. The sample obtained after 6 h of shell growth showed a complex structure consisting of disordered wires with voids at individual NW boundaries (Figs. 8.2c1,c2). This may have been caused by adatom flux shadowing from adjacent wires, resulting in tapered side facets [23]. Wire tips adopted a disordered arrangement while remaining triangular, inducing a top surface roughness exceeding several micrometers. Figures 8.2a3–c3 show planar images of typical NWs obtained from AlGaAs shell growth times of 30 min, 2 h, and 6 h, respectively. The diameters increased over time while preserving the regular hexagons with sharp {110} facets [65]. These results show the coalescence among neighboring NWs produce fully buried structures.

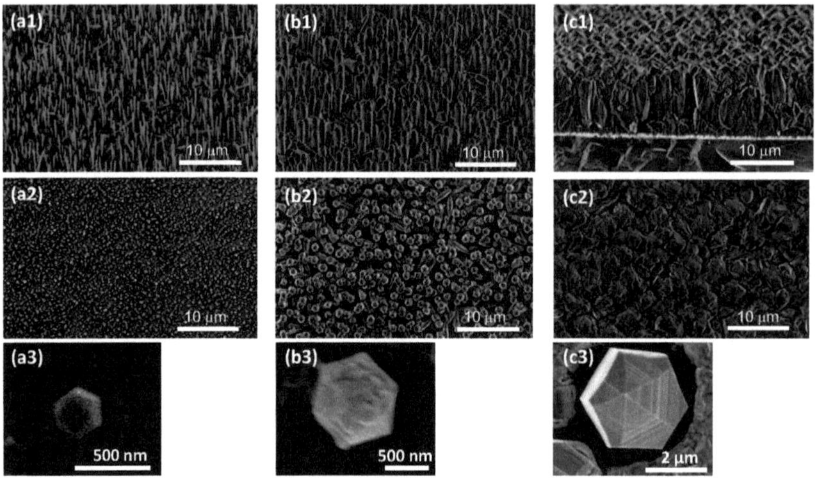

Figure 8.2 GaAs NWs buried in the AlGaAs shell layer. 45°-tilted and planar ensembles, along with planar SEM images of single NWs obtained after shell growth times of (a1–a3) 30 min, (b1–b3) 2 h, and (c1–c3) 6 h.

Figure 8.3 shows the planar SEM images of Figs. 8.2(a2–c2) digitally processed to exhibit the surface coverage in these images. In Figs. 8.3a–c, the red areas are the wire-covered areas for the wire ensembles obtained after shell growth times of 30 min, 2 h, and 6 h corresponding to Figs. 8.2a2,b2,c2, respectively.

Figure 8.4 shows the dependence of surface coverage on wire diameter. To obtain strong optical emission from the wire ensembles, the aggregate volume of the wires is expected to be larger. For the ensemble sample comprising 1 μm-diameter wires with shells grown for 2 h, the surface coverage is about 40% and the wire density is about 5×10^3 cm^{-2}, as seen in Fig. 8.3b.

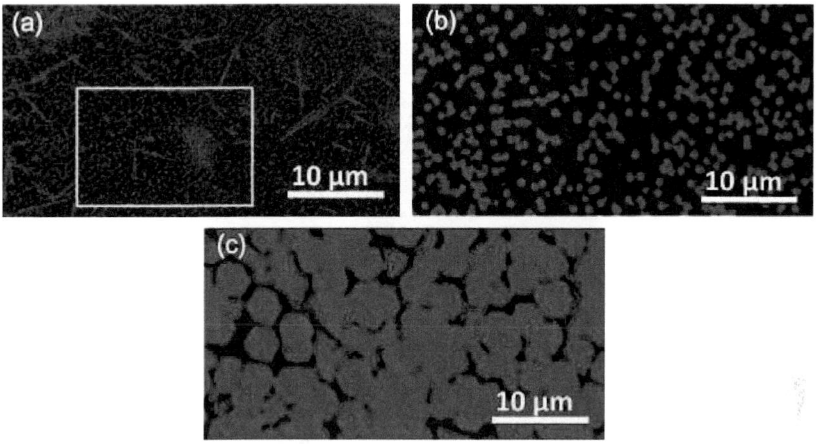

Figure 8.3 Processed planar SEM images of Figs. 8.2a2–c2. Red areas indicate wire-covered areas for the wire ensembles obtained after shell growth times of (a) 30 min, (b) 2 h, and (c) 6 h.

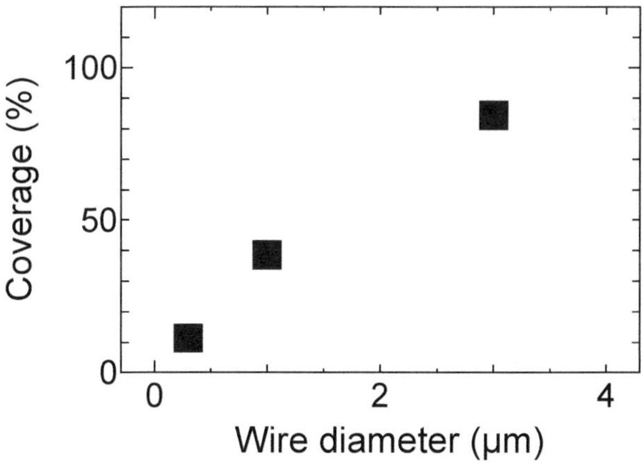

Figure 8.4 Variation of surface coverage for different wire diameters.

Figure 8.5 shows planar SEM images of the sample with an AlGaAs shell growth time of 2 h. As shown in the two enlargements in this figure, some of the wires, which have diameters of ~1 μm, have coalesced with adjacent wires. Surrounding individual wires normally have regular hexagonal shape. However, the coalesced wires show irregular structures, as has been reported for GaN NWs showing elongated ensembles [66, 67].

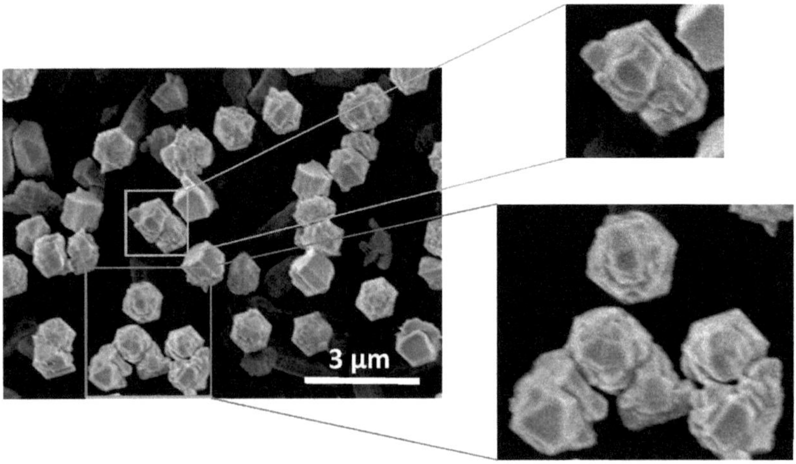

Figure 8.5 Planar SEM images of a sample with an AlGaAs shell growth time of 2 h. The areas marked with yellow and green squares, enlarged at right, show coalescence. Reprinted with permission from [66]. Copyright 2016 WILEY-VCH Verlag GmbH & Co. KGaA.

We examined the coalescence of the initial GaAs cores by selectively etching the NWs whose AlGaAs shells were grown for 2 h. Al-rich AlGaAs shells are selectively etched by a hydrofluoric acid solution, which was applied to the above NW samples to etch out the AlGaAs shell layer preserving the GaAs core. Figures 8.6a,b show 45°-tilted SEM images of the NWs with AlGaAs shell growth time of 2 h before and after HF etching, respectively. After etching, the AlGaAs shells at the bottoms and middles of the NWs were clearly etched off. The tips of the wires were not etched because of some oxide complexes that formed during the hydrofluoric etching, but they are beyond the scope of this study. Mostly, as clearly shown in the figure, there were many thin wires at the bottom of the remaining tips of the thick NWs with a diameter of about 1 µm, which seemed to support the tips. This result clearly indicates that the wires coalesced during growth.

To investigate how the crystallographic structure changed during growth, we performed XRD. Figure 8.7 shows XRD results from various samples: a reference GaAs NW with only a GaAs core of 100 nm diameter and a GaAs NW with a diameter of 300 nm, which was grown under the same conditions except without Al flux and with the growth time controlled to obtain the intended 300 nm diameter of the shells.

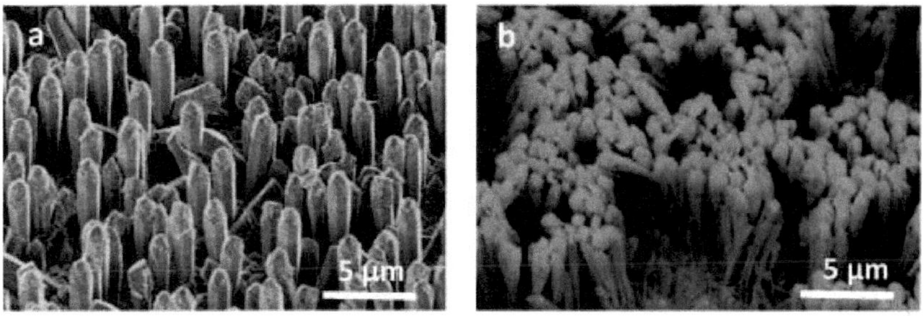

Figure 8.6 45°-tilted SEM images of the NWs with AlGaAs shell growth times of 2 h (a) before and (b) after HF etching. Reprinted with permission from [66]. Copyright 2016 WILEY-VCH Verlag GmbH & Co. KGaA.

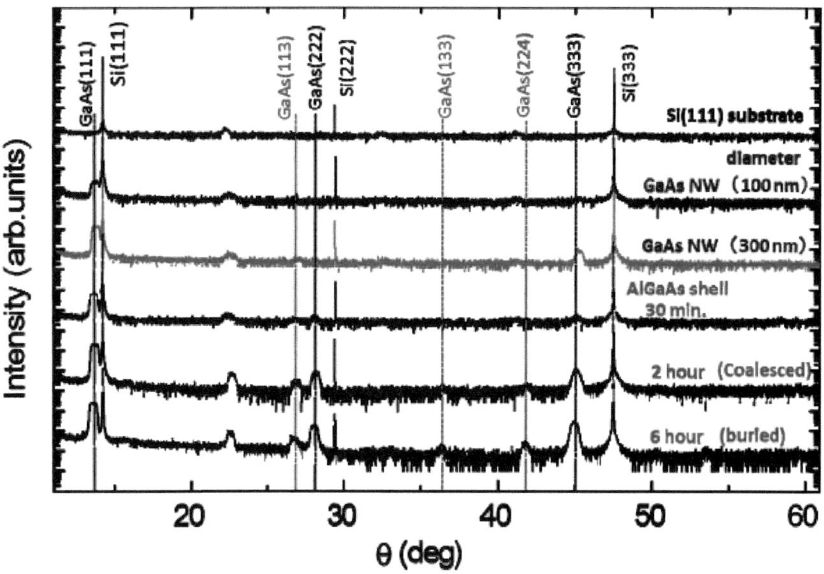

Figure 8.7 X-ray diffraction curves, from top to bottom, for the Si substrate, initial GaAs NWs with 100 nm diameter before AlGaAs shell formation, GaAs NWs with 300 nm diameter as a reference, and samples of NWs with a 100 nm GaAs core and an AlGaAs shell grown at times of 30 min, 2 h, and 6 h. Note that the peak positions of GaAs and AlAs cannot be resolved in this broad plot.

The Si substrate shows strong, sharp diffraction peaks originating from Si(111) and Si(333), associated with a weak Si(222) peak and a broad peak at 23°, an artifact from the system that appeared for all the curves in Fig. 8.5. For the thin GaAs core with a 100 nm diameter, we start to observe the GaAs(111) peak associated with the GaAs NWs. When the thickness of the GaAs NWs increased to 300 nm, the GaAs(111) peak strengthened and the GaAs(333) peak appeared. For the sample with an AlGaAs shell, there were also GaAs(111), (333), and (222) peaks. We could not resolve the GaAs and AlAs diffraction peaks for this series of samples because they have very similar lattice constants and because of peak broadening induced by variations in the inclinations of the NWs. For the coalesced NWs with an AlGaAs shell, we observed asymmetry in the (113), (133), and (224) diffraction peaks for GaAs and in those of AlGaAs. Those peaks become prominent by extending the shell width to the samples with growth times of the shells. This result suggests that the coalescence of wires induced the formation of a polycrystalline phase.

Based on these results, Fig. 8.8 summarizes the coalescence phenomena and formation mechanism of the fully buried structures. At first, the AlGaAs shells of the NWs grow individually, preserving the single crystals until they begin to coalesce with adjacent wires and become polycrystalline. The polycrystalline AlGaAs shell further develops laterally, forming energetically stable {110} facets, then becomes a large hexagonal rod that contain a certain number of GaAs cores. The polycrystalline AlGaAs shell continues to grow, eventually forming the fully buried structure.

Figure 8.9 shows an illustration of the suggested crystalline structure in the buried system. At first, the GaAs/AlGaAs core–shell structure develops with a single crystalline phase. The structure becomes rhomboidal because of the adatom diffusion and shadowing of the ensembles [68]. After occasional coalescence between adjacent wires, the crystal structure becomes polycrystalline and further develops [69–71]. Before coalescence, symmetric lattice planes appeared, as shown in the XRD curve in Fig. 8.7. The formation of the polycrystalline structure generates asymmetric lattice planes, and the intensity increases as their volume increases from the extension of the AlGaAs shell.

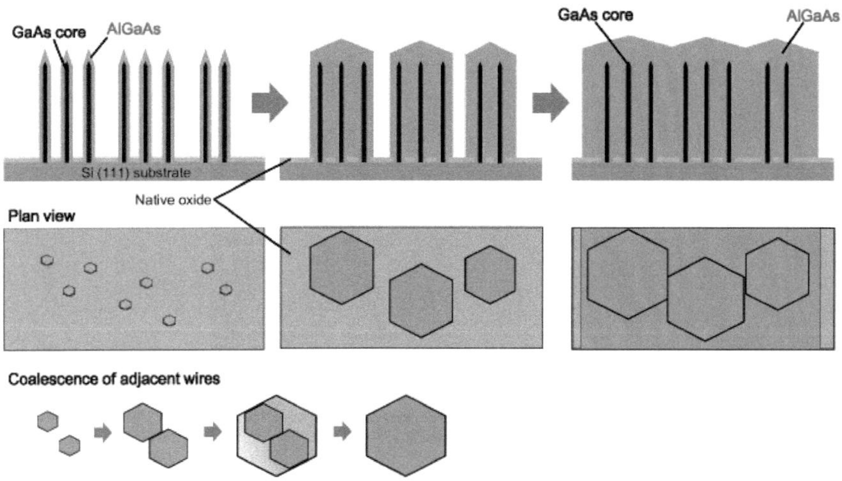

Figure 8.8 Schematic showing the formation mechanism of GaAs NWs buried in an AlGaAs layer. Reprinted with permission from [66]. Copyright 2016 WILEY-VCH Verlag GmbH & Co. KGaA.

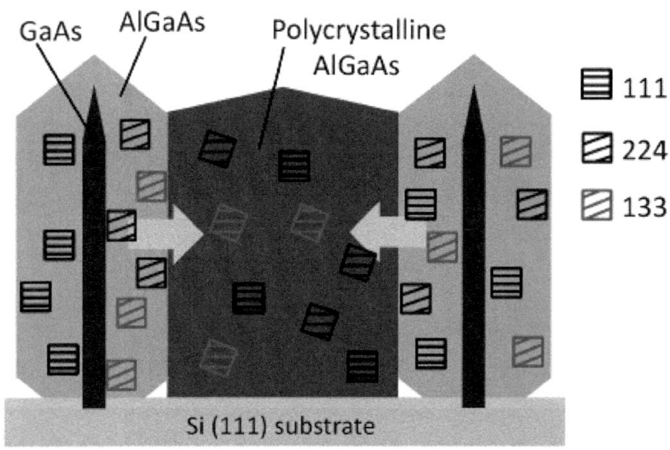

Figure 8.9 Suggested crystalline structure for the fully buried structure.

In summary, we grew GaAs/AlGaAs core–shell NWs by using molecular beam epitaxy to form a fully buried structure through extension and coalescence of the AlGaAs shell. The GaAs wire core became buried in the AlGaAs shell as the shell's diameter increased, forming a fully planar structure. The growth mechanism and its structural characteristics were studied by investigating the growth of various samples. At first, the AlGaAs shells of the NWs grew individually, preserving the single crystals until they coalesced with adjacent wires, forming the polycrystalline structure. The lateral growth of the AlGaAs shell promoted the formation of energetically stable facets. The coalesced wires then became large hexagonal rods containing a certain number of GaAs cores. The polycrystalline AlGaAs shell continued to extend, eventually forming the fully buried structure.

8.3 Wet Oxidation

We subjected the Al-rich AlGaAs/GaAs system to wet oxidation to integrate oxide functions into the semiconductor, as has been done for vertical-cavity surface emitting lasers [16–18]. The wet oxidation transforms the Al-rich AlGaAs into the amorphous oxide $AlGaO_x$, resulting in high refractive index contrast between the materials before oxidation as well as current confinement because of the insulating oxides [33]. Because of these advantages, monolithically synthesized GaAs NWs buried in a coalesced AlGaAs shell have great potential for future functional devices [11]. In the work reported here, we buried GaAs NWs in the AlGaAs shells by expanding the shell diameter to form fully planar structures. The formation mechanism and the crystalline feature of the system were investigated.

8.3.1 Wet Oxidation Process

Wet oxidation of Al-rich AlGaAs is a well-established technique for its application to semiconductor optical devices, especially for vertical cavity surface emitting laser as described in the Section 8.1 [28, 72]. The large selectivity between AlGaAs and GaAs provides a discretely controlled insulating film monolithically multilayered with GaAs. Further, the large refractive index contrast with GaAs realizes strong optical confinement [33]. The wet oxidation was conducted for the NW samples for 2 h in a furnace at 370 or 435°C [33]. Water was kept at 90°C in a bubbler, and the extracted steam was transmitted into the furnace using N_2 gas.

The Al-rich AlGaAs shell selectively underwent oxidation because its oxidation rate differed significantly from that of the GaAs core [42]. Hence, we can obtain semiconductor–oxide composite GaAs/AlGaO_x or fully oxidized GaO_x/AlGaO_x by simply tuning the oxidation temperature

8.3.2 Selective Oxidation of GaAs/AlGaAs Core–Shell Nanowires

The synthesized GaAs/AlGaAs core–shell NWs underwent wet oxidation for 2 h, as shown in Fig. 8.10 [42]. The wires were selectively or nonselectively oxidized by varying the oxidation temperatures between 370 and 435°C because the GaAs core and Al-rich AlGaAs shell displayed large differences in their oxidation rates [33]. At 370°C, the selective oxidation solely affected the AlGaAs shell while preserving the GaAs core. Conversely, under nonselective conditions (above 435°C), the entire structure was oxidized to produce GaO_x/AlGaO_x NWs.

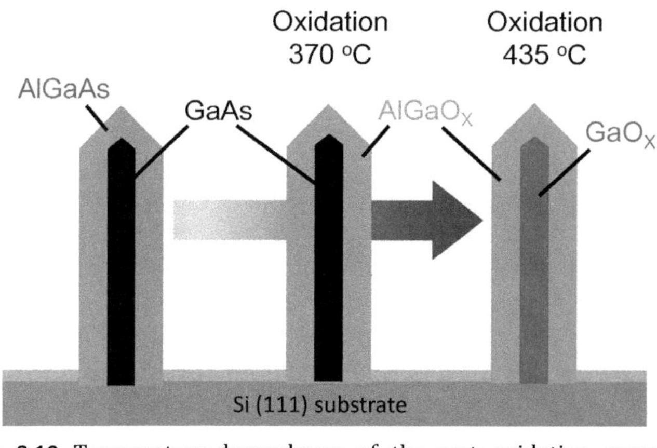

Figure 8.10 Temperature-dependence of the wet oxidation procedure. Reprinted with permission from [42]. Copyright 2014 American Chemical Society.

Figure 8.11a shows a BF-STEM image of a single GaAs/AlGaAs core–shell NW after oxidation at 435°C. The GaAs core and the AlGaAs shell were grown for 15 and 30 min, respectively. The wire was slightly bent because of the slicing step required during sample preparation.

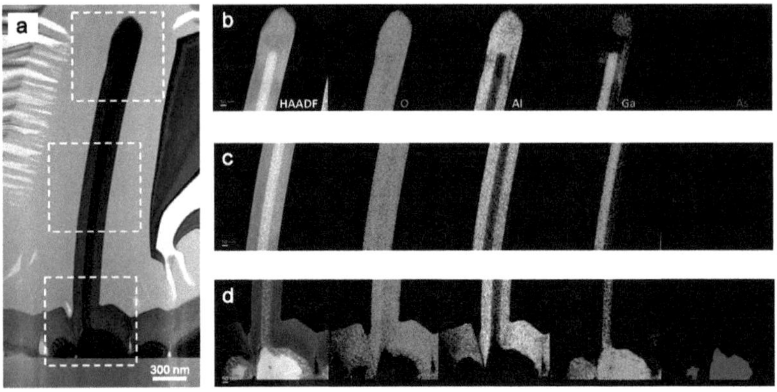

Figure 8.11 (a) BF-STEM image of a single GaAs/AlGaAs core–shell NW after oxidation at 435°C. (b–d) HAADF-STEM and EDX elemental mapping images for the areas delimited by the dashed squares in (a). The intensity of As is enhanced because the signal is close to the EDX detection limit. Reprinted with permission from [42]. Copyright 2014 American Chemical Society.

The wire consists of a 100 nm-diameter GaAs core surrounded by a wide (approximately 100 nm) AlGaAs shell. Small crystals observed on the sample surface coalesced with the NW bases but exhibited negligible effects on the upper portion of the NW. HAADF-STEM and EDX elemental mapping images (Figs. 8.11b–d) showed strong intensities for Ga and O_2 at the core. Conversely, Al and O_2 presented strong intensities in the shell. Arsenic generated a weak signal inside the wires, except at the bottom, where its intensity approached the detection limit. This indicates that the wire was oxidized by exchanging As with O_2 throughout the structure, preserving the doping-level As at a density of approximately 10^{19} cm^{-3}. The oxidation progressed inward on the wires and the upper portion of the surface crystals exposed to air, consistent with the formation of core–shell GaO_x/$AlGaO_x$ heterostructures. Note that NWs were found easier to oxidize than the surface crystals. Under the same reaction conditions, GaAs did not undergo oxidation on planar samples [33]. However, exposing a large surface area to the oxidizing steam enhanced the oxidation. Therefore, to selectively oxidize AlGaAs over GaAs, the reaction was performed at a lower temperature.

Figure 8.12a shows the HAADF-STEM and EDX elemental mapping images for the same NW as above but oxidized at a reduced temperature of 370°C. Unlike for the NW oxidized at 435°C, As and Ga showed strong intensities at the core upon oxidation. Conversely, Al and O exhibited strong intensities at the shell. This demonstrates that the shell was selectively oxidized at 370°C to provide the GaAs/$AlGaO_x$ heterostructure NW. The diffraction pattern of the shell presented a diffused ring-like pattern (Fig. 8.12b), consistent with its amorphous nature. In contrast, the core showed a clear diffraction spot, indicative of the crystalline structure of GaAs. Therefore, wet oxidation selectively produced a semiconductor/amorphous oxide NW at 370°C. Overall, this approach provided oxide/semiconductor or oxide/oxide heterostructure NW by adjusting the process conditions.

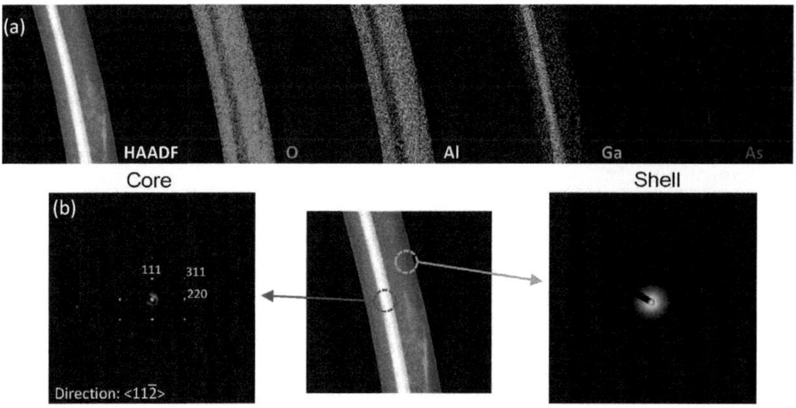

Figure 8.12 (a) HAADF and EDX elemental mapping images of NW after oxidation at 370°C. (b) Diffraction patterns of wire core (left) and shell (right) at the indicated spots. Reprinted with permission from [42]. Copyright 2014 American Chemical Society.

8.3.3 Oxidation of Buried GaAs/AlGaAs Heterostructure Nanowires

We synthesize buried GaAs/AlGaAs heterostructure NWs as discussed in Section 8.2. As schematically depicted in Fig. 8.13, freestanding NWs are expected to be buried under other shells if their shell diameters expand upon growth time increase. When the entire structure is sufficiently buried by low-refractive-index oxides, the structure itself may exhibit attractive optical properties, such as photonic crystal effects, because of the refractive index contrast between the scale of NWs and the oxides with their separations equaling several hundreds of nanometers [52, 73].

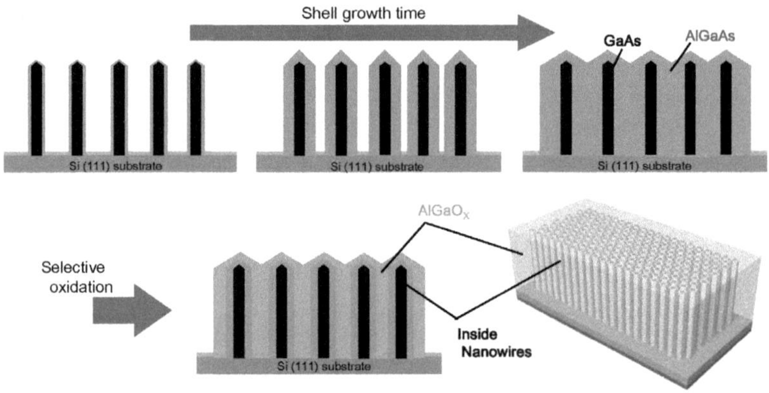

Figure 8.13 Formation procedure of GaAs NWs buried in an oxide layer. Reprinted with permission from [42]. Copyright 2014 American Chemical Society.

Figures 8.14a–c show post-oxidation cross-sectional TEM images of the sample composed of coalesced wires (Figs. 8.2c1,c2). The images presented a quarried part of the entire sample because the thickness of the finely sliced sample was smaller than 200 nm for the measurement. Grain boundaries, which may result from the complex NW growth and coalescence at large shell width, were observed (Fig. 8.14, dashed lines). Specifically, rhomboidal regions, which displayed heights and widths of approximately 8 and 1.5 μm, respectively, were detected within the buried structure. These regions may be considered as initial GaAs/AlGaAs core–shell NWs unaffected by adjacent NWs. Their structure may stem from the shadowing of the adatoms observed during the growth of dense and thick NWs [68]. In addition, the lower portion of the rhomboidal region exhibited preserved single-crystalline areas containing 90 nm-diameter GaAs NWs (Fig. 8.14c). The EDX mapping of this rhomboidal region revealed that oxygen showed strong intensity in the upper portion, but As was predominant in the bottom portion (Fig. 8.14d), suggesting that the As initially present in the AlGaAs shell was exchanged for oxygen by wet oxidation at 370°C.

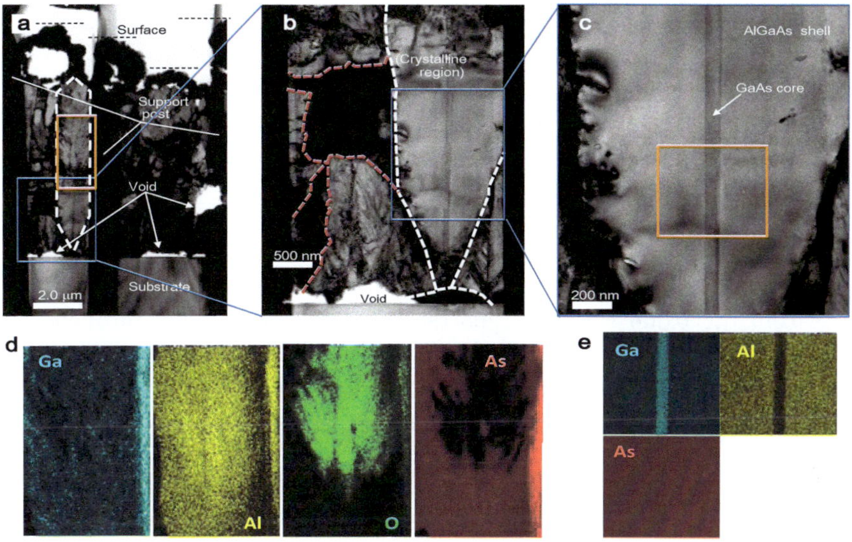

Figure 8.14 (a–c) Cross-sectional TEM images of the sample exhibiting coalesced shells after oxidation. Dashed lines are guides for the eye indicating interfaces between different crystallites within the structure. Images (b) and (c) are enlargements of areas marked by the blue rectangles. (d) Summarized EDX images of the area delimited by the orange box in (a). (e) Summarized EDX images of the area delimited by the orange box in (c). The image shows a quarried part of the entire sample because the thickness of the finely sliced sample was below 200 nm for the measurement. Reprinted with permission from [42]. Copyright 2014 American Chemical Society.

Considering the finely sliced nature of the sample, these results indicate that wet oxidation was suitable for this region and affected the 5 μm uppermost portion of the sample. This oxidation preserved the freestanding GaAs/AlGaAs core–shell NWs within the structure (Figs. 8.14c,e). The strong oxide intensity observed in the rhomboidal region (Fig. 8.14d) suggests that the oxidation may have progressed from the surface, grain boundaries, and voids. Consequently, $AlGaO_x$ oxides can confine the AlGaAs shell-surrounded GaAs core. EDX images revealed that crystalline GaAs/AlGaAs core–shell heterostructures were preserved within the system. A more precise structural control over surface, boundary, and void configurations may open the door to advanced functions relying on electrical and optical confinement.

8.4 Optical Characteristics

Our main and initial motivation for the synthesis of the semiconductor/oxide composite NWs is to employ the structure for opto-electronic devices, as described in Section 8.1. The low refractive index of oxides compared to conventional semiconductors provides dynamic control of the system, as utilized in the developed vertical cavity surface emitting lasers. Further, we have found characteristic emissions over the entire visible spectrum from the oxide NWs, as reported in [42]. Here, mainly through investigations of cathodoluminescence and photoluminescence, we discuss the NW characteristics, the origin of the luminescence, and possible future applications.

8.4.1 Individual Nanowire

To elucidate the optical properties of the composite NW series, we performed cathodoluminescence measurements of the GaAs/$AlGaO_x$ heterostructure. Specifically, we investigated GaAs/$Al_{0.9}Ga_{0.1}As$ NWs after oxidation at 370°C (see Fig. 8.12). Results are summarized in Figs. 8.15a,b. Two NWs were transferred from the original Si substrate to a carbon-coated supporting film. The contrast in the NW in Fig. 8.15a (the panchromatic CL image) reflects the core–shell structure of the NWs. The upper right corner of Fig. 8.15a shows the spectrum acquired from the 1×1 μm^2 area involving the NW. The spectrum exhibits a sharp peak at 340 nm and a broad peak extending from approximately 400 to 700 nm, probably constituting several overlapped peaks. This spectrum is similar to previously reported spectra of Al_2O_3 nanoparticles [43–46]. As seen in Fig. 8.15a, the shell layer emitted strong luminescence at 342 and 423 nm. Furthermore, the observed emissions were localized, consistent with luminescence sourced from nanostructured oxides. This suggests the formation of shell oxides with luminescent properties. In the literature, the observed spectral range varies from violet to green-yellow [43–46], possibly reflecting the different arrangements of molecules responsible for the luminescence [43]. In our case, luminescence is observed from ultraviolet (300 nm) to near infrared (700 nm) wavelengths, including the broadened tail of the spectrum. This broad emission might be sourced from the small Ga constituent within the amorphous $AlGaO_x$ shell, as seen in Fig. 8.12, and the wide nanoscale particle-size distribution of the shell [45].

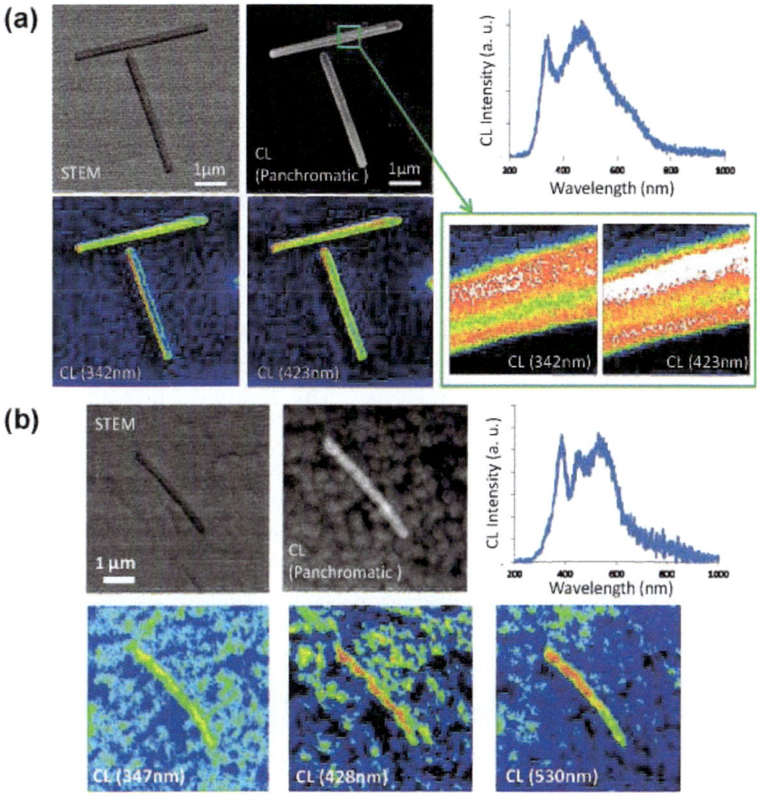

Figure 8.15 Cathodoluminescence (CL) of $GaAs/AlGaO_x$ NWs. (a) Back-scattered electron image, panchromatic CL image, cathodoluminescence spectra acquired from the 1 × 1 μm^2 area involving the NW, and monochromatic CL images at 342 nm and 423 nm. Enlarged images of the area delineated by the green square in the CL image are shown at the bottom-right for 342 nm and 423 nm. (b) CL of another $GaAs/AlGaO_x$ NW fabricated from the same sample: Back-scattered electron image, panchromatic CL image, CL spectrum acquired from the 1 × 1 μm^2 area involving the NW, monochromatic cathodoluminescence images at 347 nm, 728 nm, and 530 nm. Warmer colors in the color-imaged intensity plots indicate stronger intensity. Regions of lowest and highest (saturation) intensity are shown in black and white, respectively. Reprinted with permission from [42]. Copyright 2014 American Chemical Society.

Notably, the spectral features, especially the broad peak in the visible range, vary among the NWs, being strongest at approximately 530 nm (a green wavelength) in some cases. Figure 8.15b illustrates the cathodoluminescence characteristics of another NW obtained from the same sample as Fig. 8.12. This NW exhibits strongest emission intensity in the green region of the visible spectrum (around 530 nm).

On the other hand, the GaAs peak around 850 nm was not observed for all the NWs in Figs. 8.15a,b. Negligible GaAs luminescence is expected from the weak nature of the unpassivated 100 nm GaAs core [37]. In the present study, the passivation effect is inhibited by the nonoptimized structure and oxidation conditions. Instead, the optical quality of the core GaAs can be degraded by the irradiation of oxygen or water vapor during the oxidation process, corresponding to the GaAs core/$AlGaO_x$ shell interface in our case. The luminescence efficiency of the core GaAs could be improved by additional passivation such as an AlGaAs multi-shell passivation layer with low Al concentration [37]. Optimizing the luminescence efficiency of the proposed GaAs-oxide composite system is beyond the scope of this study, but combining the optical characteristics of oxides and the compound semiconductor GaAs could be considered in future work.

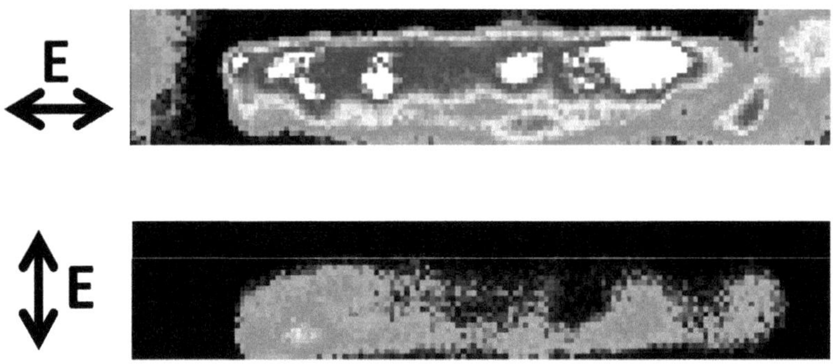

Figure 8.16 Polarization dependent CL image for a GaAs/$AlGaO_x$ NW measured at wavelength 450 nm. Warmer colors in the color-imaged intensity plots indicate stronger intensity.

Figure 8.16 shows polarization dependent CL images for a GaAs/$AlGaO_x$ NW measured at a wavelength of 450 nm. The measurement was carried out using a polarized CL system [74]. As seen in the figure, the light is strongly polarized along the length directions. Hence, the wire may have applications to polarized light sources.

8.4.2 Microwire Ensembles

In this section, we report on the GaAs/AlGaO$_x$ wire ensembles having 1 mm diameter with a large surface coverage of 40%, which were shown in Figs. 8.2b, 8.3b, 8.5, and 8.6. We investigate the mechanisms at the origin of their light emission in the visible range.

The structural characteristics of these samples were investigated using SEM. Cathodoluminescence (CL) measurements were performed at 8 K with a Zeiss ULTRA55 field-emission SEM equipped with a Gatan monoCL4 system using a photomultiplier for detection. The samples on Si substrates were mounted on a tilted sample stage before being introduced into the SEM chamber. For the acquisition of both CL spectra and CL images, the acceleration voltage of the electron beam was set to 5 kV and the beam current to about 1 nA. Continuous-wave μ-photoluminescence (μ-PL) experiments were carried out using a HeCd laser for excitation (λ = 325 nm). The laser beam with a power of about 1 mW was focused on a 1 μm-diameter spot on the sample surface using a microscope objective with a numerical aperture of 0.65. The emitted light was collected using the same objective and then directed to a monochromator (focal length 80 cm, 600 lines per mm grating) followed by a charge-coupled device (CCD) camera. The samples were mounted on a cold finger cryostat to measure PL at 10 and 300 K.

Figures 8.17a,b show bird's-eye view SEM and panchromatic CL images, respectively, taken at 8 K on an ensemble of GaAs/AlGaO$_x$ NWs with a diameter of 1 μm and a surface coverage of 40%. The one-to-one correspondence between the SEM and panchromatic CL images in Figs. 8.17a,b demonstrates that all GaAs/AlGaO$_x$ NWs emit light. To determine whether this signal originates from the GaAs core or from the AlGaO$_x$ shells, we show in Fig. 8.17c a CL spectrum taken at 8 K on the NW ensemble shown in Figs. 8.17a,b. The light emission extends from the near-infrared (1.5 eV) to the ultraviolet range (4 eV) and is the strongest at energies of about 2.6 eV. This broad emission arises probably from deep centers, as previously reported [42, 43]. No emission peak at about 1.51, 1.52 eV associated with the GaAs core could be resolved, a finding presumably resulting from the degradation of the optical efficiency of the GaAs core by the oxidation process. The $AlGaO_x$ emission is strongly quenched with increasing temperature, and no CL signal is detected at room temperature.

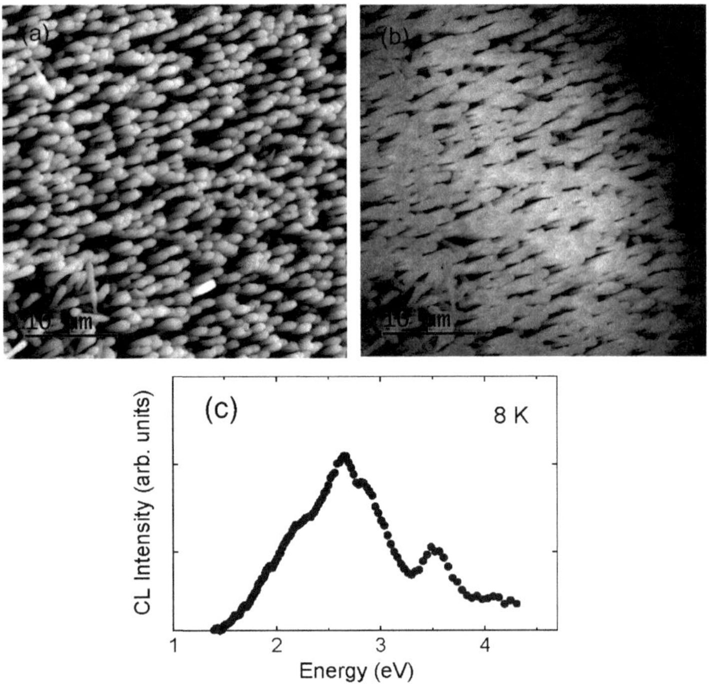

Figure 8.17 Tilted (a) SEM and (b) panchromatic CL images taken on an ensemble of $GaAs/AlGaO_x$ NWs at 8 K. The lower contrast at the right edge of (b) originates from the inhomogeneous photon collection efficiency of our CL system at large fields of view. (c) CL spectrum of the ensemble at 8 K. Reprinted with permission from [66]. Copyright 2016 WILEY-VCH Verlag GmbH & Co. KGaA.

Figures 8.18a,b show PL spectra measured at 10 and 300 K on the same $GaAs/AlGaO_x$ core/shell NW sample as in Figs. 8.2b, 8.3b, 8.5, and 8.6. PL spectra taken under the same conditions on the as-grown $GaAs/Al_{0.9}Ga_{0.1}As$ NWs prior to the wet oxidation treatment are also shown for comparison. The PL at 10 K from the as-grown NWs is dominated by a band centered at 1.502 eV that we attribute to the recombination of excitons in the GaAs core. This energy is about 13 meV lower than the value expected for the excitonic bandgap of strain-free GaAs, which may indicate that light emission occurs preferentially at polytypic NW segments [75, 76]. The weaker transitions detected between 1.6 and 2.4 eV may stem from exciton recombination at Ga-rich inclusions within the $Al_{0.9}Ga_{0.1}As$ shells [77]. With regard to the high Al content in the shells, the latter may have naturally oxidized, and some of the transitions in the 1.6 to 2.4 eV range can also originate from the deep centers observed in Fig. 8.17c for the NWs with intentionally oxidized shells.

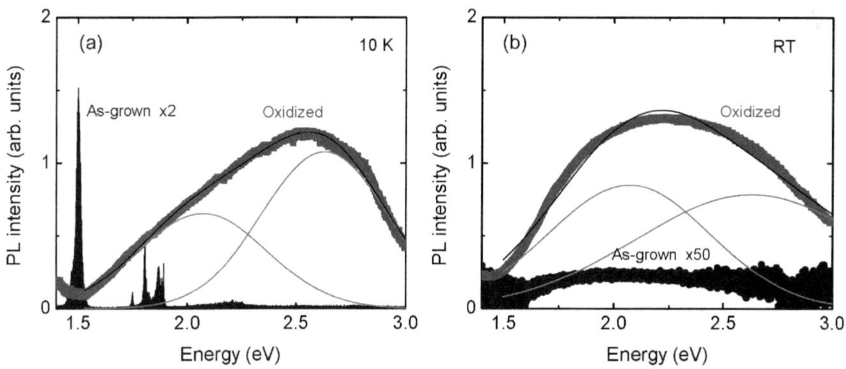

Figure 8.18 Photoluminescence spectra taken on ensembles of as-grown GaAs/(Al,Ga)As (blue) and oxidized GaAs/$AlGaO_x$ core–shell NWs (red) at (a) 10 K and (b) 300 K. The intensities in (a) and (b) are comparable. The black lines are the result of a fit with two Gaussians (grey lines) centered at 2.07 and 2.63 eV. Reprinted with permission from [66]. Copyright 2016 WILEY-VCH Verlag GmbH & Co. KGaA.

In agreement with the CL experiments in Fig. 8.17c, we do not detect any PL from the GaAs core for the oxidized NWs at 10 K. Instead, one detects a broad signal that covers the entire visible spectral range between 1.5 and 3 eV and that peaks at 2.55 eV, which we attribute to light emission from the $AlGaO_x$ shells. The PL from the $AlGaO_x$ shells presents an asymmetric line shape. A deconvolution of the spectrum using two Gaussians yields at least two bands centered at 2.07 and 2.63 eV, respectively.

Crystalline Al_2O_3 show numerous emission lines in the visible and ultraviolet ranges [78]. These lines have been attributed to F centers, which are related to the presence of anion vacancies. Amorphous AlO_x films also show PL in the visible and ultraviolet ranges [46, 79]. Although the observed emission bands are much broader than that reported for crystalline Al_2O_3 films, they are also thought to originate from F centers [79]. Consequently, we also ascribe the band centered at 2.07 and 2.63 eV to F centers in the $AlGaO_x$ shells of our NWs. However, in view of the significant broadening of these bands, it seems to be difficult to get precise knowledge on the microscopic origin of these centers based on PL experiments. In particular, the inhomogeneous compositions of the NW shells before and after selective oxidation (see the transitions between 1.6 and 2.4 eV in Fig. 8.18a) that are due to alloy fluctuations in the (Al,Ga)As shells lead to a broadening and/or a shift of the transitions related to F centers [46].

With increasing temperature from 10 to 300 K, the PL intensity from the GaAs core for the as-grown NW samples is quenched drastically. This decrease in the radiative efficiency between 10 and 300 K is due to a decrease in the nonradiative lifetime combined with an increase in the radiative lifetime, as discussed in Ref. 31. In contrast, the spectrally integrated PL intensity from the $AlGaO_x$ shells for the oxidized NWs increases with temperature and is two times more intense at 300 K than at 10 K (Fig. 8.18). An increasing PL intensity with increasing temperature has already been reported in Ref. 86 for Si-rich AlO_x nanoscale systems and has been ascribed to the thermal activation of additional radiative channels. Between 10 and 300 K, the PL peak energy shifts from 2.55 to 2.21 eV. As it is usually assumed that the PL energies of trap sites in AlO_x do not depend on temperature [79], we attribute this redshift to the fact that the increase in PL intensity for the band at 2.07 eV is larger than for the band at 2.63 eV. This observation agrees well with the temperature dependence reported for Si rich AlO_x layers [79].

The difference in the evolution of the luminescence intensity between 10 and 300 K in the CL and PL experiments is striking. While the PL intensity increases with increasing temperature (Figs. 8.18a,b), the CL intensity is strongly quenched, and no signal could be detected at 300 K. These two opposite behaviors are the consequence of the different nature of generated carriers in CL and PL experiments. On the one hand, as schematically depicted in Fig. 8.19, the high-energy electron irradiation is generated in the shell electrons and holes that can recombine radiatively, leading to light emission at the bandgap energy of $AlGaO_x$, or become trapped at color centers, giving rise to the bands at 2.07 and 2.63 eV. Alternatively, free electron–hole pairs can also diffuse to the surface or to the GaAs core of the NW, where they recombine nonradiatively. As shown in Ref. 84, for GaAs/(Al,Ga)As core–shell NWs, the charge-carrier transfer rate from the shell to the core increases with temperature. Since recombination in the core of GaAs/$AlGaO_x$ core/shell NWs is mostly nonradiative at room temperature (Fig. 8.18), the quenching of the CL signal with increasing temperature arises from the increased transfer rate of charge carriers from the shell to the GaAs core or to the NW sidewalls. On the other hand, laser excitation at 325 nm is likely to be below the bandgap of $AlGaO_x$ (Fig. 8.19). Generation of free electron–hole pairs in PL experiments can thus be excluded, and photogenerated carriers occupy exclusive sites such as the F centers, giving rise to the emission bands at 2.07 and 2.63 eV [12]. Charge carrier transfer from these sites to the core or to the surface is hindered, and the PL intensity shows a much weaker dependence on temperature, in agreement with our observations in Fig. 8.18.

In conclusion, the broad visible light emission observed for GaAs/$AlGaO_x$ core/shell NWs arises from recombination at F centers in the $AlGaO_x$ shells. At low temperatures, the emission from the F centers can be detected after resonant excitation in PL or above bandgap excitation in CL. While the PL intensity from the F centers slightly increases with increasing temperature, their CL intensity shows a significant quenching. The latter decrease is a result of the thermally activated transfer of carriers from the $AlGaO_x$ shell to the GaAs core and to the NW sidewalls, where they recombine nonradiatively. Although the broad emission from the F centers is promising for white light emission, the reduction of the density of nonradiative defects in the shell and of the transfer efficiency of carriers to the GaAs core are prerequisite to the realization of efficient white light emitters based on GaAs/$AlGaO_x$ core/shell NWs.

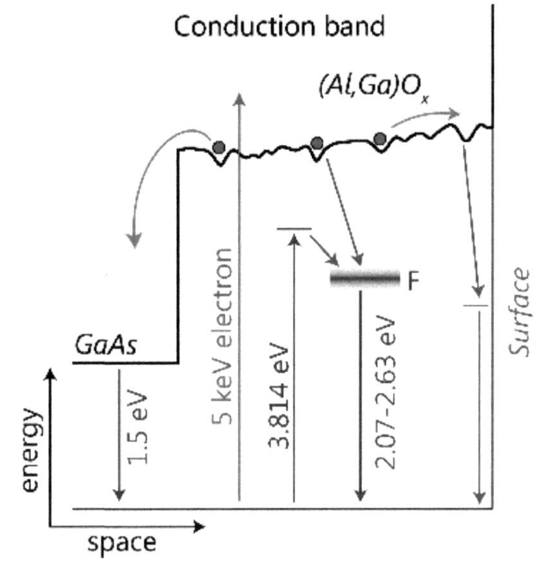

Figure 8.19 Schematic of the carrier generation and recombination process in GaAs/AlGaO$_x$ core/shell NWs. The thick black line shows the bottom of the conduction band. Electron irradiation in CL experiments generates free carriers that can bind to F centers, giving rise to the emission bands in the visible (blue arrows), or transfer to the core or the surface or the NW (green arrows) where they recombine. With increasing temperature, the transfer of carriers to the core or the surface is facilitated. In PL, excitation at 325 nm generates carriers directly at F centers. Reprinted with permission from [66]. Copyright 2016 WILEY-VCH Verlag GmbH & Co. KGaA.

8.5 Summary

We have proposed a composite NW material system consisting of compound semiconductors and oxides. The NWs were synthesized by molecular beam epitaxial growth and subsequent wet oxidation. Selective oxidation applied to the GaAs/AlGaAs core–shell NWs grown on silicon substrates produced GaAs/AlGaO$_x$ NWs. Nonselective oxidation conditions produced GaO$_x$/AlGaO$_x$ NWs. The oxidized amorphous AlGaO$_x$ shell produced luminescence related to oxygen over the entire visible spectrum. The wire core became buried in the shells when the diameter of the shell increased. The micro scale wire ensemble exhibited luminescence from the (Al,Ga)O$_x$ shells, found to be spatially homogeneous at low temperatures. With increasing temperature, the cathodoluminescence intensity strongly decreases, while the photoluminescence intensity is observed to increase as the temperature increases from 10 to 300 K. Together with the energy of the luminescence bands, these opposing dependencies demonstrate that the emission stems from F centers in the $(Al,Ga)O_x$ shells. These μ-wires are promising structures for obtaining white light emission on a Si platform. To realize efficient white light emitters, it is essential to prevent the transfer of carriers to the GaAs core and the shell surface. The system would provide extended functions for both semiconductor and oxide-materials systems, possibly utilized for light emitters, as well as solar cells in which we can utilize their absorbance. Further, through the combination of other compound semiconductors and oxides, the system could be extended to applications such as photo-catalysis and thermoelectric systems.

Acknowledgments

We thank Pierre Corfdir, Uwe Jahn, and Oliver Brandt for their invaluable discussions and the photoluminescence and cathodoluminescence investigations on the wire ensembles, and Goh Yoshida and Satoshi Shimomura for X-ray diffraction measurements as well as the collaboration from Masato Morifuji, Yoshiaki Araki, Hiroaki Goto, Namsoo Ahn, Hideaki Hibi, Yoshihiko Akamatsu, Kohei Nishioka, and Kyohei Takada for the series of epitaxial growth and nanowire study. This work was partly supported by KAKENHI (No. 16H05970 and 23686004) from the Japan Society for the Promotion of Science, a research scholarship from the Kato Foundation for Promotion of Science, the Kurata Memorial Hitachi Science and Technology Foundation, and the Murata Science Foundation.

References

1. Yang, P., Yan, R., and Fardy, M. (2010). Semiconductor nanowire: What's next?, *Nano Lett.*, **10**, pp. 1529–1536.
2. Mieszawska, A. J., Jalilian, R., Sumanasekera, G. U., and Zamborini, F. P. (2007). The synthesis and fabrication of one-dimensional nanoscale heterojunctions, *Small,* **3**, pp. 722–756.

3. Thelander, C., Agarwal, P., Brongersma, S., Eymery, J., Feiner, L., Forchel, A., Scheffler, M., Riess, W., Ohlsson, B., Gosele, U., and Samuelson, L. (2006). Nanowire-based one-dimensional electronics, *Mater. Today*, **9**, pp. 28–35.
4. Gudiksen, M. S., Lauhon, L. J., Wang, J., Smith, D. C., and Lieber, C. M. (2002). Growth of nanowire superlattice structures for nanoscale photonics and electronics, *Nature*, **415**, pp. 617–620.
5. Ozel, T., Bourret, G. R., Schmucker, A. L., Brown, K. A., and Mirkin, C. A. (2013). Hybrid semiconductor core-shell nanowires with tunable plasmonic nanoantennas, *Adv. Mater.*, **25**, pp. 4515–4520.
6. Xia, X., Tu, J., Zhang, Y., Wang, X., Gu, C., Bing Zhao, X., and Fan, H. J. (2012). High-quality metal oxide core/shell nanowire arrays on conductive substrates for electrochemical energy storage, *ACS Nano*, **6**, pp. 5531–5538.
7. Gomathi, A., Vivekchand, S. R. C., Govindaraj, A., and Rao, C. N. R. (2005). Chemically bonded ceramic oxide coatings on carbon nanotubes and inorganic nanowires, *Adv. Mater.*, **17**, pp. 2757–2761.
8. Kolmakov, A., and Moskovits, M. (2004). Chemical sensing and catalysts by one-dimensional metal-oxide nanostructures, *Ann. Rev. Mater. Res.*, **34**, pp. 151–180.
9. Sze, S. M. (1981). *Physics of Semiconductor Devices* (John Wiley & Sons, New York).
10. Zhuang, X., Ning, C. Z., and Pan, A. (2012). Composition and bandgap-graded semiconductor alloy nanowires, *Adv. Mater.*, **24**, pp. 13–33.
11. Dillen, D. C., Kim, K., Liu, E.-S., and Tutuc, E. (2014). Radial modulation doping in core-shell nanowires, *Nat. Nanotech.*, **9**, pp. 116–120.
12. Yao, J., Yan, H., and Lieber, C. M. (2013). A nanoscale combing technique for the large-scale assembly of highly aligned nanowires, *Nat. Nanotech.*, **8**, pp. 329–335.
13. Tomioka, K., Yoshimura, M., and Fukui, T. (2012). A III–V nanowire channel on silicon for high-performance vertical transistors, *Nature*, **488**, pp. 189–192.
14. Chen, R., Tran, T. -T. D., Ng, K. W., Ko, W. S., Chuang, L. C, Sedgwick, F. G., and Chang-Hasnain, C. (2011). Nanolasers grown on silicon, *Nat. Photonics*, **5**, pp. 170–175.
15. Mårtensson, T., Svensson, C. P. T., Wascaser, B. A., Larsson, M. W., Seifert, W., Deppert, K., Gustafsson, A., Wallenberg, L. R., and

Samuelson, L. (2004). Epitaxial III–V nanowires on silicon, *Nano Lett.*, **4**, pp. 1987–1980.

16. Hersee, S. D., Sun, X., and Wang, X. (2006). The controlled growth of GaN nanowires, *Nano Lett.*, **6**, pp. 1808–1811.

17. Tomioka, K., Motohisa, J., Hara, S., Hiruma, K., Fukui, T. (2010). GaAs/AlGaAs core multishell nanowire-based light-emitting diodes on Si, *Nano Lett.*, **10**, pp. 1639–1644.

18. Bjork, M. T., Thelander, C., Hansen, A. E., Jensen, L. E., Larsson, M. W., Wallenberg, L. R., and Samuelson, L. (2004). Few-electron quantum dots in nanowires, *Nano Lett.*, **4**, pp. 1621–1625.

19. Lauhon, L. J., Gudiksen, M. S., Wang, D., and Lieber, C. M. (2002). Epitaxial core–shell and core–multishell nanowire heterostructures, *Nature*, **420**, pp. 57–61.

20. Kamiya, T., and Hosono, H. (2010). Material characteristics and applications of transparent amorphous oxide semiconductors, *NPG Asia Mater.*, **2**, pp. 15–22.

21. Nomura, K., Ohta, H., Takagi, A., Kamiya, T., Hirano, M., and Hosono, H. (2004). Room-temperature fabrication of transparent flexible thin-film transistors using amorphous oxide semiconductors, *Nature*, **432**, pp. 488–492.

22. Wang, Z. L., and Song, J. (2006). Piezoelectric nanogenerators based on zinc oxide nanowire arrays, *Science*, **312**, pp. 242–246.

23. Choi, Y. C., Kim, W. S., Park, Y. S., Lee, S. M., Bae, D. J., Lee, Y. H., Park, G. -S., Choi, W. B., Lee, N. S., and Kim, J. M. (2000). Catalytic growth of β-Ga_2O_3 nanowires by arc discharge, *Adv. Mater.*, **12**, pp. 746–750.

24. Maximenko, S. I., Mazeina, L., Picard, Y. N., Freitas, Jr., J. A., Bermudez, V. M., and Prokes, S. M. (2009). Cathodoluminescence studies of the inhomogeneities in Sn-doped Ga_2O_3 nanowires, *Nano Lett.*, **9**, pp. 3245–3251.

25. Chen, P.-H., Hsieh, C.-H., Chen, S.-Y, Wu, C.-H., Wu, Y.-J., Chou, L.-J., and Chen, L.-J. (2010). Direct observation of Au/Ga_2O_3 peapodded nanowires and their plasmonic behaviors, *Nano Lett.*, **10**, pp. 3267–3271.

26. Lin, J., Huang, Y., Bando, Y., Tang, C., Li, C., and Golberg, D. (2010). Synthesis of In_2O_3 nanowire-decorated Ga_2O_3 nanobelt heterostructures and their electrical and field-emission properties, *ACS Nano*, **4**, pp. 2452–2458.

27. Richters, J.-P., Voss, T., Kim, D. S., Scholz, R., and Zacharias, M. (2008). Enhanced surface-excitonic emission in ZnO/Al_2O_3 core-shell nanowires, *Nanotechnology*, **19**, p. 305202.

28. Choquette, K. D., Geib, K. M., Ashby, C. I. H., Twesten, R. D., Blum, O., Hou, H. Q., Follstaedt, D. M., Hammons, B. E., Mathes, D., and Hull, R. (1997). Advances in selective wet oxidation of AlGaAs alloys, *IEEE Sel. Top. Quantum Electron.*, **3**, pp. 916–926.

29. Lear, K., Choquette, K., Schneider, R., Kilcoyne, S., and Geib, K. (1995). Selectively oxidised vertical cavity surface emitting lasers with 50% power conversion efficiency, *Electron. Lett.*, **31**, pp. 208–209.

30. Huffaker, D. L., Deppe, D. G., Kumar, K., and Rogers, T. J. (1994). Native oxide defined ring contact for low threshold vertical cavity lasers, *Appl. Phys. Lett.*, 1994, **65**, pp. 97–99.

31. Lou, Y., and Hall, D. C. (2008). Low-loss nonselectively oxidized $AlxGa_{1-x}As$ heterostructure waveguides, *Appl. Phys. Lett.*, **93**, pp. 261111-1-3.

32. Luo, Y., and Hall, D. C. (2005). Nonselective Wet Oxidation of AlGaAs Heterostructure Waveguides Through Controlled Addition of Oxygen, *IEEE J. Selec. Top. Quant. Electron.*, **11**, p. 1284.

33. Hirai, Y., Yamada, T., Kondow, M., and Ishikawa, F. (2012). Characterization of the oxide film obtained by wet oxidation of Al-rich AlGaAs, *Jpn. J. Appl. Phys.*, **51**, pp. 02BG10-1-4.

34. Wallentin, J., Ek, M., Vainorious, N., Mergenthaler, K., Samuelson, L., Pistol, M. -E., Wal-lenberg, L. R., and Borgstrom, M. T. (2013). Semiconductor-oxide heterostructured nanowires using postgrowth oxidation, *Nano Lett.*, **13**, pp. 5961–5966.

35. Li, J., An, L., Lu, C., and Liu, J. (2006). Conversion between Hexagonal GaN and β-Ga_2O_3 nanowires and their electrical transport properties, *Nano Lett.*, **6**, pp. 148–152.

36. Dai, X., Zhang, S., Wang, Z., Adamo, G., Liu, H., Huang, Y., Couteau, C., and Soci, C. (2014). GaAs/AlGaAs nanowire photodetector, *Nano Lett.*, **14**, pp. 2688–2693.

37. Mayer, B., Rudolph, D., Schnell, J., Morkotter, S., Winnerl, J., Treu, J., Muller, K., Bracher, G., Abstreiter, G., Koblmüller, G., and Finley, J. J. (2013). Lasing from individual GaAs-AlGaAs core-shell nanowires up to room temperature, *Nat. Commun.*, **4**, pp. 2931-1-7.

38. Zheng, C., Wong-Leung, J., Gao, Q., Tan, H. H., Jagadish, C., and Etheridge, J. (2013). Polarity-driven 3-fold symmetry of GaAs/AlGaAs core multishell nanowires, *Nano Lett.*, **13**, pp. 3742–3748.

39. Funk, S., Li, A., Ercolani, D., Gemmi, M., Sorba, L., and Zardo, I. (2013). Crystal phase induced bandgap modifications in AlAs nanowires probed by resonant Raman spectroscopy, *ACS Nano*, **7**, pp. 1400–1407.

40. Fickenscher, M., Shi, T., Jackson, H. E., Smith, L. M., Yarrison-Rice, J. M., Zheng, C., Miller, R., Etheridge, J., Wong, B. M., Gao, Q., Deshpande, S., Tan, H. H., and Jagadish, C. (2013). Optical, structural, and numerical investigations of GaAs/AlGaAs core–multishell nanowire quantum well tubes, *Nano Lett.*, **13**, pp. 1016–1022.

41. Li, A., Ercolani, D., Lugani, L., Nasi, L., Rossi, R., Salviati, G., Beltram, R., and Sorba, L. (2011). Synthesis of AlAs and AlAs–GaAs core–shell nanowires, *Cryst. Growth Des.*, **11**, pp. 4053–4058.

42. Hibi, H., Yamaguchi, M., Yamamoto, N., and Ishikawa, F. (2014). Selective synthesis of compound semiconductor/oxide composite nanowires, *Nano Lett.,* **14**, pp. 7024–7030.

43. Anjiki, A., and Uchino, T. (2012). Visible photoluminescence from photoinduced molecular species in nanometer-sized oxides: Crystalline Al_2O_3 and amorphous SiO_2 nanoparticles, *J. Phys. Chem. C*, **116**, pp. 15747–15755.

44. Chen, Y.-T., Cheng, C.-L., and Chen, Y.-F. (2008). Giant white and blue light emission from Al_2O_3 and ZnO nanocomposites, *Nanotechnology*, **19**, p. 445707.

45. Yu, Z. Q., Chang, D., Cun, Li, Zhang, N., Feng, Y. Y., and Dai, Y. Y. (2011). Blue photoluminescent properties of pure nanostructured γ-Al_2O_3, *J. Mater. Res.,* **16**, pp. 1890–1893.

46. Ortiz, A., Alonso, J. C., Pankov, V., and Albarran, D. (1999). Violet-blue photoluminescence in aluminium oxide films prepared by ultrasonic spray pyrolysis, *J. Lumin.*, **81**, p. 45.

47. Tomioka, K., Motohisa, J., Hara, S., Hiruma, K., and Fukui, T. (2010). GaAs/AlGaAs core multishell nanowire-based light-emitting diodes on Si, *Nano Lett.*, **10**, p. 1639.

48. Elaine, L., Woong, K., and Yang, P. (2008). Vertical nanowire array-based light emitting diodes, *Nano Res.*, **1**, p. 123.

49. Claudon, J., Bleuse, J., Malik, N. S., Bazin, M., Jaffrennou, P., Gregersen, N., Sauvan, C., Lalanne, P., and Gérard, J.-M. (2010). A highly efficient single-photon source based on a quantum dot in a photonic nanowire, *Nat. Photon.*, **4**, pp. 174–177.

50. Wallentin, J., Anttu, N., Asoli, D., Huffman, M., Åberg, I., Magnusson, M. H., Siefer, G., Fuss-Kailuweit, P., Dimroth, F., Witzigmann, B., Xu, H. Q., Samuelson, L., Deppert, K., and Borgström, M. T. (2013). InP nanowire array solar cells achieving 13.8% efficiency by exceeding the ray optics limit, *Science*, **339**, pp. 1057–1060.

51. Patriarche, G., Glas, R., Tchernycheva, M., Sartel, C., Largeau, L., Harmand, J. –C., Cirlin, G. E. (2008). Wurtzite to zinc blende phase

transition in GaAs nanowires induced by epitaxial burying, *Nano Lett.*, **8**, pp. 1638–1643.

52. Mascoli, D., Gerace, D., Andreani, L. (2011). Q-factor optimization for TM-like modes in pillar-based photonic crystal cavities with planar slot waveguides, *Photon. Nanostr. Fundam. Appl.*, **9**, p. 63.

53. Paek, J. H., Nishiwaki, T., Yamaguchi, M., and Sawaki, N. (2009). Catalyst free MBE-VLS growth of GaAs nanowires on (111)Si substrate, *Phys. Stat. Solidi (c)*, **6**, pp. 1436–1440.

54. Araki, Y., Yamaguchi, M., Ishikawa, F. (2013). Growth of dilute nitride GaAsN/GaAs heterostructure nanowires on Si substrates, *Nanotechnology*, **24**, p. 065601.

55. Ahn, N., Araki, Y., Kondow, M., Yamaguchi, M., and Ishikawa, F. (2014). Effects of growth interruption, As and Ga fluxes, and nitrogen plasma irradiation on the molecular beam epitaxial growth of GaAs/GaAsN core-shell nanowires on Si(111), *Jpn. J. Appl. Phys.*, **53**, p. 065001.

56. Fontcuberta i Morral, A., Colombo, C., Abstreiter, G., Arbiol, J., and Morante, J. R. (2008). Nucleation mechanism of gallium-assisted molecular beam epitaxy growth of gallium arsenide nanowires, *Appl. Phys. Lett.*, **92**, p. 063112.

57. Zhang, J., Neave, J. H., Dobson, P. J., and Joyce, B. A. (1987). Effects of diffraction conditions and processes on rheed intensity oscillations during the MBE growth of GaAs, *Appl. Phys. A*, **42**, pp. 317–326.

58. Däweritz, L., and Hey, R. (1990). Reconstruction and defect structure of vicinal GaAs(001) and $Al_xGa_{1-x}As$(001) surfaces during MBE growth, *Surf. Sci.*, **236**, pp. 15–22.

59. Honda, M., and Yamamoto, N. (2014). High-Q band edge mode of plasmonic crystals studied by cathodoluminescence, *Appl. Phys. Lett.*, **104**, p. 081112.

60. Watanabe, H., Honda, M., and Yamamoto, N. (2014). Size dependence of band-gaps in a one-dimensional plasmonic crystal, *Opt. Express*, **22**, pp. 5155–5165.

61. Takeuchi, K., and Yamamoto, N. (2011). Visualization of surface plasmon polariton waves in two-dimensional plasmonic crystal by cathodoluminescence, *Opt. Express*, **19**, pp. 12365–12374.

62. Mandl, B., Stangl, J., Hilner, E., Zakharov, A. A., Hillerich, K., Dey, A. W., Samuelson, L., Bauer, G., Deppert, K., and Mikkelsen, A. (2010). Growth mechanism of self-catalyzed group III–V nanowires, *Nano Lett.*, **10**, pp. 4443–4449.

63. Plante, M. C., and LaPierre, R. R. (2008). Control of GaAs nanowire morphology and crystal structure, *Nanotechnology*, **19**, p. 495603.

64. Chen, C., Plante, M., Fradin, C., and LaPierre, R. J. (2006). Layer-by-layer and step-flow growth mechanisms in GaAsP/GaP nanowire heterostructures, *Mater. Res.*, **21**, pp. 2801–2809.

65. Colombo, C., Spirkoska, D., Frimmer, M., Abstreiter, G., and Fontcuberta i Morral, A. (2008). Ga-assisted catalyst-free growth mechanism of GaAs nanowires by molecular beam epitaxy, *Phys. Rev. B*, **77**, p. 155326.

66. Ishikawa, F., Corfdir, P., Jahn, U., Brandt, O. (2016). (Al,Ga)Ox microwire ensembles on Si exhibiting luminescence over the entire visible wavelength range, *Adv. Opt. Mater.*, DOI: 10.1002/adom.201600695.

67. Brandt, O., Fernández-Garrido, S., Zettler, J. K., Luna, E., Jahn, U., Chèze, C., and Kaganer, V. M. (2014). Statistical analysis of the shape of one-dimensional nanostructures: Determining the coalescence degree of spontaneously formed GaN nanowires, *Cryst. Growth Des.*, **14**, pp. 2246–2253.

68. Sartel, C., Dheeraj, D. L., Jabeen, F, J. C., and Harmand, J. (2010). Effect of arsenic species on the kinetics of GaAs nanowires growth by molecular beam epitaxy, *J. Cryst. Growth*, **14**, pp. 2073–2077.

69. Consonni, V., Knelangen, M., Jahn, U., Trampert, A., Geelhaar, L., and Riechert, H. (2009). Effects of nanowire coalescence on their structural and optical properties on a local scale, *Appl. Phys. Lett.*, **95**, p. 241910.

70. Grossklaus, K. A., Banerjee, A., Jahangir, S., Bhattacharya, P., and Millunchick, J. M. (2013). Misorientation defects in coalesced self-catalyzed GaN nanowires, *J. Cryst. Growth*, **371**, pp. 142–147.

71. Jenichen, B., Brandt, O., Pfüller, C., Dogan, P., Knelangen, M., Trampert, A. (2011). Macro- and micro-strain in GaN nanowires on Si(111), *Nanotechnology*, **22**, p. 295714.

72. Sarzala, R. P., and Nakwaski, W. (2006). Separate-confinement-oxidation vertical-cavity surface-emitting laser structure, *J. Appl. Phys.*, **99**, p. 123110.

73. Birowosuto, M. D., Yokoo, A., Zhang, G., Tateno, K., Kuramochi, E., Taniyama, H., Takiguchi, M., and Notomi, M. (2014). Movable high Q nanoresonators realized by semiconductor nanowires on a Si photonic crystal platform, *Nat. Mater.*, **13**, pp. 279–285.

74. Yamamoto, N., García de Abajo, F. J., Myroshnychenko, V. (2015). Interference of surface plasmons and Smith-Purcell emission probed

by angle-resolved cathodoluminescence spectroscopy, *Phys. Rev. B*, **91**, p. 125144.

75. Jahn, U., Lähnemann, J., Pfüller, C., Brandt, O., Breuer, S., Jenichen, B., Ramsteiner, M., Geelhaar, L., and Riechert, H. (2012). Luminescence of GaAs nanowires consisting of wurtzite and zinc-blende segments, *Phys. Rev. B*, **85**, p. 045323.
76. Graham, A. M., Corfdir, P., Heiss, M., Conesa-Boj, S., Uccelli, E., Fontcuberta i Morral, A., and Phillips, R. T. (2013). *Phys. Rev. B*, **87**, p. 125304.
77. Corfdir, P., Küpers, H., Lewis, R. B., Flissikowski, T., Grahn, H. T., Geelhaar, L., and Brandt, O. (2016). Exciton dynamics in GaAs/(Al,Ga)As core-shell nanowires with shell quantum dots, *Phys. Rev. B*, **94**, p. 155413, arXiv:1603.01111.
78. Evans, B. D., Pogatshnik, G. J., and Chen, Y. (1994). Optical properties of lattice defects in α-Al_2O_3, *Nucl. Instr. Meth. B*, **91**, pp. 258–262.
79. Vergara Hernandez, E., Torchynska, T. V., Jedrzejewski, J., and Balberg,I. (2014). Transformation of photoluminescence and Raman scattering spectra of Si-rich Al_2O_3 films at thermal annealing, *Phys. B*, **453**, pp. 107–110.

Chapter 9

GaAs/$SrTiO_3$ Core–Shell Nanowires

Xin Guan and José Penuelas

Université de Lyon,
Institut des Nanotechnologies de Lyon (UMR5270/CNRS),
Ecole Centrale de Lyon, 36 avenue Guy de Collongue, Ecully, 69134, France

jose.penuelas@ec-lyon.fr

The growth of $SrTiO_3$ on self-catalyzed GaAs nanowires was studied. Prior to the growth of the $SrTiO_3$ shell, the GaAs nanowires were protected using an arsenic capping–decapping procedure in order to prevent the uncontrolled oxidation and the contamination of the nanowire surface during the transfer from the III–V to the oxide molecular beam epitaxy reactor. To validate the efficiency of this method, the surface chemistry of GaAs nanowires was thoroughly studied. $SrTiO_3$ shell was then constructed via a two-step growth method, showing most of the perovskite structure is oriented with respect to the GaAs lattice. A detailed discussion of the morphological, chemical and structural properties of such core–shell nanowires can be found in this chapter.

9.1 Introduction

The growth of semiconductor nanowires (NWs) has reached a certain maturity level and many devices for electronics, photonics,

Novel Compound Semiconductor Nanowires: Materials, Devices, and Applications
Edited by Fumitaro Ishikawa and Irina Buyanova

ISBN 978-981-4745-76-5 (Hardcover), 978-1-315-36440-7 (eBook)
www.panstanford.com

and biology have been made in recent years. In order to modify or improve the functional properties of these nano-objects, coupling heterogeneous materials with complementary features is promising. Different morphologies can be considered: axial junction, radial junction (core–shell NWs) and NWs with nanoparticles anchored on their facets.

Compared to a classic stacking of thin film, the NW geometry has specificities that make these hybrid structures valuable:

- The cost is lowered for reducing the amount of material needed for the manufacturing of certain devices.
- Due to their small diameter, it is possible to produce NWs relaxed and without defects on substrates with high lattice mismatch.
- The support exhibits a high specific surface area, which is of strong interest in the field of catalysis for instance.
- NWs are relatively flexible and hence their integration into flexible substrate is relevant; otherwise this flexibility enables nanoelectromechanical (NEMS) type applications or any other piezoelectric device.

Despite the strong interest of hybrid core–shell NWs, only few studies reported the growth of such heterostructures. Most of the reported GaAs-based core/shell NWs are made of similar III–V semiconductor shell like AlGaAs [1, 2], GaInP [3, 4], etc., which are easily grown but cannot meet the need for multifunctional devices. Integration of heterogeneous crystalline materials such as nitride [5], silicon [6, 7], silicide [8, 9], or metals [10] on GaAs NWs has been achieved recently. Owing to their wide-ranging properties that are complementary to those of semiconductors, functional oxides with perovskite structure are promising. The epitaxial integration of heterogeneous shell materials, especially functional oxides, on the GaAs NW core could lead to the fabrication of multifunctional devices by coupling III–V semiconductor properties with oxides properties, such as ferroelectricity [11, 12] and piezoelectricity [13]. However, achieving an epitaxial growth in semiconductor/perovskite oxide heterostructures is challenging due to the rapid oxidation of classical semiconductors such as Si or III–V when exposed to air or O_2 partial pressure [14, 15]. The GaAs NWs oxidation can occur during two critical steps: the transfer from the III–V reactor to the oxide reactor and the beginning of the oxide growth when the pristine GaAs surface is exposed to a partial pressure of O_2. To prevent such phenomenon, two strategies have been used: the reversible As-cap method and the two-step $SrTiO_3$ growth.

An uncontrolled oxidation and contamination of the III–V NW facets during the transfer is inevitable, since most of the III–V epitaxy reactors do not permit sample transfer under ultrahigh vacuum (UHV) conditions between separate reactors for growing most of metals, oxides, or semiconductors. Wet chemical treatment [16] and atomic hydrogen [7, 17] etching are usually applied to obtain clean GaAs NW facets before implementing the growth of shell. However, such strategies have several drawbacks like introducing contaminations from solvents and surface modifications caused by etching. To preserve the surface structure and chemistry of NWs, a contamination-free and reversible surface protection such as reversible arsenic (As)-cap method [18] is preferable.

To prevent the oxidation during the beginning of the $SrTiO_3$ growth, Mc Kee et al. [19] reported that surface passivation of the Si substrate with half a monolayer of Sr allows the growth of $SrTiO_3$ thin films by molecular beam epitaxy (MBE). More recently, $SrTiO_3$ was successfully grown on GaAs substrates using similar strategy [20, 21]. Integration of $SrTiO_3$ on the facets of GaAs NWs would pave the way for the fabrication of semiconductor (core)/functional oxides (shell) NWs and to the exploration of their properties.

9.2 The Epitaxial Growth of GaAs NWs on Si (111) Substrate

NWs were grown on n-doped Si (111) substrates by MBE. In order to remove surface contaminations, the epi-ready Si substrates were ultrasonic cleaned for 5 min in ethanol and acetone, respectively, before being introduced into UHV, while the native oxides were conserved. The substrate was then outgassed at 200°C under UHV for a few minutes. GaAs NWs were grown in a MBE reactor dedicated to III–V semiconductors equipped with a Ga Knudsen cell and a valved cracked As source. The substrate temperature was controlled by a calibrated thermocouple and checked by a pyrometer. During growth, the sample holder was continuously rotated to enhance growth homogeneity. The MBE reactor is equipped with reflection high energy electron diffraction (RHEED). Self-catalyzed GaAs NWs were grown by the vapor–liquid–solid (VLS) method using Ga droplets as catalyst [22–26]. Firstly, the Si substrate temperature was increased to 530°C and one nominal monolayer of Ga was deposited at a deposition rate of 1.4 Å/s on the Si substrate. At this temperature, the deposition leads to the formation of Ga liquid droplets on the native-oxides-covered surface. These droplets locally decompose the native oxide and create pinholes [25] via either of following equations [27, 28]:

$$SiO_2 + 4Ga \rightarrow Si + 2Ga_2O \tag{9.1}$$

$$Ga + SiO_2 \rightarrow Ga{:}Si + O_2 \tag{9.2}$$

Ga atom can easily accumulate in the defect site and then ensure the VLS NWs growth. Afterwards, the substrate temperature was increased up to 610°C to initiate the growth of GaAs NWs. The Ga deposition rate was 2.1 Å/s in units of equivalent 2D GaAs growth rate as measured from RHEED oscillations during Ga-limited growth and the As_4 beam equivalent pressure was set at 3.3×10^{-6} Torr, which corresponds to a deposition rate of 3 Å/s as measured during As-limited growth [29]. The Ga shutter and the As valve were opened at the same time. After 15 min, the Ga shutter was closed and the temperature of sample holder was decreased down to room temperature under identical As_4 atmosphere in order to stop the axial growth of NW by crystallizing the Ga droplet as well as to avoid GaAs decomposition. A representative SEM image of the sample is shown in Fig. 9.1a. The NWs are vertical, with an approximate length of 1.0 μm. The NW density is about 7 NWs/μm^2. Nearly all the GaAs NWs have a uniform diameter of 50 nm with scarcely tapering effect. The Ga catalyst is no longer visible at the NW tip. GaAs nanocrystals are also observed between the NWs. The SEM top view image of the NWs shows that the NWs have perfect hexagonal cross section. The six facets can be indexed as {1–10} crystal planes, which is common for III–V NWs grown on (111) oriented substrates. Moreover, the lateral surfaces are systematically aligned with the in-plane Si{1–10} crystal plane. This morphology of the GaAs is in accordance with the following epitaxial relationship:

GaAs[111]//Si[111] and GaAs[1–10]//Si[1–10], in correspondence with XRD result (discussed later).

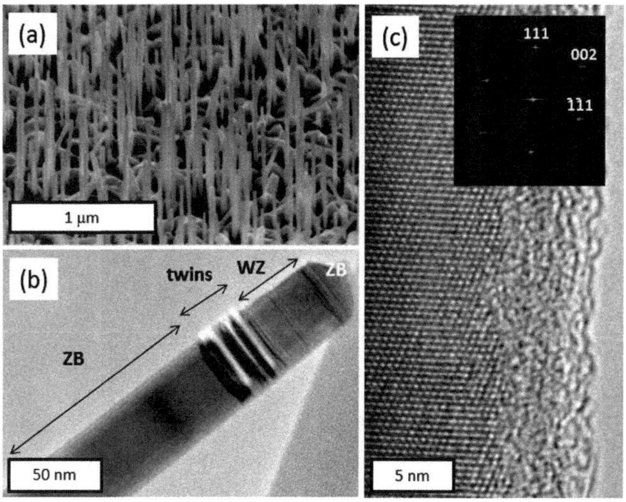

Figure 9.1 SEM of the GaAs NWs grown on Si(111) substrate (a). TEM images of a single GaAs NW showing its morphology and multi-twinned domains close to the NWs tip (b), lattice resolved TEM image of a GaAs NWs, with its Fast-Fourier transform (FFT) in inset (c). Reprinted with permission from [30]. Copyright (2016) American Chemical Society.

TEM images of GaAs NWs are shown in Fig. 9.1b,c. The GaAs NW in Fig. 9.1b is quite representative, which exhibits four different domains. From the bottom to the top, the first one is a zinc-blende (ZB) phase with very few structural defects. The second domain, about 30 nm in length, is located at approximate one hundred nm away from the NW tip and contains a large number of twining defects along the growth axis and several short wurtzite (WZ) segments. The following one is made of a pure WZ region of around 50 nm length. And finally the NW ends up to a very thin ZB structure corresponding to the consumption of the Ga catalyst droplet by As_4 flux during the cooling down procedure. Such a phase sequence has already been reported [31–33] and is supposed to be a consequence of an evolution of the Ga droplet shape during the VLS NW growth. Figure 9.1c shows a high-resolution TEM (HRTEM) image of a GaAs NW in the pure ZB region (the inset shows its FFT), the NW growth axis is [111] and the zone axis is [1–10]. The lattice parameter is about

5.62 Å, approximately equal to that of bulk GaAs (5.65 Å). A thin amorphous layer is observed on the surface of NW, which corresponds to the GaAs oxide formed by the uncontrollable oxidation during the transfer to the microscope. It would be critical for the growth of the epitaxial crystalline shell to avoid the formation of such amorphous oxide layer, which will be discussed later.

For getting more information on lattice matching, XRD measurement was carried out. Figure 9.2a shows the reciprocal space mapping of the GaAs NWs grown on Si(111), measured around the Si(111) reflection and the GaAs(111) reflection. The Si(111) peak is intense and narrow, typical for a substrate. Whilst the GaAs(111) peak located at 27.32° is narrow along the radial direction ($\theta/2\theta$) and larger along the rocking angle (omega) which is in accordance with slightly tilted epitaxial NWs. The GaAs (111) peak (Fig. 9.2b) is in good agreement with bulk GaAs, showing the full relaxation of the GaAs NWs. The lattice parameter is calculated via the Scherrer equation, and the average crystal size is 49 nm.

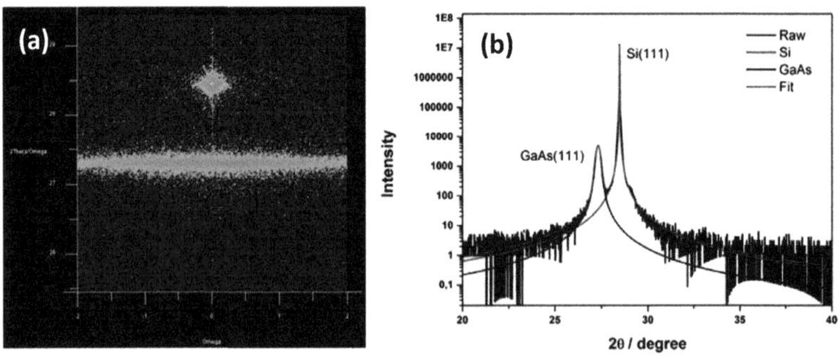

Figure 9.2 (a) The reciprocal space mapping and (b) $\theta/2\theta$ scan of GaAs NW array on Si (111) substrate. Reproduced from Ref. [34] with permission from The Royal Society of Chemistry.

9.3 The Reversible As Cap Method Against the Uncontrolled Oxidation of NW Surface

9.3.1 Surface Oxidation of GaAs and the Reversible As Cap Method

A relatively clean and abrupt interface is required as the purpose of constructing epitaxial core/shell NWs. As aforementioned, an amorphous layer of oxides is observed around the GaAs NW, which could be the obstacle of epitaxial shell growth and the further performance improvement. The study on the composition of oxidation of GaAs NW will be necessary for searching for an appropriate oxidation-proof method. Obviously, a contamination-free and reversible surface protection would be preferable to well preserve the NW surface structure and chemistry.

XPS measurements were performed for the analysis of GaAs surface oxides. To obtain a reference spectrum, the newly fabricated GaAs NW array was transferred directly to the XPS chamber under UHV, without exposure to air, which means no contamination or oxidation during the whole process. As shown in Fig. 9.3, row I, the As $3d_{5/2}$ peak is at 41.0 eV and the Ga $3d_{5/2}$ peak at 19.1 eV, which corresponds to the positions and the stoichiometry expected for bulk GaAs [35, 36]. The O 1s core level at 533 eV merely originates from the native SiO_x oxide on the surface of the Si (111) substrate. In order to observe the NW oxidation, two GaAs NW samples were kept in ambient air at room temperature (RT) without any other chemical or physical procedure, one for 24 h (Fig. 9.3, row II) and the other for as long as 1 month (Fig. 9.3, row III). After 24 h air-exposure, the As 3d core level (Fig. 9.3, row II) exhibits a supplementary doublet at 44.3 eV, which is attributed to As_2O_3 [35]. The Ga 3d core level is also modified and fitted with two more peaks at 20.75 and 20.30 eV, in accordance with the Ga $3d_{3/2}$ and $3d_{5/2}$ of Ga_2O_3, [37], respectively. The O 1s peak (Fig. 9.2c, row II) is broadened and asymmetrical due to the appearance of two other oxides, namely As_2O_3 and Ga_2O_3 [38]. When the exposure (to air) time was prolonged to 1 month, the intensity of As(As_2O_3) and Ga(Ga_2O_3) peaks is notably increased (Fig. 9.3b, row III). The intensity of O 1s core level, related to As_2O_3 and Ga_2O_3, increases as well in proportion to the air exposure time.

The thermal stability of the oxides was probed by measuring the Ga 3d and As 3d core levels during annealing inside the XPS analysis chamber. The sample appears to be stable under UHV at a temperature below 500°C. The As 3d intensity of As oxides abruptly decreases at 500°C, while a new component appears at around 41 eV related to GaAs. No significant evolution of the Ga oxide component is observed, which indicates a better stability of the Ga oxides compared to the As ones. This thermal stability difference between As and Ga oxides makes the simple heating treatment not suitable: considering the difficulty on the decomposition and evaporation speed control, it is hard to maintain the identical chemical structure of GaAs surface, except for the change on NWs' diameter and length. The reversible As-cap method is proposed for having been implemented to protect thin films [18, 39–42] and some one-dimensional NWs [43–46]. This method is clearly appropriate for the MBE facility.

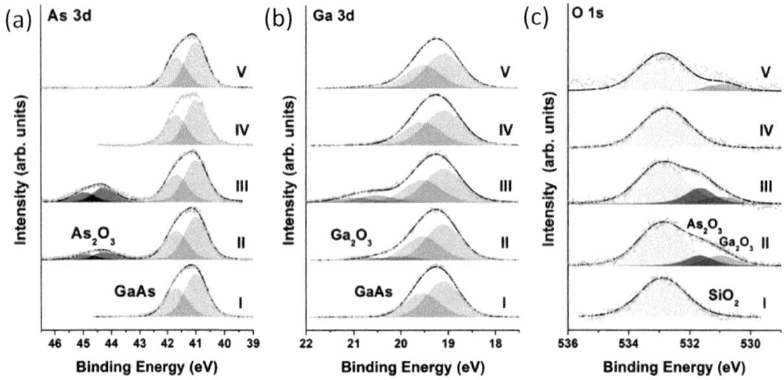

Figure 9.3 XPS spectra of As 3d core level (a), Ga 3d (b) and O 1s (c) of different NW arrays. I, as-grown GaAs NWs without contact with ambient air; II, GaAs NWs without As-cap stored in ambient air for 24 h; III, GaAs NWs without As-cap stored in ambient air for 1 month; IV, GaAs NWs with As-cap kept in the ambient air for 24 h and then decapped; V, GaAs NWs with As-cap kept in the ambient air for 2 months and then decapped. Reproduced from Ref. [34] with permission from The Royal Society of Chemistry.

The protective As capping layer was grown on the GaAs NWs prior to the growth of the shell by the deposition of As_2 within the same MBE chamber for the GaAs NW growth. The As capping procedure was initiated once the substrate had cooled to 160°C, thus avoiding As layer desorption from the NWs at higher temperature. Instead of the smooth surface of a typical GaAs NW without oxidation (Fig. 9.4b), the newly fabricated surface of As-capped NW is rough and amorphous confirmed via TEM image and RHEED pattern (Fig. 9.4f). NWs' diameter is thickened from 50 to 100 nm, given a 2 h-capping (Fig. 9.4a,b,e,f). EDX elemental mapping (Fig. 9.5) further confirms the formation of the As cap layer which perfectly wrapped the core. Comparing with the As and Ga 3d core level of pristine GaAs NW showing in Fig. 9.4d, the As 3d peak in Fig. 9.4h slightly shifts toward higher binding energy (0.3 eV), because elemental As is detected rather than As in GaAs whose photoelectron mean free path is shorter than the thickness of the As layer. The Ga 3d core level has not been detected for the same reason (Fig. 9.4h).

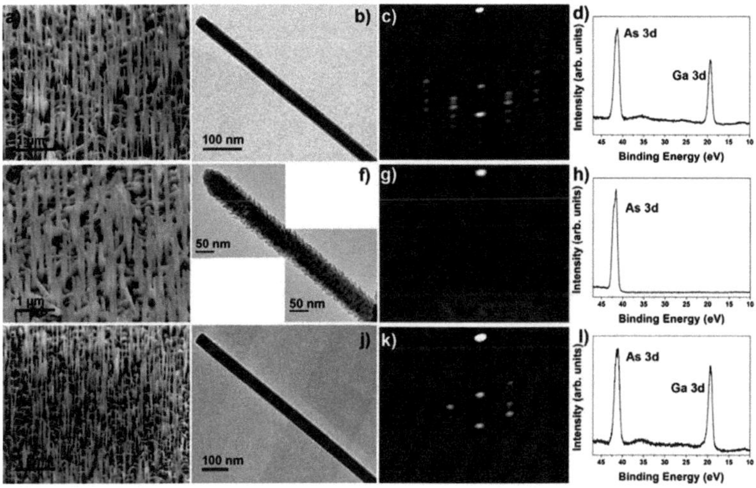

Figure 9.4 SEM image, TEM image, RHEED patterns and XPS spectra of: (a–d) as-grown GaAs NWs, (e–h) NWs after As capping and (i–l) NWs after As decapping under UHV. Reproduced from Ref. [34] with permission from The Royal Society of Chemistry.

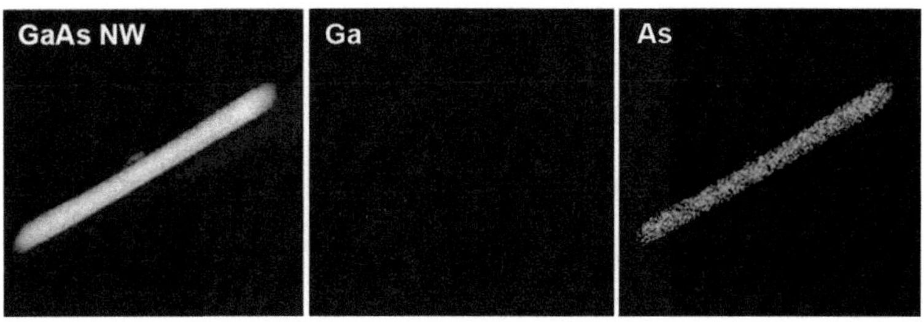

Figure 9.5 STEM image and EDX elemental mapping of the As-capped GaAs NW. Reproduced from Ref. [34] with permission from The Royal Society of Chemistry.

The heating treatment under UHV was applied to the capped NW array inside the MBE chamber for removing the As layer, termed in "decap" procedure. The process was monitored by in situ

RHEED. The RHEED pattern changed gradually from a continuous bright background (Fig. 9.4g) to a spotty pattern (Fig. 9.4k) while progressively increasing the temperature from RT to 440°C. The final pattern (Fig. 9.4k) is almost the same as Fig. 9.4c, revealing highly twinned ZB and WZ crystal structure, consistent with the TEM results. It also shows that the capping layer was completely removed and that the GaAs crystal structure was not modified after annealing. The TEM image (Fig. 9.4j) shows that the NW regained the smooth facets and returned to the initial diameter of 50 nm. Neither NW structure nor morphology changes occurred after the As decap process. Furthermore comparing the Ga 3d and As 3d core level spectra of as-grown GaAs NWs, capped NWs and decapped NWs (Fig. 9.4d,h,l), the disappearance and re-emergence of Ga peak also reveals the reversibility of the As cap method.

To show that the oxidation-proof procedure is necessary and effective, the reversible As-cap method was implemented right after the growth of GaAs NWs. Then, the capped sample was taken out of the MBE chamber and stored in the ambient air at RT for 24 h without any other chemical or physical treatment and finally reintroduced into the MBE chamber. Then the As layer was removed by thermal treatment. The moderate heating did not destroy the GaAs NW structure, as shown in Fig. 9.4j-l. The XPS spectra of the decapped NW sample (Fig. 9.3, row IV) are very similar to those of as-grown GaAs NWs (Fig. 9.3, row I). Neither As nor Ga oxides are detected. O 1s peak is also symmetrical, revealing its sole origin from native SiO_x on the Si substrate. When the air-exposure time was extended to as long as 2 months and other procedures were kept the same, only a little amount of Ga oxide was detected (Fig. 9.3, row V), less than the quantity detected in Fig. 9.3, row II, which demonstrates that the oxidation-proof As capping method is efficient.

9.3.2 The Impact on the Structure of AlGaAs Shell

To further demonstrate the impact of the uncontrolled oxidation on the shell structure, the growth of an AlGaAs shell was investigated on three different GaAs NW samples. The first one corresponds to as-grown GaAs NWs without exposure to ambient air, the second one to GaAs NWs stored in ambient air for 45 days and the third one to As-capped GaAs NWs stored in ambient air at RT for 40 days and then decapped under UHV. For the first NW array, following the typical growth of the GaAs NWs in the MBE chamber, an AlGaAs shell was grown on the surface of the GaAs core right after the temperature of the sample holder was reduced from 610 to 515°C under As_4 atmosphere, referred to as "in situ grown GaAs/AlGaAs NWs." The As_4 crystallized the Ga catalyst on the tip of the NWs, inhibiting the axial growth. TEM images (Fig. 9.6a,b) reveal the uniform growth of the AlGaAs shell on the GaAs NW facets. The HRTEM picture in Fig. 9.6b clearly shows continuous uniform lattice fringes of the shell material. The interplanar spacing is equal to 3.23 Å, which corresponds to the (111) planes of AlGaAs alloys. No interfacial layer between these two materials is observed. Through the FFT analysis of the selected area around GaAs/AlGaAs interface, only one set of patterns is obtained, indicating that AlGaAs shell was epitaxially grown on the GaAs core.

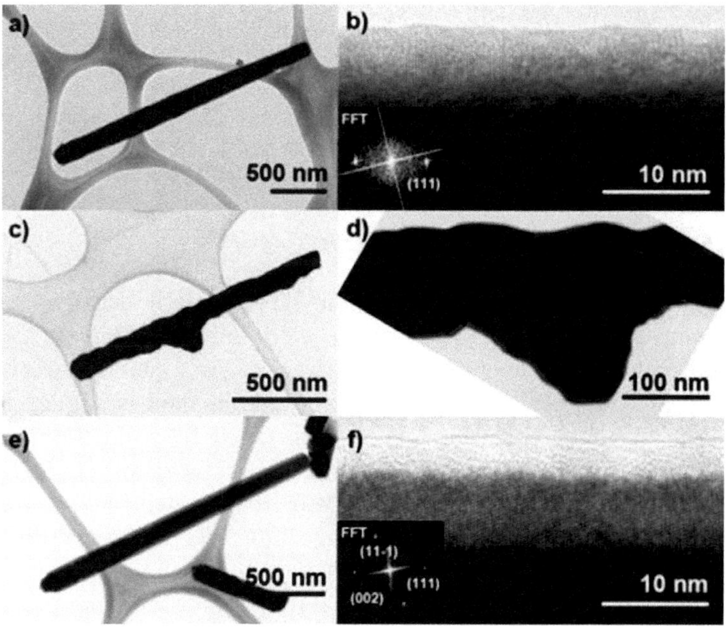

Figure 9.6 TEM and HRTEM images of an in situ grown GaAs/AlGaAs NW (a, b), an oxidized-GaAs/AlGaAs NW (c, d) and an As-capped-decapped-GaAs/AlGaAs NW (e, f). Reproduced from Ref. [34] with permission from The Royal Society of Chemistry.

In the case of the oxidized GaAs NW array, the newly formed surface of core/shell NWs is prominently roughened (Fig. 9.6c). The rough surface is due to the abundant crystal defects in the entire shell, as shown in the HRTEM picture in Fig. 9.6d. These results prove a non-epitaxial relationship between oxidized GaAs and AlGaAs and the detrimental effect of the oxidized interface on the morphology of NWs. This kind of structural disorder is clearly detrimental to the properties of the nanostructures [47]. Finally, for the third one, the As capped GaAs NW array, it was preserved under the ambient atmosphere for 40 days and then introduced back into the MBE chamber to remove the oxidation-proof As-cap under UHV, and to subsequently construct the core/shell NWs using the same procedure. As seen in Fig. 9.6e, NWs with smooth facets were also obtained. The HRTEM image and its corresponding FFT pattern (Fig. 9.6f) demonstrate the epitaxial growth, comparable to the results obtained from the in situ grown GaAs/AlGaAs NWs. During this soft decap process, no structure changes occur based on the RHEED pattern. It should be noted that oxidized-GaAs NWs are also heated from room temperature to 440°C in order to maintain a same heating history with the decapped ones.

9.4 GAAs (Core)/SrTiO$_3$ (Shell) NWs

9.4.1 The Growth of GaAs/SrTiO$_3$ NWs

Confirmed by the above discussion, the profound harm of the uncontrolled oxidation is revealed. And the reversible As cap method is proven effective and can be used as a general method for protecting NWs from oxidation. So this method is applied to the construction of heterogeneous core/shell NWs, GaAs/SrTiO$_3$ NWs, for assuring a relatively clean and abrupt interface (Fig. 9.7a). The SrTiO$_3$ shell was grown in a MBE reactor dedicated to the growth of functional oxides. O$_2$ injection was controlled in a differentially pumped pre-chamber connected to the reactor via a butterfly valve. In order to avoid the formation of an amorphous oxide on the GaAs NW facets, the two-step growth process is used as proposed by Niu et al. for growing SrTiO$_3$ films on Si substrates [48]. The first step (Fig. 9.7c,d) consisted of depositing an amorphous SrTiO$_3$ buffer layer at 350°C under low

O_2 partial pressure of 5×10^{-8} Torr in order to obtain a relatively sharp interface. Annealing at 550°C under UHV for 15 min led to the crystallization of this $SrTiO_3$ buffer layer (Fig. 9.7e,f). In the second step (Fig. 9.7g), further $SrTiO_3$ growth is performed at higher temperature (550°C) under higher O_2 partial pressure (1×10^{-7} Torr). It should be noted that probably thanks to the $SrTiO_3$ buffer layer that is grown at low temperature and acts as a passivation film, the GaAs/$SrTiO_3$ interface is preserved. After $SrTiO_3$ deposition at low temperature, the RHEED pattern does not show any change except a brighter background, typical for an amorphous layer (Fig. 9.7d). However, rings appeared during the annealing due to the $SrTiO_3$ crystallization without preferential orientation (Fig. 9.7f). During the second step of $SrTiO_3$ growth, rings become more intense (Fig. 9.7h). All the rings observed in Fig. 9.7h are in agreement with bulk $SrTiO_3$.

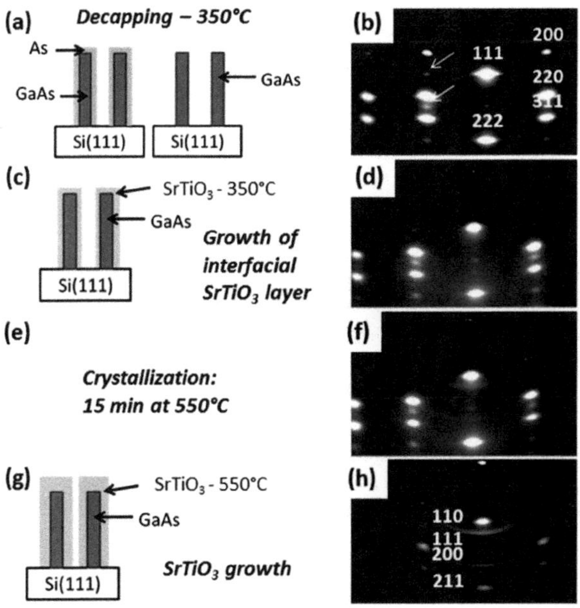

Figure 9.7 Illustration of the capping/decapping and of the two-step $SrTiO_3$ growth process (a, c, e, g). RHEED pattern measured along [1–10] directions: at 350°C after the decapping of the As protective shell (b), after the growth of $SrTiO_3$ buffer during 15 min at 350°C (d), after the annealing at 550°C during 15 min (f), after the growth of $SrTiO_3$ during 15 min at 550°C (h). Reprinted with permission from [30]. Copyright (2016) American Chemical Society.

In order to investigate the impact of the growth temperature on the structural and morphological properties, a series of sample was prepared. The one-step deposition of 40 nominal monolayers of $SrTiO_3$ under low O_2 partial pressure (5×10^{-8} Torr) was performed for comparison with the two-step method. Crystal structure analyses reveal that at low growth temperature (350°C), $SrTiO_3$ is mainly amorphous, while at 620°C the $SrTiO_3$ is crystalline but strongly dewetted. The composition of the NWs was studied by EDX elemental mapping as shown in Fig. 9.8a. The results clearly show the presence of the $SrTiO_3$ shell wrapping the GaAs core. When compared to the rest of the NW, the tip exhibits a $SrTiO_3$ enrichment. This is probably the consequence of the verticality of the NWs and of the geometry of the MBE reactor that induced a preferential growth on the tip compared to the facets. However, radial growth is also evidenced without a tapering effect. A slight composition variation is visible along the cross section of Fig. 9.8c that could be due to the shadowing effect. In the case of the NW array with good verticality, the arrival of vapor atoms at the sidewalls may be restricted due to the high wire densities.

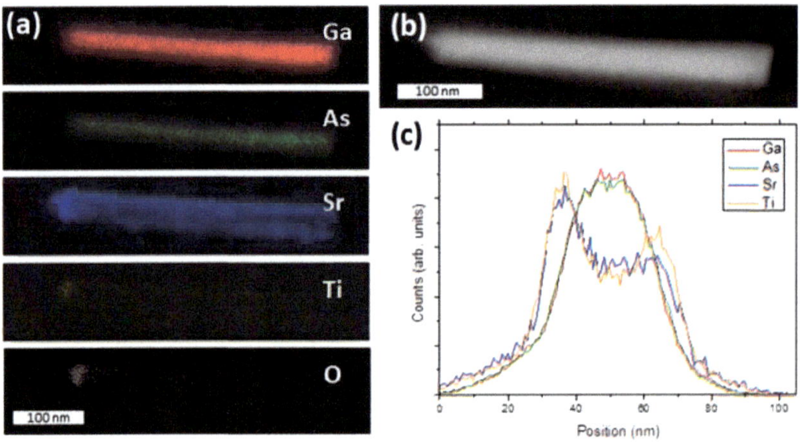

Figure 9.8 Two-dimensional element mapping of Ga, As, Sr, Ti, and O, respectively, for a single nanowire (a) and its corresponding scanning transmission electron microscopy image for a spot size of 1.5 nm (b). Line-scan performed along the nanowire diameter (c). Reprinted with permission from [30]. Copyright (2016) American Chemical Society.

9.4.2 Structure, Morphology, and Chemistry at the GaAs/SrTiO$_3$ Interface

Figure 9.9 shows TEM images of the GaAs NWs after growth of SrTiO$_3$. The NW surface is relatively rough because SrTiO$_3$ seems to form agglomerated nanocrystals or facetted shell. The GaAs surface is well covered by SrTiO$_3$ even if some NWs exhibit regions free of SrTiO$_3$ probably due to the shadowing effect during the oxide growth as discussed above. The total equivalent thickness of SrTiO$_3$ was 10 nm. As the NW array is with a density of 7 NWs/μm^2, 45 nm in diameter and 1.0 μm in length, the exposed GaAs lateral surface is equivalent to 98% of the 2D surface. Assuming the growth of a two-dimensional SrTiO$_3$ layer on GaAs as reported, [49] the SrTiO$_3$ shell thickness should be about 5 nm, in accordance with the TEM image.

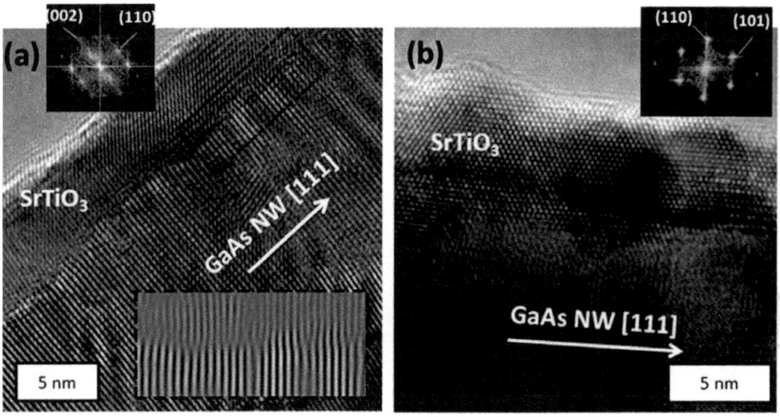

Figure 9.9 TEM images showing an abrupt (a) and an amorphous (b) GaAs core/SrTiO$_3$ shell NW interface. Reprinted with permission from [30]. Copyright (2016) American Chemical Society.

In Fig. 9.9a, some regions clearly exhibit Moiré fringes due to the overlapping of SrTiO$_3$ and GaAs lattices. SrTiO$_3$ nanocrystals are unambiguously identified by measuring their lattice spacing. Moreover, the chemical composition of the nanocrystals measured by EDX mapping is in accordance with SrTiO$_3$ stoichiometry. Most of the SrTiO$_3$ layer appears to be oriented with respect to the GaAs lattice. Measurements performed on the spots of the upper FFT in inset of Fig. 9.9a (2.29 and 2.76 Å) are in good agreement with [002] and [110] reflections of $SrTiO_3$. The alignment of $SrTiO_3$ [110] with GaAs [111] results in a high mismatch that induces the formation of interfacial dislocation as shown in the filtered image of the interface (insert of Fig. 9.9a). The interface is abrupt but not perfectly flat, which could be the consequence of GaAs/$SrTiO_3$ interfacial reaction or micro-faceting of the GaAs NW (note that {111} micro-facets have been reported [50]). Figure 9.9b shows another interface between a $SrTiO_3$ nanocrystal and the GaAs NW core. Although the interface is not abrupt and seems to be partially amorphous, some parts of the $SrTiO_3$ layer are epitaxially grown on the GaAs lattice. The $SrTiO_3$ lattice is clearly identified, the FFT of the lattice shows three series of spots related to {110} planes (the zone axis is [–111]). From this image, it can be deduced that the $SrTiO_3$ [1–12] is aligned with the NW growth direction which is GaAs [111]. If GaAs NW facets were {1–10} planes as reported, [51] $SrTiO_3$ [–111] would be aligned with GaAs [11–2]. Such alignment could minimize the mismatch to as small as 2.3% along $SrTiO_3$ [1–12]//GaAs [222] and $SrTiO_3$ [–111]//GaAs [11–2]. Another alignment, GaAs [1–10]//$SrTiO_3$ [110], can also be easily deduced. Figure 9.10 shows a possible model of the interface between $SrTiO_3$ nanocrystals and GaAs (1–10) facet in accordance with the lattice matching in Fig. 9.9b. To summarize, both sharp and amorphous interface are discovered in the as-prepared NW array. As for a classical two-dimensional growth, the amorphous interfacial layer can be formed after the $SrTiO_3$ growth, which is compatible with epitaxial growth under oxygen partial pressure. However, some NWs exhibit polycrystalline areas, particularly at the NW tip.

XPS measurements were performed at Soleil synchrotron facility after the $SrTiO_3$ growth, with a photon energy of 750 eV and a resolution of about 70 meV [52]. Figure 9.11 shows As 3d, Ga 3d, Sr 3d, Ti 2p, and O 1s core levels before annealing. The Ti 2p core level is made of a single spin-orbit doublet (Ti^{4+}) with a spin-orbit splitting of 5.73 eV [53–55]. The Sr 3d core level exhibits a main spin-orbit doublet with Sr $3d_{5/2}$ at 133.2 eV and a spin-orbit splitting of 1.74 eV as in bulk $SrTiO_3$ [54–56] and two other spin-orbit doublets at 133.8 and 134.5 eV for Sr $3d_{5/2}$. These latter peaks have been related to the presence of stacking faults and SrO, respectively [55, 56]. The formation of SrO in MBE-grown $SrTiO_3$ is attributed to the Sr segregation at the surface and at the interface during growth. The Ga 3d core level consists of a spin-orbit doublet corresponding to GaAs with Ga $3d_{5/2}$ at 18.9 eV and a spin-orbit splitting of 0.4 eV, [57] and an intense component at 20.1 eV corresponding to Ga oxides. Contribution from O 2s and Sr 4p core levels is also reported. The As 3d peak position at 44.8 eV corresponds to As oxide, and not to the As 3d doublet (spin orbit of 0.69 eV), which would appear at around 41 eV between Sr 4s/Ti 3p peaks and As oxide peak. Therefore, an oxidized GaAs has been formed either at the GaAs/$SrTiO_3$ interface during $SrTiO_3$ growth or at the GaAs NW surface after growth due to a partially inefficient passivation of GaAs NWs. The core level O 1s also shows three components at 532.15, 530.94 and 529.74 eV attributed to $SrTiO_3$, As oxide and Ga oxide.

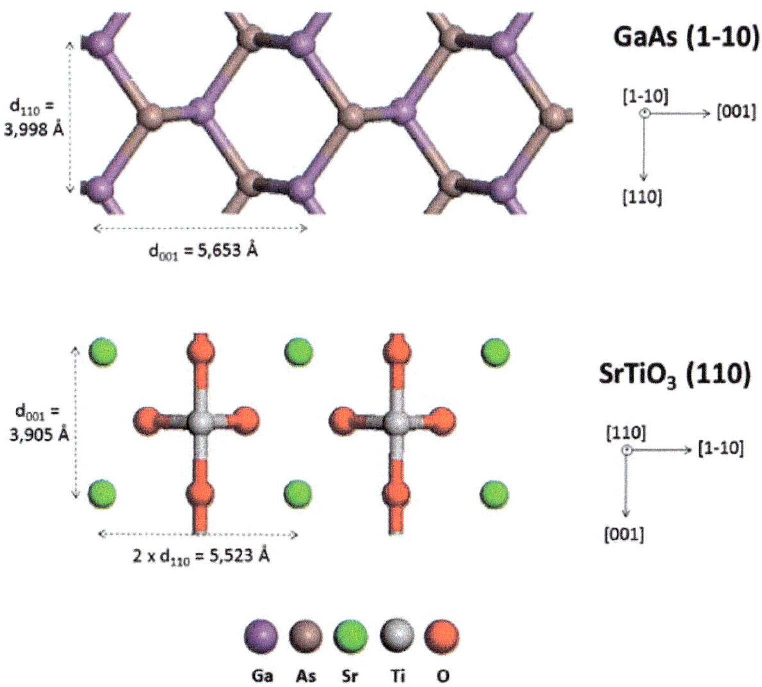

Figure 9.10 Ball-and-stick model showing the GaAs(1–10) plane and the $SrTiO_3$ (110) plane. Reprinted with permission from [30]. Copyright (2016) American Chemical Society.

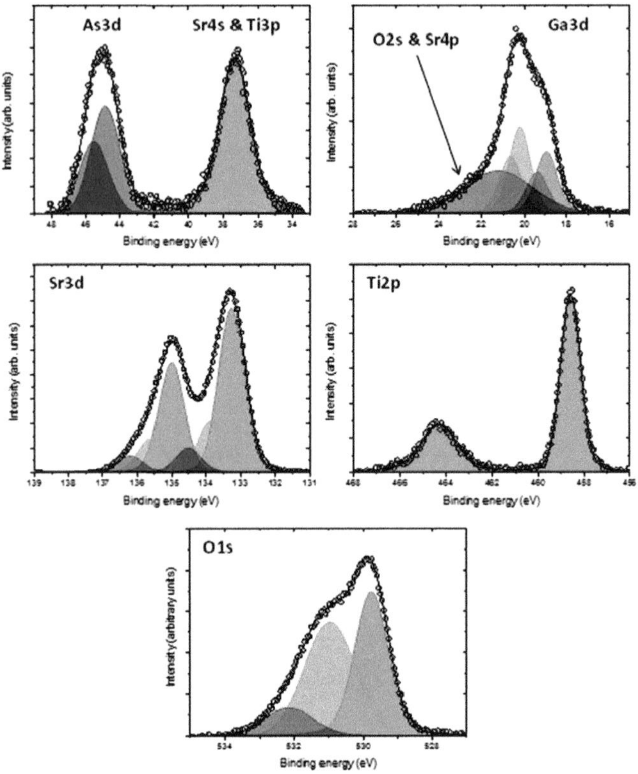

Figure 9.11 XPS core levels of $GaAs/SrTiO_3$ NWs measured at room temperature. The photon energy was 750 eV. Reprinted with permission from [30]. Copyright (2016) American Chemical Society.

Figure 9.12a shows the PL spectrum of GaAs NWs after As cap/decap procedure. The peak located at 1.43 eV is in agreement with the emission energy of the bulk GaAs ZB phase NWs at RT [58]. Then, the two-step $SrTiO_3$ deposition method was applied to the decapped GaAs NWs under UHV. As shown in Fig. 9.12b, light emission of GaAs has been barely detected, while a rather low broad band at about 1.25–1.3 eV originates from defects of the parasitic 2D layer. It should be noted that similar spectra were obtained even though $SrTiO_3$ shell was grown at lower temperature. The suppression of PL for this decapped-$GaAs/SrTiO_3$ core/shell NW array could be the consequence of interface defects caused by oxidation during the $SrTiO_3$ growth (despite the moderate growth temperature and oxidation pressure used during the first growth step) or inter-diffusion of the metallic species (Sr, Ti) into the NWs. To avoid the occurrence of poor light emission property of GaAs core, an AlGaAs passivating shell around the GaAs was grown. The shape of PL peak of GaAs/AlGaAs (Fig. 9.12c) is quite similar to Fig. 9.12a, but 5 times more intense, showing the efficiency of passivation through this wider-band gap material. Finally, GaAs/AlGaAs core/shell NWs were capped with As layer and then transferred into the MBE chamber dedicated to functional oxides. After the As-decap procedure, $SrTiO_3$ shell was deposited by the two-step method. An evident strong peak arises at 1.43 eV (Fig. 9.12d), in accordance with result of the sample before $SrTiO_3$ growth (Fig. 9.12c). From time-resolved PL measurements we extracted a lifetime of 300 and 209 ps for GaAs/AlGaAs and GaAs/AlGaAs/$SrTiO_3$, respectively (Fig. 9.13). It is well clarified that the AlGaAs passivation can avoid the introduction of non-radiative recombination centers at the GaAs/$SrTiO_3$ interface during the fabrication of the shell.

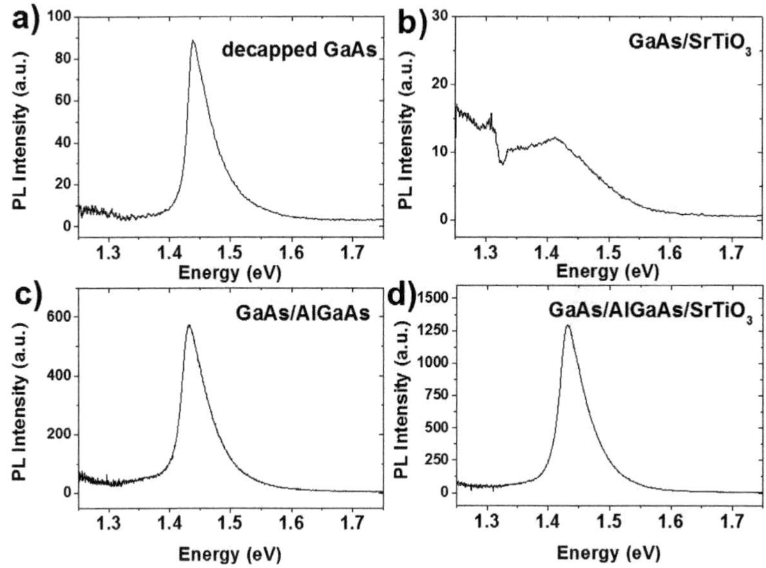

Figure 9.12 Photoluminescence spectra of decapped GaAs NW array (a), GaAs/$SrTiO_3$ core/shell NW array (b), GaAs/AlGaAs core/shell NW array (c) and GaAs/AlGaAs/$SrTiO_3$ core/shell NW array (d). For sample (b) and (d), the $SrTiO_3$ was grown using the two-step method. The measurements were performed at 300 K. Reprinted with permission from [30]. Copyright (2016) American Chemical Society.

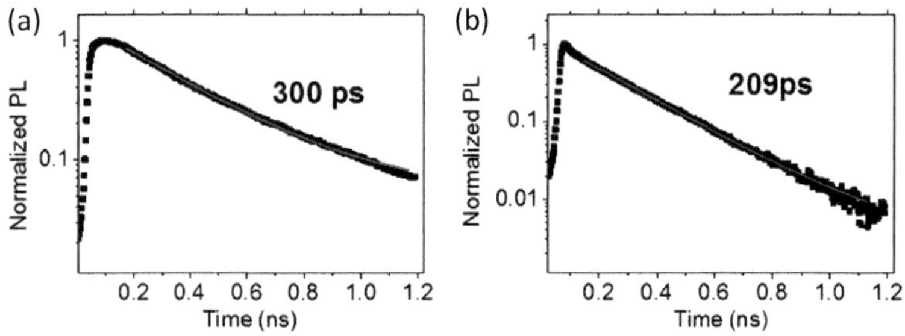

Figure 9.13 Time-resolved PL measurements for GaAs/AlGaAs NWs (a) and GaAs/AlGaAs/SrTiO$_3$ NWs. The lifetime was extracted from monoexponential decay fitting (red line). Measurements were performed at 300 K. Reprinted with permission from [30]. Copyright (2016) American Chemical Society.

Acknowledgments

This work has been partly funded by the French Agence Nationale pour la Recherche (ANR), project COSCOF ANR 12JS10 00301. Chinese Scholarship Council (CSC) is acknowledged for its financial support. The authors thank the CLYM and NanoLyon for access to equipment and acknowledge the contribution of the "Heteroepitaxy and Nanostructures" group at Lyon Institute of Nanotechnology for the growth of III–V and perovskite oxides by molecular beam epitaxy. N. Chauvin is acknowledged for the photoluminescence measurements. The authors thank F. Sirotti and F. Silly from TEMPO beamline at Synchrotron SOLEIL for their help with real-time x-ray photoelectron spectroscopy measurements and N. P. Blanchard for his help with transmission electron microscopy.

References

1. Kauko, H., Zheng, C., Zhu, Y., Glanvill, S., Dwyer, C., Munshi, A., Fimland, B., van Helvoort, A., and Etheridge, J. (2013). Compositional analysis of GaAs/AlGaAs heterostructures using quantitative scanning transmission electron microscopy, *Appl. Phys. Lett.*, **103**, p. 232111.
2. Songmuang, R., Giang, L. T. T., Bleuse, J., Den Hertog, M., Niquet, Y. M., Dang, L. S., and Mariette, H. (2016). Determination of the optimal shell thickness for self-catalyzed GaAs/AlGaAs core–shell nanowires on silicon, *Nano Lett.*, DOI: 10.1021/acs.nanolett.5b03917.

3. Sköld, N., Karlsson, L. S., Larsson, M. W., Pistol, M. E., Seifert, W., Trägårdh, J., and Samuelson, L. (2005). Growth and optical properties of strained GaAs–$Ga_xIn_{1-x}P$ core–shell nanowires, *Nano Lett.*, **5**, pp. 1943–1947.

4. Mariani, G., Scofield, A. C., Hung, C. H., and Huffaker, D. L. (2013). GaAs nanopillar-array solar cells employing in situ surface passivation, *Nat. Commun.*, **4**, p. 1497.

5. Chen, S. L., Chen, W. M., Ishikawa, F., and Buyanova, I. A. (2015). Suppression of non-radiative surface recombination by N incorporation in GaAs/GaNAs core/shell nanowires, *Sci. Rep.*, **5**, p. 11653.

6. Conesa-Boj, S., Boioli, F., Russo-Averchi, E., Dunand, S., Heiss, M., Ruffer, D., Wyrsch, N., Ballif, C., Miglio, L., and Fontcuberta i Morral, A. (2014). Plastic and elastic strain fields in GaAs/Si core-shell nanowires, *Nano Lett.*, **14**, pp. 1859–1864.

7. Conesa-Boj, S., Dunand, S., Russo-Averchi, E., Heiss, M., Ruffer, D., Wyrsch, N., Ballif, C., and Fontcuberta i Morral, A. (2013). Hybrid axial and radial Si-GaAs heterostructures in nanowires, *Nanoscale*, **5**, pp. 9633–9639.

8. Jenichen, B., Hilse, M., Herfort, J., and Trampert, A. (2015). Real structure of lattice matched GaAs–Fe_3Si core–shell nanowires, *J. Cryst. Growth*, **410**, pp. 1–6.

9. Hilse, M., Herfort, J., Jenichen, B., Trampert, A., Hanke, M., Schaaf, P., Geelhaar, L., and Riechert, H. (2013). GaAs–Fe_3Si core–shell nanowires: Nanobar magnets, *Nano Lett.*, **13**, pp. 6203–6209.

10. Rüffer, D., Huber, R., Berberich, P., Albert, S., Russo-Averchi, E., Heiss, M., Arbiol, J., i Morral, A. F., and Grundler, D. (2012). Magnetic states of an individual Ni nanotube probed by anisotropic magnetoresistance, *Nanoscale*, **4**, pp. 4989–4995.

11. Dubourdieu, C., Bruley, J., Arruda, T. M., Posadas, A., Jordan-Sweet, J., Frank, M. M., Cartier, E., Frank, D. J., Kalinin, S. V., and Demkov, A. A. (2013). Switching of ferroelectric polarization in epitaxial $BaTiO_3$ films on silicon without a conducting bottom electrode, *Nat. Nanotechnol.*, **8**, pp. 748–754.

12. Wang, Y., Ganpule, C., Liu, B., Li, H., Mori, K., Hill, B., Wuttig, M., Ramesh, R., Finder, J., and Yu, Z. (2002). Epitaxial ferroelectric $Pb(Zr, Ti)O_3$ thin films on Si using $SrTiO_3$ template layers, *Appl. Phys. Lett.*, **80**, p. 97.

13. Park, K. I., Son, J. H., Hwang, G. T., Jeong, C. K., Ryu, J., Koo, M., Choi, I., Lee, S. H., Byun, M., and Wang, Z. L. (2014). Highly-efficient, flexible piezoelectric PZT thin film nanogenerator on plastic substrates, *Adv. Mater.*, **26**, pp. 2514–2520.

14. Lukeš, F. (1972). Oxidation of Si and GaAs in air at room temperature, *Surf. Sci.*, **30**, pp. 91–100.

15. Contour, J. P., Massies, J., Fronius, H., and Ploog, K. (1988). An XPS study of the passivating oxide layer produced on GaAs (001) substrate by heating in air above 200°C, *Jpn. J. Appl. Phys.*, **27**, p. L167.

16. Alekseev, P. A., Dunaevskiy, M. S., Ulin, V. P., Lvova, T. V., Filatov, D. O., Nezhdanov, A. V., Mashin, A. I., and Berkovits, V. L. (2015). Nitride surface passivation of GaAs nanowires: Impact on surface state density, *Nano Lett.*, **15**, pp. 63–68.

17. Hjort, M., Lehmann, S., Knutsson, J., Timm, R., Jacobsson, D., Lundgren, E., Dick, K., and Mikkelsen, A. (2013). Direct imaging of atomic scale structure and electronic properties of GaAs wurtzite and zinc blende nanowire surfaces, *Nano Lett.*, **13**, pp. 4492–4498.

18. Karpov, I., Venkateswaran, N., Bratina, G., Gladfelter, W., Franciosi, A., and Sorba, L. (1995). Arsenic cap layer desorption and the formation of GaAs (001) c (4× 4) surfaces, *J. Vac. Sci. Technol. B*, **13**, pp. 2041–2048.

19. McKee, R., Walker, F., and Chisholm, M. (1998). Crystalline oxides on silicon: The first five monolayers, *Phys. Rev. Lett.*, **81**, pp. 3014–3017.

20. Meunier, B., Bachelet, R., Grenet, G., Botella, C., Regreny, P., Largeau, L., Penuelas, J., and Saint-Girons, G. (2016). The role of titanium at the SrTiO$_3$/GaAs epitaxial interface, *J. Cryst. Growth*, **433**, pp. 139–142.

21. Liang, Y., Curless, J., and McCready, D. (2005). Band alignment at epitaxial SrTiO$_3$–GaAs(001) heterojunction, *Appl. Phys. Lett.*, **86**, p. 082905.

22. Fauzia, J., Vincenzo, G., Silvia, R., and Faustino, M. (2008). Self-catalyzed growth of GaAs nanowires on cleaved Si by molecular beam epitaxy, *Nanotechnology*, **19**, p. 275711.

23. Ramdani, M. R., Harmand, J. C., Glas, F., Patriarche, G., and Travers, L. (2012). Arsenic pathways in self-catalyzed growth of GaAs nanowires, *Cryst, Growth Des.*, **13**, pp. 91–96.

24. Priante, G., Ambrosini, S., Dubrovskii, V. G., Franciosi, A., and Rubini, S. (2013). Stopping and resuming at will the growth of GaAs nanowires, *Cryst. Growth Des.*, **13**, pp. 3976–3984.

25. Matteini, F., Tütüncüoglu, G. Z., Potts, H., Jabeen, F., and Fontcuberta i Morral, A. (2015). Wetting of Ga on SiO$_x$ and its impact on GaAs nanowire growth, *Cryst. Growth Des.*, **15**, pp. 3105–3109.

26. Dubrovskii, V., Xu, T., Álvarez, A. D., Plissard, S., Caroff, P., Glas, F., and Grandidier, B. (2015). Self-equilibration of the diameter of Ga-catalyzed GaAs nanowires, *Nano Lett.*, **15**, pp. 5580–5584.

27. Madsen, M. H., Aagesen, M., Krogstrup, P., Sørensen, C., and Nygård, J. (2011). Influence of the oxide layer for growth of self-assisted InAs nanowires on Si (111), *Nqnoscale Res. Lett.*, **6**, pp. 1–5.

28. Fontcuberta i Morral, A., Colombo, C., Abstreiter, G., Arbiol, J., and Morante, J. (2008). Nucleation mechanism of gallium-assisted molecular beam epitaxy growth of gallium arsenide nanowires, *Appl. Phys. Lett.*, **92**, p. 063112.

29. Rudolph, D., Hertenberger, S., Bolte, S., Paosangthong, W., Spirkoska, D. E., Döblinger, M., Bichler, M., Finley, J. J., Abstreiter, G., and Koblmüller, G. (2011). Direct observation of a noncatalytic growth regime for GaAs nanowires, *Nano Lett.*, **11**, pp. 3848–3854.

30. Guan, X., Becdelievre, J., Meunier, B., Benali, A., Saint-Girons, G., Bachelet, R., Regreny, P., Botella, C., Grenet, G., Blanchard, N. P., Jaurand, X., Silly, M. G., Sirotti, F., Chauvin, N., Gendry, M., and Penuelas, J. (2016). GaAs core/$SrTiO_3$ shell nanowires grown by molecular beam epitaxy, *Nano Lett.*, **16**, pp. 2393–2399.

31. Kim, Y. H., Park, D. W., and Lee, S. J. (2012). Gallium-droplet behaviors of self-catalyzed GaAs nanowires: A transmission electron microscopy study, *Appl. Phys. Lett.*, **100**, p. 033117.

32. Yu, X., Wang, H., Lu, J., Zhao, J., Misuraca, J., Xiong, P., and von Molnár, S. (2012). Evidence for structural phase transitions induced by the triple phase line shift in self-catalyzed GaAs nanowires, *Nano Lett.*, **12**, pp. 5436–5442.

33. Plissard, S., Dick, K. A., Larrieu, G., Godey, S., Addad, A., Wallart, X., and Caroff, P. (2010). Gold-free growth of GaAs nanowires on silicon: Arrays and polytypism, *Nanotechnology*, **21**, p. 385602.

34. Guan, X., Becdelievre, J., Benali, A., Botella, C., Grenet, G., Regreny, P., Chauvin, N., Blanchard, N. P., Jaurand, X., Saint-Girons, G., Bachelet, R., Gendry, M., and Penuelas, J. (2016). GaAs nanowires with oxidation-proof arsenic capping for the growth of an epitaxial shell, *Nanoscale*, **8**, pp. 15637–15644.

35. Negrila, C., Logofatu, C., Ghita, R., Cotirlan, C., Ungureanu, F., Manea, A., and Lazarescu, M. (2008). Angle-resolved XPS structural investigation of GaAs surfaces, *J. Cryst. Growth*, **310**, pp. 1576–1582.

36. Ambrosini, S., Fanetti, M., Grillo, V., Franciosi, A., and Rubini, S. (2011). Self-catalyzed GaAs nanowire growth on Si-treated GaAs(100) substrates, *J, Appl. Phys.*, **109**, p. 094306.

37. Ishikawa, T., and Ikoma, H. (1992). X-ray photoelectron spectroscopic analysis of the oxide of GaAs, *Jpn. J. Appl. Phys.*, **31**, p. 3981.

38. Razek, N., Otte, K., Chasse, T., Hirsch, D., Schindler, A., Frost, F., and Rauschenbach, B. (2002). GaAs surface cleaning by low energy hydrogen ion beam treatment, *J. Vac. Sci. Technol. A*, **20**, pp. 1492–1497.

39. Gallagher, M., Prince, R., and Willis, R. (1992). On the atomic structure and electronic properties of decapped GaAs (001)(2 × 4) surfaces, *Surf. Sci.*, **275**, pp. 31–40.

40. Jiang, X., Wang, R., Van Dijken, S., Shelby, R., Macfarlane, R., Solomon, G., Harris, J., and Parkin, S. (2003). Optical detection of hot-electron spin injection into GaAs from a magnetic tunnel transistor source, *Phys. Rev. Lett.*, **90**, p. 256603.

41. Jiang, X., Wang, R., Shelby, R., Macfarlane, R., Bank, S., Harris, J., and Parkin, S. (2005). Highly spin-polarized room-temperature tunnel injector for semiconductor spintronics using MgO (100), *Phys. Rev. Lett.*, **94**, p. 056601.

42. Wistey, M., Bank, S., Yuen, H., Goddard, L., Gugov, T., and Harris Jr, J. (2005). Protecting wafer surface during plasma ignition using an arsenic cap, *J. Vac. Sci. Technol. B*, **23**, pp. 1324–1327.

43. Xu, T., Dick, K. A., Plissard, S., Nguyen, T. H., Makoudi, Y., Berthe, M., Nys, J. P., Wallart, X., Grandidier, B., and Caroff, P. (2012). Faceting, composition and crystal phase evolution in III–V antimonide nanowire heterostructures revealed by combining microscopy techniques, *Nanotechnology*, **23**, p. 095702.

44. Capiod, P., Xu, T., Nys, J. P., Berthe, M., Patriarche, G., Lymperakis, L., Neugebauer, J., Caroff, P., Dunin-Borkowski, R. E., Ebert, P., and Grandidier, B. (2013). Band offsets at zincblende-wurtzite GaAs nanowire sidewall surfaces, *Appl. Phys. Lett.*, **103**, p. 122104.

45. Potts, H., Friedl, M., Amaduzzi, F., Tang, K., Tütüncüoglu, G., Matteini, F., Alarcon Lladó, E., McIntyre, P. C., and Fontcuberta i Morral, A. (2016). From twinning to pure zincblende catalyst-free InAs(Sb) nanowires, *Nano Lett.*, **16**, pp. 637–643.

46. Díaz Álvarez, A., Xu, T., Tütüncüoglu, G. Z., Demonchaux, T., Nys, J. P., Berthe, M., Matteini, F., Potts, H. A., Troadec, D., and Patriarche, G. (2015). Nonstoichiometric low-temperature grown GaAs nanowires, *Nano Lett.*, **15**, pp. 6440–6445.

47. Ishikawa, F., Akamatsu, Y., Watanabe, K., Uesugi, F., Asahina, S., Jahn, U., and Shimomura, S. (2015). Metamorphic GaAs/GaAsBi heterostructured nanowires, *Nano Lett.*, **15**, pp. 7265–7272.

48. Niu, G., Vilquin, B., Penuelas, J., Botella, C., Hollinger, G., and Saint-Girons, G. (2011). Heteroepitaxy of $SrTiO_3$ thin films on Si (001) using different growth strategies: Toward substratelike qualitya, *J. Vac. Sci. Technol. B*, **29**, p. 041207.

49. Louahadj, L., Le Bourdais, D., Largeau, L., Agnus, G., Mazet, L., Bachelet, R., Regreny, P., Albertini, D., Pillard, V., and Dubourdieu, C. (2013). Ferroelectric Pb (Zr, Ti)O_3 epitaxial layers on GaAs, *Appl. Phys. Lett.*, **103**, p. 212901.

50. Mariager, S. O., Sørensen, C. B., Aagesen, M., Nygård, J., Feidenhans, R., and Willmott, P. (2007). Facet structure of GaAs nanowires grown by molecular beam epitaxy, *Appl. Phys. Lett.*, **91**, p. 083106.

51. Capiod, P., Xu, T., Nys, J., Berthe, M., Patriarche, G., Lymperakis, L., Neugebauer, J., Caroff, P., Dunin-Borkowski, R., and Ebert, P. (2013). Band offsets at zincblende-wurtzite GaAs nanowire sidewall surfaces, *Appl. Phys. Lett.*, **103**, p. 122104.

52. Bergeard, N., Silly, M., Krizmancic, D., Chauvet, C., Guzzo, M., Ricaud, J., Izquierdo, M., Stebel, L., Pittana, P., and Sergo, R. (2011). Time-resolved photoelectron spectroscopy using synchrotron radiation time structure, *J. Synchrotron Radi.*, **18**, pp. 245–250.

53. Liang, Y., Wei, Y., Hu, X., Yu, Z., Droopad, R., Li, H., and Moore, K. (2004). Heteroepitaxy of $SrTiO_3$ on vicinal Si (001): Growth and kinetic effects, *J. Appl. Phys.*, **96**, pp. 3413–3416.

54. Cai, H., Wu, X., and Gao, J. (2009). Effect of oxygen content on structural and transport properties in $SrTiO_{3-x}$ thin films, *Chem. Phys. Lett.*, **467**, pp. 313–317.

55. Wang, Z., Yang, F., Zhang, Z., Tang, Y., Feng, J., Wu, K., Guo, Q., and Guo, J. (2011). Evolution of the surface structures on $SrTiO_3$ (110) tuned by Ti or Sr concentration, *Phys. Rev. B*, **83**, p. 155453.

56. Szot, K., Speier, W., Breuer, U., Meyer, R., Szade, J., and Waser, R. (2000). Formation of micro-crystals on the (100) surface of $SrTiO_3$ at elevated temperatures, *Surf. Sci.*, **460**, pp. 112–128.

57. Ley, L., Pollak, R., McFeely, F., Kowalczyk, S. P., and Shirley, D. (1974). Total valence-band densities of states of III–V and II–VI compounds from x-ray photoemission spectroscopy, *Phys. Rev. B*, **9**, p. 600.

58. Breuer, S., Pfüller, C., Flissikowski, T., Brandt, O., Grahn, H. T., Geelhaar, L., and Riechert, H. (2011). Suitability of Au- and self-assisted GaAs nanowires for optoelectronic applications, *Nano Lett.*, **11**, pp. 1276–1279.

Part III

Devices and Applications

Chapter 10

Ga(In)N Nanowires Grown by Molecular Beam Epitaxy: From Quantum Light Emitters to Nanotransistors

Žarko Gačević and Enrique Calleja
ISOM—ETSIT, Universidad Politécnica de Madrid,
Avda. Complutense s/n, 28040 Madrid, Spain
gacevic@isom.upm.es

This chapter summarizes the recent progress in the fabrication of GaN nanowires (NWs) and their employment in photonic and electronic nanodevices. In the first section, a brief overview of the early discovery of GaN NWs and their most relevant properties is given. The second section first summarizes advantages and drawbacks of the two most common GaN NW fabrication methods, the bottom-up approach via self-assembled (SA) heteroepitaxy and the top-down approach via selective etching. Then, a detailed description of the mechanisms which drive the GaN NW formation by selective area growth (SAG) homoepitaxy is given. The convenience of SAG to obtain NWs with exceptional crystal quality and in a highly controlled manner (site, size and orientation control) is also addressed. The third and the fourth sections provide insights into the realization of GaN-NW-based photonic and electronic devices, such as quantum

Novel Compound Semiconductor Nanowires: Materials, Devices, and Applications
Edited by Fumitaro Ishikawa and Irina Buyanova

ISBN 978-981-4745-76-5 (Hardcover), 978-1-315-36440-7 (eBook)
www.panstanford.com

light emitters (single photon sources, SPSs) and field effect nanotransistors. The fabrication and characterization of a SPS, relying on a pencil-like InGaN/GaN dot-in-a-wire structure is presented. These sources emit linearly polarized single photons over a wide spectral range and are suitable for high-temperature operation. Their realization is an important step toward quantum communications. Next, the fabrication and characterization of a metal-semiconductor field effect transistor (FET), relying on a GaN:Si NW is given. The NWs are processed horizontally, employing a semi-cylindrical top Ti/Au Schottky gate. The final NW-FETs (non-planar devices) are characterized by an improved channel electrostatic control. Their realization is an important step toward "3D computer" architectures. Finally, the performances of the hereby presented SPS and nanoFET are systematically compared to similar devices realized via different fabrication methods within the III-nitride material system.

10.1 Introduction

In 1964 R. S. Wagner and W. C. Ellis reported on a vapor–liquid–solid (VLS) mechanism to obtain Si single crystals with a quasi one-dimensional (1D) morphology (whiskers) [50]. Half a century later, their report is widely considered as the birth of the research field on 1D nanocrystals, commonly referred to as nanowires (NWs). In the 1990s, a wide variety of novel methods and materials for NWs synthesis was discovered, boosting thus the research on NW-based electronic and photonic nanodevices.

In parallel with the growing interest for NWs, the early 1990s brought groundbreaking results in the field of wide-band gap III-nitride materials. The wide choice of III-nitride direct band gaps (from ultraviolet to near infrared) and the high InGaN emission efficiency in the near-ultraviolet/blue spectral range, made this ternary compound a material of choice for the realization of white light-emitting diodes (LEDs) and blue light emitters [41]. In addition, the unique properties of GaN made this binary material very suitable for transistors operating over a wide range of temperatures, frequencies and powers [38]. Despite significant technological improvements over the last two decades, the lack of cheap substrates compatible with the growth of III-nitrides with high crystal quality still remains one of the major issues for further technological improvements. Somewhat fortunately, in the late 1990s several groups discovered that the growth of GaN with extraordinary crystal quality was possible on a variety of commercial substrates (such as silicon and sapphire) in the form of self-assembled (SA) NWs, achieved via a simple catalyst-free growth method (bottom-up) [42, 55]. While the catalyst-free approach enables an exceptional chemical purity, the high NW surface-to-volume ratio, allowing for an efficient strain relaxation, yields GaN NWs with a virtually perfect crystal structure.

The random character of the GaN NW SA heteroepitaxy is inconvenient to control NW site, size (diameter and height) and orientation (tilt and twist) and results in the difficulty to process NW arrays into devices. To overcome these highly undesirable drawbacks, several groups developed an alternative approach for GaN NW growth, commonly referred to as selective area growth (SAG) homoepitaxy, with drastically improved controllability [3, 30, 44]. The novel fabrication methods have further boosted the research on GaN-NW-based photonics and electronics. Apart from the highly desirable outstanding crystal quality, GaN NWs offer a quasi 1D morphology, which is also convenient for a wide variety of applications. In the case of nanophotonics, the NWs high surface-to-volume ratio favors elastic (vs. plastic) strain relaxation, preventing the material structural deterioration and resulting in high electron–hole recombination efficiency. Combined with the NW 1D morphology, which facilitates photon guiding and enhances light extraction, the NWs result very attractive building blocks for next-generation classical and quantum light emitters. As for nanoelectronics, the NWs high surface-to-volume ratio enables novel designs, such as semi-cylindrical top-gate and surrounding-gate transistors. These novel non-planar (3D) architectures are promising for increased on-chip transistor densities. In addition, the NWs exceptionally flat sidewalls generate very low surface scattering resulting in high carrier mobility, desirable for high-frequency applications. Combined together, these features could lead to a change in the architecture paradigm resulting in "3D computers" operating at exceptionally high frequencies. They would extend Moore's law well beyond the limits imposed by the planar technology and would pave the way for further improvements in computing power.

10.2 GaN Nanowire Fabrication

10.2.1 Bottom-Up versus Top-Down Approach

Starting from the first reported mechanism for NW fabrication (vapor–liquid–solid) half a century ago [50], a tremendous progress has been made in the NWs engineering. The nanotechnology fabrication methods are commonly divided into two groups, referred to as bottom-up and top-down approaches [54]. The bottom-up approach starts with individual atoms (and/or molecules) which, via chemical bonding, build up into a desired nanostructure. In the specific case of GaN NWs, the process starts with a GaN nucleation seed ("0D crystal") which then, due to preferential GaN growth along the polar (0001 or $000\bar{1}$) crystal direction "elongates" into a GaN NW (1D). The bottom-up growth is usually performed by metal-organic vapor phase epitaxy (MOVPE) or molecular beam epitaxy (MBE) (Fig. 10.1a). The top-down approach, on the other hand, starts from a three-dimensional (3D) GaN layer which is then converted into an array of GaN NWs (1D) by selective etching (Fig. 10.1b).

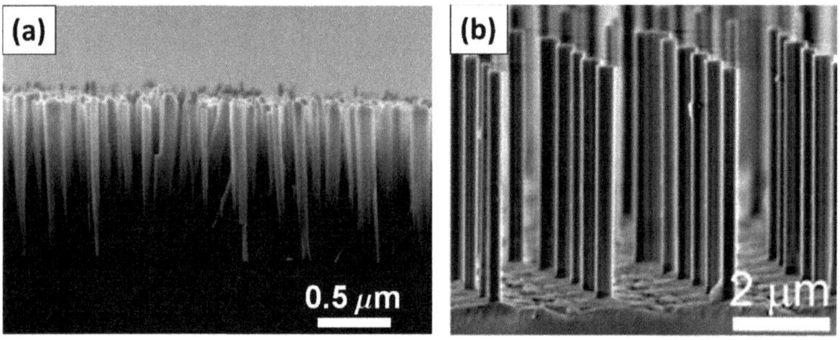

Figure 10.1 GaN NWs fabricated by (a) bottom-up approach by SA heteroepitaxy on a Si(111) substrate and (b) top-down approach, by lithographic dry etch followed by an anisotropic wet etch (adapted from Gačević et al. and Li et al.) [17, 35].

Both bottom-up and top-down approaches have advantages and drawbacks, so that, the selection of either one depends on the targeted application. Concerning the NW chemical purity and crystal quality, much better results are usually obtained by the bottom-up approach, since the ultra high vacuum enables an exceptional chemical purity and the NW strain-free growth results in an exceptional crystal quality. The top-down fabrication is, on the other hand, initiated from a 3D GaN layer, usually of lower quality due to dislocations and other extended defects, which is further subjected to surface damage and contamination during nano-imprinting/etching steps. Concerning the NW fabrication controllability, i.e., site, size and orientation control, much better results are usually obtained by the top-down approach, since it relies on the use of geometrically perfect masking. SA growth is inherently a random process from its initiation (islands nucleation) leading to a strong dispersion in NW heights and diameters. In addition, when SA epitaxy is performed on Si substrates the formation of a thin (2 nm) SiN layer on its surface leads to a weak epitaxial relation between the GaN nanocrystals and the underlying Si substrate. Note that the SiN interlayer was "traditionally" considered amorphous [11]; however, recently there are numerous indications about its actual "weak crystallinity" [22]. This, in turn, leads to a dispersion of nanocrystals orientation (tilt and twist), which is further concomitant with electrical and optical properties dispersion (aside from the difficulty for planar processing of NWs arrays).

To combine the advantages of the above-mentioned bottom-up and top-down approaches, significant efforts have been made to develop SAG homoepitaxy. Nowadays, this technique has been successfully used to grow GaN NWs by both MOVPE and MBE techniques and on both GaN/sapphire(0001) and GaN/Si(111) templates [3, 30, 44]. The technique itself relies on three key technological features: (i) a bottom-up approach to obtain high NW chemical purity and crystal quality, (ii) the use of nanohole lithographic masks to control the initial nucleation, achieving NW site and size (diameter and height) control and (iii) an homoepitaxial process (i.e., direct nucleation of GaN islands on the underlying GaN template) to control the NW orientation (tilt and twist) [17]. The SAG technique opens the door to fabricate a wide variety of uniform and scalable NW-based photonic and electronic devices, not available by the former SA heteroepitaxy. These devices include arrays of classical or quantum light nano-emitters (nano-LEDs and SPSs), photonic crystals and vertical surrounding-gate nanoFETs.

10.2.2 SAG Homoepitaxy: NWs Growth Mechanisms

This section reviews the driving mechanisms for GaN NW SAG homoepitaxy, studied from a series of samples grown by MBE [18].

A series of six GaN NWs samples (designated as A to F) was grown in a RIBER Compact 21 MBE system, equipped with a radio-frequency plasma nitrogen source and two standard Knudsen cells for Ga. The growths were performed on 3.3 μm-thick commercial GaN/sapphire (0001) templates (Lumilog). Prior to the MBE growth, a 7 nm-thick Ti layer was deposited on the GaN templates covered with polystyrene nanobeads by e-beam evaporation and further transformed into a nanohole mask (colloidal lithography). The nanoholes follow a regular hexagonal arrangement with 600 nm pitch and 500 nm diameter. Within the MBE growth chamber and prior to epitaxial growth, the Ti mask is converted into a stable TiN one, by exposing it to the plasma nitrogen flux at elevated temperatures. For all samples (A to F) the impinging fluxes and the growth temperature were kept constant (Φ_{Ga} = 15 nm/min, Φ_N = 4 nm/min and T_g = 820°C). The growth time, on the other hand, was varied in steps as t = 30, 60, 90, 120, 180, and 300 min to assess the different stages of the NW formation process. The morphological characterization was performed by scanning electron microscope (SEM, FEI Inspect F50). Further technical details can be found elsewhere [18].

The SAG of GaN NWs is a process which can be divided in two growth stages. Stage I (Fig. 10.2a–c) is driven by SAG kinetics and consists of: (i) initial nucleation of small GaN islands (crystal seeds), typically close to the nanohole edge, (ii) coalescence onset, along the nanohole inner perimeter (ring-like structure) and (iii) full coalescence, resulting in a single nanocrystal per nanohole. In stage II (Fig. 10.2d–f), driven by free-surface-energy minimization, the formed nanocrystal undergoes a morphological evolution from: (iv) a cylindrical-like shape to (v) a dodecagonal shape, ending with a thermodynamically stable (vi) hexagonal shape. Detailed descriptions of these steps are:

(i) Nucleation. During the initial growth stage, the adatoms impinge on the bare surface of the GaN template covered with a TiN nanohole mask. Under SAG conditions, GaN nucleates within the nanoholes (exclusively) and preferentially along their inner perimeter (Fig. 10.2a). During this nucleation step, the number of nucleation seeds progressively increases until their density saturates.

(ii) Coalescence onset. By enlarging their size the nucleation seeds start coalescing to form a ring-like structure. This ring-like structure starts to grow towards the nanohole center, Fig. 10.2b. The number of seeds (per given nanohole) progressively decreases, until eventually a single GaN nanocrystal is formed (Fig. 10.2b–c).

(iii) Full coalescence. The newly formed nanocrystal, with still irregular shape (Fig. 10.2c), continues to extend until it covers the entire nanohole (Fig. 10.2d).

(iv) Cylindrical-like shape. If the SAG proceeds in a perfectly isotropic way, at the end of stage I the GaN nanocrystal shape follows the cylindrical one, that is, the shape of the hosting nanohole (Fig. 10.2d). The cylindrical shape, "imposed" to a wurtzite nanocrystal, contains many energetically "high-cost" non-polar facets and is thus thermodynamically unstable. It is worth noticing that, since not all facets evolve simultaneously, not all nanocrystals (but most of them) show a cylindrical shape (some show already distinct facets).

(v) Dodecagonal shape. The surface energy of *m*- and *a*-planes is lower than that of other non-polar crystal planes. During the nanocrystal lateral growth the free surface of these other non-polar crystal planes is reduced (elimination of energetically "high-cost" nonpolar planes) promoting the formation of *a*- and *m*-crystal facets. Consequently, a dodecagonal (*6a*/*6m*) nanocrystal is formed (Fig. 10.2e).

(vi) Hexagonal shape. Being the surface energy of *m*-planes somewhat lower than that of *a*-planes, the GaN nanocrystal morphology evolves toward a thermodynamically stable hexagonal shape. Consequently, the free surface of energetically favorable *m*-facets gradually expands at the cost of the energetically less favorable *a*-facets (Figs. 10.2e,f). Once the thermodynamically stable hexagonal (*6m*) shape is reached, the GaN nanocrystal ends its morphological evolution.

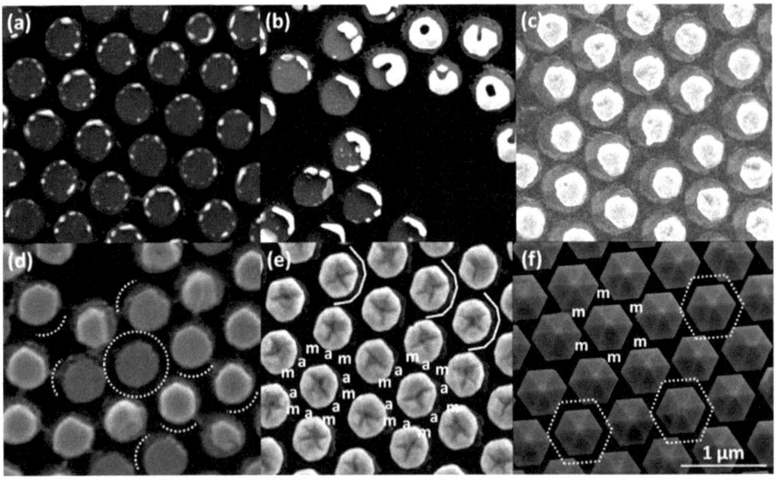

Figure 10.2 Top SEM view of samples A to C and D to F, corresponding to stages I and II, respectively. Stage I: (a) (30 min) Nucleation; (b) (60 min) Coalescence onset; and (c) (90 min) Full coalescence. Stage II: (d) (120 min) Cylindrical-like shape; (e) (180 min) Dodecagonal shape; (f) (300 min) Hexagonal shape (adapted from Gačević et al.) [18].

Model for growth stage I

The growth conditions which promote SAG lead to GaN nucleation within the nanoholes, exclusively. The driving mechanism for SAG is mainly attributed to different Ga adatom and/or GaN nuclei interaction with the two different underlying surfaces. Under steady-state growth conditions, the two surfaces (amorphous TiN and crystalline GaN) are at the same temperature. The much weaker Ga adatom interactions with the amorphous TiN (vs. crystalline GaN) leads to a higher Ga desorption from TiN (vs. GaN), as well as a higher decomposition rate of the GaN seeds that may eventually nucleate on TiN. Consequently, for fixed impinging Ga and N fluxes there is a growth temperature range ("growth window") within which the GaN-on-GaN growth rate is positive (>0) whereas the GaN-on-TiN growth rate is zero (0), resulting in growth selectivity. In addition, the weak Ga adatom interaction with the amorphous TiN enhances the adatom diffusion on the mask which in turn results in a net flux of Ga adatoms (in addition to the directly impinging one) towards the nanoholes (preferential nucleation domains). The peripheral part of the nanohole thus behaves as the preferential area for initial GaN nucleation, as experimentally observed (Figs. 10.2a–b). This result is in qualitative agreement with other nucleation scenarios reported for SAG of III-nitride nanostructures that point to nanoholes periphery (i.e., "next-to-discontinuity region") as the preferential nucleation domain [27, 43, 51].

Model for growth stage II

The final NW size slightly surpasses the size of the underlying nanohole (see Fig. 10.2f). This, however, is not due to a GaN nucleation on the mask (which is hindered by the SAG kinetics) but due to the NW morphological evolution toward the shape which minimizes its total energy per unit volume [46].

A GaN nanocrystal is composed of atoms which, as a function of their energy, can be roughly divided into four groups:

(i) Bulk atoms, located inside the nanocrystal
(ii) Interface atoms, located at or close to the underlying interface
(iii) Surface atoms, located at or close to the free crystal facets
(iv) Edge atoms, located at or close to the crystal edges (where the crystal facets meet)

For the analysis of different energetic contributions (i–iv), the energy per unit volume of an infinite defect- and strain-free GaN crystal is taken as a reference ($E_{\text{bulk}} = 0$). With respect to this value, the energy of atoms may be altered due to either strain or proximity to the crystal boundaries interface, surfaces or edges. The total energy per unit volume of a growing nanocrystal thus can be written as [46]:

$$E_{\text{total}} = E_{\text{strain}} + E_{\text{interface}} + E_{\text{surface}} + E_{\text{edge}}. \tag{10.1}$$

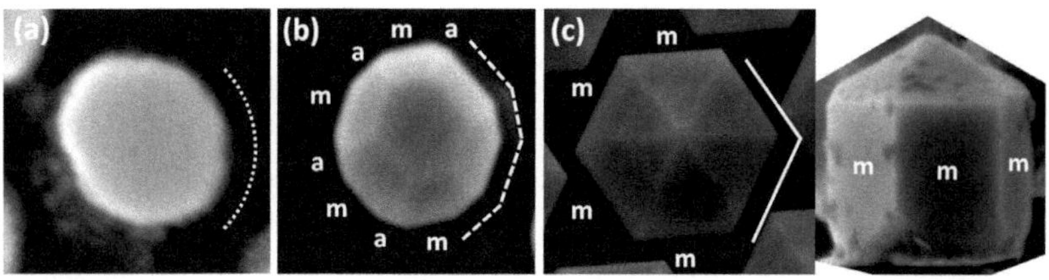

Figure 10.3 Stage II: SEM images of: (a) cylindrical-like, (b) dodecagonal, and (c) hexagonal shape. The relevant crystal facets are indicated for clarity (adapted from Gačević et al.) [18].

The difference in total free energy of two distinct morphologies is likely the driving mechanism for the observed shape transition, that is: $\boldsymbol{E}_{\text{total}}$ (*cylindrical*) > $\boldsymbol{E}_{\text{total}}$ (*dodecagonal*) > $\boldsymbol{E}_{\text{total}}$ (*hexagonal*). Among the four terms contributing to $\boldsymbol{E}_{\text{total}}$ in Eq. (10.1), the energetically dominant one will drive the transition.

Recall that the SAG proceeds homoepitaxially yielding strain-free nanocrystals; the interface between the GaN NWs and the GaN substrate does not store free energy, that is, $E_{\text{strain}} = E_{\text{interface}} = 0$. In addition, the free surface energy per unit volume of a growing nanocrystal (with diameter d and total volume V) is proportional to the total area of its free surface (S):

$$E_{\text{surface}} \sim S \cdot \frac{1}{V} \sim \frac{1}{d}. \tag{10.2}$$

Similarly, the total edge energy per unit volume is proportional to the total edge length (L):

$$E_{\text{edge}} \sim L \cdot \frac{1}{V} \sim \frac{1}{d^2}. \tag{10.3}$$

By increasing the crystal diameter ($d\nearrow$) the contribution of the edge energy vanishes in respect to that of the free-surface energy. Recently, Consonni et al. [11] pointed out that for GaN nanocrystals in the 5 nm size range, the contribution of the edge energy can be significant (thus comparable to contributions of strain and surface energy); in the present case, however, the NWs have nearly two orders of magnitude larger diameter (~500 nm), so that the contribution of the edge energy can be safely neglected (in respect to the free-surface one). A hierarchy in the magnitude of the contributing energy terms for the considered nanocrystal in formation is thus defined as

$$E_{\text{surface}} >> E_{\text{edge}} > E_{\text{interface}} = E_{\text{strain}} = 0. \tag{10.4}$$

yielding

$$E_{\text{total}} \sim E_{\text{surface}} = \frac{1}{V}\sum \gamma_j S_j, \tag{10.5}$$

where γ_j is the surface energy per unit area of the j-th crystal facet and S_j is its corresponding area.

At the onset of growth stage II, the GaN nanocrystal starts its morphological evolution from a cylindrical-like shape (containing virtually all non-polar facets) toward the energetically favorable dodecagonal and hexagonal shapes (Figs. 10.3 and 10.4a). Figure 10.4b summarizes the evolution of the γ value of the GaN nanocrystal outer shell as a function of its shape and rotational coordinate ϕ (due to its six-fold periodicity, the $\gamma(\phi)$ function is featured only over 0 to 60° domain—the periodicity is a direct consequence of the spatial six-fold symmetry of the wurtzite GaN).

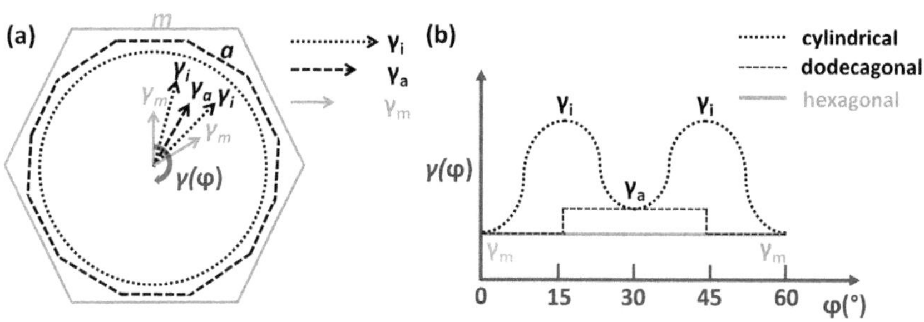

Figure 10.4 (a) Schematic presentation of the nanocrystal shape morphological evolution: cylindrical-like, dodecagonal ($6a/6m$) and hexagonal ($6m$) shape. (b) As the morphological evolution proceeds, the $\gamma(\phi)$ evolves: from an initial sine-wave-like function, to an intermediate step-like function, finishing as a constant function (adapted from Gačević et al.) [18].

Note that the $\gamma(\phi)$ function (ϕ = 0–360°) has 12 local minima, six of them in the direction of m-planes, labeled as γ_m ($\phi_m = 0° + k \cdot 60°, k = 0, 1, ..., 5$) and six of them in the direction of a-planes, labeled as γ_a ($\phi_a = 30° + k \cdot 60°, k = 0, 1, ..., 5$). Similarly, the function also exhibits 12 local maxima. The crystal planes at which $\gamma(\phi)$ maximizes (the energetically "most costly" non-polar facets) are labeled as intermediate i-planes, being the corresponding function maxima labeled as γ_i ($\phi_i \sim 15° + k \cdot 30°$, $k = 0, 1, ..., 11$, Fig. 10.4a,b) [18].

The existence of 12 local maxima of $\gamma(\phi)$ is identified as the driving mechanism for the temporal nanocrystal dodecagonalization. Namely, the fact that $\gamma(\phi)$ function has its maxima in 12 crystal directions implies that the initial cylindrical-like nanocrystal will tend to minimize the free surface of the energetically "high-cost" i-facets, by means of a faster growth rate along these 12 directions (in respect to other non-polar directions), leading to a dodecagonal $6a/6m$ nanocrystal shape

(Figs. 10.3 and 10.4). Similarly, the existence of six minima of the $\gamma(\phi)$ function is identified as the driving mechanism for the final NW hexagonalization; being the formation of free *m*-facets energetically preferential in respect to free *a*-facets ($\gamma_m < \gamma_a$) the GaN nanocrystal continues its evolution from the *6a/6m* dodecagonal to the 6*m* hexagonal form (Figs. 10.3 and 10.4). From this point on, the NW vertical growth proceeds while keeping the thermodynamically stable hexagonal shape.

10.2.3 Discussion: SA Heteroepitaxy versus SAG Homoepitaxy

Figure 10.5 shows top and lateral SEM images of samples A (SA) and B (SAG) [17]. The SA heteroepitaxy is characterized by a random self-nucleation mechanism and a poor epitaxial relation with the Si(111) substrate through the SiN interlayer. This leads to: (i) a random NW positioning (no site control), (ii) significant dispersion in NW size (diameter and height) and (iii) misorientation (tilt and twist). A relatively high dispersion in NWs size and orientation is evident in the SEM images of Fig. 10.5A. The right-hand insets in Fig. 10.5 represent the corresponding reflection high-energy electron diffraction (RHEED) pattern, taken in situ along the (11–20) azimuth. The annular RHEED pattern, a fingerprint of SA heteroepitaxy, is a confirmation of a poor crystallographic uniformity of the analyzed ensemble.

The SAG homoepitaxy (Fig. 10.5B) is, on the other hand, characterized by ordering and a full epitaxial relationship with the GaN/sapphire template. The combination of these two desired properties leads to NWs (i) site control, (ii) low size dispersion and (iii) high orientation uniformity. The RHEED in this case, also taken in situ along the (11–20) azimuth, reveals a bright spotty pattern, a signature of 3D electrons diffraction in a single-crystal, a fingerprint of the high crystallographic uniformity of the analyzed ensemble.

SEM images in Fig. 10.5 clearly show that the SAG homoepitaxy leads to a dramatically improved uniformity. To quantify it, a statistical analysis of the NWs diameters over a 2.2 × 2.2 μm^2 area of SEM images (top-view images featured in Fig. 10.5) was done using free software from Nanotec. As expected the analysis reveals a huge improvement on NW diameter uniformity from SA to SAG

NWs, characterized by standard deviations of $\sigma_D(A) \sim 45\%$ and $\sigma_D(B) \sim 6\%$ respectively. Notice that the statistical broadening in sample A is mainly attributed to the random nature of the initial nucleation and further NWs coalescence. A slight statistical broadening in SAG NWs is mainly attributed to the size dispersion of the polystyrene nanobeads (estimated at ~5%) used for the fabrication of the colloidal Ti nanohole mask [17].

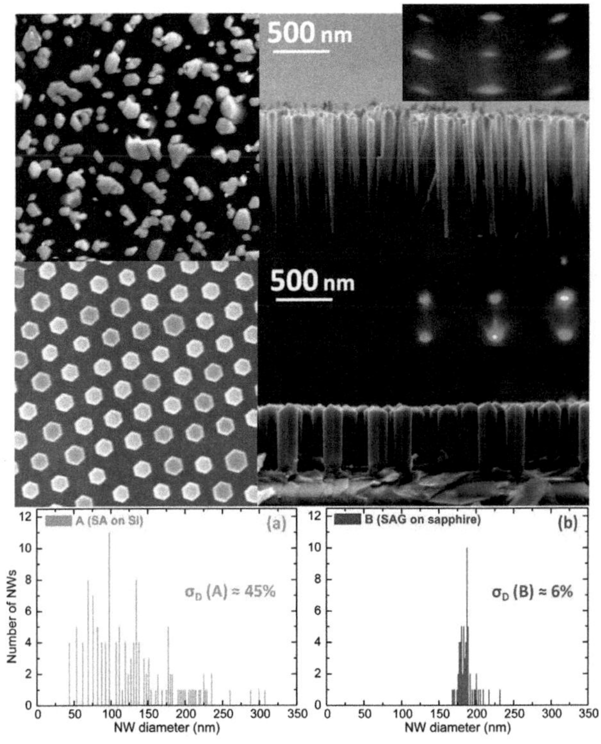

Figure 10.5 (A-B) Top and lateral SEM images from samples A and B. The insets represent the corresponding RHEED patterns, taken in situ along the (11–20) azimuth. Panels a and b show statistical analyses of NWs diameters, from the SEM images in panels A and B. The number of analyzed NWs is: 132 and 60, respectively. The diameter standard deviations σ_D are estimated to be ~45% and 6%, respectively (adapted from Gačević et al.) [17].

Consonni and co-workers reported that in SA heteroepitaxial growth of GaN NWs on a Si(111) substrate, the NW formation follows a two-step process [11]. The first step (nucleation) occurs under a weak epitaxial relation due to the presence of the SiN interlayer which impedes direct GaN-to-Si crystal-to-crystal contact. It is initially driven by adatom surface kinetics and results in GaN nucleation in the form of spherical caps at random sites. The second step (morphological evolution) is, on the other hand, driven by total energy per unit volume minimization. At a certain (critical) size, the spherical cap undergoes a morphological transition into a thermodynamically stable hexagonal prism. From this point on, the NWs vertical growth proceeds preserving the stable hexagonal form. In contrast, the initial nucleation in the case of SAG homoepitaxy is facilitated by the adatom surface kinetics and determined by the nanohole lithographic mask; it is the nanohole diameter which determines the onset of the second growth stage, during which the nucleation seed undergoes morphological transition toward thermodynamically stable hexagonal shape. Finally, recall that in the SAG case the growth proceeds homoepitaxially, i.e., under strong epitaxial constraint, thus the NW orientation is directly inherited from the crystallographic orientation of the GaN pseudosubstrate crystal planes. The combination of previous highly desired factors explains the high controllability over SAG homoepitaxy (versus SA heteroepitaxy), more precisely over NWs' nucleation site, size and orientation.

Concerning the very initial phase of SA heteroepitaxy, notice that the crystal accretion (a gradual accumulation of the material around the nucleation center) is usually suggested as the mechanism which drives the initial nucleation seed (spherical cap) formation [11]. In contrast, the SAG GaN NWs do not develop from a single nucleation seed (Fig. 10.2), but from coalescence of several ones, being thus prone to defect generation [18]. In addition, because of the direct SAG homoepitaxy on the GaN/sapphire template, dislocations already existing in it may propagate to the growing NWs. Although dislocations may generate as well when growing GaN NWs heteroepitaxially by SA on Si(111) (to accommodate strain), their density and occurrence on a given NW would be much lower than for homoepitaxial SAG (in this case, obviously, depending significantly on the GaN/sapphire template quality, but commonly in the $\sim 10^9$ cm^{-2} density range). These two intrinsic differences between the homo- and heteroepitaxy might lead to a significantly lower crystal quality of homoepitaxial SAG GaN NWs.

To investigate crystal quality of SAG GaN NWs, a detailed structural analysis of the NW bottom and top parts has been performed (Fig. 10.6) [1, 17]. High-resolution transmission electron microscopy (HRTEM) data from the NW bottom part (Fig. 10.6a) reveals a dense network of dislocations, which are either inherited from the underlying template or created during the initial SAG growth stage (nucleation seeds coalescence, Fig. 10.2). The generated dislocations "wander around" within the NW until they reach its free surface, where they eventually get annihilated. This mechanism is commonly referred to as dislocation filtering [45]. The crystal structure of the NW top part (Fig. 10.6b) is, on the other hand, found free from crystal defects. High angle annular dark field (HAADF), selective area electron diffraction (SAED) and annular bright field (ABF) images reveal a virtually perfect wurtzite crystal structure. The dramatic improvement of the NW crystal quality from its bottom toward its top part is mainly attributed to the dislocations filtering effect.

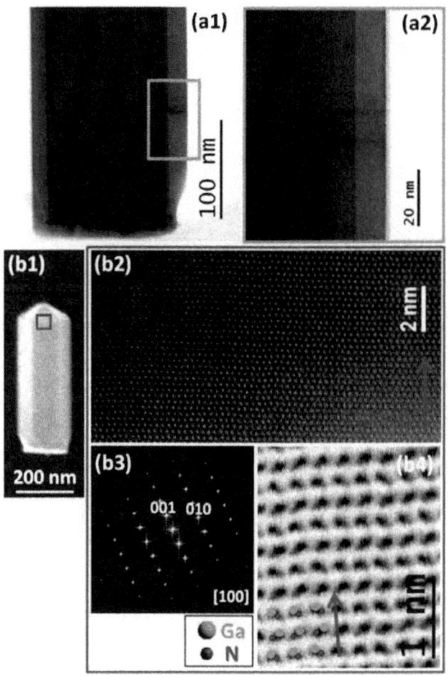

Figure 10.6 (a1–2) HRTEM images of a homoepitaxial SAG GaN NW showing initial dislocations and the mechanism for their filtering. (b1–4) SEM, HAADF, SEAD and ABF images, respectively, taken at the NW top part, reveal a virtually perfect wurtzite structure (adapted from Aseev et al. and Gačević et al.) [1, 17].

Bearing in mind that TEM provides very local structural information from the studied sample, low-temperature PL measurements (~10 K) on NWs from samples A and B (SA and SAG, respectively, see Fig. 10.5) have been performed and compared to that from the commercial GaN/sapphire template. To avoid interfering PL signals from the GaN buffer and/or parasitic GaN material nucleated on the mask, the studied NWs were transferred to optically inactive bare Si substrates (Fig. 10.7A,B); in the PL experiment about 10^6 NWs were probed together. Before starting the PL data analysis, it is worth to recall the expected features from a nearly perfect crystal:

(i) intense near-band-edge emission
(ii) narrow near-band-edge emission
(iii) no below-band-gap emission (other than phonon replica)

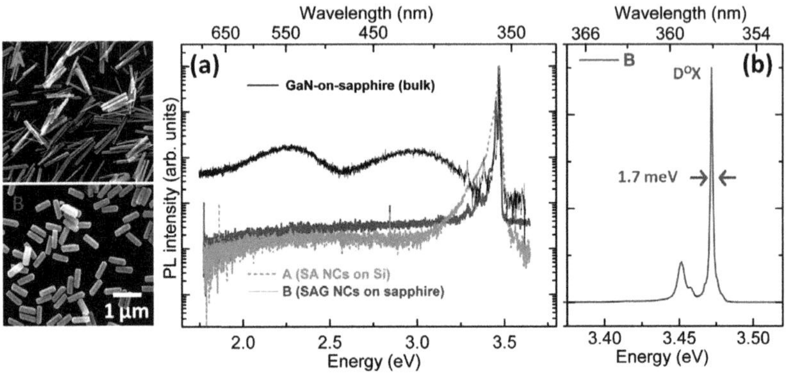

Figure 10.7 SEM images of NWs transferred to optically inactive Si substrate from samples A (SA GaN NWs) and B (SAG GaN NWs). (a) Normalized PL spectra of samples A and B compared to that of the commercial GaN/sapphire template (Lumilog). PL spectra recorded on both NW types are characterized by intense and narrow near-band-edge and no below-band-gap emission. (b) A donor-bound-exciton (D_0X) line as narrow as 1.7 meV is recorded from a bunch of ~10^6 random NWs from sample B, further confirming the material's outstanding crystal quality (adapted from Gačević et al.) [17].

Considering the GaN/sapphire template, due to the high lattice mismatch the generation of a dense network of threading dislocations at the hetero-interface is expected, typically in the 10^9 cm^{-2} density range. Consequently, the PL spectrum from the GaN template shows a significant below-band-gap emission

(Fig. 10.6a), most remarkably the yellow band. On the other hand, PL spectra from GaN NWs is characterized by a very intense near-band-edge emission (feature (i)) and is dominated by a narrow donor-bound exciton (D_0X) line, at 3.471 eV (Fig. 10.6b). The D_0X line-width is about 1.7 meV for sample B (feature (ii)), a similar value to the narrowest lines reported for SA GaN NCs [15, 42]. Finally, below-band-gap emission bands are not present in the GaN NWs (feature (iii)).

The systematic comparison of SA and SAG GaN NWs clearly points to the exceptional potential of the SAG method to develop nanodevices from NW arrays (also from single NWs). The key advantages of the method can be summarized in (i) site-, (ii) size-(diameter and height), and (iii) orientation- (tilt and twist) control. The following two sections address the use of GaN NWs to fabricate quantum light emitters and field effect nanotransistors.

10.3 Ordered InGaN/GaN NWs as Arrays of Quantum Light Emitters

10.3.1 III-Nitride Quantum Dots as Single Photon Sources

The achievement of highly efficient GaN-based classical light emitters (light-emitting diodes and laser diodes) has driven a revolution in the market of solid-state lighting [41]. In the case of classical light emitters the number of emitted photons (in a given time interval) follows a statistical distribution. On the other hand, their quantum counterparts, commonly referred to as single photon sources (SPSs), generate individual photons (or a pair of photons) in response to an external demand (trigger). When triggered, an ideal SPS emits a single photon with a probability of 100% [7, 58].

Among different approaches to develop SPSs, the one based on semiconductor quantum dots (QDs) has attracted significant attention [7]. The main advantages of this approach are the potential for excellent SPS controllability (emission wavelength, lifetime, photon polarization and operating temperature) and the compatibility with the well established semiconductor technology. A significant boost to semiconductor SPS research was given in

2000 when the first optically pumped SPS was reported on single InAs QD (grown by Stranski-Krastanov, SK) [37]. The exploitation of InAs/InP QDs further allowed single-photon operation at telecommunication wavelengths (~1.55 μm) compatible with silica-fiber-based optical communications [48].

As for III-nitride semiconductors, it is very interesting to note that the key features which make them a material of choice for classical light emission also qualify them as exceptional candidates for quantum light emitters, operated over a wide spectral range and at high temperatures. The benefits of the III-nitrides highly varying band gaps have already been partially exploited to achieve SPSs with emission wavelength tunable over the 280–620 nm range [10, 12–14, 23, 24, 28, 29, 31, 34, 49, 53, 57].

Next, the benefits of their high band offsets have also been exploited for strong exciton confinement which allows single photon operation at room temperature and even in hot environments (at 350 K) [13, 23, 24]. A certain degree of single photon polarization control has also been achieved via manipulation of the emitting QD shape [49]. Finally, electrically triggered SPSs have also been reported using QDs in both columnar and bulk device morphologies [13, 14].

Many recent advances in the field of III-nitride SPSs have been achieved using QDs embedded in NWs [10, 12–14, 23, 24, 28, 34, 49, 53, 57] obtained by both top-down and bottom-up approaches. In the latter case, thanks to their high crystal quality, GaN NWs can be used to host QDs with exceptional crystal structure. In addition, the growth of GaN NWs can be site-controlled and they can be easily removed from their native substrates for individual spectral characterization. These features are highly desirable for SPS integration into photonic nanodevices.

In this section, a brief review of the recent progress in the field of GaN NW-based quantum light emitters is given. The most relevant SPS designs, those relying on QDs embedded in 3D bulk and 1D NW crystals, are commented. The SPS fabrication is also addressed in respect to top-down vs. bottom-up approaches. The specific fabrication details and optical characterization of SPSs grown by SAG MBE are described. Finally a systematic comparison between different SPS types is given.

10.3.2 III-Nitride Dot-in-a-Wire SPSs: Top-Down versus Bottom-Up Approach

The first report of a III-nitride SPS by Kako et al. in 2006 was based on SK GaN/AlN QDs grown by MOVPE and emitting at ~350 nm [29]. The fingerprint for SPS operation is obtained from second order photon correlation measurements $g^{(2)}(\tau)$ using the Hanbury Brown and Twiss (HBT) setup [21]. The $g^{(2)}(\tau)$ function describes the joint probability of detecting the arrival of a photon at time t and of another one at time $t + \tau$. For a SPS the $g^{(2)}(\tau)$ function exhibits a pronounced "antibunching" dip, ideally reaching a zero-value at zero-time delay: $g^{(2)}(0) = 0$. The GaN/AlN QD SPS developed by Kako et al., showed a clear antibunching behavior $g^{(2)}(0) = 0.30$, with the single-photon emission character up to 200 K [29]. Such temperature stability was attributed to the high GaN/AlN band offsets which prevented carrier escape. That high-temperature stability is difficult to achieve by other III–V materials and this early result strongly suggested that the nitrides SPSs could achieve highly desirable room temperature operation.

The fabrication process reported by Kako et al. relied on GaN SK QDs, a method which is characterized by an intrinsically low controllability. The GaN SK QDs were embedded in 3D compact layers, thus not quite convenient for individual QD spectroscopy either.

In the last few years several groups developed dot-in-a-wire nitride-based SPSs following either a top-down (Fig. 10.8a), or bottom-up (Fig. 10.8b–d) approaches. Zhang et al. (Fig. 10.8a) reported quantum light generation from ordered arrays of GaN nanopillars containing InGaN quantum disks. These SPSs were fabricated in a three-step procedure, consisting of MOCVD growth of a single InGaN QW on a GaN template, capped with a thin GaN layer [57]. The planar structure was then patterned via electron beam lithography and subsequently etched into pillars by inductively coupled plasma reactive-ion etching. The finished device contains multiple arrays of ordered nanopillars, each pillar containing a single InGaN nanodisk as the QD active region. Deshpande et al. fabricated p-i-n GaN/InGaN/GaN NW SPSs (Fig. 10.8b) grown by SA PA-MBE on Si(111) substrates [14]. Similar to Zhang et al., each individual NW contains a single

InGaN nanodisk as active region. Holmes et al. reported on ordered arrays of GaN/AlGaN NWs with a single GaN nanodisk as active region for single photon emission (Fig. 10.8c) [23]. They were grown by MOCVD on AlN templates covered with a nanohole SiO_2 mask (patterned by electron beam lithography and reactive ion etching), resulting in ordered arrays of pencil-like NWs. Similarly, Chernysheva et al. reported ordered arrays of pencil-like InGaN/GaN NWs, each one containing a single InGaN nanodisk [10]. The pencil like morphology favors the formation of nanodisks including both polar and semipolar facets (Fig. 10.8d). The growth was performed by plasma-assisted MBE SAG on GaN/sapphire templates covered with a nanohole Ti mask (colloidal lithography). The following section provides a detailed insight into the fabrication and optical characterization of this particular SPS structure. Then, a systematic comparison of most relevant III-nitride quantum light sources reported to date is addressed in the discussion section.

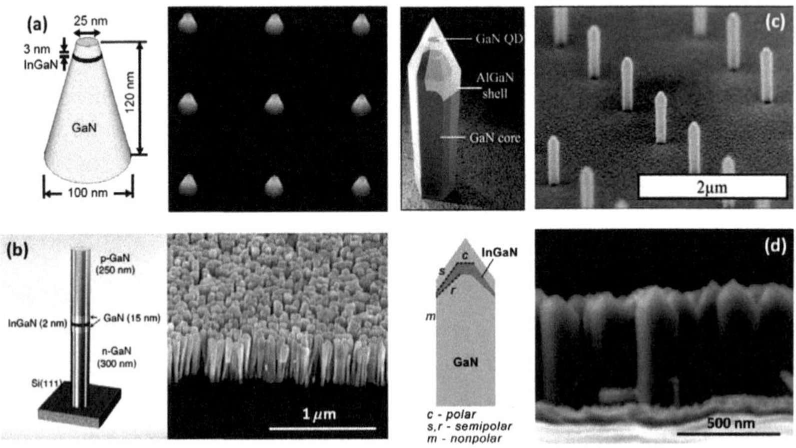

Figure 10.8 Schematics and SEM images of III-nitride SPSs fabricated by top-down (a) and bottom-up (b–d) approaches. (a) An ordered array of InGaN dot in GaN nanopillar SPSs, patterned via electron beam lithography and etched by inductively coupled plasma reactive-ion etching. (b) InGaN/GaN dot-in-a-wire SPSs, fabricated by SA MBE. (c) GaN/AlGaN dot-in-a-wire SPS fabricated by SAG MOVPE (d) InGaN/GaN dot-in-a-wire SPSs fabricated by SAG MBE (adapted from Zhang et al., Desphande et al., Holmes et al. and Chernysheva et al.) [10, 14, 23, 57].

10.3.3 InGaN/GaN NW SPS Grown by SAG MBE

The samples were grown in a RIBER Compact 21 MBE system equipped with a radio-frequency plasma nitrogen source and standard Knudsen cells, two for Ga and one for In [20]. The growth was performed on commercial GaN/sapphire templates covered by a ~7 nm-thick Ti mask with a matrix of hexagonal ordered nanoholes (300 nm pitch and 180 nm diameter). The structure growth proceeded in three steps: (i) initial pencil-like GaN NW, (ii) InGaN quantum disk and (iii) top GaN capping. To achieve selectivity, step (i) was performed at high temperature (~820°C) resulting in ordered GaN NWs with pencil-like morphology. To allow In incorporation and the formation of InGaN quantum disks the growth temperature was decreased to ~620°C. Finally the structure was capped with GaN at the same growth temperature.

Figure 10.9a,b shows a top SEM image of a representative sample area and a bright field SEM (BF-STEM) micrograph of an individual NW. Figure 10.9a confirms the ordered NW growth, with highly uniform diameters and crystallographic tilts. The STEM micrograph in Fig. 10.9b reveals the shape and size of each of the three NW sections.

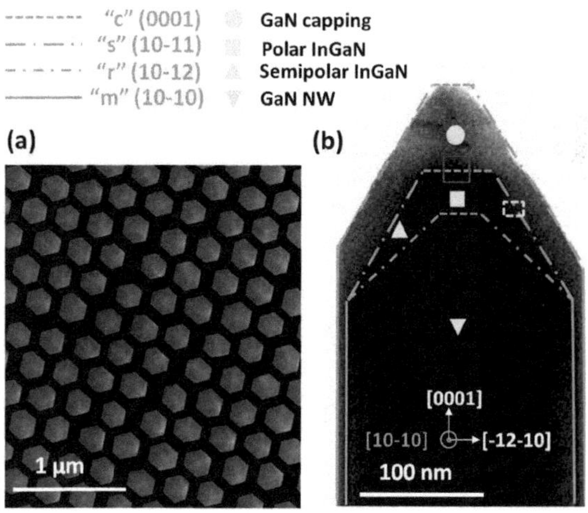

Figure 10.9 (a) Top SEM image of a sample area revealing an ordered hexagonal array of pencil-like SPSs. (b) TEM micrograph of a single NW revealing its structure, i.e., different sections (adapted from Gačević et al.) [20].

The initial GaN NW shows a truncated pyramidal top, built out of a polar *c* facet and six semipolar *r* facets (*c*-6*r* shape). During the second growth step, the InGaN nanodisk grows basically on the polar *c* and semipolar *r* facets (with a negligible radial growth, on the nonpolar *m* facets). When starting the second growth step (InGaN nanodisk) the growth temperature and III/N ratio are both significantly reduced producing a change of the energetically preferential NW top shape, from a *c*-6*r* to a *c*-6*s* one, being the *s* facet another semipolar plane (Fig. 10.9b). During the third step (GaN cap) the growth temperature is kept unchanged, whereas the III/N ratio is further slightly reduced (due to In cell shutter closing); this change does not affect the preferential form of the NW top section, as seen in Fig. 10.9b [20].

Figure 10.10 shows a chemical mapping of the same NW, performed by energy dispersive X-ray spectroscopy (EDS). In the EDS experiment, a single lying NW is probed by a high-energy electron beam, which penetrates perpendicularly to the growth direction and excites the NW tip at different heights (Fig. 10.10a). Figure 10.10b shows the spatial distribution of the In-$L\alpha 1$ line intensity. A moderate intensity of the In-$L\alpha 1$ line at positions corresponding to semipolar InGaN, indicates a moderate In content on these planes (estimated at ~10%). The In-$L\alpha 1$ line intensity rises significantly when polar InGaN gets excited, revealing a significantly higher In content of the polar InGaN (estimated at ~20%). Finally, the In-$L\alpha 1$ intensity drops to zero when the GaN capping is probed indicating zero In content (for further experimental and technical details see Gačević et al.) [20].

The strong difference of the In content from semipolar to polar InGaN planes leads to different emission wavelength (color). Figure 10.10b shows a direct nanoscale correlation of the annular dark-field (ADF) STEM image and the CL wavelength image of a single NW. The CL exhibits a dominant emission centered at 360 nm, attributed to GaN bulk emission. The InGaN nanodisk clearly shows two emission bands, the first one, centered at 400 nm, originates from the semipolar section, whereas the second one, centered at 500 nm, originates from the polar top one.

Micro-PL (μPL) measurements were performed with the laser beam focused to a ~1.5 mm spot probing simultaneously around 30 standing NWs (a native sample) [10]. The results confirm the same emission bands (GaN bulk, semipolar InGaN and polar InGaN) as evidenced in the spectrum shown in Fig. 10.11a. In addition, the spectrum reveals sharp and intense Lorentzian QD-like emission lines, most likely induced by intra-disk indium clustering. These QD-like centers are actually the single photon emission centers and are discussed later.

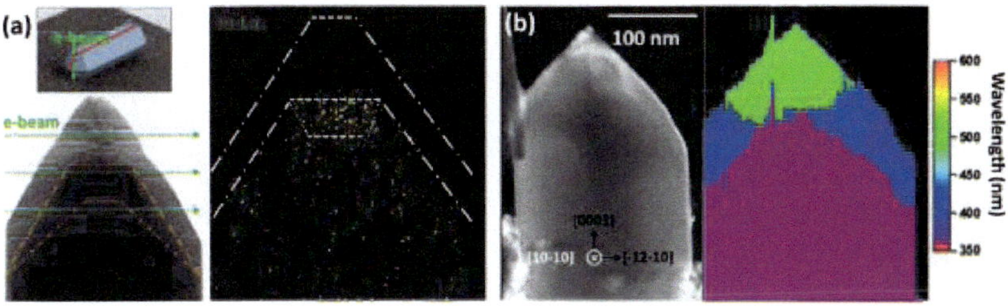

Figure 10.10 (a) A NW probing is performed as shown in the schematics (left), with the e-beam penetrating perpendicularly to the central c-axis (green arrow) at different points. The spatial distribution of the In-Lα1 line intensity yields information about In incorporation on different facets of the InGaN nanodisk. The In content is found significantly higher in the polar than in the semipolar section. (b) ADF STEM image of a single NW combined with a high spatially resolved CL mapping (at 15 K). The images reveal that the green/violet luminescence originates from the polar/semipolar InGaN, whereas the UV luminescence comes from the core GaN (the wavelength is color coded, adapted from Gačević et al.) [20].

Figure 11.11b shows polar InGaN emission from a series of samples. The light originating from the polar InGaN nanodisk section (Fig. 10.11b) can be tuned across the blue-green spectral emission range (from 2.80 to 2.35 eV) only by changing the growth temperature and/or impinging In/Ga fluxes, while keeping the same nanostructure design (which is the one shown in Fig. 10.9b).

For quantum optics experiments, the individual NWs were assessed. μ-PL spectrum from single NW is presented in Fig. 10.12a. The intense, narrow lines observed within the polar InGaN emission band indicate that In fluctuations across this nanodisk polar section produce quantum confinement of excitons in all three spatial directions, that is, 0D quantum states. The linewidth of these QD-like emissions, found in 350–600 μeV range, is similar to values reported for high-quality SK InGaN QDs [31].

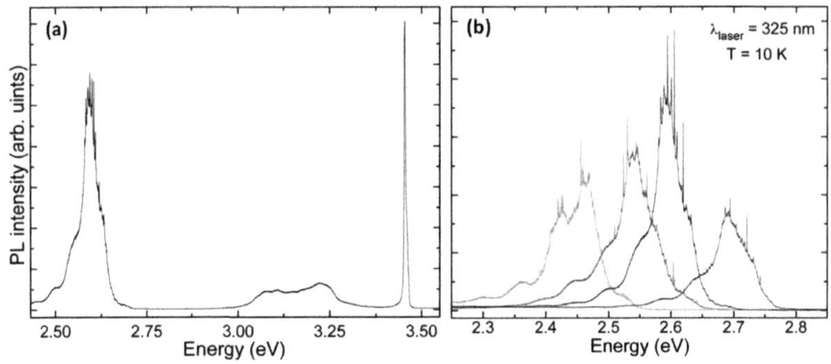

Figure 10.11 (a) μ-PL spectrum from an ensemble of NWs. (b) Enlarged μ-PL spectra from the polar sections of InGaN nanodisks. The featured spectra were taken on four different samples. These samples contain NWs with varied indium content and identical design, featured in Fig. 10.10b (adapted from Chernysheva et al.) [10].

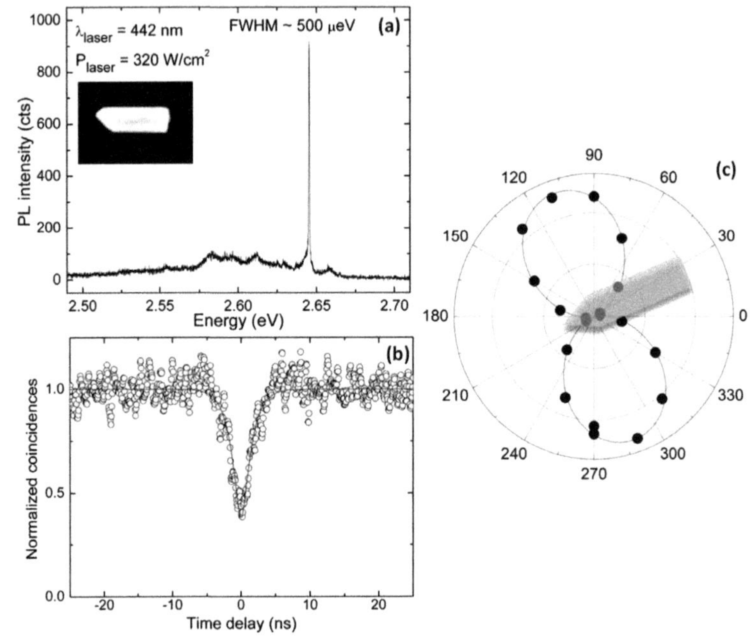

Figure 10.12 (a) μ-PL spectrum of a single (dispersed) NW. (b) Normalized coincidence counts for the QD emission peak in (a) showing the anti-bunching dip. (c) Angular polarization distribution of the QD-like emission, showing strong linear polarization perpendicular to the NW c-axis (adapted from Chernysheva et al.) [10]. The lines in panels b and c are fits to appropriate equations, details of which can be found elsewhere [10].

The QD-like emission "spikes" are observable up to 80 K (not shown). The exceptionally narrow emission line with Lorentzian shape strongly suggests that these sharp PL lines are due to recombination of electrons and holes confined at 0D quantum objects. The linear dependence of the integrated PL peak intensity on the excitation power density (not shown), further suggests single exciton recombination (i.e., not charged exciton or biexciton recombination).

To test the nature of light originating from these emission centers (classical vs. quantum), second order coherence $g^{(2)}$ measurements (intensity autocorrelation) were performed on selected NWs. As a selection criteria, the spectral width (ΔE < 1 meV) and emission intensity (signal-to-total counts ratio $\rho = S/(S + B) > 0.75$, where S and B stand for the main signal and background counts, respectively) were used. Figure 10.12b shows a μPL spectrum of one selected QD, hosted within the polar section of the InGaN nanodisk. The histogram shows a pronounced antibunching from a reduction of coincidences at zero time delay. After correction to account for the instrument time response, the second order photon correlation at zero-time-delay was estimated at $g^{(2)}(0) = 0.20 \pm 0.04$. This value was further corrected to account for background contamination (originating from the photons not emitted by the analyzed emission center, but which eventually get detected by the HBT detectors) yielding a final $g^{(2)}(0) = 0.06 \pm 0.04$ value, which is far below the 0.5 two-photon threshold. The second order coherence measurements thus unambiguously prove the quantum character of the emitted light.

Figure 10.12c summarizes the optical polarization measurements performed on the same QD emission peak. The emitted light has a high degree of linear polarization (DLP), estimated at 90% (note that high linearly polarized single photons are desirable for quantum communications). The high degree of in-plane linear polarization (>70%) with no preferential crystallographic direction has been found for all studied intra-disk InGaN QDs (not shown). These findings suggest the random asymmetry of the probed QDs, which agrees well with the random In fluctuations being the driving mechanism for their formation.

10.3.4 Discussion

The data in Table 10.1 summarize the most relevant parameters of III-nitrides-based SPSs reported so far [10, 12–14, 23, 24, 28, 29, 31, 34, 49, 53, 57]. Note first that, concerning the fabrication technique, the bottom-up approach is much more frequently used than the top-down one. There are several factors which make the top-down approach less efficient for SPS applications. First, due to a very small Bohr radius (the typically cited value of which is ~3 nm for III-nitride material system) the nitride QDs must be very small to host 0D excitons, which is at or below the limit of the existing lithographic techniques. Second, the crystal quality of QDs acting as SPSs is a highly critical issue. Since the crystal quality of bulk GaN layers, from which the top-down etching proceeds, is significantly lower than that of 1D NWs (bottom-up approach), the crystal quality of the resulting top-down nanostructures will be much lower, as well. Combined together, these two factors favor the use of bottom-up approaches, even though being more complicated from a technological point of view.

Table 10.1 The summary of the most relevant parameters of GaN-based SPSs reported so far [10, 12–14, 23, 24, 28, 29, 31, 34, 49, 53, 57]

Reference	Material/ structure type	Emission wavelength	Best $g^{(2)}(0)$	Max reported operating T	Single photon polarization
S. Kako et al. (2006)	GaN/AlN SK QDs	~355 nm	0.30	200 K	—
Kremling et al. (2012)	InGaN/GaN SK QDs	~426 nm	0.23	50 K	Linear DLP ~ 80%*
S. Desphande et al. (2013)	InGaN/GaN dot in SA NW**	~426 nm	0.16	10 K	Linear DLP ~ 70%
S. Desphande et al. (2013)	InGaN/AlGaN dot in SA NW	~450 nm	0.20	200 K	Linear DLP ~ 78%
L. Zhang et al. (2013)	InGaN disk in etched GaN nanopillars***	~420 nm	0.18	90 K	Linear DLP = 84%
M. Holmes et al. (2014)	GaN/AlGaN dot in SAG NW	~285 nm	0.12	300 K	—

Reference	Material/ structure type	Emission wavelength	Best $g^{(2)}(0)$	Max reported operating T	Single photon polarization
S. Kako et al. (2014)	GaN/AlN SK QDs****	~336 nm	0.05	100 K	—
S. Desphande et al. (2014)	InGaN/GaN SK QDs**	~600 nm	0.11	280 K	—
S. Lazić et al. (2015)	InGaN/GaN dot in SAG NW	435–525 nm	0.06	80 K	Linear DLP = 92%
E. Chernysheva et al. (2015)	InGaN/GaN dot in SAG NW	435–525 nm	0.06	80 K	Linear DLP > 90%
C. H. Teng et al. (2015)	InGaN disk in etched GaN nanopillars***	~420 nm	0.26	10 K	Linear DLP = 60–80%
T. Yamamoto et al. (2016)	InGaN/GaN dot in SAG NW	560–600 nm	0.40	20 K	Linear DLP ~ 90%
M. Holmes et al. (2016)	GaN/AlGaN dot in SAG NW	~292 nm	0.34	350 K	—

*Averaged value for 16 analyzed QDs.
**Electrically driven.
***Top-down.
****Cubic GaN QDs.

It is also worth to notice that most III-nitride SPSs rely on wurtzite (hexagonal) QDs, although, some results on zincblende (cubic) QDs have also been reported [28]. The wurtzite nitrides are polar materials in which an internal electric field is created at the polar (and semipolar) heterostructure interfaces. This leads to an intra-dot electron–hole separation and thus a higher exciton lifetime and slower SPS operation. The motivation behind the use of zincblende (nonpolar) nitrides (Kako et al.) is to significantly suppress the internal electric field [28]. The absence of internal electric fields within zincblende QDs leads to a higher electron–hole wavefunction overlap, thus a faster electron–hole recombination and SPS operation. However, the zincblende QDs require a more complicated fabrication process, due to the intrinsic instability of this crystal phase in the III-nitride material system.

III-nitride SPSs are most commonly based on SK QDs and dots-in-wire structures (both bottom-up). The advantages of the latter ones are typically a higher crystal quality of the host material (1D NW versus 3D bulk layer) and their potential for posterior integration into nanophotonic circuits. Due to a wide range of III-nitrides band gaps, they are the material of choice to cover a wide emission range (from 200 to 1700 nm). Indeed, SPS operation with III-nitrides has already been reported over a wide wavelength range (285–620 nm). Lazić et al. and Chernysheva et al. demonstrated that the SPS wavelength can be relatively easily tuned over the blue and green spectral ranges just by increasing the In content of the active region while keeping the same nanostructure design [10, 34]. The higher success achieved at shorter emission wavelengths (UV and visible) versus longer ones (infrared) is likely due to higher InGaN emission efficiency for low (<40%) vs. high (>40%) indium content (the same result is also widely observed for III-nitride quantum wells [48]). It is worth to notice that high-In content emitters could provide single photon emission at both 1.3 mm and 1.55 mm telecommunication wavelengths, opening their use and integration with existing silica-fiber communication systems.

Similarly, the possibility to form QDs with very high band offsets using III-nitrides allows SPS operation at high temperatures. As already mentioned in Section 10.3.2, the very first reported GaN-based SPS exhibited proper operation up to 200 K [29]. In recent years room-temperature operation was demonstrated on both GaN/AlGaN and InGaN/GaN QDs [13, 23]. Actually, Holmes et al. reported that GaN/AlGaN system is appropriate for operation even in hot environments (up to 350 K) [24].

Finally, a high degree of linear polarization (DLP) is systematically observed for all III-nitride QDs (>60%) [10, 12, 14, 31, 34, 49, 53, 57]. Notice that, concerning III-nitrides, the strong valence band mixing, high internal electric fields (wurtzite) and strain anisotropy all lead to a strongly anisotropic physical system [2]. The system anisotropy is, very likely, also enabled by the QD shape randomness expected to be particularly strong in the case of InGaN QDs, due to proneness of indium to clustering. The combination of the previous factors likely results in a high system anisotropy, leading to high degree of linear optical polarization, as systematically observed in numerous experiments.

10.4 GaN Nanowire Fabrication

10.4.1 GaN NWs as Field-Effect Nanotransistors

Recent designs of commercial devices, such as FinFETs and trigate transistors have broken the standard planar (2D) approach and stepped into the area of non-planar (3D) transistor architectures. The main motivation for 3D architectures is a further improvement in electrostatic control of the transistor channel and a possibility of transistors vertical stacking (3D computers) [16, 32, 33].

Nanowire field-effect transistors (NW-FETs) are promising candidates to further boost the progress in device scaling, making use of non-planar approaches. Several key factors make semiconductor NWs attractive building blocks for 3D architectures. Their outstanding crystal quality combined with smooth lateral surfaces (in particular in the case of bottom-up NWs) provides high carrier mobility, in comparison with "top-down" nanodevices with a similar size [52]. They can be fabricated with a wide range of aspect ratios and their electrical properties can be precisely controlled. Finally, they can be processed both horizontally and vertically [36].

Among many different materials (Si, Ge, InAs, ZnO, SnO_2, In_2O_3, etc.) the GaN NWs have attracted significant research attention, since they can be easily fabricated by bottom-up approach with a virtually perfect crystal structure, as already commented. GaN is an environment-friendly material, exhibiting a high thermal stability and a high breakdown electric field. Similar to many other semiconductor NWs, due to surface Fermi-level pinning (and depending very much on the doping) GaN NWs exhibit a lateral depletion space charge layer (from the surface toward the NW core). Consequently, depending on the NW diameter and doping, conducting (or fully depleted) channels along its core can exist [8].

In this section, the recent progress in the field of GaN NW transistors is briefly reviewed, commenting on different fabrication approaches (bottom-up vs. top-down) and processing methods (back-gate vs. top-gate or surrounding-gate). Then a detailed description of top-gate GaN NW metal-semiconductor (MES) FETs fabrication and electrical characterization is given. Four different transistor working regimes are identified and explained. Detailed insights into background mechanisms which drive each of the regimes, as well as the transistor equivalent large-signal models for each of them are given. Finally, the results obtained by different fabrication methods are systematically compared and discussed.

10.4.2 Fabrication and Processing Approaches: Bottom-Up versus Top-Down

First GaN NW-based transistors were reported in 2002, using NWs grown by laser assisted catalytic (bottom-up) method, further processed into Si back-gate configuration (Fig. 10.13a) [26]. Devices exploiting similar back-gate designs were then reported by several other groups. The back-gate GaN NW FETs are typically "normally-ON" transistors [9, 39]. The pinch off capability of these devices, however, remained an issue since very high negative gate bias (typically tens of volts) was necessary to fully deplete the NW conductive channel. Such a high gate voltage leads to a giant electric field in the thin gate oxide which can result in the device breakage. To overcome this problem, several groups reported on the fabrication of semi-cylindrical top-gate NW FETs (Fig. 10.13b) [5, 6, 19]. This configuration offers significantly improved gate/channel proximity, leading to a better channel electrostatic control, i.e., reduced gate threshold $|V_{TH}|$ and increased transconductance g_m. In 2011 Blanchard et al. reported top-gate GaN NW-MESFETs exploiting MOVPE grown NWs (also bottom-up), where the gate oxide was completely removed, providing excellent gate-channel proximity and leading to threshold voltages as low as V_{TH} = −2.6 V [5]. The NWs had diameters in the 210–470 nm range and n-type doping (Si) in 4×10^{17}–4×10^{18} cm^{-3} range, yielding a high drain-current density J_{DS} > 18 kA/cm^2. Further improvements on GaN NW-MESFETs were reported by Gačević et al. with V_{TH} = −0.7 V and maximum drain-source current density of J_{DS} > 78 kA/cm^2. GaN NWs were grown by MBE and processed with a semi-cylindrical top-gate configuration [19]. The V_{TH} and J_{DS} improvements were mainly attributed to a reduced NW diameter, down to 90 nm (hence, further improved gate-channel proximity) and increased doping level $N_D = 7 \times 10^{18}$ cm^{-3} (also Si doped).

The previous studies were all based on single GaN NWs transferred from the substrate onto an insulating material and then processed into functional nanotransistors. Although this approach has proven excellent for fundamental research, it is not acceptable for real industrial applications and mass production. In 2016 Yu et al. reported vertical GaN transistors based on vertically aligned NWs with surrounding gates (Fig. 10.13c), where the vertical NWs were obtained by a top-down etching with inductively coupled plasma reactive ion etching (ICP RIE) from MOVPE grown compact layers [56]. Seven NWs were processed into a vertical MOSFET. Compared to bottom-up growth methods, the top-down one has significant advantages in terms of doping control (doping of 3D compact layers has been more extensively studied than that of 1D NWs), device geometry and scalability. The change in gate technology has enabled normally-off transistors (V_{TH} = 1.2 V) with significantly improved maximum drain-source current output I_{DS} ~ 3 mA (vs. ~10 μA typically obtained on single NW transistors).

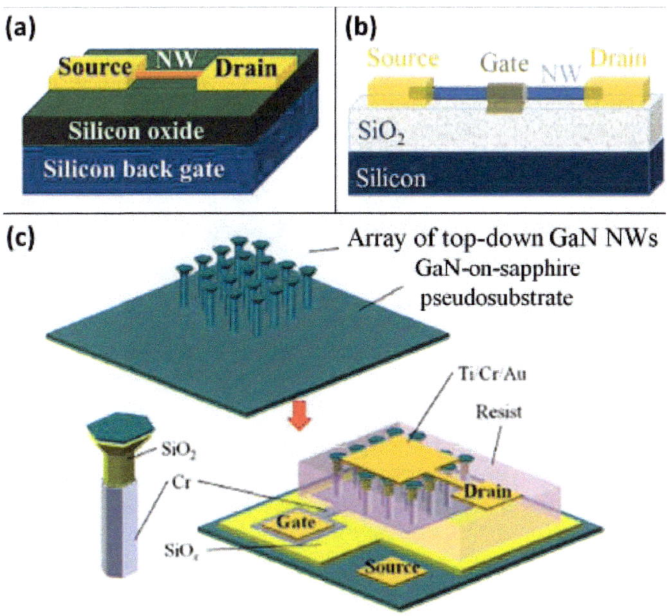

Figure 10.13 Schematics of (a) back-gate, (b) semi-cylindrical top-gate, and (c) vertical surrounding-gate GaN NW-based transistors, adapted from Huang et al., Gačević et al., and Yu et al. [19, 26, 56].

10.4.3 A Semi-Cylindrical Top-Gate GaN NW MESFET

Fabrication. SA GaN NWs were heteroepitaxially grown on a 2 inch Si(111) substrate by PA-MBE, with a growth temperature of 800°C and impinging fluxes of Φ_{Ga} ~ 4 nm/min and Φ_N ~ 10 nm/min [19]. The NWs were n-type doped by keeping the temperature of Si cell at 1000°C. The donor concentration in NWs was estimated to be ~7×10^{18} cm^{-3} from standard capacitance–voltage measurements performed on GaN compact layers grown with the same growth rate and under the same doping conditions. Prior to NW-MESFET fabrication the NWs were dispersed on a ~500 nm-thick insulating SiO_2 layer, thermally grown on a host Si(001) substrate. The as-grown sample was dipped in isopropanol in an ultrasonic bath (sonication). The isopropanol droplets were dispersed over SiO_2 surface, previously heated to ~100°C. The isopropanol evaporates rapidly leaving single lying NWs behind (Fig. 10.14a). The NW-MESFET is then fabricated in two steps: (a) source/drain contacting and (b) gate fabrication.

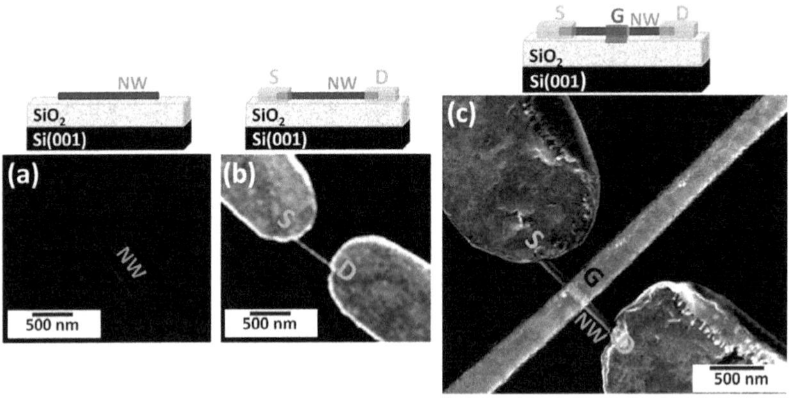

Figure 10.14 SEM images and schematics of the GaN NW-MESFET fabrication steps, adapted from Gačević et al. [19]. (a) Single NW spotted on a SiO_2 surface. (b) GaN NW with metallic contacts (source (S) and drain (D)) fabricated at its edges. (c) Fully fabricated GaN NW-MESFET with a metallic gate (G).

(a) Source/drain contacts were deposited on individual sufficiently long NWs (Fig. 10.14a). A thin polymeric resist is deposited over the SiO_2 surface by standard *spin coating*. Second, the resist is exposed to *electron beam lithography* to pattern the source and drain contacts. Third, Ti(~20 nm)/Au(~100 nm) metal layers are deposited over the sample surface by *electron beam evaporation*. Fourth, standard *lift-off* is performed to remove the resist and the metal deposited on it. Finally, the contacts are *thermally annealed* to achieve ohmic behavior (Fig. 10.14b).

(b) The gate fabrication follows a similar procedure, the only significant difference being the very last *thermal annealing* step. The gate was fabricated by depositing a metal layer (Ti 15 nm/Au 100 nm) directly on the GaN NW (Fig. 10.14c). In this case, to obtain a Schottky gate contact, the device was not thermally annealed after the final lift-off step. The fabrication steps and parameters used are shown in Fig. 10.14a–c and Table 10.2, respectively [19].

Table 10.2 GaN NW MESFET fabrication parameters (adapted from Gačević et al.) [19]

NanoFET	Fabrication parameters
Substrate	500 nm-thick SiO_2 thermally grown on Si(001)
GaN nanowire	Diameter/length: 90 nm/1.2 μm Doping: uniformly doped with Si (donor concentration estimated at $N_D \sim 7 \times 10^{18}$ cm^{-3})
Source/Drain contacts (ohmic)	Metal: Ti 20 nm/Au 100 nm Thermal annealing: at 450°C for 5 min
Gate contact (Schottky)	Dielectrics: No Metal: Ti 15 nm/Au 100 nm Thermal annealing: no

Electrical characterization. To obtain the transistor transfer curves (I_D (V_G)), the V_D voltage was kept at three different constant values: 0.5, 2.5, and 4.5 V. The V_G voltage was varied from −1.5 to 0.0 V ($V_G \leq V_S$) and then further from 0.0 to 0.5 × V_D ($V_G \leq 0.5 \times V_D$), represented by solid and dashed lines in Fig. 10.15a, respectively. Gate voltage was kept always sufficiently low to ensure the Schottky diode in reverse bias (the details are explained hereafter). The NW MESFET, working in a normally-ON mode shows a drastic decrease of drain current when V_G is reduced below −0.7 V. This value is identified as the transistor threshold voltage (V_{TH}).

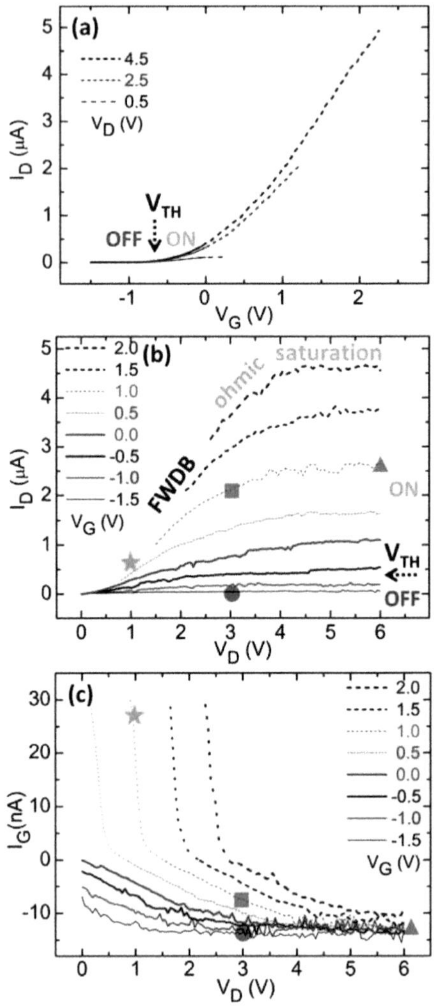

Figure 10.15 Electrical characterization of the GaN NW MESFET with a semi-cylindrical top Ti/Au gate. (a) The transfer curves confirm a strong modulation of the NW conductivity by V_G, the threshold voltage being V_{TH} = −0.7 V. (b) The NW-MESFET output curves $I_D(V_D)$ and (c) gate current curves $I_G(V_D)$, reveal three standard working regimes: (A) OFF (circle), (B) ON-ohmic (square) and (C) ON-saturation (triangle), for negative gate biases; an additional (D) forward bias FWDB working regime (star), which is considered prohibited is identified for positive gate biases (adapted from Gačević et al.) [19].

To determine the transistor output curves $I_D(V_D)$, V_G was first kept at constant non-positive values: –1.5 to 0.0 V, whereas the V_D was varied from 6 to 0 V (solid lines in Fig. 10.15b). Then, V_G was increased to positive values: 0.0–2.0 V, whereas V_D was varied from 6.0 V downwards, keeping it always sufficiently high as to prevent Schottky junction forward bias (dashed lines in Fig. 10.15b). Simultaneously, the gate current (I_G) was recorded (Fig. 10.15c).

As long as the V_G is kept below the threshold (V_G = –1.5, –1.0 V) the drain current remains in the nA range (OFF mode), in agreement with results in Fig. 10.15a. Once V_G surpasses the threshold (V_G = –0.5, 0.0 V) the I_D approaches the μA range (ON mode). As long as the $V_G \leq 0.0$ V, the I_G remains negative (solid lines in Fig. 10.15c), confirming exclusively reverse-bias Schottky operation. The positive V_G, on the other hand, can lead to either reverse- or forward-bias operation; this is directly confirmed by measuring both negative and positive I_G currents, dashed and dotted lines in Fig. 10.15c, respectively. The transistor output curves reveal three standard FET operation regimes: OFF, ON-ohmic and ON-saturation, for negative Schottky bias. An additional regime, which is considered prohibited, is observed for positive V_G, when the Schottky is in forward-bias operation.

Transistor working regimes. For clear understanding of the four identified transistor regimes, the output curves featured in Fig. 10.15b are linked to the formation of different conductive paths across the processed NW, as shown in Fig. 10.16. In all schematics, the NW surface depletion is neglected. The depletion is considered at ohmic contacts, that is, at source and drain (a very thin depletion region) and at Schottky contact, that is, at gate (a considerably thicker depletion region, Fig. 10.16). The transistor regimes are identified as follows:

(A) OFF (circle: V_G = –1.5 V, V_D = 3.0 V). The high reverse Schottky junction bias causes an extension of the depletion region, leading eventually to a full NW depletion (pinch-off). There is no drain-to-source conductive channel (open circuit) and the transistor current is zero (I_D = 0).

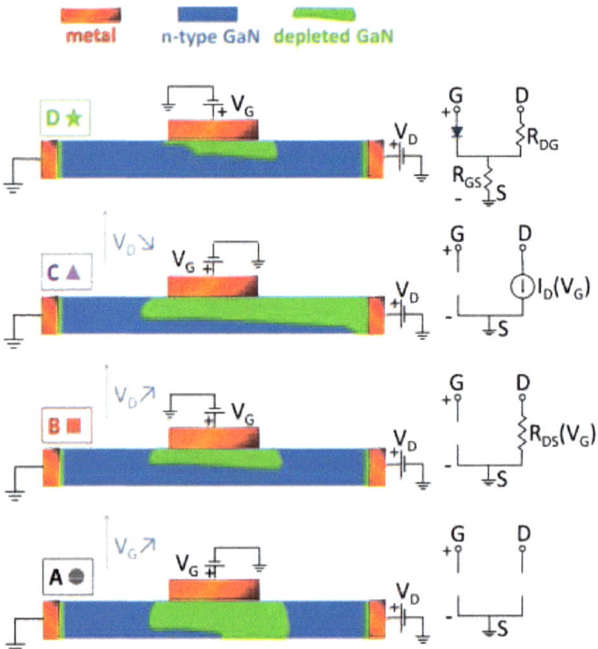

Figure 10.16 (Left) The schematics relate the observed transistor regimes A-D (see Fig. 10.15b) with conductive paths across the NW volume. V_G and V_D values vary over the ranges shown in Fig. 10.15b. The vertical arrows indicate the corresponding voltage increase/decrease (for example, when moving from A to B in the output curve diagram, Fig. 10.15b, only V_G is increased, whereas V_D is kept constant). (Right) Approximate large signal models.

(B) ON-ohmic (square: V_G = 1.0 V, V_D = 3.0 V). As V_G increases (from −1.5 to +1.0 V) the decreasing Schottky reverse bias reduces the depletion region creating a conductive channel. The NW resistance (R_{DS}) depends mainly on the conductive channel cross section which is a function of V_G. For a fixed V_G, the transistor channel behaves as a resistor, the resistivity of which is controlled by V_G.

(C) ON-saturation (triangle: V_G = 1.0 V, V_D = 6.0 V). As V_D increases (from 3.0 to 6.0 V) the increasing potential along the NW provokes a depletion region extension toward the drain, leading eventually to a full depletion on the drain side. The transistor channel resistivity is no more controlled by V_D and the transistor current is fixed almost exclusively by V_G.

(D) Schottky junction at the conduction onset (star: $V_G = 1.0$ V, $V_D = 1.0$ V). As V_D decreases (from 3.0 to 1.0 V) the decreasing potential along the NW reduces the Schottky junction reverse polarization leading to a thinner depletion region. At a certain point, the Schottky junction goes to forward bias and the current starts to flow through the junction. The nanodevice does not behave as a transistor anymore.

Transistor large signal model. For further clarity, the four different transistors regimes (A–D) have been approximated by equivalent large signal circuit models (see Fig. 10.16, right).

(A) In OFF regime the depletion region extends across a significant part of the NW volume. There is no conductive channel along the NW, thus the drain–source path is modeled with an open circuit. The Schottky diode is in a reverse bias, thus the gate–source path is also modeled with an open circuit (Fig. 10.16, right).

(B) In ON-ohmic regime the depletion region is retracted enabling formation of the conductive channel, thus the drain–source path is modeled with a resistor. The Schottky diode remains in a reverse bias.

(C) In ON-saturation regime the depletion region extends toward the drain. The transistor current is no more controlled by V_D but fixed by V_G, thus the drain–source path is modeled with a current source. The Schottky diode remains in a reverse bias.

(D) In FWD-bias regime, both drain-source and gate-source paths are conductive. The former one is modeled by a resistor, whereas the latter one is modeled by a Schottky diode in forward bias (see Fig. 10.16 right). The equivalent large signal model facilitates determination of the gate voltage limit (V_{GL}) between the Schottky reverse and forward-bias operation (see Fig. 10.16 right, for clarity) $V_{GL} = V_D \times R_{GS}/(R_{DG} + R_{GS})$. The V_{GL} value has been experimentally estimated (see Fig. 10.15e) at: $V_{GL} = \sim 0.65 \times V_D$, for the present NW-MESFET. Note that the hereby determined V_{GL} value justifies the maximum gate voltage $V_G \leq 0.5 \times V_D$ used to obtain the transistor transfer curves in Fig. 10.15a; under this condition the transistor is kept out of the prohibited FWD-bias regime.

10.4.4 Discussion

Table 10.2 compares the most relevant parameters of the GaN NW-MESFETs reported to date. The back-gate transistors show systematically lower performance compared to top-gate ones [9, 26, 39]. Their threshold voltages (V_{TH}) are negative and typically in the range of tens of volts while saturation voltages are high and transconductances remain below 0.5 μS. Transistors with semi-cylindrical top-gate geometry improve the gate/channel proximity reducing significantly the threshold and saturation voltages (typically few volts) and improving transconductance (typically in the μS range) [5, 6, 19]. When the top gate oxide is removed (MES), the gate and the channel are brought even closer, leading to a further reduction of the threshold voltage, best reported values being as low as −2.6 V and −0.7 V [5, 19]. In order to adequately compare two similar MESFET structures their sizes have to be taken into consideration, that is, their physical parameters have to be properly rescaled by the NW diameters. In that sense, the normalized transconductance (defined as: g_m/d) and current density (defined as: $J_{DS} = 4I_{SD}/d^2\pi$) allow also for transistors comparison with different designs. Comparing results reported by Blanchard et al. and Gačević et al. (threshold voltage −2.6 V vs. −0.7 V, normalized transconductance 4–9 mS/mm vs. 22 ms/mm and maximum saturation current >18 kA/cm^2 vs. >78 kA/cm^2) suggests that smaller NW diameters help to improve the channel electrostatic control (yielding lower threshold and higher normalized transconductance), whereas higher doping enables a higher transistor current density [5, 19]. It is important to notice that since the NW is usually depleted close to its surface, the actual diameter of the NW conductive channel is unknown, thus for quantitative estimations the NW full diameters were used (leading to underestimated g_m/d and J_{DS} values). Notice also that the normalized transconductance reported by Gačević et al. (although likely underestimated) compares well with the 20–50 mS/mm values typically reported for thin film planar GaN MESFETs [4, 25, 47].

The fabrication of vertical surrounding-gate GaN transistors allowed increasing the transistor current output from μA to mA range [56]. This dramatic improvement is achieved mainly because of a strongly enlarged device cross section by more than 200 times (in the device reported by Gačević et al. the current flows through a single NW with d = 90 nm, whereas in the device reported by Yu et al. the current flows through seven NWs each with d = 500 nm). Once the transistor physical parameters are rescaled (considering their geometry), for adequate comparison, it is found that the vertical design overperforms the horizontal one. The transistor current density increase (78 vs. 240 kA/cm^2) might be attributed to a further (slight) increase in doping (see Table 10.3). As for normalized transconductance (22 vs. 125 mS/mm), the improvement is mainly attributed to the fully surrounding gate, which is expected to control the channel pinch-off more efficiently than the semi-cylindrical one.

10.5 Summary

The progress in III-nitride materials achieved over the last 20 years had a significant impact in the field of solid-state lighting and high-frequency/high-power electronics. The discovery in the late 1990s of GaN NWs grown by SA heteroepitaxy and having virtually perfect crystal structure suggest nowadays their use as building blocks for a next-generation photonic and electronic nanodevices. One of the main obstacles to develop GaN NW-based photonics and electronics was a poor controllability, inherent to the SA heteroepitaxy. In this chapter, the progress made on controllability of GaN NW fabrication has been summarized, with emphasis on SAG homoepitaxy achieved by PA–MBE; the formation mechanisms, morphological, structural and optical properties of SAG NWs have been systematically addressed. The development of quantum light emitters and non-planar nanotransistors, both relying on GaN NWs, has been discussed in detail. For SPSs a significant progress has been recently made by the use of ordered arrays of bottom-up NWs, whereas for nanotransistors, relevant results have been achieved exploiting both bottom-up and top-down approaches. These achievements may represent very important initial steps toward the next-generation of GaN-NW-based quantum photonics and 3D electronics, which might revolutionize our view on modern communications and computation.

Table 10.3 The most relevant parameters of GaN NWs processed into non-planar FETs, reported in the literature [5, 6, 9, 19, 26, 39, 56] (adapted from Gačević et al. [19])

Reference	NW diameter/doping Transistor type Gate technology	Best V_{TH} [V]	MAX g_m [μS] g_m/d [mS/mm][a]	MAX I_D [μA]/V_{DS} [V] J_{DS} [kA/cm^2][b]
Y. Huang et al. (2002)	$d \sim 18$ nm/unintentional MOSFET Back Si gate	−8.0 V	0.4 μS	2 μA/>8 V
H. Y. Cha et al. (2006)	$d \sim 33$ nm/unintentional MOSFET Back Si gate	−12.5 V	<0.5 μS	4 μA/20 V
H. Y. Cha et al. (2006)	$d \sim 33$ nm/unintentional MOSFET Top semi-cylindrical Ni-Au gate	−4.0 V	<1.0 μS	4 μA/4 V
A. Motayed et al. (2007)	$d \sim 95$–195 nm/unintentional MOSFET Back Si gate	−30 V	<0.1 μS	<1 μA/>30 V
P. Blanchard et al. (2008)	$d \sim 210$–470 nm/$N_D = 4 \times 10^{17} - 4 \times 10^{18}$ cm^{-3} MESFET Top semi-cylindrical Pt-Au gate	[−15.1, −2.6[c]] V	1.7[c]–3.4[d] μS 4–9 mS/mm	12 μA /~4 V[e] >18 kA/cm^2
P. Blanchard et al. (2012)	$d \sim 210$–470 nm/$N_D = 4 \times 10^{17} - 4 \times 10^{18}$ cm^{-3} MOSFET Fully conformal Ni-Au gate	[−12.0, −4.0] V	10.0 μS	45 μA/>6 V
Gačević et al. (2016)	$d = 90$ nm/$N_D \sim 7 \times 10^{18}$ cm^{-3} MESFET Top semi-cylindrical Ti–Au gate	−0.7 V	2.0 μS 22 mS/mm	5.0 μA/4.5 V >78 kA/cm^2
Yu et al., (2016)	$d = 500$ nm/$N_D \sim 10^{18}$–10^{19}cm^{-3} Vertical MOSFET Surrounding Cr gate	+1.2 V	~500 μS 125 mS/mm	3.0 mA /7 V >240 kA/cm^2

[a]Due to a strong influence of the presence of gate oxide on transconductance, the normalized transconductances are omitted for single NW MOSFET structures.

[b]Due to a strong influence of the presence of gate oxide on NW surface depletion and, thus, effective conductive channel diameter, the estimated saturation current are omitted for single NW MOSFET structures.

[c]NW with $d = 210$ nm.

[d]NW with $d = 360$ nm.

[e]NW with $d = 290$ nm.

Acknowledgments

Žarko Gačević acknowledges fruitful collaborations with Pavel Aseev on GaN NW growth, with Ekaterina Chernisheva and Dr. Snežana Lazić on GaN NW quantum light emitters, and with David López Romero on GaN NW nanotransistors.

References

1. Aseev, P., Gačević, Ž., Torres-Pardo, A., González-Calbet, J. M., and Calleja, E. (2016). Improving optical performance of GaN nanowires grown by selective area growth homoepitaxy: Influence of substrate and nanowire dimensions. *Appl. Phys. Lett.,* 108(253109), pp. 1–5.
2. Bardoux, R., Guillet, T., Gil, B., Lefebvre, P., Bretagnon, T., Taliercio, T., Rousset, S., and Semond, F. (2008). Polarized emission from GaN/AlN quantum dots: Single-dot spectroscopy and symmetry-based theory. *Phys. Rev. B,* 77(235315), pp. 1–13.
3. Bengoechea-Encabo, A., Barbagini, F., Fernández-Garrido, S., Grandal, J., Ristić, J., Sánchez-García, M. A., Calleja, E., Jahn, U., Luna, E., and Trampert, A. (2011). Understanding the selective area growth of GaN nanocolumns by MBE using Ti nanomasks. *J. Cryst. Growth,* 325, pp. 89–92.
4. Binari, S. C., Kruppa, W., Dietrich, H. B., Kelner, G., Wickenden, A. E., and Freitas, J. A. (1997). Fabrication and characterization of GaN FETs. *Solid-State Electronics,* 41, pp. 1549–1554.
5. Blanchard, P. T., Bertness, K. A., Harvey, T. E., Mansfield, L. M., Sanders, A. W., and Sanford, N. A. (2008). MESFETs made from individual GaN nanowires. *IEEE Trans. Nanotechnol.,* 7, pp. 760–765.
6. Blanchard, P. T., Bertness, K. A., Harvey, T. E., Sanders, A. W., Sanford, N. A., George, S. M., and Seghete, D. (2012). MOSFETs made from GaN nanowires with fully conformal cylindrical gates. *IEEE Trans. Nanotechnol.* 11, pp. 479–482.
7. Buckley, S., Rivoire, K., and Vučković, J. (2012). Engineered quantum dot single-photon sources. *Rep. Prog. Phys.,* 75(126503), pp. 1–27.
8. Calarco, R., Stoica, T., Brandt, O., and Geelhaar, L. (2011). Surface-induced effects in GaN nanowires. *J. Mater. Res.,* 26, pp. 2157–2168.
9. Cha, H.-Y., Wu, H., Chandrashekhar, M., Choi, Y. C., Chae, S., Koley, G., and Spencer, M. G. (2006). Fabrication and characterization

of pre-aligned gallium nitride nanowire field-effect transistors. *Nanotechnology,* 17, pp. 1264–1271.

10. Chernysheva, E., Gačević, Ž., García-Lepetit, N., van der Meulen, H. P., Müller, M., Bertram, F., Veit, P., Torres-Pardo, A., González-Calbet, J. M., Christen, J., Calleja, E., Calleja, J. M., and Lazić, S. (2015). Blue-to-green single photons from InGaN/GaN dot-in-a-nanowire ordered arrays. *Euro. Phys. Lett.,* 111(24001), pp. 1–6.
11. Consonni, V., Hanke, M., Knelangen, M., Geelhaar, L., Trampert, A., and Riechert, H. (2011). Nucleation mechanisms of self-induced GaN nanowires grown on an amorphous interlayer. *Phys. Rev. B,* 83(035310), pp. 1–8.
12. Deshpande, S., Das, A., and Bhattacharya, P. (2013). Blue single photon emission up to 200 K from an InGaN quantum dot in AlGaN nanowire. *Appl. Phys. Lett.,* 102(161114), pp. 1–5.
13. Deshpande, S., Frost, T., Hazari, A., and Bhattacharya, P. (2014). Electrically pumped single-photon emission at room temperature from a single InGaN/GaN quantum dot. *Appl. Phys. Lett.,* 105(141109), pp. 1–5.
14. Deshpande, S., Heo, J., Das, A., and Bhattacharya, P. (2013). Electrically driven polarized single-photon emission from an InGaN quantum dot in a GaN nanowire. *Nat. Commun.,* 4(1675), pp. 1–8.
15. Dogan, P., Brandt, O., Pfüller, C., Bluhm, A.-K. Geelhaar, L., and Riechert, H. (2011). GaN nanowire templates for the pendeoepitaxial coalescence overgrowth on Si (111) by molecular beam epitaxy. *J. Cryst. Growth,* 323, pp. 418–421.
16. Ferain, I., Colinge, C., and Colinge, J.-P. (2010). Multigate transistors as the future of classical metal-oxide-semiconductor field-effect transistors. *Nature,* 479, pp. 310–316.
17. Gačević, Ž., Bengoechea-Encabo, A., Albert, S., Torres-Pardo, A., González-Calbet, J. M., and Calleja, E. (2015). Crystallographically uniform arrays of ordered (In)GaN nanocolumns. *J. Appl. Phys.,* 117(035301), pp. 1–12.
18. Gačevicì, Ž., Gómez-Sánchez, D., and Calleja, E. (2015). Formation mechanisms of GaN nanowires grown by selective area growth homoepitaxy. *Nano Lett.,* 15, pp. 1117–1121.
19. Gačević, Ž., López-Romero, D., Juan Mangas, T., and Calleja, E. (2016). A top-gate GaN nanowire metal–semiconductor field effect transistor with improved channel electrostatic control. *Appl. Phys. Lett.* 108(033101), pp. 1–5.

20. Gačević, Ž., Vukmirović, N., García-Lepetit, N., Torres-Pardo, A., Müller, M., Metzner, S., Albert, S., Bengoechea-Encabo, A., Bertram, F., Veit, P., Christen, J., González-Calbet, J. M., and Calleja, E. (2016). Influence of composition, strain, and electric field anisotropy on different emission colors and recombination dynamics from InGaN nanodisks in pencil-like GaN nanowires. *Phys. Rev. B*, 93(125436), pp. 1–11.

21. Hanbury Brown, R., and Twiss, R. Q. (1956). A test of a new type of stellar interferometer on Sirius. *Nature*, 178, pp. 1046–1048.

22. Hestroffer, K., Leclere, C., Cantelli, V., Bougerol, C., Renevier, H., and Daudin, B. (2012). In situ study of self-assembled GaN nanowires nucleation on Si(111) by plasma-assisted molecular beam epitaxy. *Appl. Phys. Lett.*, 100(212107), pp. 1–4.

23. Holmes, M. J., Choi, K., Kako, S., Arita, M., and Arakawa, Y. (2014). Room-temperature triggered single photon emission from a III-nitride site-controlled nanowire quantum dot. *Nano Lett.*, 14, pp. 982–986.

24. Holmes, M. J., Kako, S., Choi, K., Arita, M., and Arakawa, Y. (2016). Single photons from a hot solid-state emitter at 350K. *ACS Photonics*, 3, pp. 543–546.

25. Hong, S. J., Chapman, P., Krein, P. T., and Kim, K. (2006). Selective-area growth and fabrication of recessed-gate GaN MESFET using plasma-assisted molecular beam epitaxy. *Phys. Status Solidi A—Appl. Mater. Sci.*, 203, pp. 1872–1875.

26. Huang, Y., Duan, X., Cui, Y., and Lieber, C. M. (2002). Gallium nitride nanowire nanodevices. *Nano Lett.*, 2, pp. 101–104.

27. Ishizawa, S., Sekiguchi, H., Kikuchi, A., and Kishino, K. (2007). Selective growth of GaN nanocolumns by Al thin layer on substrate. *Phys. Status Solidi B*, 244, pp. 1815–1819.

28. Kako, S., Holmes, M., Sergent, S., Bürger, M., As, D. J., and Arakawa, Y. (2014). Single-photon emission from cubic GaN quantum dots. *Appl. Phys. Lett.*, 104(011101), pp. 1–4.

29. Kako, S., Santori, C., Hoshino, K., Götzinger, S., Yamamoto, Y., and Arakawa, Y. (2006). A gallium nitride single-photon source operating at 200K. *Nat. Mater.*, 5, pp. 887–892.

30. Kishino, K., Sekiguchi, H., and Kikuchi, A. (2009). Improved Ti-mask selective-area growth (SAG) by rf-plasma-assisted molecular beam epitaxy demonstrating extremely uniform GaN nanocolumn arrays. *J. Cryst. Growth*, 311, pp. 2063–2068.

31. Kremling, S., Tessarek, C., Dartsch, H., Figge, S., Höfling, S., Worschech, L., Kruse, C., Hommel, D. Forchel, A. (2012). Single photon emission from InGaN/GaN quantum dots up to 50 K. *Appl. Phys. Lett.*, 100(061115), pp. 1–3.

32. Kuhn, K. J. (2011). Moore's crystal ball: Device physics and technology past the 15 nm generation. *Microelectron. Eng.*, 88, pp. 1044–1049.

33. Larrieu, G., and Han, X.-L. (2013). Vertical nanowire array-based field effect transistors for ultimate scaling. *Nanoscale*, 5, pp. 2437–2441.

34. Lazić, S., Chernysheva, E., Gačević, Ž., Garcíýa-Lepetit, N., van der Meulen, H. P., Müller, M., Bertram, F., Veit, P., Christen, J., Torres-Pardo, A., González-Calbet, J. M., Calleja, E., and Calleja, J. M. (2015). Ordered arrays of InGaN/GaN dot-in-a-wire nanostructures as single photon emitters. *Proc. SPIE*, 9363.

35. Li, Q., Wang, G. T., Wright, J., Xu, H., Luk, T.-S., Brener, I., Figiel, J., Cross, K., Crawford, M.-H., Lee, S. R., and Koleske, D. D. (2012). Nanofabrication of tunable nanowire lasers via electron and ion-beam based techniques. Sandia report SAND2012–7665.

36. Lu, W., Xie, P., and Lieber, C. M. (2008). Nanowire transistor performance limits and applications. *IEEE Trans. Electron Devices*, 55, pp. 2859–2876.

37. Michler, P., Kiraz, A., Becher, C., Schoenfeld, W. V., Petroff, P. M., Zhang, L., Hu, E., and Imamoglu, A. (2000). A quantum dot single-photon turnstile device. *Science*, 290, pp. 2282–2285.

38. Mishra, U. K., Shen, L., Kazior, T. E., and Wu, Y.-F. (2008). GaN-based RF power devices and amplifiers. *Proc. IEEE*, 96, pp. 287–305.

39. Motayed, A., Vaudin, M., Davydov, A. V., Melngailis, J., He, M., and Mohammad, S. N. (2007). Diameter dependent transport properties of gallium nitride nanowire field effect transistors. *Appl. Phys. Lett.*, 90(043104), pp. 1–3.

40. Mukai, T., Yamada, M., and Nakamura, S. (1999). Characteristics of InGaN-based UV/blue/green/amber/red light-emitting diodes. *Jpn. J. Appl. Phys.*, 38, pp. 3976–3981.

41. Nakamura, S., and Fasol, G. (1997). *The Blue Laser Diode* (Springer, Berlin).

42. Sánchez-García, M. A., Calleja, E., Monroy, E., Sánchez, F. J., Calle, F., Muñoz, E., and Beresford, R. J. (1997). The effect of the III/V ratio and substrate temperature on the morphology and properties of

GaN- and AlN-layers grown by molecular beam epitaxy on Si(111), *J. Cryst. Growth,* 183, pp. 23–30.

43. Schumann, T., Gotschke, T., Limbach, F., Stoica, T., and Calarco, R. (2011). Selective-area catalyst-free MBE growth of GaN nanowires using a patterned oxide layer. *Nanotechnology,* 22(095603), pp. 1–6.

44. Sekiguchi, H., Kishino, K., and Kikuchi, A. (2008). Ti-mask selective-area growth of GaN by RF-plasma-assisted molecular-beam epitaxy for fabricating regularly arranged InGaN/GaN nanocolumns, *Appl. Phys. Express,* 1(124002), pp. 1–3.

45. Sekiguchi, H., Kishino, K., and Kikuchi, A. (2010). Formation of InGaN quantum dots in regularly arranged GaN nanocolumns grown by rf-plasma-assisted molecular-beam epitaxy. *Phys. Status Solidi C,* 7, pp. 2374–2377.

46. Shchukin, V. A., and Bimberg, D. (1999). Spontaneous ordering of nanostructures on crystal surfaces. *Rev. Mod. Phys.,* 71, pp. 1125–1171.

47. Shin, M. W., and Trew, R. J. (1995). GaN MESFETs for high-power and high-temperature microwave applications. *Electron. Lett.,* 31, pp. 498–500.

48. Takemoto, K., Takatsu, M., Hirose, S., Yokoyama, N., Sakuma, Y., Usuki, T., Miyazawa, T., and Arakawa, Y. (2007). An optical horn structure for single-photon source using quantum dots at telecommunication wavelength. *J. Appl. Phys.,* 101(081720), pp. 1–5.

49. Teng, C. H., Zhang, L., Hill, T. A., Demory, B., Deng, H., and Ku, P.-C. (2015). Elliptical quantum dots as on-demand single photons sources with deterministic polarization states. *Appl. Phys. Lett.,* 107(191105), pp. 1–4.

50. Wagner, R. S., and Ellis, W. C. (1965). Vapor-liquid-solid mechanism of single crystal growth. *Appl. Phys. Lett.,* 4, pp. 89–90.

51. Winden, A., Mikulics, M., Stoica, T., Von Der Ahe, M., Mussler, G., Haab, A., Grützmacher, D., and Hardtdegen, H. (2013). Site-controlled growth of indium nitride based nanostructures using metalorganic vapour phase epitaxy. *J. Cryst. Growth,* 370, pp. 336–341.

52. Xiang, J., Lu, W., Hu, Y., Wu, Y., Yan, H., and Lieber, C. M. (2006). Ge/Si nanowire heterostructures as high-performance field-effect transistors. *Nature,* 441, pp. 489–493.

53. Yamamoto, T., Maekawa, M., Imanishi, Y., Ishizawa, S., Nakaoka, T., and Kishino, K. (2016). Photon correlation study of background suppressed single InGaN nanocolumns. *Jpn. J. Appl. Phys.,* 55(04EK03), pp. 1–5.

54. Yang, P., Yan, R., and Fardy, M. (2010). Semiconductor nanowires: What's next?, *Nano Lett.*, 10(5), pp. 1529–1536.
55. Yoshizawa, M., Kikuchi, A., Mori, M., Fujita, N., and Kishino, K. (1997). Growth of self-organized GaN nanostructures on Al_2O_3 (0001) by RF-radical source molecular beam epitaxy, Japan. *J. Appl. Phys.*, 36, pp. L459–L462.
56. Yu, F., Rümmler, D., Hartmann, J., Caccamo, L., Schimpke, T., Strassburg, M., Gad, A. E., Bakin, A., Wehmann, H.-H., Witzigmann, B., Wasisto, H. S., and Waag, A. (2016). Vertical architecture for enhancement mode power transistors based on GaN nanowires. *Appl. Phys. Lett.,* 108(213503), pp. 1–5.
57. Zhang, L., Teng, C.-H., Hill, T. A., Lee, L-K., Ku, P.-C., and Deng, H. (2013). Single photon emission from site-controlled InGaN/GaN quantum dots. *Appl. Phys. Lett.,* 103, p. 192114.
58. Zhu, T., and Oliver, R. A. (2016). Nitride quantum light sources. *Euro. Phys. Lett.,* 113(38001), pp. 1–7.

Chapter 11

InP-Related Nanowires for Light-Emitting Applications

Kenichi Kawaguchi

Fujitsu Laboratories Ltd., 10-1 Morinosato-Wakamiya, Atsugi 243-0197, Japan

k_kawaguchi@jp.fujitsu.com

InP-related materials are suitable for creating optical devices in near-infrared regions because of their band gap energy and variety of alloys for heterostructures. By applying the features of these materials to InP nanowires (NWs), small optical devices that overcome the limits of conventional film devices are expected to be developed. In particular, devices formed in the shape of vertical NWs have an advantage of producing arrays or integrating with other devices. In this chapter, fabrication techniques of InP-related NWs for light-emitting applications will be presented.

11.1 Introduction

InP-related materials have been developed as constituents of optical devices in near-infrared regions, especially for telecom applications. For instance, lasers, optical amplifiers, and detectors

Novel Compound Semiconductor Nanowires: Materials, Devices, and Applications
Edited by Fumitaro Ishikawa and Irina Buyanova

ISBN 978-981-4745-76-5 (Hardcover), 978-1-315-36440-7 (eBook)
www.panstanford.com

have been produced in thin-film devices using metal-organic vapor-phase epitaxy (MOVPE) [1–11]. In such thin-film devices, heterostructures composed of multiple layers are used for generating functionalities. For light-emitting applications, quantum structures such as quantum wells (QWs) or quantum dots (QDs) are used as an active region for efficient radiative processes. Conductive semiconductor layers with impurities are formed for controlling carrier transport properties. In addition to the crystal growth, a metal contact process has been developed for efficient carrier injection.

Vertical InP NWs are interesting as a building block for next-generation optical devices [12–15]. Since the NWs have small in-plane sizes, the NWs have the advantages of creating small devices and device integration. In order to develop NW light-emitting devices, equivalent fabrication techniques established in thin-film devices are inevitably needed for the NWs.

In this chapter, our studies of InP-related NWs for light-emitting applications in the near-infrared region are reviewed. First, the fabrication technique of heterostructures on InP NWs is given. The InP NWs are grown via the vapor–liquid–solid (VLS) method, and the heterostructures are formed in the radial direction. Next, position-controlled growth of vertical NWs is described. This technique is required for the metal contact process that creates electrically driven devices. Then, crystal growth of the device structure is discussed using position-controlled NWs. Then, the technology of device fabrication is described. The device characteristics are also shown. Furthermore, crystal growth of vertical InP NWs on Si substrates, which provides an opportunity of expanding application field, is discussed.

11.2 Radial Heterostructures Using Randomly Located InP Nanowires

11.2.1 Configuration of Heterostructures

The introduction of heterostructures is necessary for controlling carrier transport or optical confinement in semiconductor NWs. As one interesting feature of NWs, the NW is allowed to have two heterostructure configurations, in the axial direction and the radial direction. A schematic view of the configuration is shown in Fig. 11.1. In the axial heterostructure (Fig. 11.1a), the size of each material segment is defined by the NW diameter and layer thickness. Therefore, an extremely small segment can be achieved when NWs with a small diameter and a thin layer are used. For light-emitting applications, such structure can be applied to single photon sources that need QD states by three-dimensional quantum confinement. The configuration of the radial heterostructure (Fig. 11.1b) is unique, since the growth direction of a NW and that of the stacking layers is orthogonal. The volume of the sheath layer is proportional to the surface area of the NWs. This means that the volume of the active region can be increased by increasing the NW length even if the in-plane area is limited. For optical device applications, such a feature is advantageous in increasing optical gains. In the following sections, we focus on the radial heterostructure.

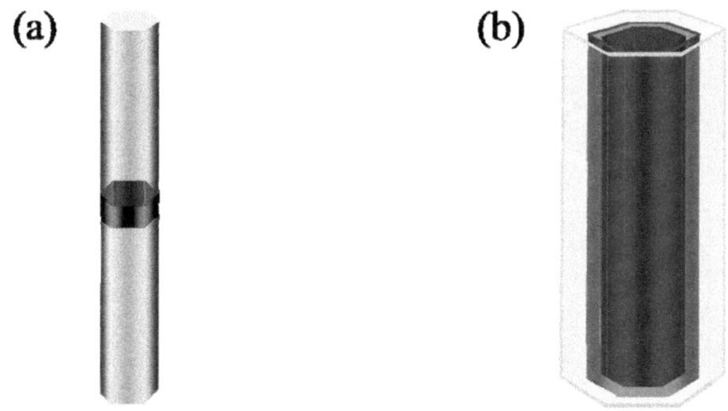

Figure 11.1 Two types of nanowire heterostructure: (a) axial heterostructure and (b) radial heterostructure.

11.2.2 Controlling Growth Direction for Radial Heterostructure on InP Nanowires

InP NWs can be grown via VLS mode using catalytic metal particles such as Au particles. Au particles formed by different methods are allowed to be used. Aerosol particles, colloidal particles, and pieces of films cut by nanoscale patterning have been reported to be effective catalysts. The NW growth proceeds underneath the metal catalysts, and the formed InP NWs generally have a catalyst on the top of them. For the fabrication of radial heterostructures, a changing growth direction is necessary after the NW growth.

In MOVPE growth, the lateral growth ratio is generally increased by increasing the growth temperature or V/III ratios. However, the strong catalytic effect of Au disturbs the formation of the radial heterostructure in some cases. Scanning electron microscope (SEM) images of before and after InAsP shell growth on an InP NW with Au catalyst are shown in Fig. 11.2. After the InAsP shell growth, the NW diameter of the original part was increased by the radial growth. However, strong axial growth was observed at the same time. From the EDX analysis of the structure, it was revealed that the top part has a higher As content than the radial growth part, and that As-rich materials favorably grow in the axial direction. Therefore, removing Au particle is necessary to suppress axial growth during the radial growth step.

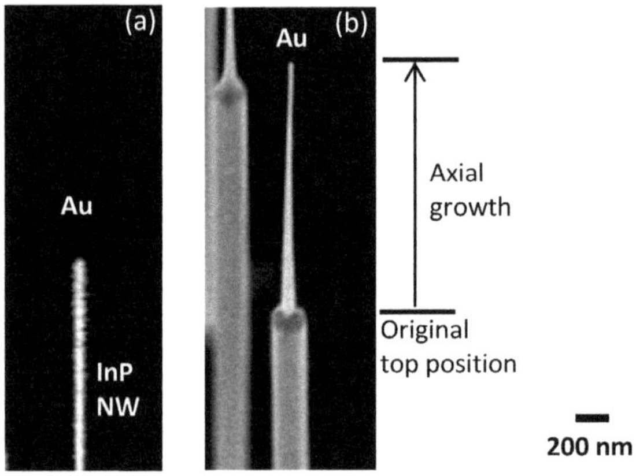

Figure 11.2 (a) InP NW core and (b) InP NW with InAsP shell grown without removing Au catalyst.

Au particles on InP NWs were selectively removed by a two-step wet chemical process. First, the In-containing part is removed by H_2SO_4. Then, residual Au parts are removed using solution of a mixture of KI and I_2. After the removal of Au particles, the anomalous axial growth is well suppressed [16].

11.2.3 InAs Radial Quantum Heterostructure on InP Nanowires

In the previous section, the technique of controlling the growth direction for radial heterostructures was described. In this section, the fabrication of a specific heterostructure, InP/InAs, is shown [17]. InP/InAs is useful as a heterostructure for light-emitting applications. Since InAs has a narrow band gap of 0.36 eV in bulk form, it covers a wide wavelength region from the telecom application region to the gas detection application region.

InAs/InP is a strained material system, and InAs grows in Stranski–Krastanow (S–K) mode on the planar InP(001) substrate [18, 19]. Below critical thickness, InAs grows in a film form. When the critical thickness is exceeded, InAs self-assembly forms QDs [20, 21]. However, the growth behavior of InAs on the sidewall of InP NWs was unknown, and therefore, it was investigated using InP NW with different crystal structures.

First, InAs growth on the sidewall of pure wurtzite (WZ) InP NWs was investigated. InP NWs were grown at 420°C, with $\chi_{TMI} = 6.4 \times 10^{-6}$ and a V/III ratio of 950. The pure-WZ phase was achieved using in situ S-doping with $\chi_{H2S} = 1.4 \times 10^{-4}$. A high-resolution transmission electron microscope (TEM) image is shown in Fig. 11.3. An InAs layer with a nominal thickness of 8 ML was grown at 460°C with $\chi_{TMI} = 6.4 \times 10^{-6}$ and arsine (AsH_3) at a V/III ratio of 46. Interestingly, a smooth WZ-InAs QW layer is observed. It was also confirmed that the smooth surface of the InAs layer was maintained even when the InAs film thickness was increased by a factor of two. Strain-driven S–K mode QD formation was not observed on the surface of perfect WZ InP NWs even for an InAs layer thickness exceeding the critical layer thickness on zinc-blende (ZB) InP(001) substrate by more than a factor of two. The NW side facets have a low surface energy, which results in a long migration length of adatoms [22]. In such a situation, when a step is created by random two-dimensional nucleation, the adatoms seem to be consumed for the rapid lateral growth, which is considered to develop a smooth layer and enhances the metastable layer thickness.

Next, crystal growth of InAs on the sidewall of defective InP NWs was investigated. InP NWs purposely designed with stacking faults were grown by switching the H_2S flow such that S-doped segments consisting of a perfect WZ crystal phase were separated by undoped segments consisting of WZ with stacking faults. A high-resolution TEM image of the stacking-fault segment is shown in Fig. 11.4. InAs QD formation was observed on the InP NW with stacking-fault segments, in contrast to the smooth overgrowth observed on the sidewall of the pure WZ InP NWs. Note that the crystal structure of InAs QDs matches with that of the adjacent InP NW core, indicating that the QD has grown epitaxially on the NW core. The shape of the QD was asymmetric. No specific facets were observed. Such shape invokes the different formation mechanism of QDs from the S–K mode.

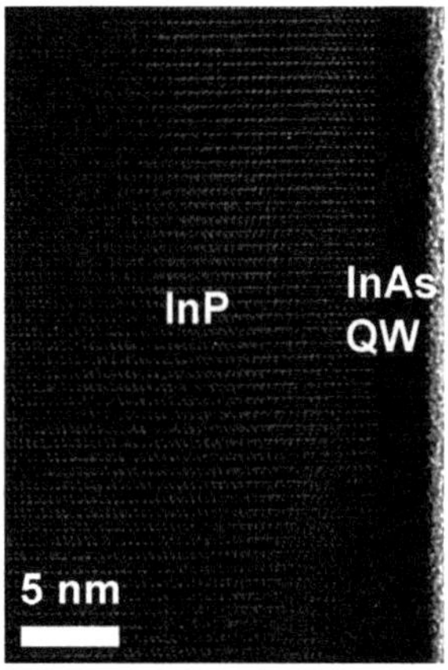

Figure 11.3 Radial InAs QW grown on the sidewall of pure WZ InP NW.

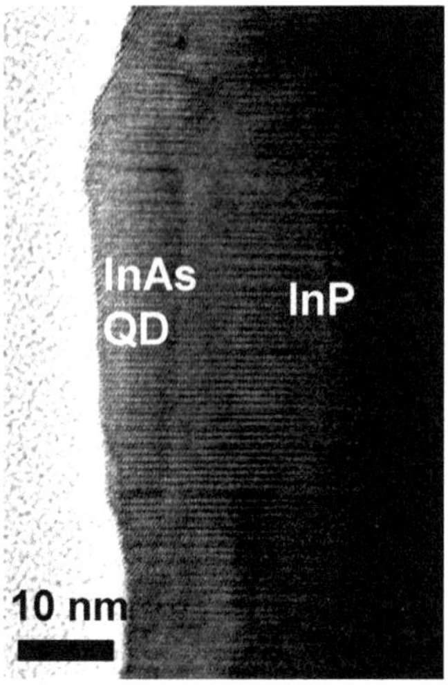

Figure 11.4 InAs QD grown on the sidewall of InP NW with stacking faults.

For the WZ NWs with {1–100} sidewall facets with designed stacking faults, steps are created on the NW sidewall at the intersection of the WZ and the ZB segments with {111} facets [23]. The surface steps act as a physical migration barrier of adatoms. In addition to the barrier of surface steps, the surface energy of ZB {111} facets larger than that of WZ {1–100} facets may modulate the adatom migrations [24]. Such preferential adatom trapping is considered to enhance the adatoms gathering on the stacking-fault region and enhance three-dimensional nucleation.

Both the radial InAs QW and QDs were found to be optically active. The surface of InAs quantum structures was covered by InP capping layers with a radial thickness of approximately 100 nm with a growth temperature of 460°C, and then photoluminescence (PL) characterization of a single NW at 5 K was performed for the NWs transferred to Si substrate. The pure WZ NW with an InAs QW showed an emission with a peak energy at 0.83 eV with a full width at half maximum (FWHM) of 20 meV. On the other hand, the NW with stacking-fault segments showed a broad luminescence of which the highest peak was located at around 0.78 eV, attributed to an emission from the InAs layer including QDs. The transitions attributed to the QD are lower in energy than those of the QW, since the height (the main quantization axis) of the QDs was larger than the QW thickness. The FWHM was 120 meV, which was larger than the typical values of 50–80 meV observed in the ensemble of S–K InAs QDs grown on InP(001) [25]. The QDs on the NW might have a larger size distribution, compared with self-assembly-formed QDs [25, 26]. Also, the co-existence of the WZ and ZB (stacking faults) crystal phase, which brings different transition energies depending on the size of ZB segments and/or stacking-fault density [27–29], is considered to contribute to the spectral broadening. Crystal-phase controlled InP NWs and radial InAs quantum structures will expand the variety of light-emitting structures.

11.3 Position-Controlled Growth of InP Nanowires

11.3.1 Demand for Position Control Method and Geometry for Light-Emitting Application

VLS NWs have been widely studied using randomly deposited Au particles because of the ease of experimental preparation. Such method is suitable for studying fundamental crystal growth of VLS NWs and their crystal properties. Some NWs can be used for devices in the lateral configuration, with a fabrication procedure employing rearrangement of NWs by transferring. In view of practical applications, vertical NW devices are more interesting, as discussed above. For vertical NW devices using as grown NWs, however, NW position must be defined precisely. In addition to the position of one NW, the separation of each NW is important for light-emitting application. A schematic view of arrayed vertical NW devices is shown in Fig. 11.5. Each optical source consists of one NW and is individually driven as an independent component. In such application, the separation of each NW should be designed based on two aspects. One comes from the cross talk of lights, which means that optical rays from neighboring devices should be delivered to the destination without mixing. The other comes from electrodes. Required space for the electrodes depends on the line width and pad size of metals. Generally speaking, the separation of the NWs should be in the range of 10 μm or more. In this situation, the NW density is extremely low, and the occupation ratio is only in the order of 10^{-4}. In the following, position-controlled growth of NWs with such low densities will be presented [30].

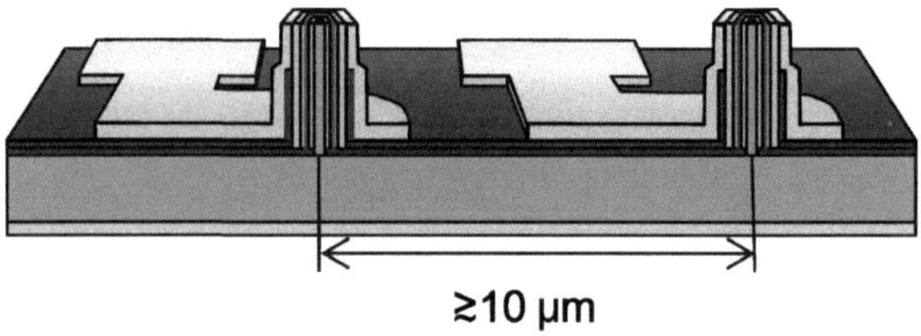

Figure 11.5 Schematic view of arrayed vertical NW devices with individual electrodes.

11.3.2 Fabrication Procedure of Position-Controlled InP Nanowires

A schematic view of fabrication steps of position-controlled NWs is shown in Fig. 11.6. Patterned substrates are fabricated using electron beam (EB) lithography and a lift-off process. First, a SiO_2 film with a typical thickness of 50 nm is deposited on an

InP(111)B substrate. Then, an EB resist is coated on the SiO_2 surface, and two-dimensionally arrayed circular holes with a typical diameter of 100 nm are developed on the resist using EB lithography. Arrayed openings are formed on SiO_2 by wet chemical etching. The size of holes in the SiO_2 film becomes slightly larger than that of the EB resist, which makes it easier to perform the lift-off process. Then, Au disks with a typical thickness of 40 nm are set in the openings, by deposition and a lift-off process.

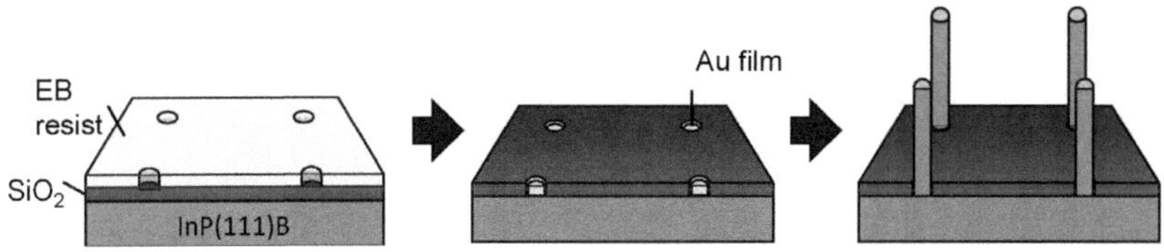

Figure 11.6 Fabrication procedure of position-controlled InP NWs.

The NW growth is performed using MOVPE. Prior to the NW growth, thermal treatment in PH_3 ambient is performed at around 600°C. This step is effective for cleaning substrates and converting the Au film pieces to Au-In single particles. InP NWs are grown at around 400°C, which is a similar growth temperature to that used for the VLS growth based on random Au catalysts. As discussed below, the introduction of HCl gas during the NW growth was found to be effective as a method of controlling the shape of position-defined InP NWs.

11.3.3 Growth of Position-Controlled InP Nanowires

Using the procedure described above, vertical InP NWs were grown with V/III ratios in a range of 10^3–10^4. In order to enhance the formation of the WZ crystal phase, and thus suppress the formation of the ZB crystal phase in the InP NWs, S-doping by H_2S gas was introduced during NW growth [31, 32]. Typical InP NW grown on patterned InP substrate is shown in Fig. 11.7a. The pattern pitch was 10 μm. TMIn flow rate was 0.036 ccm, and the growth time was 5 min. A vertical VLS-InP NW on which catalytic particles were retained was formed. However, the anomalous growth at the root of the NW was observed. The relationship between the axial growth rate of the vertical NWs and anomalous growth ratios, and TMIn flow rate is summarized in Fig. 11.7b. The growth rate changes almost linearly with the TMIn flow rate, which means a group-III supply limiting growth regime. This behavior is basically the same as the conventional VLS growth on maskless substrates under the growth condition of sufficient V/III ratios [33]. This growth rate dependence indicates that the incorporation rate of In into Au catalysts and crystallization to InP simply depends on the amount of TMIn material, due to the high V/III ratios. The growth rate on the InP patterned substrate with SiO_2 masks was approximately 5 times higher than that on InP substrates without SiO_2 masks. The enhanced growth rate is a result of the contribution of diffused material from the mask regions. As the TMIn flow rate increased, the ratio of anomalous growth increased rapidly. This indicates that excess group-III materials are supplied to the openings. In MOVPE, two material diffusion processes, i.e., surface diffusion and vapor-phase diffusion, contribute to the growth. Since the vapor-phase diffusion process has large diffusion lengths that have an order of magnitude in the range of 10^1 to 10^2 μm [34, 35], it is considered that a large amount of supplied material on the mask region can reach the mask openings where the active bonds of the semiconductor surface enhance chemical reactions for crystal growth. The shape of anomalous growth structures is similar to NWs, implying that the excess materials formed In droplets and that the In droplets acted as self-catalysts and enhanced VLS growth [36, 37]. The accumulation of excess group-III materials suggests that the incorporation rate into Au catalysts is finite.

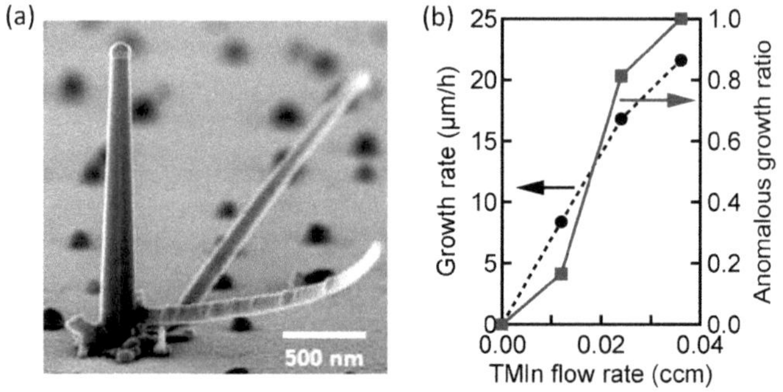

Figure 11.7 (a) Vertical InP NW with anomalous growth and (b) Growth rate and anomalous growth ratio vs. TMIn flow rate. Reproduced from *Appl. Phys. Lett.*, **104**(063102) (2014), with the permission of AIP Publishing.

The introduction of HCl gas was found to be effective as a means of eliminating the excess materials supplied via vapor-phase diffusions. SEM images of InP NWs with HCl flow are shown in Fig. 11.8. The pattern pitch is 10 μm. Vertical InP NWs with completely suppressed anomalous growth were obtained, as shown in Fig. 11.8a.

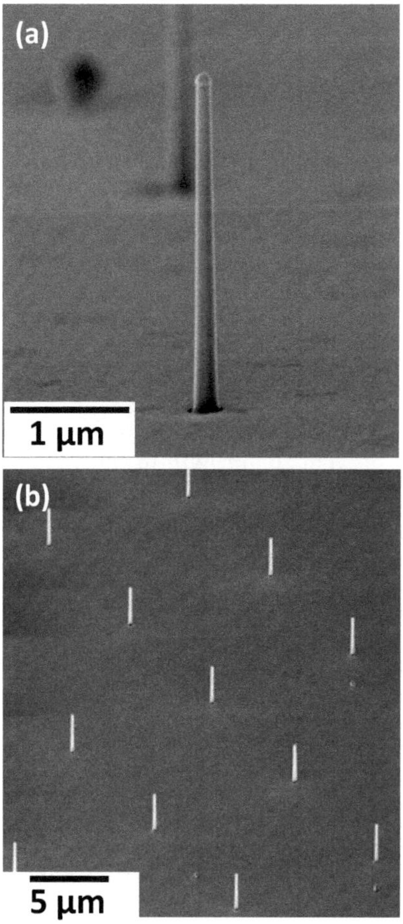

Figure 11.8 SEM images of shape-controlled InP NWs with (a) close view and (b) wide view. Reproduced from *Appl. Phys. Lett.*, **104**(063102) (2014), with the permission of AIP Publishing.

By introducing HCl during NW growth, the following process is considered to occur. First, the HCl gas reacts with the excess

In vapor species, which forms InCl molecules. Then, the InCl molecules get desorbed easily from the SiO_2-mask surface due to their high vapor pressure [38]. In fact, HCl has been reported to be sufficiently reactive even at low temperatures necessary for NW growth [39, 40]. Since InCl has the least heat of vaporization compared with other indium chlorides such as $InCl_2$ and $InCl_3$ [41], most of the indium chloride molecules are expected to be desorbed as InCl. Other indium chlorides might form simultaneously. However, it is not believed to affect the macroscopic mechanism discussed above.

In addition to the suppression of anomalous growth of inclined NW-like structures, the tapering of vertical VLS-NWs along their axis was also found to decrease with an increasing HCl flow rate. The tapering angles changed from 5.1° to 1.5° when the HCl flow rate was changed from 0 to 0.012 ccm. A similar effect was reported in the VLS growth of InP NWs using randomly dispersed Au particles on maskless InP substrates [39, 40]. The results show that the lateral growth component was removed in the vapor phase, while axial growth was maintained by the driving force of the catalytic particles. The created InCl molecules might prevent the growth in the vapor–solid mode that dominates lateral growth. The introduction of HCl also decreases the axial growth rate slightly. However, the reduction ratio was only 20% for an HCl flow rate of 0.012 ccm, and from the perspective of practical application, the axial growth rate is sufficiently high.

A wide-angle view of the NW arrays grown with HCl gas flow is shown in Fig. 11.8b. Two-dimensionally arrayed InP NWs with controlled position and shape were obtained.

Using the same growth condition, VLS-InP NWs were found to grow with suppressed anomalous growth for wider pattern pitches such as 100 μm, as shown in Fig. 11.9a. The lengths of shape-controlled InP NWs grown on patterned substrates with different pitches are summarized in Fig. 11.9b. NWs grown without HCl are also investigated for comparison (not shown here). It is noted that NWs grown without HCl supply are accompanied by anomalous growth, as shown in Fig. 11.7a. The length of InP NWs grown without the introduction of HCl is almost independent of the pattern pitch, which could be because the axial growth rate is regulated by the incorporation rate into Au catalysts. This trend is maintained when the HCl gas is introduced, which indicates that HCl does not affect the activity of Au catalytic particles.

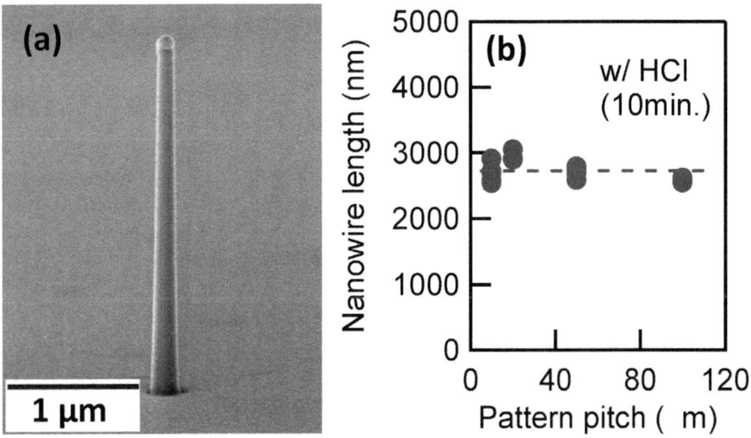

Figure 11.9 (a) SEM image of InP NW grown on 100 μm-pitch pattern. Reproduced from *Appl. Phys. Lett.*, **104**(063102) (2014), with the permission of AIP Publishing. (b) Relationship between NW length and pattern pitch.

11.4 Two-Dimensional Array of Radial Heterostructures

11.4.1 Structural Properties

Position-defined VLS-InP NWs were realized using mask-patterned substrates and HCl gas introduction. The next step for the crystal growth of device structures is forming radial heterostructure NW arrays using position-controlled InP NWs as a template. Therefore, single InP/InAsP/InP QWs were radially grown on the sidewall of the NWs [42]. 3 μm-long InP NWs with a 10 μm pitch were used. The procedure for growing radial heterostructures on position-defined NWs is basically the same as that on the randomly grown NWs: After the InP NW growth, Au catalysts were removed by wet-chemical etching. Then, the sample was reloaded into the reactor, and the radial structures were grown. Figure 11.10a shows an SEM image of a NW with a radial QW. The NW diameter is seen to increase due to the lateral growth. The NW has <1–100>

side facets. A cross-sectional TEM image of the radial QWs is shown in Fig. 11.10b. The observation direction is the <11–20> direction, which is parallel to the <110> direction of ZB InP substrate. The contrast between InAsP QW layer and InP barrier layers can be clearly seen. The InP/InAsP/InP QW structure grew epitaxially on the sidewall of the WZ-InP, which resulted in the pure WZ crystal phase.

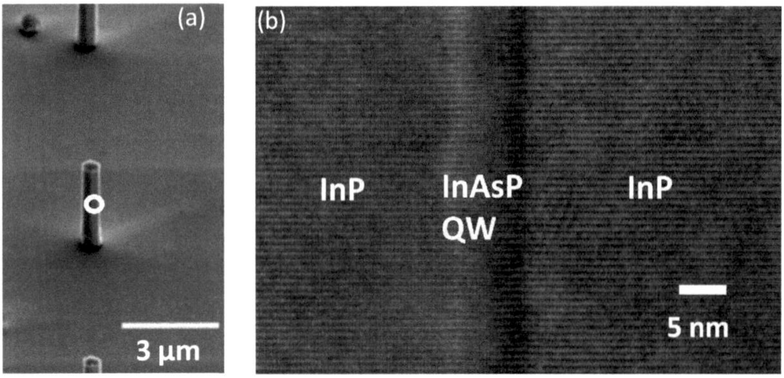

Figure 11.10 (a) SEM image and (b) TEM image of InP NW with radial InP/InAsP/InP QW.

11.4.2 Optical Properties

Optical properties of single NWs in array can be investigated using micro-PL setup, as shown in Fig. 11.11a. A PL-intensity mapping image of the radial InP/InAsP/InP NW array detected at room temperature is shown in Fig. 11.11b. The investigated area is the middle of the array with a size of 1 × 1 mm^2. The detection wavelength was 1350 nm. Two-dimensional arrayed PL patterns coming from the radial QWs were clearly observed in the 1.3 μm wavelength region at room temperature. Each bright spot comes from a single NW, and the intensity of the interval of the spots drops into the background level, since the neighboring NWs are sufficiently separated from each other. Some intensity variations are seen among the NWs. This is considered to be because of the combination of structural fluctuation of radial QWs as well as the variation of focusing. The obtained NW array, however, has achieved sufficient quality as a light-emitting structure for demonstrating light-emitting devices.

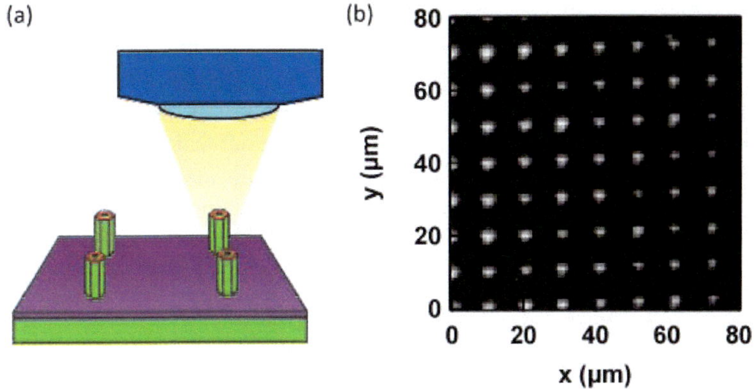

Figure 11.11 (a) Optical characterization geometry for individual NWs of arrayed sample (b) Mapping image of PL intensity for arrayed NWs with radial InP/InAsP/InP QW.

11.5 MOVPE Growth of Radial p-i-n Nanowires

For radial light-emitting structures, a p-i-n diode structure should also be created in the radial direction for efficient carrier injection. In our InP NW growth, S-doping is employed for enhancing WZ crystal formation. The NW core can be used as an n-type conductivity layer. Therefore, a radial p-i-n structure is obtained if a p-type shell is formed. In this section, p-type InP shell growth on the radial light-emitting NW structure is reviewed [43].

p-type InP shells were grown with in situ Zn doping. DEZn, which is a common for MOVPE growth, was used as a Zn source. The Zn incorporation can be evaluated using optical characterization or secondary ion mass spectrometry. From the preliminary work on the p-type shell growth, it was found that Zn can be doped into InP shells [44], covering the concentration that can be used as a p-type layer of diodes. However, Zn is known to be a diffusive element in InP [45–48]. In MOVPE film growth, the Zn doping concentration saturates at around 2×10^{18} cm^{-3}, and furthermore Zn doping causes strong diffusion via the kick-out process. The diffused Zn atoms deteriorate the active region underneath. Therefore, the optical properties of the radial light-emitting structure after p-type shell growth were investigated.

It was found that the amount of doped Zn atoms strongly affected the optical properties even though no significant difference was observed in the NW shape. PL spectra of p-i-n NWs with various *D*/III ratios for p-InP shell growth are shown in Fig. 11.12. As seen in the spectrum of the reference sample with an undoped shell (*D*/III = 0), the luminescence from the radial InAsP QW layer in this series was 1.4 μm. Due to the structural fluctuation of the QW layer, the spectral width is relatively broad, but it does not affect the purpose of this work.

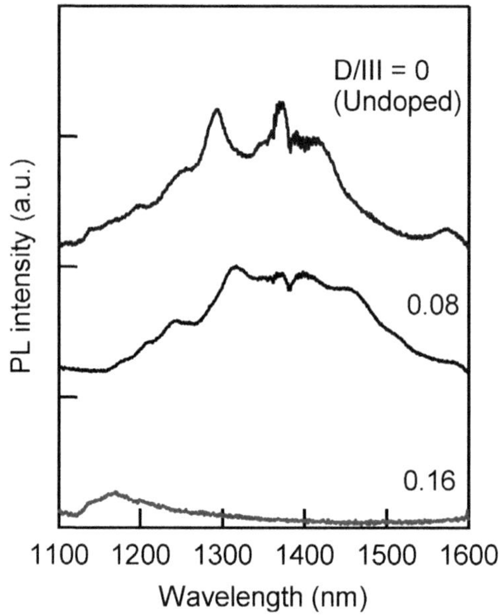

Figure 11.12 PL spectra of p-i-n NW with different p-type shell doping. Copyright (2015) The Japan Society of Applied Physics.

At the highest *D*/III ratio of 0.16, a strong reduction in PL intensity was observed. Since the growth condition we used here was estimated to be close to the saturation concentration for InP films, the excess Zn atoms have already diffused into the active region, which reduced the internal quantum efficiency of the radiative recombination.

Meanwhile, the PL intensity was maintained for the *D*/III ratio of 0.08, in which the estimated Zn concentration was ~10^{18} cm^{-3}. Thus, it was found that a range satisfying both conductivity and luminescence exists for the Zn doping of InP shells, and p-i-n radial NWs applicable to device fabrication were obtained.

11.6 Light-Emitting Device with p-i-n Nanowires

11.6.1 Fabrication Procedure of NW Light-Emitting Devices

The actual device structure we fabricated is shown in Fig. 11.13; 10 × 10 two-dimensionally arrayed NWs are used for the device. The pitch of the NWs is 10 μm, and therefore, the effective area of light emission is approximately 10^4 μm^2. It is noted that a mask pattern with 10 × 10 arrayed openings is used as one device region and that there are no sacrificed NWs during the device fabrication process. A common p-electrode is fabricated on the 10 × 10 two-dimensional NW array. Light emissions are designed to be extracted from the top of the NWs.

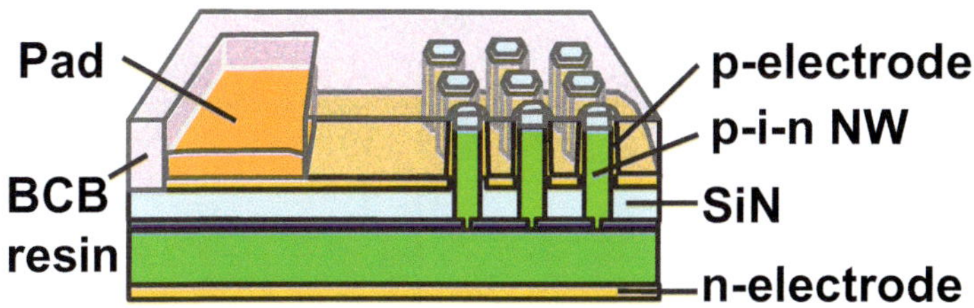

Figure 11.13 Schematic view of NW light-emitting device Copyright (2015) The Japan Society of Applied Physics.

The fabrication steps of NW light-emitting devices are illustrated in Fig. 11.14. This procedure satisfies the structure with p-type sidewall electrodes and light extraction from the top of the NWs.

(a) A SiN passivation film with a nominal thickness of 1 μm is deposited. Owing to the difference in deposition rate between the horizontal and vertical directions, a thinner SiN layer is deposited on the NW sidewall than that on the top of the NWs and the substrate surface.

(b) The sidewall part of the SiN film is selectively removed by wet control etching. SiN caps on top of the NWs act to minimize process damage during later metal removal.

(c) Ti (130 nm)/Au (100 nm) metal layers for the p-type electrode are deposited on the entire surface, by sputtering.

(d) Au pads for probe access are fabricated by electric gilding on the part of p-type metal layers.
(e) p-Type leads are fabricated by reactive-ion etching (RIE).
(f) The NWs are buried by benzocyclobutene (BCB) resin to protect the NWs during the removal of metals on the top of the NWs.
(g) Using RIE, p-type metals covering the top of the NWs are removed to extract light emission.
(h) AuGe (40 nm)/Au (200 nm) metal layers were deposited on the back surface as the n-electrode, and contact holes were created on the pads by RIE.

It is noted that this fabrication process is applicable to NWs with a maximum length of approximately 5 μm, which makes it a widely applicable process for various kinds of NWs.

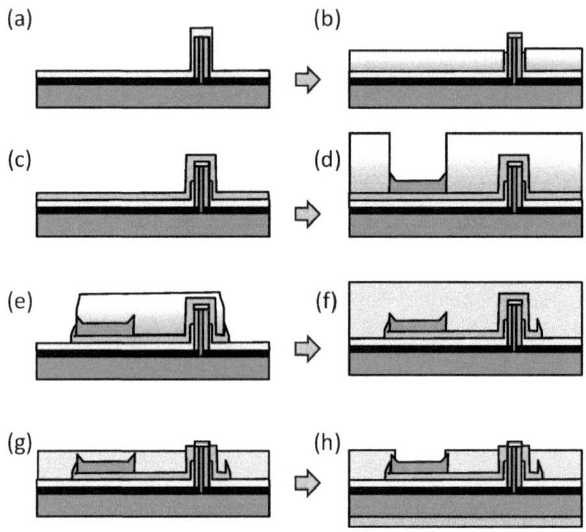

Figure 11.14 Fabrication procedure of NW light-emitting device.

11.6.2 Device Characteristics

The current-voltage (I–V) curve is shown in Fig. 11.15. Current rectification due to the p-i-n structure was clearly observed. The ideality factor (n) of this diode was approximately 4, which was deduced by fitting with a simple diode characteristic model given by

$$I = I_0\left(\exp\left(\frac{q(V - IR_s)}{nkT}\right) - 1\right), \tag{11.1}$$

where I_0 is the reverse saturation current, q is the elementary charge, R_S is the series resistance, k is the Boltzmann constant, and T is the absolute temperature. This n value is relatively large compared with the value for the ideal recombination current regime of 2, which suggests the existence of a complex of series resistances rather than one R_S. The series resistance R_S was 42 Ω. A single NW diode with the same radial p-i-n structure was investigated as a reference (not shown). The series resistance of the single NW diode was estimated to be 1.4 kΩ. A smaller R_S for the arrayed device, owing to the parallel conduction in plural NWs, was confirmed. Regarding the resistance per NW, however, that of the arrayed NW device is higher. It can be improved by increasing the uniformity of the NWs.

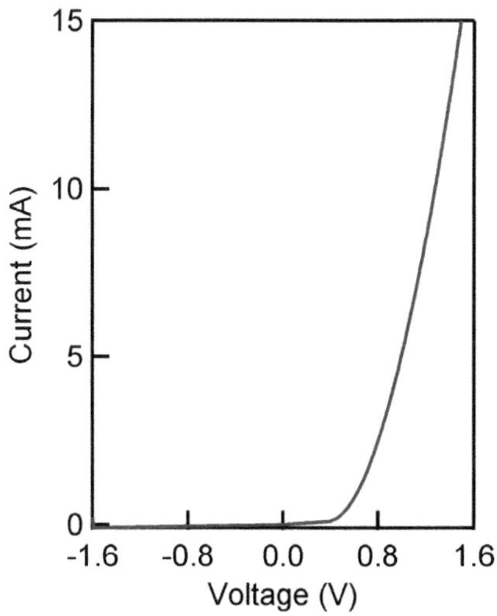

Figure 11.15 I–V curve of radial p-i-n NW light-emitting device. (Copyright (2015) The Japan Society of Applied Physics).

An electroluminescence (EL) spectrum at room temperature is shown in Fig. 11.16. The injected current was 15 mA. A PL spectrum taken from the sample before the contact process is also indicated as a comparison. The EL peak wavelength corresponds well to the PL peak wavelength, which indicates that heterostructures are maintained during the device fabrication process. The EL intensity will be improved by further optimization of the active region.

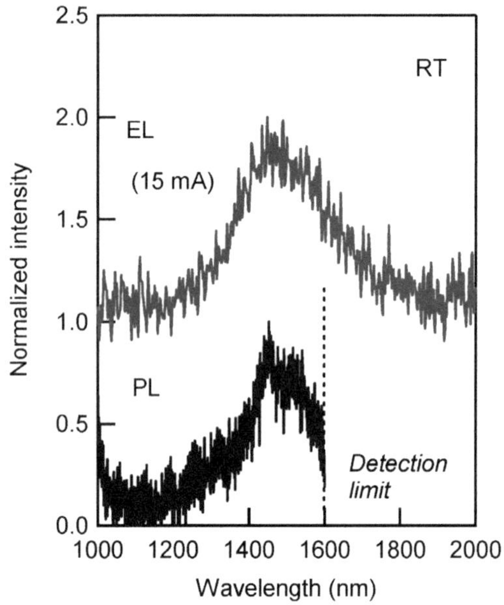

Figure 11.16 EL spectrum of NW light-emitting device. PL spectrum is also shown as comparison. Copyright (2015) The Japan Society of Applied Physics.

11.7 Nanowire Light-Emitting Structures on Si Substrates

11.7.1 Demand for NW Light-Emitting Devices on Si Substrates

Until the previous section, the study of InP-related NW light-emitting devices was performed on InP substrates. From the viewpoint of applications, the implementation of such VLS InP NWs onto Si substrates is important, since it will pave the way for the realization of monolithic integration with other functional devices such as silicon photonics and/or CMOS. For NW growth, however, changing substrates is expected to cause some differences. In the growth technique, the most crucial thing is the selection of metal catalysts. Although Au is one of the best metal catalysts for the VLS growth of InP NWs, Au forms deep-level centers when it is incorporated into Si [49]. Therefore, the use of Au is not appropriate for NW growth on Si substrates if we plan to integrate the NW devices monolithically with other devices in the future. As a growth technique of VLS-InP NWs without Au catalysts, self-catalyzed VLS growth, which uses In metal particles formed by group-III sources, has been reported on bare Si substrates [37, 50]. However, position-controlled InP NWs on Si substrates using self-catalyzed growth have not been investigated yet, and applying the essence of the growth technique developed on patterned InP substrate with Au catalysts is considered to be helpful for the growth on Si.

In this section, the growth of InP-related heterostructure NWs started from the self-catalyzed growth of position-defined InP NWs on SiO_2-mask-patterned Si substrates is described [51]. During the NW growth, shape-controlling techniques developed on InP substrates are employed. Optically active radial NW heterostructures will be shown as a result.

11.7.2 Growth of Radial Hetero NWs on Si Substrates Using Self-Catalyzed VLS Method

Patterned growth substrates consisted of Si(111) substrates covered with 50 nm-thick SiO_2 films that contained openings with a diameter of 100 nm and a periodicity of 10 μm. MOVPE growth was realized with a reactor pressure of 150 Torr. After thermal annealing at 730°C in H_2 ambient, the patterned Si substrates were cooled down to the growth temperature, and then InP NWs were grown by supplying TMIn and PH_3 simultaneously. A low growth temperature of 400°C and a high TMIn flow rate of 0.05 ccm were used for the InP NW growth, aiming at forming In catalytic particles at the beginning of the growth and keeping VLS growth mode triggered by the self-catalytic particles. The growth condition in this work produced growth rates several times higher than our previous works on InP NWs with Au catalysts on InP substrates. The introduction of HCl and H_2S gases was investigated for controlling the NW shape, which was effective for growing WZ-InP NWs on patterned InP substrates, as shown in previous sections.

SEM images of typical undoped InP NWs without and with HCl (0.016 ccm) gas introduction are shown in Fig. 11.17. The growth time was 8 min. In Fig. 11.17a, a vertical, pyramidal NW was seen to be formed on the mask opening. The tapering shape comes from the radial direction during axial growth. Since the bottom part experiences longer source supply time than the top part does, the tapered structure was formed.

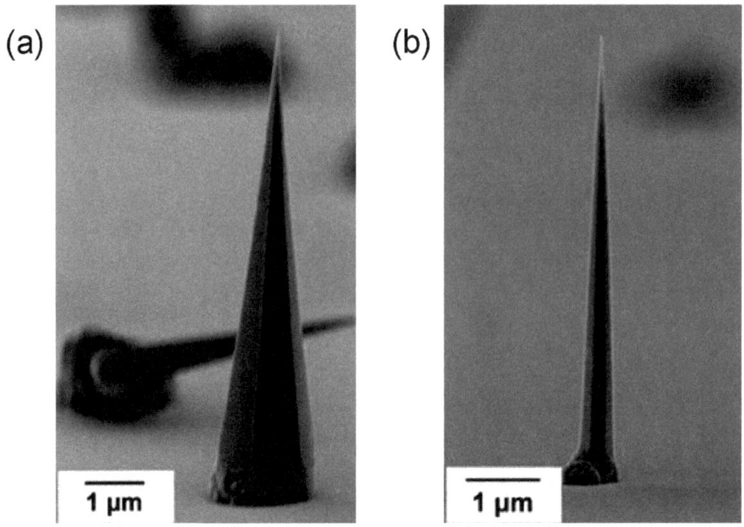

Figure 11.17 SEM images of InP NW grown on Si (a) without and (b) with HCl. Reproduced from *Appl. Phys. Lett.*, **106**(012107) (2015), with the permission of AIP Publishing.

As shown in Fig. 11.17b, a narrow InP NW compared with that in Fig. 11.1a was obtained by the introduction of HCl. The diameter at the middle and the length of the NW are approximately 250 nm and 6 μm, respectively. The mechanism of the decrease in diameter is considered to be the same as that discussed in the works on InP substrates. That is, HCl reacts with excess components of In species and forms chlorides, which reduced the lateral growth rate. It is noted that self-catalytic particles are stable enough to keep the axial growth even though HCl is reactive for In metals as well.

HCl flow rate dependence of tapering angles and NW growth rates is summarized in Fig. 11.18. As the HCl flow rate increased, the tapering angle monotonically decreased. This trend indicates that HCl flow rate can be used as a controlling parameter of the diameter of self-catalyzed InP NWs on Si as well as that of Au-catalyst VLS InP NWs on InP substrates. Although HCl preferentially reacts with excess In species at low HCl flow rates, it also affects the axial growth components when the reaction with most excess species is completed. The NW length drastically decreased for the high HCl flow rate of 0.021 ccm. As seen in Fig. 11.17b, the pyramidal shape still remained, which was defined by the residual lateral growth.

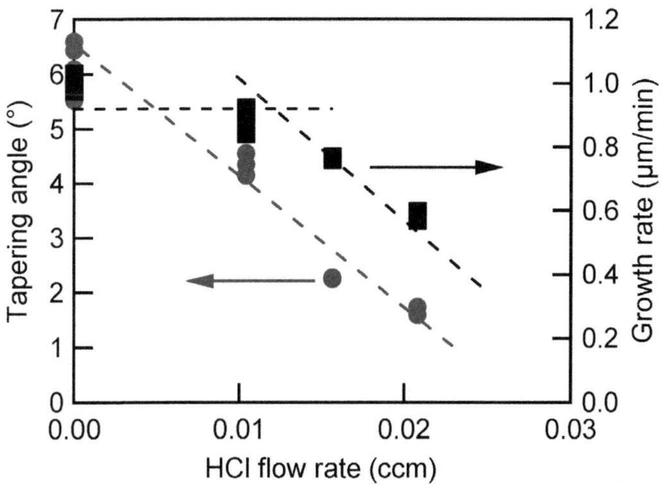

Figure 11.18 HCl flow rate dependence of tapering angle and growth rate. Reproduced from *Appl. Phys. Lett.*, **106**, 012107 (2015), with the permission of AIP Publishing.

The introduction of H_2S gas, which has the effect of enhancing the formation of the WZ crystal phase, was found to change the shape of self-catalyzed NWs as well. An SEM image of typical InP NW grown with H_2S gas introduction in addition to HCl gas flow is shown in Fig. 11.19. The flow rate of the H_2S and HCl is 0.03 and 0.016 ccm, respectively. Total growth time was 6.5 min, and an undoped segment was firstly grown for 90 s as an initial layer. The self-catalyzed InP NW shape changed to have a straight region in the middle of the NW.

The difference in crystal structure is considered to affect the shape of the NWs. That is, the smooth sidewall surface dominantly formed by the [1–100] facets of the WZ crystal phase gives a larger migration length of adatoms on the sidewall, which decreases the lateral growth rate of the bottom part. As a result, a straight region with a smaller diameter was formed in the middle of the NW. It is noted that the migration of adatoms might be also enhanced by the surface passivation of S atoms [52, 53]. Thus, controlling the shape of self-catalyzed InP NWs on patterned Si substrates was demonstrated using a combination of HCl and H_2S gases. Although the tapering of the top part still remained due to the small catalytic particle size, it will be shown that the top shape can be controlled by the successive radial growth.

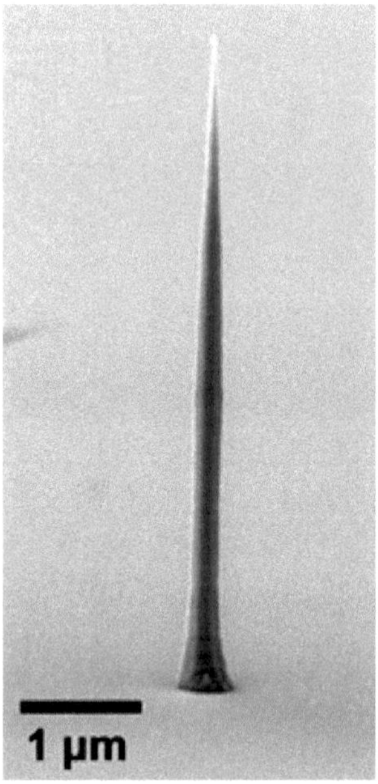

Figure 11.19 SEM image of self-catalyzed InP NW on Si substrate with shape-controlled by HCl and H_2S. Reproduced from *Appl. Phys. Lett.*, **106**(012107) (2015), with the permission of AIP Publishing.

Using the shape-controlled InP NWs formed by a combination of HCl and H_2S gas treatment, radial hetero NWs on Si substrates were fabricated. The growth sequence was as follows: After the growth of InP NW cores, the substrate was heated up to a temperature for radial growth in PH_3 ambient. Catalytic particles transform to InP crystal during this step. Then, the radial InP/InAsP/InP structure was grown successively. The first InP layer has the effect of a buffer layer after growth interruption. It also reduces the tapering of the NW core and forms smoother sidewall facets. A cross-sectional scanning transmission electron microscope (STEM) image of the NW at the straight part is shown in Fig. 11.20a. The radial InAsP QW clearly surrounds the whole sidewall, shown as a white contrast.

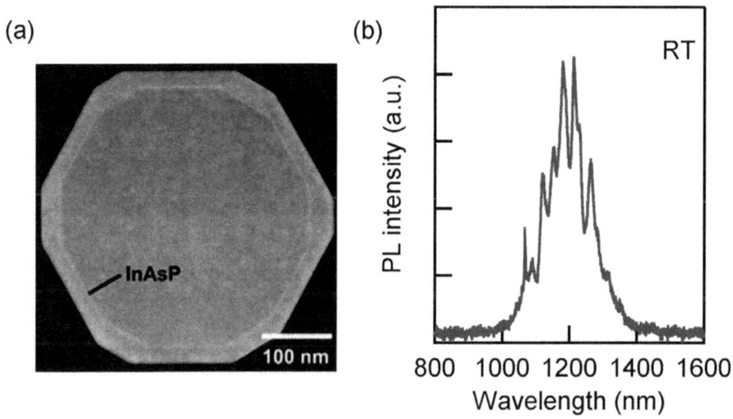

Figure 11.20 (a) STEM image of radial InP/InAsP/InP QW. (b) Room-temperature PL spectrum. Reproduced from *Applied Physics Letters*, **106**(012107) (2015), with the permission of AIP Publishing.

A room-temperature PL spectrum of a single InP NW with a radial InAsP QW on Si substrate is shown in Fig. 11.20b. PL coming from the radial InP/InAsP/InP QW structure was clearly observed. The PL intensity of this NW is comparable with the similar structure on InP substrates, implying that non-radiative recombination centers related to crystal defects induced by hetero epitaxy on Si substrates are negligible. InP luminescence around 900 nm was not observed, which indicates that radiative recombination effectively occurred in the QW region. The peak wavelength of the envelope of the spectrum is approximately 1.2 μm, and the spectral width at half maximum is as large as 200 nm. The broad peak width is considered to be due to the fluctuation of QW thickness, and the alloy composition also might affect it to some degree. Sharp multiple peaks were observed with a separation of approximately 31 nm. This is considered to be the Fabry–Perot resonance between the top and bottom of the NW [54]. The separation corresponds to a cavity length of

6.5 μm, which is reasonably consistent with the actual NW length. Further study on the optical properties related to the NW cavity will be reported elsewhere.

11.8 Summary

In this chapter, the development of the fabrication techniques of InP-related NWs for light-emitting applications was reviewed. A two-dimensional InP NW array with radial heterostructures was achieved based on the position-defined growth on patterned InP substrate with Au catalyst. A fabrication technique for a vertical NW device was developed, and room-temperature EL was also demonstrated. Moreover, the potential for VLS InP NWs on a Si platform was discussed. By the further optimization of these techniques, compact arrayed optical sources which are suitable for next-generation optoelectronics will be realized in the future.

Acknowledgments

The works reported in this chapter were partly supported by Project for Developing Innovation Systems of the Ministry of Education, Culture, Sports, Science and Technology (MEXT), Japan, and partly performed within the Nanometer Structure Consortium at Lund University.

References

1. Duchemin, J. P., Hirtz, J. P., Razeghi, M., Bonnet, M., and Hersee, S. D. (1981). GaInAs and GaInAsP materials grown by low pressure MOCVD for microwave and optoelectronic applications, *J. Cryst. Growth*, **55**, pp. 64–73.
2. Sugou, S., Kameyama, A., Miyamoto, Y., Furuya, K., and Suematsu, Y. (1984). Conditions for OMVPE growth of GaInAsP/InP crystal, *Jpn. J. Appl. Phys.*, **23**, pp. 1182–1189.
3. Nelson, A. W., Cole, S., Wong, S., Harlow, M. J., Devlin, W. J., Wake, D., Rodgers, P. M., and Robertson, M. J. (1986). High performance, long wavelength opto-electronic components by atmospheric pressure MOVPE, *J. Cryst. Growth*, **77**, pp. 579–590.
4. Nelson, A. W., Moss, R. H., Regnault, J. C., Spurdens, P. C., and Wong, S. (1985). Double heterostructure and multiquantum-well lasers

at 1.5–1.7 μm grown by atmospheric pressure MOVPE, *Electron. Lett.*, **21**, pp. 329–331.

5. Koren, U., Miller, B. I., Su, Y. K., Koch, T. L., and Bowers, J. E. (1987). Low internal loss separate confinement heterostructure InGaAs/InGaAsP quantum well laser, *Appl. Phys. Lett.*, **51**, pp. 1744–1746.
6. Yamada, H., Sasaki, T., Takano, T., Numai, T., Kitamura, M., and Mito, I. (1988). Low threshold operation of 1.55 μm GaInAsP/InP DFB-BH LDs entirely grown by MOVPE on InP gratings, *Electron. Lett.*, **24**, pp. 147–149.
7. Moseley, A., Urquhart, J., and Riffat, J. (1988). High-speed GaInAs/InP multiquantum well avalanche photodiodes grown by atmospheric-pressure MOCVD, *Electron. Lett.*, **24**, pp. 313–315.
8. Magari, K., Okamoto, M., and Noguchi, Y. (1991). 1.55 μm polarization-insensitive high-gain tensile-strained-barrier MQW optical amplifier, *IEEE Photon. Technol. Lett.*, **3**, pp. 998–1000.
9. Streubel, K., Wallin, J., Landgren, G., Öhlander, U., Lourdudoss, S., and Kjebon, O. (1994). Importance of metalorganic vapor phase epitaxy growth conditions for the fabrication of GaInAsP strained quantum well lasers, *J. Cryst. Growth*, **143**, pp. 7–14.
10. Kawaguchi, K., Matsuda, M., Ekawa, M., Yamamoto, T., Kuwatsuka, H., Sugawara, M., and Arakawa, Y. (2008). Formation of $InAs_{1-x}Sb_x$ quantum dots on vicinal InP(0 0 1) for 1.55 μm DFB laser applications, *J. Cryst. Growth*, **310**, pp. 2999–3003.
11. Kawaguchi, K., Yasuoka, N., Ekawa, M., Ebe, H., Akiyama, T., Sugawara, M., and Arakawa, Y. (2008). Growth of columnar quantum dots by metalorganic vapor-phase epitaxy for semiconductor optical amplifiers, *Jpn. J. Appl. Phys.*, **47**, pp. 2888–2892.
12. Borgström, M. T., Zwiller, V., Müller, E., and Imamoglu, A. (2005). Optically bright quantum dots in single nanowires, *Nano Lett.*, **5**, pp. 1439–1443.
13. Mohan, P., Motohisa, J., and Fukui, T. (2006). Fabrication of InP/InAs/InP core-multishell heterostructure nanowires by selective area metalorganic vapor phase epitaxy, *Appl. Phys. Lett.*, **88**, 133105 (3 pp.).
14. Tchernycheva, M., Cirlin, G. E., Patriarche, G., Travers, L., Zwiller, V., Perinetti, U., and Harmand, J.-C. (2007). Growth and characterization of InP nanowires with InAsP insertions, *Nano Lett.*, **7**, pp. 1500–1504.
15. Minot, E. D., Kelkensberg, F., van Kouwen, M., van Dam, J. A., Kouwenhoven, L. P., Zwiller, V., Borgström, M. T., Wunnicke, O., Verheijen, M. A., and Bakkers, E. P. A. M. (2007). Single quantum dot nanowire LEDs, *Nano Lett.*, **7**, pp. 367–371.

16. Kawaguchi, K., Heurlin, M., Lindgren, D., Borgström, M. T., and Samuelson, L. (2011). MOVPE growth and optical properties of wurtzite InP nanowires with radial InP/InAsP quantum wells, *23rd International Conference on Indium Phosphide and Related Materials—IPRM 2011, May 22–26, 2011, Berlin, Germany.*

17. Kawaguchi, K., Heurlin, M., Lindgren, D., Borgström, M. T., Ek, M., and Samuelson, L. (2011). InAs quantum dots and quantum wells grown on stacking-fault controlled InP nanowires with wurtzite crystal structure, *Appl. Phys. Lett.*, **99**, 131915 (3 pp.).

18. Seifert, W., Carlsson, N., Miller, M., Pistol, M.-E., Samuelson, L., and Wallenberg, L. (1996). In situ growth of quantum dot structures by the Stranski-Krastanow growth mode, *Prog. Cryst. Growth Charact. Mater.*, **33**, pp. 423–471.

19. Seifert, W., Carlsson, N., Johansson, J., Pistol, M.-E., and Samuelson, L. (1997). In situ growth of nano-structures by metal-organic vapour phase epitaxy, *J. Cryst. Growth,* **170**, pp. 39–46.

20. Schneider Jr., R. P., and Wessels, B. W. (1990). InAs-InP strained single quantum wells grown by atmospheric pressure organometallic vapor phase epitaxy, *Appl. Phys. Lett.*, **57**, pp. 1998–2000.

21. Marchand, H., Desjardins, P., Guillon, S., Paultre, J.-E., Bougrioua, Z., Yip, R. Y.-F., and Masut, R. A. (1997). Metalorganic vapor phase epitaxy of coherent self-assembled InAs nanometer-sized islands in InP(001), *Appl. Phys. Lett.*, **71**, pp. 527–529.

22. Jensen, L. E., Björk, M. T., Jeppesen, S., Persson, A. I., Ohlsson, B. J., and Samuelson, L. (2004). Role of surface diffusion in chemical beam epitaxy of InAs nanowires, *Nano Lett.*, **4**, pp. 1961–1964.

23. Algra, R. E., Verheijen, M. A., Borgström, M. T., Feiner, L.-F., Immink, G., Enckevort, W. J. P., Vlieg, E., and Bakkers, E. P. A. M. (2008). Twinning superlattices in indium phosphide nanowires, *Nature,* **456**, pp. 369–372.

24. Yamashita, T., Akiyama, T., Nakamura, K., and Ito, T. (2010). Effects of facet orientation on relative stability between zinc blende and wurtzite structures in group III–V nanowires, *Jpn. J. Appl. Phys.*, **49**, 055003 (5 pp.).

25. Carlsson, N., Junno, T., Montelius, L., Pistol, M.-E., Samuelson, L., and Seifert, W. (1998). Growth of self-assembled InAs and $InAs_xP_{1-x}$ dots on InP by metalorganic vapour phase epitaxy, *J. Cryst. Growth,* **191**, pp. 347–356.

26. Fu, Y., Ferdos, F., Sadeghi, M., Wang, S. M., and Larsson, A. (2002). Photoluminescence of an assembly of size-distributed self-assembled InAs quantum dots, *J. Appl. Phys.*, **92**, pp. 3089–3092.

27. Bao, J., Bell, D. C., Capasso, F., Wagner, J. B., Mårtensson, T., Trägårdh, J., and Samuelson, L. (2008). Optical properties of rotationally twinned InP nanowire heterostructures, *Nano Lett.*, **8**, pp. 836–841.

28. Spirkoska, D., Arbiol, J., Gustafsson, A., Conesa-Boj, S., Glas, F., Zardo, I., Heigoldt, M., Gass, M. H., Bleloch, A. L., Estrade, S., Kaniber, M., Rossler, J., Peiro, F., Morante, J. R., Abstreiter, G., Samuelson, L., and Morral, A. F. (2009). Structural and optical properties of high quality zinc-blende/wurtzite GaAs nanowire heterostructures, *Phys. Rev. B*, **80**, 245325 (9 pp.).

29. Akopian, N., Patriarche, G., Liu, L., Harmand, J.-C., and Zwiller, V. (2010). Crystal phase quantum dots, *Nano Lett.*, **10**, pp. 1198–1201.

30. Kawaguchi, K., Sudo, H., Matsuda, M., Ekawa, M., Yamamoto, T., and Arakawa, Y. (2014). Position-controlled InP nanowires with 10–100 μm pitches using Au-deposited SiO_2/InP patterned substrates, *Appl. Phys. Lett.*, **104**, 063102 (4 pp.).

31. van Weert, M. H. M., Helman, A., van den Einden, W., Algra, R. E., Verheijen, M. A., Borgström, M. T., Immink, G., Kelly, J. J., Kouwenhoven, L. P., and Bakkers, E. P. A. M. (2009). Zinc incorporation via the vapor-liquid-solid mechanism into InP nanowires, *J. Am. Chem. Soc.*, **131**, pp. 4578–4579.

32. Tuin, G. L., Borgström, M. T., Trägårdh, J., Ek, M., Wallenberg, L. R., Samuelson, L., and Pistol, M.-E. (2011). Valence band splitting in wurtzite InP nanowires observed by photoluminescence and photoluminescence excitation spectroscopy, *Nano Res.*, **4**, pp. 159–163.

33. Dick, K. A., Deppert, K., Samuelson, L., and Seifert, W. (2006). Optimization of Au-assisted InAs nanowires grown by MOVPE, *J. Cryst. Growth*, **297**, pp. 326–333.

34. Alam, M. A., People, R., Isaacs, E., Kim, C. Y., Evans-Lutterodt, K., Siegrist, T., Pernell, T. L., Vandenberg, J., Sputz, S. K., Chu, S. N. G., Lang, D. V. L. Smith, and Hybertsen, M. S. (1999). Simulation and characterization of the selective area growth process, *Appl. Phys. Lett.*, **74**, pp. 2617–2619.

35. Greenspan, J. E., Zhang, X., Puetz, N., and Emmerstorfer, B. (2000). Growth mechanisms and modeling for metalorganic chemical vapor deposition selective-area epitaxy on InP substrates, *J. Vac. Sci. Technol. A*, **18**, pp. 648–651.

36. Novotny, C. J., and Yu, P. K. L. (2005). Vertically aligned, catalyst-free InP nanowires grown by metalorganic chemical vapor deposition, *Appl. Phys. Lett.*, **87**, 203111 (3 pp.).

37. Mattila, M., Hakkarainen, T., Lipsanen, H., Jiang, H., and Kauppinen, E. I. (2006). Catalyst-free growth of In(As)P nanowires on silicon, *Appl. Phys. Lett.*, **89**, 063119 (3 pp.).

38. McNevin, S. C. (1986). Chemical etching of GaAs and InP by chlorine: The thermodynamically predicted dependence on Cl_2 pressure and temperature, *J. Vac. Sci. Technol. B*, **4**, pp. 1216–1226.

39. Borgström, M. T., Wallentin, J., Trägårdh, J., Ramvall, P., Ek, M., Wallenberg, L. R., Samuelson, L., and Deppert, K. (2010). In situ etching for total control over axial and radial nanowire growth, *Nano Res.*, **3**, pp. 264–270.

40. Borgström, M. T., Wallentin, J., Kawaguchi, K., Samuelson, L., and Deppert, K. (2011). Dynamics of extremely anisotropic etching of InP nanowires by HCl, *Chem. Phys. Lett.*, **502**, pp. 222–224.

41. Tsang, W. T., Kapre, R., and Sciortino Jr., P. F. (1994). In situ dry etching of InP using phosphorus trichloride and regrowth inside a chemical beam epitaxial growth chamber, *J. Cryst. Growth*, **136**, pp. 42–49.

42. Kawaguchi, K., Sudo, H., Matsuda, M., Ekawa, M., Yamamoto, T., and Arakawa, Y. (2014). VLS growth of position-controlled InP nanowires and formation of radial heterostructures on mask-patterned InP substrates, *MRS Proceedings*, 1659, mrsf13-1659-ss16-05.

43. Kawaguchi, K., Sudo, H., Matsuda, M., Ekawa, M., Yamamoto, T., and Arakawa, Y. (2015). Room-temperature electroluminescence from radial p–i–n InP/InAsP/InP nanowire heterostructures in the 1.5-μm-wavelength region, *Jpn. J. Appl. Phys.*, **54**, 04DN02 (5 pp.).

44. Wallentin, J., Borgström, M. T., Samuelson, L., Ekawa, M., and Kawaguchi, K. (2013). Photoluminescence study of Zn-doped wurtzite InP core-shell nanowires, *Appl. Phys. Lett.*, **102**, 032105(5 pp.).

45. Chu, S. N. G., Logan, R. A., Geva, M., and Ha, N. T. (1995). Concentration dependent Zn diffusion in InP during metalorganic vapor phase epitaxy, *J. Appl. Phys.*, **78**, pp. 3001–3007.

46. Moon, Y., Si, S., Yoon, E., and Kim, S. J. (1998). Low temperature photoluminescence characteristics of Zn-doped InP grown by metalorganic chemical vapor deposition, *J. Appl. Phys.*, **83**, pp. 2261–2265.

47. Tai, C.-Y., Seiler, J., and Geva, M. (1999). Modeling of Zn diffusion in InP/InGaAs materials during MOVPE growth, *11th International Conference on Indium Phosphide and Related Materials*, pp. 245–248.

48. Höglund, A., Castleton, C. W. M., and Mirbt, S. (2008). Diffusion mechanism of Zn in InP and GaP from first principles, *Phys. Rev. B*, **77**, p. 113201.

49. Lang, V., Grimmeiss, H. G., Meijer, E., and Jaros, M. (1980). Complex nature of gold-related deep levels in silicon, *Phys. Rev. B,* **22**, p. 3917.

50. Gao, L., Woo, R. L., Liang, B., Pozuelo, M., Prikhodko, S., Jackson, M., Goel, N., Hudait, M. K., Huffaker, D. L., Goorsky, M. S., Kodambaka, S., and Hicks, R. F. (2009). Self-catalyzed epitaxial growth of vertical indium phosphide nanowires on silicon, *Nano Lett.,* **9**, pp. 2223–2228.

51. Kawaguchi, K., Sudo, H., Matsuda, M., Takemoto, K., Yamamoto, T., and Arakawa, Y. (2015). Radial InP/InAsP/InP heterostructure nanowires on patterned Si substrates using self-catalyzed growth for vertical-type optical devices, *Appl. Phys. Lett.,* **106**, 012107(4 pp.).

52. Çakmak, M., and Srivastava, G. P. (1998). An ab initio calculation of the adsorption of H_2S onto InP(110)-(1 × 1) surface, *Appl. Surf. Sci.,* 123/124, pp. 52–55.

53. Lu, H.-L., Terada, Y., Shimogaki, Y., Nakano, Y., and Sugiyama, M. (2009). In situ passivation of InP surface using H_2S during metal organic vapor phase epitaxy, *Appl. Phys. Lett.,* **95**, 152103 (3 pp.).

54. Ding, Y., Motohisa, J., Hua, B., Hara, S., and Fukui, T. (2007). Observation of microcavity modes and waveguides in InP nanowires fabricated by selective-area metalorganic vapor-phase epitaxy, *Nano Lett.,* **7**, pp. 3598–3602.

Chapter 12

InP/InAs Quantum Heterostructure Nanowires

Guoqiang Zhang, Kouta Tateno, and Hideki Gotoh

NTT Basic Research Laboratories, NTT Corporation,
3-1, Morinosato-Wakamiya, Atsugi, Kanagawa 243-0198, Japan
zhang.guoqiang@lab.ntt.co.jp

One crucial challenge for semiconductor nanowires (NWs) has been the development of a complementary metal–oxide–semiconductor (CMOS)-compatible synthesis approach which produces semiconductor heterostructure NWs with excellent optical and electrical properties. This remains challenging mostly because gold (Au), which is widely used as a catalyst particle when NWs are synthesized with the bottom-up vapor–liquid–solid (VLS) approach [1], is not permitted in the mainstream CMOS process as Au is highly detrimental to the performance of minority carrier electronic devices [2, 3]. Here we describe the growth, structural, and optical properties of InP, InAs, and InP/InAs heterostructure NWs by developing an Au-free indium-particle-assisted VLS approach. We also present a novel approach to form the site-defined InP/InAs NWs by combining bottom-up self-assembly with top-down micro-photolithography techniques.

Novel Compound Semiconductor Nanowires: Materials, Devices, and Applications
Edited by Fumitaro Ishikawa and Irina Buyanova

ISBN 978-981-4745-76-5 (Hardcover), 978-1-315-36440-7 (eBook)
www.panstanford.com

12.1 Introduction

Semiconductor nanowires (NWs) have provided a new platform for fundamental physics research [4, 5] and have become next-generation building blocks in photonics [6, 7], electronics [8], and energy conversion [9, 10]. Semiconductor NWs are expected to play an important role as functional device elements in nanoscale electronic devices that may take us beyond the CMOS. The particle-catalyzed VLS mode [1, 11] has been widely used for NW synthesis, and shows excellent potential because it provides high controllability of size, and site, and it has a good ability to form axial and radial heterostructures. Au has been widely employed as the particle material for synthesizing various NWs. However, Au atoms are found to be incorporated into the NW body during growth, as reported for Si NW growth [12, 13]. An NW-based device may be degraded by these Au atoms, which reduce the minority diffusion length in Si NWs [14, 15] and work as nonradiative recombination centers in GaAs NWs [16]. Moreover, because the use of Au is prohibited in CMOS technology, it is extremely difficult to integrate NW-based nanoelectronic and photonic devices into the CMOS process. Au-free growth approach for semiconductor NWs is essentially important to expand the application into CMOS-compatible process and for the realization of optoelectronic integrated circuit (OEIC) and "beyond CMOS."

To overcome this problem, researchers have developed other techniques, one is selective area growth (SAG) [17–19], and the other is self-catalyzed VLS growth [20–29]. For III–V compound semiconductor NW growth, one can use one of the group III elements as the particle material and grow the NW in the VLS mode. This is called the self-catalyzed (or self-assisted) VLS growth mode. In this growth process, gallium (or indium) particles are often deposited on a substrate, and then the NW growth is initiated by introducing both III and V source materials. During growth, a gallium (or indium) particle is located at the NW tip to drive the growth. The liquidus particle enables the NW to grow as liquid phase epitaxy in a near-equilibrium condition. In 1977, Kasahara from SONY Corporation first reported the GaAs NWs grown via gallium-particle-catalyzed VLS mode for the first time [20]. They directly verified the composition of the gallium particle at the NW tip. In 2005, Novotny reported the InP NWs grown from indium particles deposited on the substrate surface [21]. Later, there have been many reports about InP, GaAs, and GaAsP NWs grown via the indium and gallium-catalyzed VLS mode [22–26, 30, 31].

The synthesis of InAs NWs in the self-catalyzed VLS mode remains difficult for long time [32–34]. InAs NWs could be grown without using Au nanoparticles, but no indium particle has been found at the NW tip [32, 34]. The presence of an indium particle at the NW tip is a straightforward evidence for the self-catalyzed VLS mode. We found that the key point is to use a high arsenic source flow and a growth temperature as low as 320–330°C [27]. This enabled us to grow InAs NWs and directly see the indium particle at the InAs NW tip after growth. This is the direct evidence that the growth is conducted in the indium-particle-catalyzed VLS mode. Grap et al. have also reported the growth of InAs NWs in the indium-particle-catalyzed VLS mode in a molecular beam epitaxy (MBE) system [35]. Recently, Gomes reported the InAs NWs vertically grown on Si substrate via the indium-particle-catalyzed VLS mode [36].

It could be possible to grow semiconductor heterostructure NWs by using a common III-group element and using it as the catalyst particle to form heterostructures in the self-catalyzed VLS mode [25, 28, 37]. Pozuelo et al. has studied InP/InSb heterostructure NWs synthesized in the self-catalyzed VLS mode [25]. However, the lattice mismatch of InSb/InP is as large as 10.4%, resulting in forming dislocations at the interface despite the high capability of lattice relaxation in the NW structure [25]. InP and InAs semiconductors have excellent optical and electronic properties and have attracted a lot of interest as regards the synthesis of their heterostructure NWs [7, 38–40]. However, the synthesis of InP/InAs heterostructure NWs are mostly made via the Au-catalyzed VLS mode, the synthesis via the self-catalyzed VLS mode remains challenging. We realized the InP/InAs heterostructure NWs via the self-catalyzed VLS mode [28]. We found that the InP/InAs heterostructure NW exhibits a coherent interface despite the 3.1% lattice mismatch [28].

In this chapter, we first describe the growth, structure, and optical properties of InAs NWs via the self-catalyzed VLS mode [27]. Then we present the InP/InAs heterostructure NWs, including growth, structural and interface analysis, tuning emission wavelength by the thickness of InAs quantum disks [28]. Finally, we describe how to make the site-defined InP/InAs heterostructure NWs by self-assembled indium particle array [29].

12.2 Experiments and Characterization Methods

We synthesized the InP/InAs NWs in a metalorganic vapor phase epitaxy (MOVPE) system. indium particles were formed on InP (111)B simply by introducing trimethyl-indium (TMIn) source material at 360°C for 15–300 s at a flow rate of 3.0 μmol/min (Fig. 12.1). Basically, the indium particles were homogeneously distributed on the substrate. The density is dependent on the migration length of indium atoms on the surface, which is significantly affected by the substrate surface temperature [29]. The indium particle size increases with the deposition time. This enables us to modify the density and size of the indium particles by the temperature and the deposition time. The temperature was then reduced to the growth temperature (320–330°C) and growth was initiated by introducing TMIn and tertiarybutylarsenic (TBA), or tertiarybutylphosphine (TBP) simultaneously.

For the InP segment growth, the flow rates of TMIn and TBP are 2.12–3.03 and 803.6 μmol/min, respectively. For the InAs segment growth, the flow rates of TMIn and TBA are 1.5–7.2 and 44.6–446 μmol/min, respectively. We also used the flow rate unit of sccm (standard cubic centimeters per minute). The mole flow rate of TMIn and TBP (or TBA) materials with 1 sccm corresponds to 1.5×10^{-7} mol/min and 4.46×10^{-5} mol/min, respectively.

The morphology of the NWs was analyzed using a scanning electron microscope (SEM, Karl Zeiss, Ultra55). A structural and compositional analysis of the NWs was performed with a TEM (JEOL, JEM2100F, operated at 200 kV) equipped with a high-angle annular dark-field (HAADF) detector and an X-ray energy dispersive spectrometer (XEDS). Cross-sectional samples for aberration-corrected scanning TEM (Cs-STEM) measurements were prepared by using a focused ion beam (FIB; SII SMI-3050SE, operated at 30 kV) system. High-resolution analysis of the heterostructures in InAs/InP heterostructure NWs was performed with a Cs-STEM (JEOL, ARM200F, operated at 200 kV).

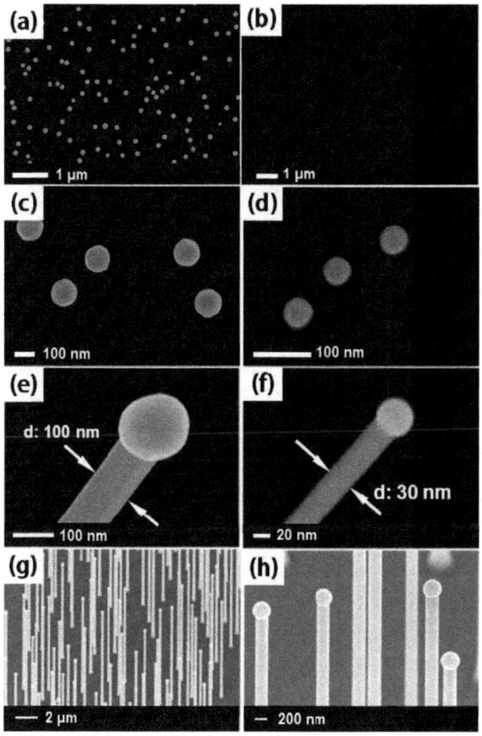

Figure 12.1 Indium particle size control by controlling deposition time and InP NWs grown from indium particles in indium-particle-assisted VLS mode. (a–d), SEM images of indium particles deposited with different time. (a) and (c), The particle diameter is 130 ± 12 nm. (b) and (d), The particle diameter is 48 ± 5 nm. The indium atoms, which are from the decomposition of the source material, migrate on the surface, accumulate into clusters, and then form indium particles. (e) and (f), SEM images of InP NWs grown from the indium particles shown in (c) and (d). The NW diameters are 100 and 30 nm, respectively. It can be clearly seen that there are indium particles at the NW tips. (g) and (h), SEM (tilt: 38°) images of InP NWs. The <111>-oriented InP were epitaxially grown on InP(111) substrate and therefore are vertically aligned on the substrate. An indium particle can be clearly seen at each NW tip.

About the photoluminescence (PL) measurement, we used a standard μ-PL setup [7]. The NW sample was optically excited using a continuous-wave diode laser operating at 640 nm. A free-space excitation technique was applied through a 0.42- and 0.50-numerical-aperture near-infrared microscope objective.

The minimum beam diameter was estimated to be 2 μm. The luminescence was collected through the same microscope objective and filtered with a dichroic beam splitter and a long-wave-pass filter (cutoff wavelength 1,064 nm). For direct observation of the emission from the NWs, we split the emission from the InAs segment and the visible illumination light using a pellicle beam splitter (cutoff wavelength 1,064 nm). The detection parts contain a visible near-infrared charge-coupled device (CCD) camera and a highly sensitive short-wave infrared InGaAs photodiode array CCD with a long-wave-pass filter (cutoff wavelength 1,100 nm) for the microscope image and the PL image, respectively. To measure the PL spectrum, we coupled the emission into the multimode fiber and directed it to a grating spectrometer with a cooled InGaAs array. The sample stage was cooled to 4 K with liquid helium for the measurement.

Raman spectroscopy has become a standard method for the non-destructive characterization of NWs due to its ability to deliver information about size, structure, carrier concentration, local density, strain and orientation. We carried out the Raman measurement by Renishaw inVia Raman Microscope. The excitation laser source is 532 nm with maximum power of 50 mW. The power for the NW measurement is normally lower than 5 mW to avoid the heating (or burning) of the NW under the strong radiation. The laser can be focused onto a single NW with a spot size of 1.2–1.5 μm by ×100 lens. The NWs were normally dispersed onto a SiO_2/Si substrate covered with Au film for Raman spectroscopy.

For the cathodoluminescence (CL) measurement, the NWs were transferred to a Si substrate covered with a thin Au film. In contrast to the conventional micro–photoluminescence with microscale resolution, the combination of CL with SEM provides nanoscale spatial resolution. We analyzed the samples with a SEM (JEOL, JSM-7800F) equipped with a CL (Gatan, MonoCL4) system. The sample stage was cooled to 8 K with liquid helium. The luminescence excited by the electron beam was collected by a parabolic mirror and guided to a grating (600 l/mm, 1000 nm), and was finally detected with a photomultiplier (Hamamatsu Photonics, R5509-73, operated at a temperature lower than 193 K).

12.3 Indium Arsenide NWs

In this section, we describe the growth, structure, and electrical properties of InAs NWs grown in the self-catalyzed VLS mode through a self-assembled indium particle. The III/V mole ratio had a significant effect on the growth and structural shape. A single InAs NW grown with a high III/V mole ratio exhibits distinct TO phonon peaks in a Raman spectroscopy observation. Furthermore, we verified that the InAs NW grown in the self-catalyzed VLS mode could function as a channel in the FET device and the undoped NW exhibits n-type conduction behavior.

12.3.1 Influence of V/III Ratio on InAs NW Growth

Figures 12.2a–e show side-view SEM images of InAs NWs grown under various TMIn and TBA flow rates. When decreasing III/V

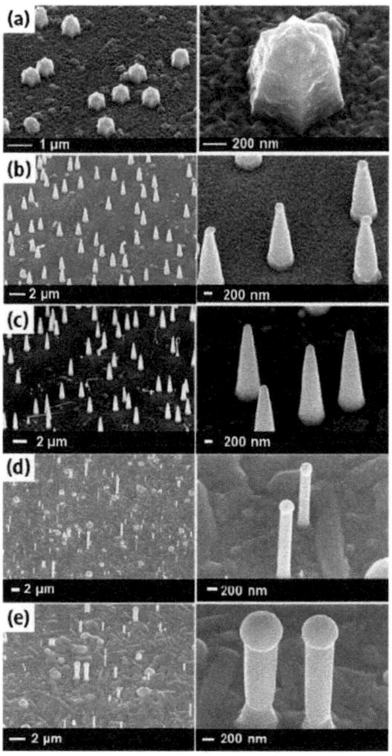

Figure 12.2 Side-view (tilt: 38°) SEM images of InAs NWs grown at 320°C under various TMIn and TBA flow rates. The III/V source mole ratios in (a), (b), (c), (d), and (e) are 0.0067, 0.027, 0.054, 0.11, and 0.16, respectively. There are two images for each condition (low magnification on the left and high magnification on the right).

mole ratio, the nanostructure exhibits a rod structure, as shown in Figs. 12.2a–e. In contrast, from Figs. 12.2b–e, the InAs NWs with a relatively long segment are vertically aligned on the substrate. Notably, there are particles at the NW tips, indicating that the NWs might be grown in the VLS mode. As the III/V mole ratio is increased, there is a continuous change in the NW shape, from a tapering shape (Figs. 12.2a–c,) to a uniform shape (Fig. 12.2d), and then a reversely tapering shape (Fig. 12.2e). In particular, there is a distinct difference between the NW tapering shapes in Figs. 12.2c,d.

12.3.2 Structural Analysis

To study the structural property of the InAs NWs, we prepared sliced samples by using a FIB system. Figure 12.3 shows TEM images of InAs NWs (Fig. 12.2c) grown under the flow rate of TMIn (32 sccm) and TBA (2 sccm). In Fig. 12.3a, there is a single NW vertically grown on the InP substrate. Notably, there is a difference in contrast in the root region of the InAs NW. The feature can be seen more clearly in Fig. 12.3d. The region is indeed a polycrystalline structure.

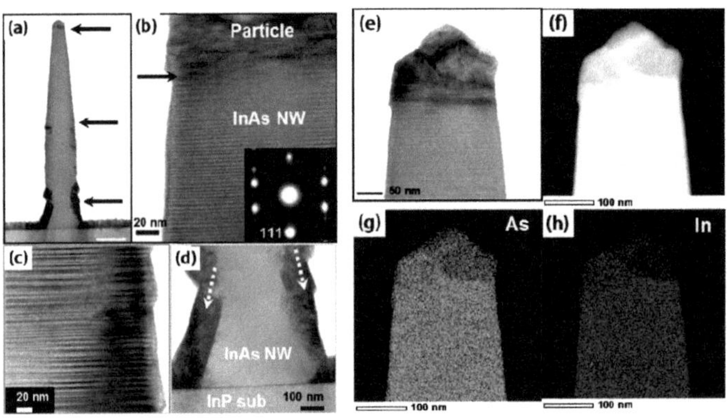

Figure 12.3 TEM images of InAs NWs grown under the III/V ratio of 0.054. (a) InAs NW on InP (111)B substrate. (b) TEM image of the region near the interface between the NW and the tip, indicated by the black arrow. The inset in (b) is the SED pattern of the InAs NW, which indicates that the NW growth direction is <111> and it has a ZB crystalline structure with stacking faults. (c) and (d) TEM images of the regions near the NW body and the InAs/InP interface, as indicated by the black arrows in (a). (e) Bright-field TEM and (f) HAADF-STEM images of the region near the tip. (g) and (h) Elemental mapping of the tip region. The scale bars in (f), (g) and (h) denote 100 nm. Compositional analysis indicates that the particle has an In:As mole ratio of 1:1.

To clarify whether the InAs nanowire was grown in the VLS mode through the Indium particle, we performed an elemental compositional mapping measurement to study the compositional difference between the particle at the tip and the NW body (Figs. 12.3e–h). The intensity fluctuation in the particle indicates a compositional inhomogeneity in the tip region. Further quantitative analysis indicates that the Indium/As mole ratio in the particle region is nearly 1:1. The results shown above indicate that the InAs NW was grown through the particle at the tip in the self-catalyzed VLS mode.

To clarify how the III/V mole ratio affects the structure and growth behavior, we also carried out a TEM analysis for the InAs NW, shown in Fig. 12.2d, grown at higher III/V mole ratio (Fig. 12.4). There is a single InAs NW vertically grown on an InP substrate and a particle is visible in the tip region. The SED pattern indicates that the growth direction of the NW is <111> and that it has a ZB crystalline structure with stacking faults. The quantitative analysis indicates that the particle at the tip is pure indium (the intensity of the As signal is below the sensitivity of the system). The fact that there is an indium particle at the NW tip indicates that an InAs NW was grown in the self-catalyzed VLS mode.

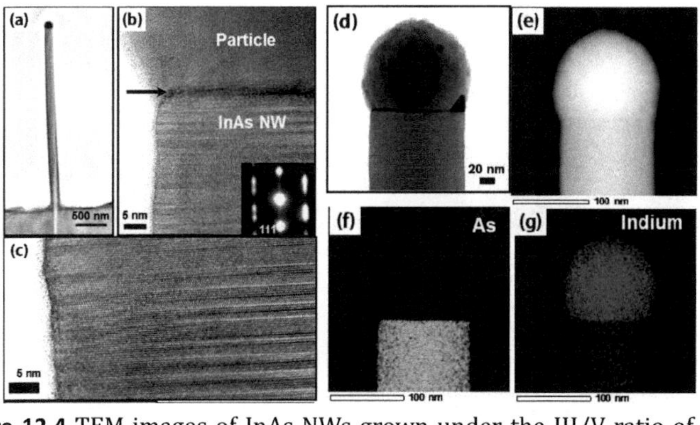

Figure 12.4 TEM images of InAs NWs grown under the III/V ratio of 0.11. (a) InAs NW on InP (111)B substrate. There is a particle at the NW tip. The black scale bar denotes 500 nm. (b) TEM image of the region near the interface between the NW and the tip, indicated by the black arrow. The inset in (b) is the SED pattern of the InAs NW, which indicates that the NW growth direction is <111> and that it has a ZB crystalline structure with stacking faults. (c) HRTEM of the InAs NW near the body side. (d) TEM and (e) HAADF-STEM images of the region near the tip. (f) and (g) Elemental mapping of the tip region. The scale bars in (e), (f) and (g) denote 100 nm. Compositional analysis indicates that the particle is pure Indium.

The TEM analysis results shown in Figs. 12.3 and 12.4 reveal that crystalline InAs NWs were grown and there were indium particles at each NW tip. This indicates that the NWs were grown in the self-catalyzed VLS mode through the indium particle. Moreover, the III/V mole ratio has a marked effect on the NW structure and the growth behavior. When the supply of indium source material was insufficient, the InAs NW has an indium-As particle at the NW tip (Fig. 12.3).

12.3.3 Influence of Temperature on InAs NW Growth

As an essential growth parameter, the influence of the growth temperature on the NW growth is also studied. Figures 12.5a,b show the InAs NW grown on InP (111)B substrates under 320 and 360°C, respectively. Both samples have the same other growth conditions including a high III/V mole ratio of 0.18. In Fig. 12.5a, InAs NWs are vertically aligned on InP substrate. There is a particle at the each NW tip. The high III/V mole ratio enables to form the NW with an indium particle at the tip and uniform

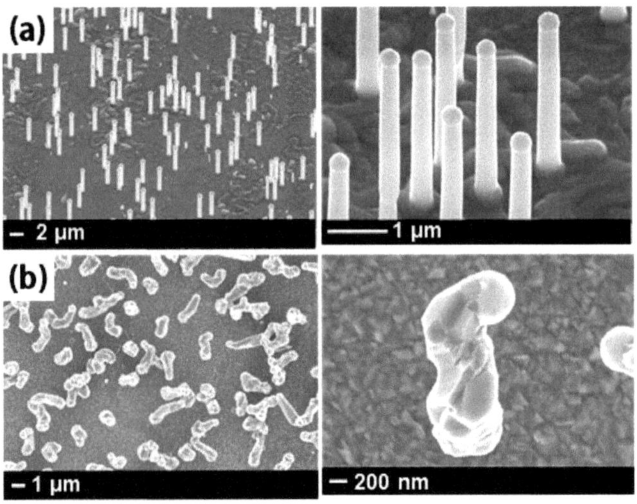

Figure 12.5 (a) Side-view (tilt: 38°) and (b) top-view SEM images of InAs nanostructures grown at (a) 320°C and (b) 360°C under the III/V ratio of 0.18. The III/V source mole ratio of the both samples is 0.18. There are two images for each condition, (low magnification on the left and high magnification on the right). In (a), there are InAs NWs vertically aligned on the substrate and there is an In particle at each NW tip.

diameter. When the temperature is increased to 360°C, we found a totally different result. In Fig. 12.5b, there is no NW formed on the substrate, only some irregular rod-like structures. High growth temperature is not helpful to grow InAs NWs in the self-catalyzed VLS mode. For the self-catalyzed growth of InAs NWs, we found that the growth temperature range was 320–340°C. Compared with the Au-catalyzed VLS growth of InAs NWs [11, 41, 42], the temperature range is quite limited. At the growth temperature as high as 400–500°C, the InAs NW could be grown in the VPE or SAG modes [32, 33, 43]. The distinct temperature difference between the self-catalyzed VLS and VPE (or SAG) may be why InAs NWs could not be found in the self-catalyzed VLS mode in their studies. Another possible reason may be the high III/V mole ratio is required for the self-catalyzed VLS growth.

12.3.4 Optical Properties

To study the lattice-vibration property of the InAs NW, we performed micro-Raman measurements on single InAs NWs. To prepare the sample, we dispersed InAs NWs onto a SiO_2 (thickness: 500 nm)/Si substrate with 200 nm Au film. By using an objective lens (×100), we obtained a spot size diameter as small as 1.2–1.5 μm, which enabled us to perform the measurement for individual NWs. Figure 12.6a shows single InAs NWs dispersed on an Au-coated SiO_2/Si substrate. The indium particle at the NW tip is visible in the figure.

Figure 12.6b shows the Raman spectrum of the InAs NW shown in Fig. 12.6a. There are two peaks; one at 217.8 cm^{-1} with a linewidth of 12.1 cm^{-1}, and the other at 234.3 cm^{-1} with a linewidth of 6.7 cm^{-1}. The peak at 217.8 cm^{-1} is assigned to the first-order transverse optic (TO) mode [44]. Compared with the longitudinal optic (LO) mode at 240.2 cm^{-1} in the bulk InAs, the peak at 234.3 cm^{-1} exhibits a redshift of 5.9 cm^{-1}. Probably, it is the low frequency branch of couple LO-phonon-plasmon (L_-) mode according to the well-established theory of interaction of LO phonons and plasmons [45]. The similar result had also been observed before in InAs NWs synthesized by a solvothermal route [46]. The distinct phonon in the Raman spectroscopy indicates the high crystalline feature of the InAs NW grown in the self-catalyzed VLS mode.

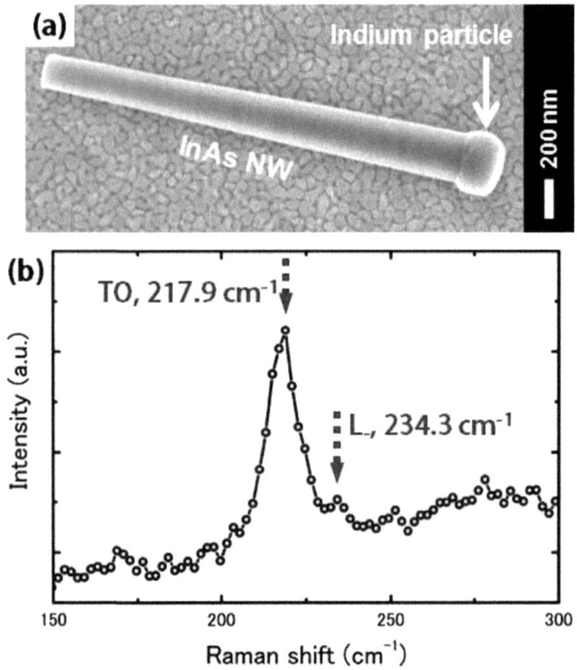

Figure 12.6 (a) SEM image of a single InAs NW grown at the III/V mole ratio of 0.11. The NWs were dispersed on a Au-coated SiO_2/Si substrate. At the NW tip, an indium particle is indicated by the white arrow. (b) Raman spectrum of the NW shown in (a). The InAs NW exhibits two distinct peaks: the TO peak at 217.8 cm^{-1} and the L_-peak at 234.3 cm^{-1}.

12.3.5 Electrical Properties

Owing to its high mobility, InAs NW shows good potential for high-speed electronic device applications. There have been numerous studies on this topic, but most of the InAs NWs were grown in the Au-catalyzed VLS mode. To examine whether the InAs NW grown in the Indium-particle-catalyzed VLS mode can function as a field effect transistor (FET) channel, we configured an InAs NW FET and studied the transport characteristics. We fabricated an InAs NW FET (Fig. 12.7a) with a gate-all-around (GAA) structure [47, 48], which enables improved electrostatic control of the channel conductance. We adopted a novel two-step gate electrode formation method, where an NW is sandwiched between the lower and upper gate metals (Fig. 12.7b).

DC electrical characterizations were performed using a semiconductor parameter analyzer at room temperature. Figure 12.7c shows typical transport characteristics for the InAs NW-FET device under a drain bias of 0.1 V. The device clearly shows the gate-voltage-dependent drain current. The InAs NW exhibits n-type conduction behavior. The inset shows the I_d–V_g characteristic on a logarithmic scale. The rectification range of the drain current is four orders of magnitude. The subthreshold slope is 500 mV/dec. The device features a peak transconductance (g_m) of 7.0 μS at a drain-source voltage of 0.1 V. The data shows n-type enhancement mode FET operation with a threshold voltage of –0.04 V. With a zero gate bias, the electron concentration of the InAs NW is estimated to be ~1 × 10^{16} cm^{-3}. Compared with the literature values of the InAs NW grown in the Au-catalyzed VLS mode (~10^{17} cm^{-3}) [49, 50], the electron concentration is rather low. The high purity of the InAs NW resulting from the Au-free VLS growth approach may facilitate the low carrier concentration.

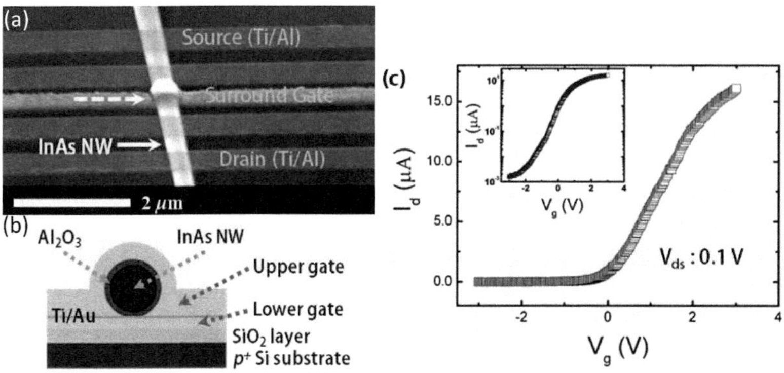

Figure 12.7 (a) SEM image of an InAs NW FET. (b) Schematic diagram of the cross-sectional structure near the surround gate electrode, where is indicated by the dashed arrow in (a). (c) Typical transfer characteristics, I_d–V_g, of the InAs-NW FET under a drain bias of 0.1 V. The inset is the I_d–V_g characteristic on a logarithmic scale, showing a rectification of 4 orders of magnitude.

With a zero gate bias, the mobility of the InAs NW is estimated to be ~1000 cm^2/Vs. Bryllert et al. reported an electron mobility of ~3000 cm^2/Vs in vertical InAs NW transistors with novel cylindrical gate structure [49], while Dayeh et al. reported a similar value of ~2740 cm^2/Vs in back-gated InAs NWFETs [50]. Compared with the mobility values, the mobility of the InAs NW grown in the self-catalyzed VLS is still low. Presumably, this may be due to the scattering caused by the stacking faults in the NW. Our InAs NWs contain high-density stacking faults (Fig. 12.7c). The mobility is also much lower than the bulk value (33,000 cm^2/Vs at 300 K). The bare side surface, which is not passivated and normally covered with oxides, scatters the carriers and consequently reduces the mobility [51].

12.4 Indium Phosphide/Indium Arsenide Heterostructure NW

In this section, we describe growth, structure and interface analysis, and optical property of multi-stacked InAs/InP heterostructure NWs. We realized the NWs by using an indium-particle-catalyzed VLS synthesis approach. The growth temperature as low as 320°C enables the formation of an atomically abrupt InP/InAs interface by suppressing the diffusion and weakening the reservoir effect in the indium droplet. The high controllability of the growth technology ensures that the luminescence can be tailored by the thickness of InAs segment and cover the 1.3–1.5 μm telecommunication window range. By using the nanoscale-spatial-resolution technology combing CL with SEM, we directly correlated the site of different-thickness InAs segments with its luminescence property in a single NW and demonstrate the InAs-thickness-controlled energy of optical emission in 1.1–1.6 μm.

12.4.1 Structural Analysis of InP/InAs Heterostructure NWs

Figure 12.8a shows a NW consisting of an InP segment with a multi-stacked heterostructure near its tip. We analyzed the region near the tip and the compositional profile along the axial direction (Figure 12.8b). The distinct contrast between the segments indicates the formation of multi-heterostructures. The compositional profile exhibits an axial composition of indium/$(InP/InAs)_3$/InP.

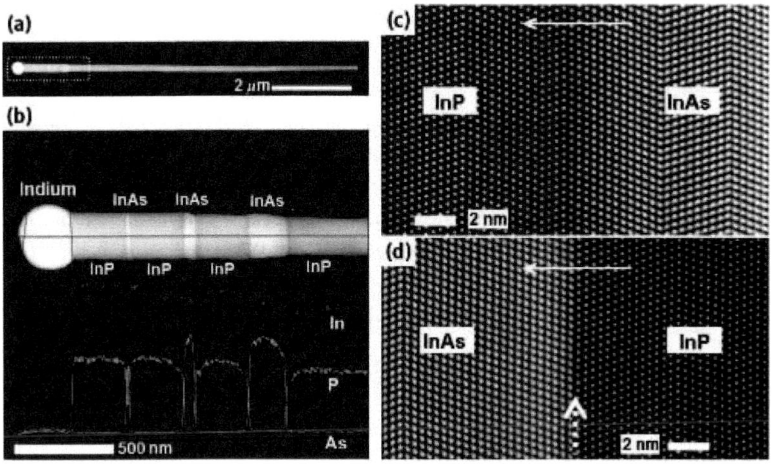

Figure 12.8 Compositional and structural analysis of InAs/InP hetero-NW. (a) HAADF-STEM image of a multi-stacked InP/InAs NW. The NW consists of three InAs segments with growth times of 5, 20, and 60 s. (b) HAADF-STEM image of the region with heterostructures near the NW tip, as indicated by the dashed rectangle in Figure 12.1a, and the compositional line scan along the axial direction. The red line in the HAADF-STEM images represents the scanning position in the NW. The red, green, and blue lines represent the indium, P, and As compositional profiles, respectively. (c) and (d) Respective HAADF-STEM image of the InAs→InP and InP→InAs heterostructures obtained with the Cs-STEM. The white arrows at the top of (c) and (d) indicate the <111> growth direction. There is a distinct difference in the contrast between the InAs and InP segments. In (c), the interface region exhibits a gradually decreasing intensity from InAs to InP, reflecting a gradual change in composition from As to P. In contrast, the intensity change in the InP→InAs heterostructure, as indicated by the dashed arrow in (d), is very sharp.

The almost 100% indium composition of the particle tip revealed by quantitative analysis indicates that the multi-stacked InAs/InP hetero-NW was grown in the indium-particle-catalyzed VLS mode. The VLS mode allows precise control of the NW length of each segment in the hetero-NW. To examine whether or not the self-catalyzed VLS growth mode also has precise thickness controllability similar to the Au-catalyzed VLS mode [52], we set the growth times of the InAs segments at 5, 20, and 60 s. The dependence of the InAs segment length on the growth time is almost linear with a growth rate of 3 nm/s.

Interface abruptness is extremely important in heterostructures to precisely engineer the carrier distribution for functional devices. To determine the abruptness of the interface formed by the self-catalyzed VLS, we analyzed the interface with a Cs-STEM system. The Cs-STEM system has a very small electron beam probe size that provides picometer resolution and is capable of distinguishing different atomic columns [53]. This makes it the most powerful tool for interface analysis with atomic layer resolution.

The materials sources in the InAs/InP hetero-NW had two different switching sequences, i.e., InP→InAs and InAs→InP, and so we analyzed both interfaces to examine the influence of the switching sequence on the interface property (Figs. 12.8c,d). In a HAADF-STEM image, the brightness of the atomic column is proportional to the atomic number. The dark and bright regions correspond to InP and InAs segments, respectively. The change in heterostructure brightness reflects the interface abruptness. The InAs→InP interface exhibits a compositionally graded layer, where an InAsP ternary composition is formed (Fig. 12.8c). The thickness of the graded layer is estimated to be 3–4 monolayers. In contrast, the InP→InAs interface exhibits a sharp contrast (Fig. 12.8d), implying a very abrupt interface.

We quantitatively analyzed the InP→InAs interface by plotting the intensity profile of the atomic column (Figs. 12.9c,d). In Fig. 12.9b, with the current contrast, the P spot is very weak because the P atom is much lighter than the As atom. Notably, there is a sharp change in the spot intensity of the V-group atoms, as indicated by the dashed arrow. The interface was analyzed quantitatively by plotting the integrated intensity profile of the spots (Figs. 12.9c,d). The As(P) exhibits an uniform intensity profile in the InAs(P) segment. In contrast, the intensity between the As and P spots changes sharply at the interface. The result shows that the InP→InAs interface is atomically abrupt.

It should be noted that a stacking fault is exactly located at the heterostructure interface. The TEM slice was prepared by technicians using an FIB system who carried out the Cs-STEM measurement. The stacking fault at the interface is there by chance. This is an interesting result that indeed needs more work to be elucidated. In our opinion the stacking fault, which is indeed a defect, may plays some role in forming the atomically abrupt interface. If further investigation proves this to result to be reproducible, it would be a very significant result and novel observation.

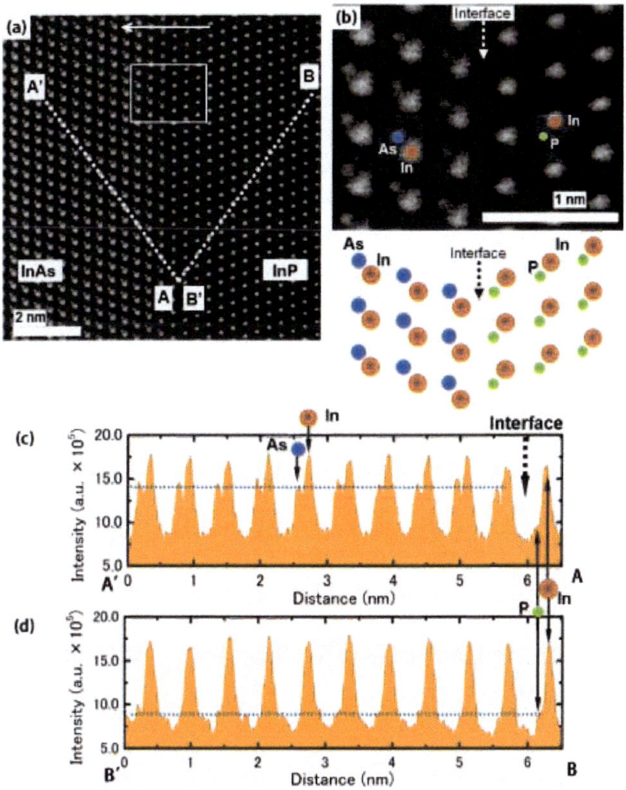

Figure 12.9 Analysis of interface abruptness with Cs-STEM. (a) and (b) High-resolution HAADF-STEM image of the InP→InAs heterostructure. The white arrow in (a) indicates the <111> growth direction. The interface region indicated by the white rectangle is enlarged in (b). A sharp change is clearly visible in the intensity of the As and P columns. The corresponding atomic model schematic is also shown here. (c) and (d) The respective intensity profiles along the A→A and B→B lines in (a). As(P) shows the uniform intensity profile in the InAs(P) segment (the dashed blue lines are guides to the eye). The intensity of As(P) shows a sharp change at the interface, as indicated by the dashed black arrow in (c). In (b-d), the blue, green, and orange balls represent As, P, and indium atoms, respectively.

An atomically abrupt heterostructure has been generally believed to be extremely difficult to realize by bottom-up VLS synthesis mainly due to the reservoir effect [54]. One effective way to suppress the reservoir effect is to use a growth temperature that is as low as possible, because solubility usually decreases with temperature. Here, the indium-particle-catalyzed VLS synthesis approach allows a very low growth temperature because of the low melting point of indium (156.6°C). The other advantage of the low growth temperature is that the atomic diffusion near the heterostructure can be considerably suppressed because it usually increases exponentially with temperature.

The switching sequence results in the InP→InAs and InAs→InP heterostructures have distinctly different interface properties since As is more soluble in indium particles than P, which consequently causes an enhanced reservoir effect. There may be other factors that have considerable influence on the interface property in addition to the reservoir effect. One is atom-dependent absorption. It has been found that additional As atoms reside predominantly on the InAs (111)B surface once the supply of As source material has been stopped. Consequently the As atoms are incorporated into the next growth layer and form a compositionally-graded layer with the supply of P source material [55, 56].

12.4.2 Tuning the Emission Wavelength of InAs Quantum Disks

A technique for forming an array of InP/InAs heterostructure NWs is very helpful in terms of integrating the NWs and making functional devices. We used the InP (111)B substrate for the growth of the vertical InAs/InP hetero-NWs (Fig. 12.10). The NWs were vertically aligned on the substrate, indicating an epitaxial relationship between the NW and the substrate (Figs. 12.10a,b). Compared with the tapering shape of the NW shown in Fig. 12.1, the NW exhibits a very uniform diameter (Figs. 12.10c,d). This is attributed to the optimization of the flow rates of the source materials and the density of the indium particles.

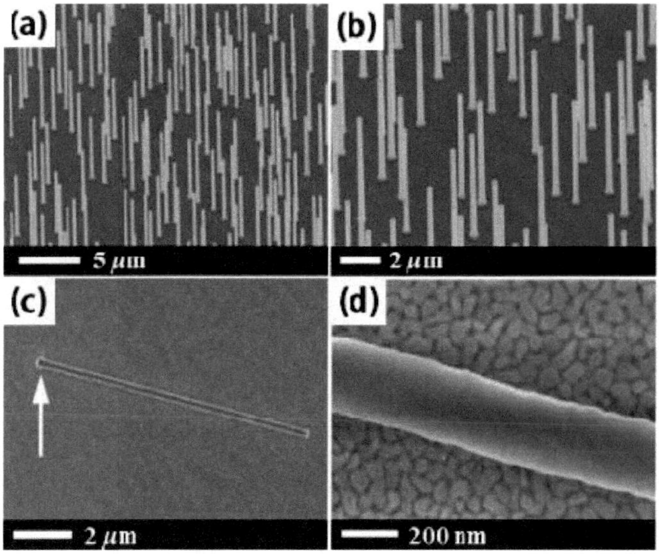

Figure 12.10 Multi-stacked InAs/InP hetero-NW. (a) and (b) SEM images (tilt: 38°) of the InP/InAs NW grown on an InP 111B substrate. The NWs are vertically aligned on the substrate, indicating an epitaxial relationship between the NW and the substrate. (c) and (d) SEM images of the NWs dispersed on a Au/Si substrate. The diameter is uniform along the growth direction. The white arrow in (c) indicates the indium particle at the NW tip.

In semiconductor technology, the heterostructure in NWs is fundamentally important for both technological and physical research and has attracted much attention in relation to optoelectronic applications. It is fundamentally important to tailor the optical and electronic properties accurately by forming a very thin layer with high thickness controllability. Consequently, it enables one to tune the luminescence wavelength, in particular, in telecommunication window (1.3–1.5 μm), through thickness control by the quantum confinement effect. We set short growth times for the InAs segment of 0.5, 1, and 1.5 sec in the NW shown in Fig. 12.10. The HAADF-STEM image of the NW (Fig. 12.11a) indicates the InP/InAs heterostructure and an indium particle at the NW tip. The three InP/InAs/InP double heterostructures in the NW are shown in Figs. 12.11b,d. The quantization effect occurs in such thin InAs layers. Figure 12.11e schematically shows the band structure of the multi-stacked InAs/InP NW. The energy gap between the ground states of electrons and holes decreases with the thickness.

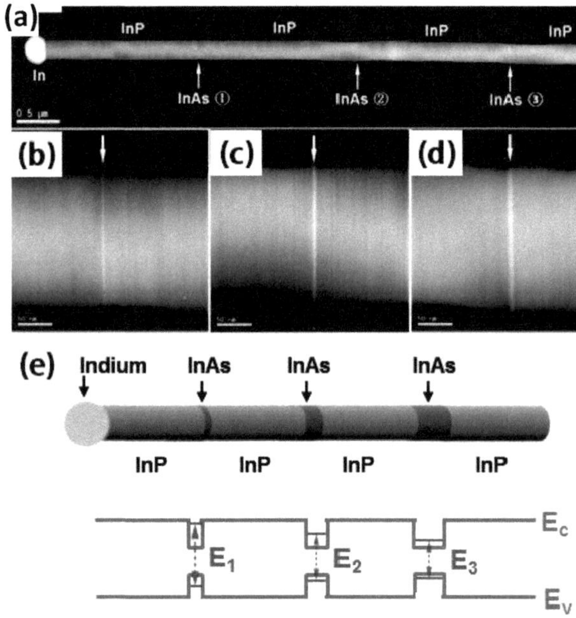

Figure 12.11 TEM characterization of single InP/InAs hetero-NW. (a) HAADF-STEM image of an InP/InAs hetero-NW with three InAs segments. The InAs segments ①, ② and ③ were grown in 0.5, 1, and 1.5 s, respectively. (b)–(d): HAADF-STEM images of the three InAs segments sandwiched between InP barriers. (e) Schematic diagrams of the composition and band structures of the InAs/InP multi-stacked NW. The red lines indicate the ground energy levels induced by the quantum confinement effect in InAs segments. The different thickness results in different energies between the ground level of electron and hole (E_1, E_2, and E_3), as indicated by the red arrows.

We then studied the optical properties of InAs segments of various thicknesses in InP NWs. Because the ground (or ground and excited) state energy is directly tuned by the thickness for the quantum confinement effect, the optical emission energy should be thickness dependent. We clarified this dependence by directly correlating the thickness of the disks with the emission energy of these InAs segments. We carried out a CL measurement in a

SEM system. The combination of CL with a SEM system enabled us to analyze the luminescence from single NWs with a high (nanometer scale) spatial resolution.

We measured a number of single NWs and found that the luminescence spectrum of a single NW consisted of four peaks (Fig. 12.12a). To identify the emission source of these luminescence peaks, we performed the monochromatic mapping of the energy of each peak (Figs. 12.12c–f). In Fig. 12.12c, there are four long bright spots, indicating that the emission at 1.4187 eV is from the InP segments. The three dark spots in Fig. 12.12c show the positions of the InAs segments. These InAs segments exhibit relatively strong individual luminescence from Figs. 12.12d–f, indicating that the InAs segments ①, ②, and ③ have corresponding emission peaks at 1.117, 0.946, and 0.826 eV, respectively. The increase in the emission energy with decreasing thickness indicates that (1) an emission energy blue shift is induced as the quantum confinement effect increases with decreasing thickness and (2) the thickness can be accurately changed by controlling the growth time.

Because the lattice mismatch of InAs/InP is as high as 3.1%, the coherent growth of the InP/InAs hetero-NW implies considerable compressive strain exists in the InAs segments. It is known that strain engineering is widely used to tailor the optical or electronic property of semiconductors [57, 58]. Noting that there are small shoulders in the high-energy side for the three peaks from InAs segments, probably this is from the recombination of light holes, which actually separated from the heavy holes in the band by the compressive strain in the InAs segment.

The linewidth of the luminescence from the InAs segments has a relatively large value, and we consider that this is induced by inhomogeneous compressive strain along the radial direction. The inhomogeneity results from the lateral lattice relaxation capability of the one-dimensional structure and it usually increases with the thickness [59]. The fact that the linewidth increases with the InAs thickness from 12 to 27 and 37 meV (Fig. 12.12a) is consistent with the above explanation. The stacking fault should also be responsible for the wide linewidth because of the documented effect of these defects on linewidth broadening [60, 61].

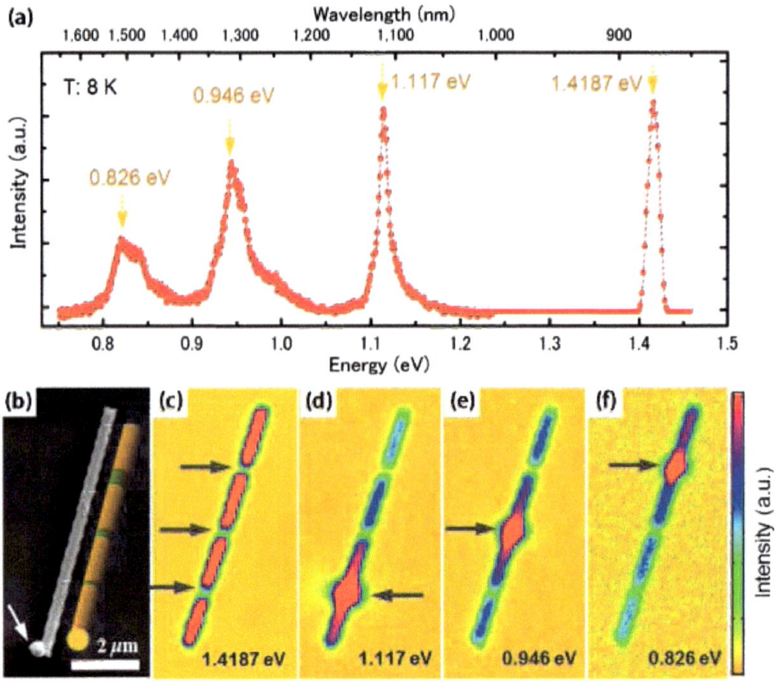

Figure 12.12 CL spectra and images of single InP/InAs hetero-NW at 8 K. (a) CL spectra of an InP/InAs NW. There are four distinct peaks located at 0.826, 0.946, 1.117, and 1.419 eV as indicated in the figure. Due to the very strong luminescence from the InP segment, the spectrum in the 1.24–1.46 eV range is normalized. (b) SEM image of an InP/InAs NW. There is an indium particle at the NW tip, as indicated by the white arrow. The inset is a schematic diagram of the NW structure. The SEM image was obtained during a monochromatic scanning CL measurement. The wave-like side face of the NW is induced by the drift of the sample stage due to the helium flow during scanning. (c)–(f) Monochromatic CL images of the NW at 1.419 eV (874 nm), 1.117 eV (1110 nm), 0.946 eV (1310 nm), and 0.826 (1500 nm), respectively. The dark spots in (c) are the positions of the InAs disks, as indicated by the arrows. The bright luminescence spots in (d–f), as indicated by the arrows, correspond to the InAs disks with the CL peaks shown in (a). The InP barrier segments also show the luminescence in the emission range of the InAs segments. Part of the carriers excited on the InP barrier will diffuse before recombination into the InAs segments, and then recombine in the photon emission with the energy of the ground (or ground and excited) states of the InAs segment.

12.5 Aligning Indium Phosphide/Indium Arsenide Heterostructure NWs by Self-Assembled Indium Particle Array

In this section, we describe a bottom-up self-assembly way to realize site-defined InP/InAs heterostructure NW array. The growth procedure we used is a two-step ("seed and grow") process whereby indium adatoms are allowed to diffuse to the open windows of an exposed InP substrate and form an indium nanoparticle array under optimized temperature and window spacing conditions. The site-defined InP NWs with a nanometer-scale diameter are then grown in the indium-particle-catalyzed VLS mode by using the indium nanoparticle array. To examine the growth feasibility of the InP/InAs heterostructure NWs with the method, we added two InAs segments with different thicknesses to the InP NW. The luminescence spots that originated from the InAs quantum disks exhibit the same pattern formed by the exposed InP windows that exist on the original substrate.

12.5.1 Strategy for Site-Defined Indium Nanoparticles

The site-defined growth of III–V NWs, whereby NWs are grown in pre-determined areas on a substrate, is a common method. It usually entails growth in gaps on an SiO_2 mask made by electron-beam or nanosphere lithography and has been achieved for GaAs [62, 63], GaN [64], InP [17] and InAs [63, 65–67]. This process is normally realized with a one-step VPE growth mechanism, and it grows nanostructures in an ordered array with the NW occupying the entire exposed substrate hole [60, 68].

Our approach consists of achieving site-defined growth with high reproducibility using a two-step "seed and grow" process. An indium particle is formed on the exposed substrate windows and NWs are then grown from the seed particles [26]. This procedure has numerous advantages. It is self-catalyzed and consequently avoids the use of Au. Also, the size of the seed particle can be adjusted during the process once it is contained in the exposed substrate window. The NW diameter can be easily controlled by varying the particle size.

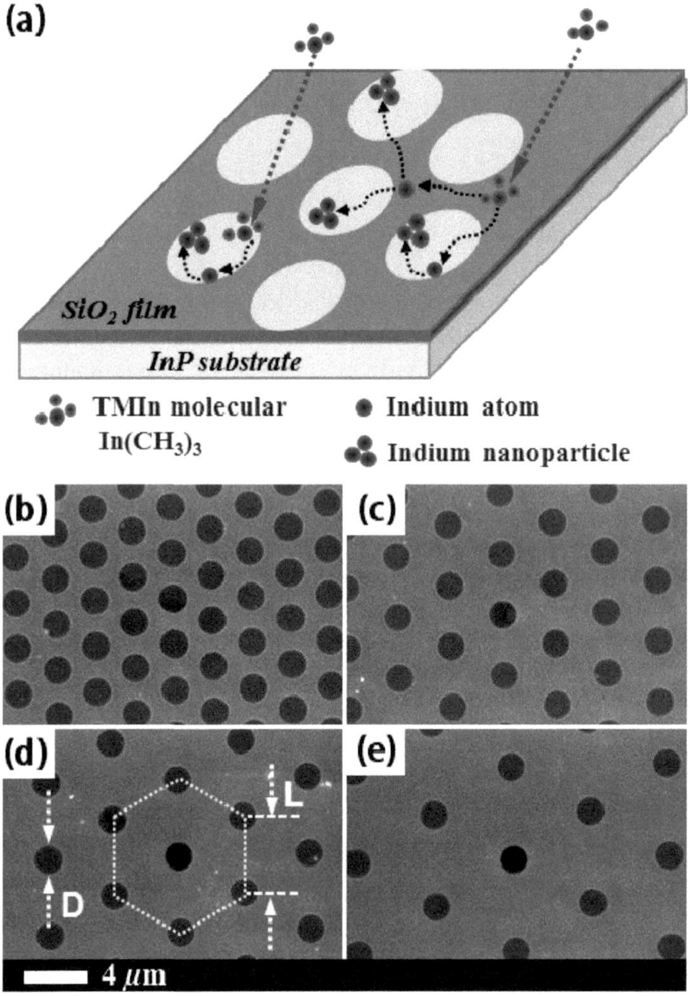

Figure 12.13 (a) Schematic diagram of absorption and decomposition of source materials, surface diffusion and accumulation of indium adatoms on SiO_2 film surface and circular open windows of exposed InP substrate. The dotted red lines indicate the absorption of the source materials on the surface. The dotted black lines indicate the indium adatom diffusion resulting from source material decomposition. The indium particles preferentially accumulate on the InP area and form nanoparticles to minimize the surface energy. (b)–(e) InP substrates with different spacings between the centers of the open windows (L). The circular window is the InP area after SiO_2 film removal. The diameters of the circular open windows, D, are all the same (2 μm). (b) L: 3 μm, (c) L: 4 μm, (d) L: 5 μm, (e) L: 6 μm.

Our procedure has two main steps: a seeding step performed by diffusion followed by a growth step in the VLS mode. The first step in the growth process involves the breakdown of trimethyl-indium into indium and its other constituents. The indium adatoms then diffuse into areas of the substrate that are not covered by the SiO_2 film. This mechanism and the substrate before absorption are shown in Fig. 12.13a. The two steps allow better control over the location and size of the grown NWs. This first step is crucial because the success of the growth step is determined by the morphology of the window region and the location of the indium droplets on the substrate. The diffusion on the substrate surface will be temperature dependent following an Arrhenius-type relationship.

At a low temperature ($T < 500°C$), adatoms are incorporated after diffusion and the competition with desorption becomes negligible. For a single atom diffusing on a 2-D substrate, using Einstein's diffusion length relation, $\lambda = \sqrt{D\tau}$, the displacement from the arrival site to the eventual incorporation site can be written as

$$\lambda = \sqrt{D\tau} = \lambda_0 \exp[-E_d/(2\kappa_B T)],$$

where τ is the mean surface diffusion time [69, 70]. In this equation, the pre-exponential factor λ_0 is the merged effective jump distance, E_d is the diffusion barrier of a single jump, and κ_B is Boltzmann's constant. This equation indicates that the diffusion length is strongly temperature dependent and simply increases with temperature. The temperature and window spacing (Figs. 12.13b–e) are therefore becoming key parameters for deciding the alignment of seed particles in the open windows of an exposed InP substrate.

12.5.2 Effect of Temperature and Window Spacing on the Alignment

The ideal seeding conditions were determined by varying the temperature and the window spacing on the substrate while keeping the deposition time constant at 10 min. The results of this test are shown in Fig. 12.14. The window diameter was kept constant at a value of 2 μm. The window diameter value was chosen intentionally because it is attainable using the conventional UV lithography.

To minimize the surface energy, indium atoms on the surface tend to accumulate and then form indium nanoparticles on the surface. If the window distance (L) is short enough, the indium atoms diffuse to the open windows of the exposed InP substrate and form indium particles inside the window; otherwise, they form indium particles on the SiO_2 film (Fig. 12.14). The influence of L is particularly marked when the temperature is fixed at 500°C. Within a single open window, the indium atoms also diffuse and finally form single or multi indium particles. Here temperature has a significant effect on the formation of a single particle in a single window (Fig. 12.15). At 400 and 440°C, multi indium particles are formed in the single window (Figs. 12.15a,b). With increasing temperature, the diffusion length increases and a single particle is therefore formed in the single open window (Figs. 12.15c,d).

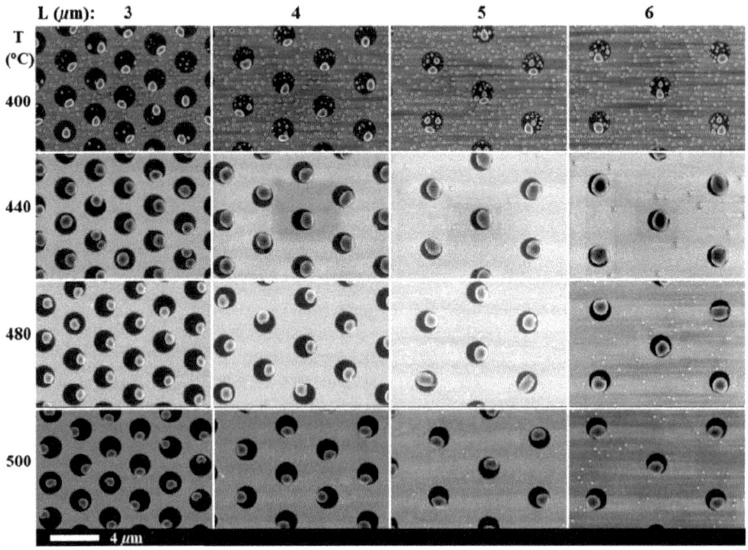

Figure 12.14 Indium particles deposited at various temperatures (400, 440, 480, and 500°C) on InP (211)B substrate with different sized open windows (L = 3, 4, 5, and 6 μm). At a low deposition temperature (400°C), indium particles are formed on the SiO_2 film. With increasing temperature, the density of the indium particles formed on the SiO_2 film decreases because of the increased diffusion length. The distance between the adjacent windows (L) has a marked effect on the indium particle distribution. With decreasing distance (L), more indium atoms can diffuse and reach the InP open window area. This also reduces the probability of indium particle formation on the SiO_2 film.

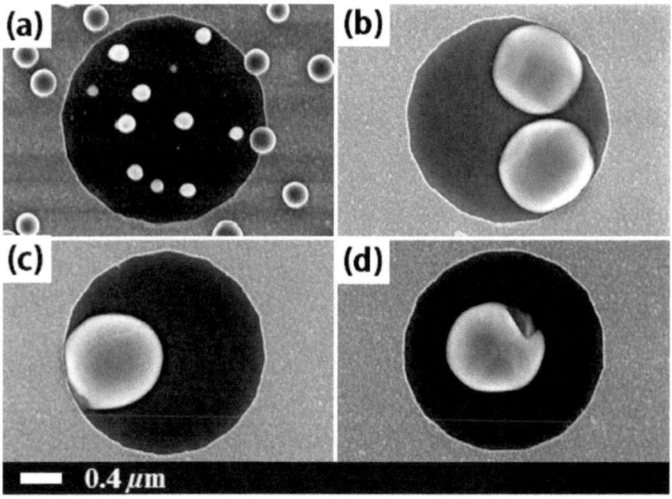

Figure 12.15 SEM images of indium particles self-assembled in a single open window under various deposition temperatures. (a) 400°C, (b) 440°C, (c) 480°C, (d) 500°C. The InP (211)B substrates have the same pattern size (D: 2 μm, *L*: 3 μm). At 400 and 440°C, multi indium particles are formed in the single window. With increasing temperature, the diffusion length increases and a single particle is therefore formed in the single open window (c and d).

The best seeding conditions were 480 and 500°C with a window spacing of 3 μm. Having all of the indium particles deposited at these higher temperatures and in the windows with closer window spacing is consistent with the temperature dependence described in the theory presented above. At these temperatures, the indium adatoms had sufficient kinetic energy to diffuse to the open windows of an exposed InP substrate, and they eventually accumulated into indium particles in the open windows.

12.5.3 Effect of the Alignment of Indium Nanoparticles on NW Growth

To study the way in which the alignment of the indium particles affects the subsequent NW growth, we compared the growth result by using substrates with indium particles deposited at 500°C, and with various window spacings (*L*) (Fig. 12.16). The dependence on the seeding step is seen clearly in this figure. For *L* values of

5 and 6 μm, indium particles present on the SiO_2 film result in irregular NW growth or have a larger diameter (Figs. 12.16c,d). For an L of 4 μm, there is very little growth on the SiO_2 film, but still some irregular growth from the window area (Fig. 12.16b). In comparison to these results, Fig. 12.16a shows a well-aligned uniform NW array, without the formation of other unwanted crystal features. Figure 12.17 is a close-up of a grown NW array showing the homogeneity of the sample. The indium particles formed on the amorphous masking film make the growth behavior more complicated and uncontrollable.

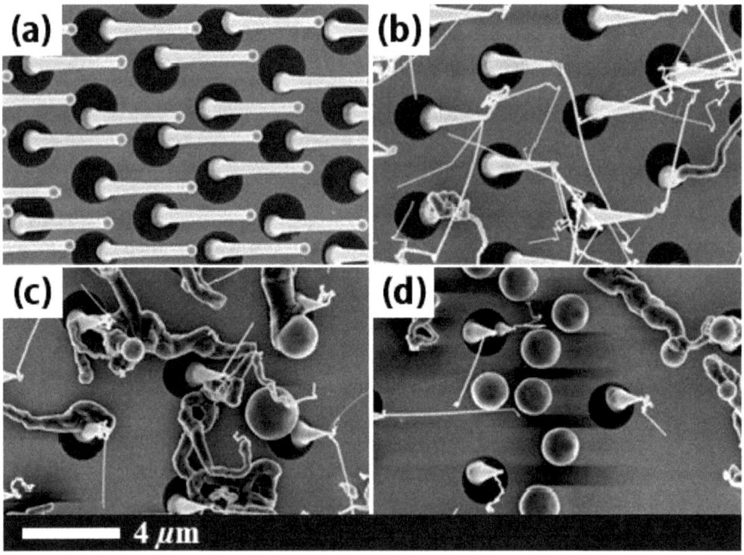

Figure 12.16 Top-view SEM images of InP NWs grown from indium particles deposited at 500°C formed on an InP (211)B substrate with the same D (2 μm) and various distances (L) between the open windows. (a) L: 3 μm, (b) L: 4 μm, (c) L: 5 μm, (d) L: 6 μm. Due to the additional indium particles on the SiO_2 film, irregular growth occurred (b, c, and d). In (a), NWs are well formed on the window area, indicating that an L of 3 μm is critical as regards forming an NW array. <111>-oriented NWs are epitaxially grown on an InP (211)B substrate, and so the NWs are inclined toward the substrate.

The other issue which may partially be responsible for the irregular growth is the relatively large window spacings (Figs. 12.16b–d). Because the growth temperature (320°C) is much lower than the deposition temperature (500°C) of indium particles, the diffusion length of indium atoms is actually much reduced during the NW growth. This may result in the additional indium particles formed on the masking film and consequently causes the irregular NW growth (Figs. 12.16b,c) and large indium particles with growth (Fig. 12.16d). The result indicates the critical significance of the alignment of the seed particle in the open window and the window spacing distance for the controllable formation of a uniform NW array.

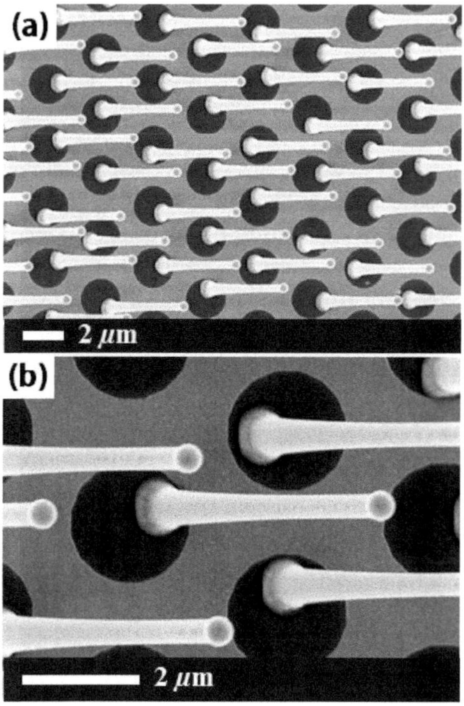

Figure 12.17 Top-view SEM images of InP NWs grown from indium particles deposited at 500°C formed on an InP (211)B substrate with a D of 2 μm and an *L* of 3 μm. The roots of all the NWs are located in the window region, indicating the seeding from the indium particles. Because the NWs were grown in the indium-particle-assisted VLS mode, there is an indium particle at each NW tip after growth.

12.5.4 InP/InAs Heterostructure NW Array

InP/InAs heterostructure NWs have a number of potential applications in optical and electronic devices [39, 40, 61]. By tuning the thickness of the InAs in InP NWs, we can modify the luminescence over a wide range, in particular, including the telecommunication window [40]. The realization of such a heterostructure in an InP NW with a site-defined function is promising as regards integrating the NW into the CMOS process.

To examine the growth feasibility of InP/InAs heterostructure NWs with the above method, we add two InAs segments with different growth time. Figure 12.18a shows the InP/InAs heterostructure NWs grown on an InP (211)B substrate with pattern size of D (2 μm) and *L* (3 μm). The NWs are inclined toward the substrate because of the <111> growth direction. The two InAs quantum disks were grown with 1 s and 2 s, respectively. This therefore results in the different thickness. Because of the quantum confinement effect, the luminescence wavelength should be tuned by the thickness (Fig. 12.18b). In Fig. 12.18b, the red lines schematically indicate the ground energy levels of electrons and holes induced by the quantum confinement effect in InAs quantum disks. The varied thickness results in different energies between the ground levels of the electrons and holes (E_1 and E_2), as indicated by the red arrows.

We then studied the optical properties of the InAs quantum disks. We used a micro-PL setup [7] to analyze the luminescence from the InAs quantum disks. NWs inclined toward the surface of an InP (211)B substrate ensure the efficient absorption of the incident laser for the excitation and collection of luminescence from the same lens. Thus, we were able to perform the optical analysis without dispersing the NWs to other substrates. This approach also enables us to obtain luminescence images and confirm the spatial distribution of emission sources. This is why we used an InP (211)B substrate for the NW array growth and subsequent analysis of the optical properties.

Figure 12.18c shows a spatial PL image when changing the lens magnification and laser spot size. We positioned an 1100 nm long-pass-filter (LPF) in front of the camera, to filter out the InP's emission leaving only the emission originating from the InAs quantum disks. The luminescence spots are clearly seen in all the images in Fig. 12.18c. We placed an LPF of 1100 nm in front of the camera, so the bright luminescence spots in the images originate from the InAs quantum disks instead of the InP segment. Most of embedded InAs quantum disks exhibit luminescence except for certain weak spots, which may result from the imperfections produced in the fabrication process. The disposition of these sources is the same hexagonal shape formed by the exposed InP windows that exists on the original substrate (Fig. 12.17a). This indicates that the InAs segments are indeed grown in the site-defined InP NWs, as shown in Fig. 12.18a. The result suggests that the method for the site-defined InP NWs is applicable to the site control of InP/InAs heterostructure NWs.

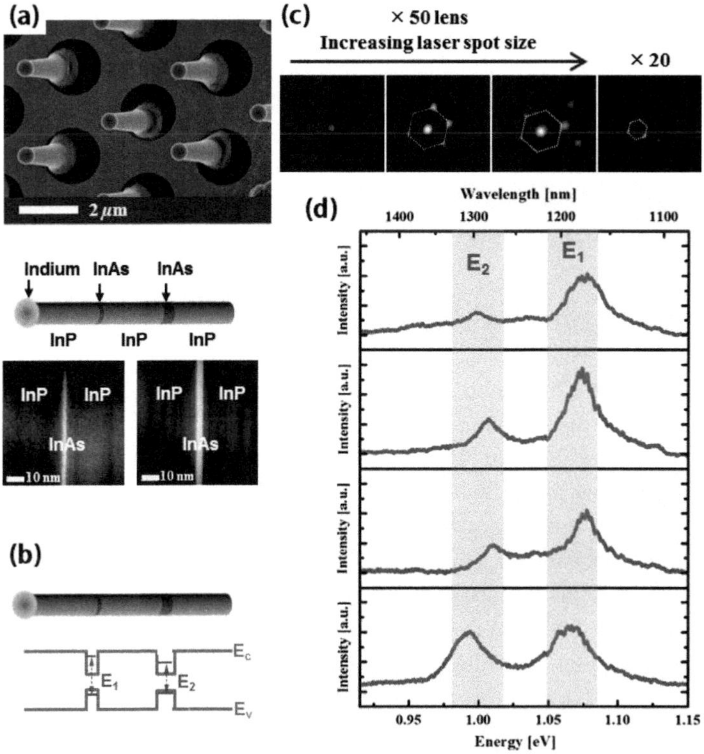

Figure 12.18 SEM image (top view), schematic diagram and HAADF-STEM images of InP/InAs heterostructure NWs grown on InP (211)B. The two InAs quantum disks grown with varied time result in the different thickness of 2.7±1 nm and 3.7±1 nm, respectively. (b) The NW band structure diagram of the InP/InAs NW. (c) Spatial images obtained during luminescence for different lens magnifications and laser spot sizes. (d) Spectroscopy of 4 single NWs selected from the sample. The excitation laser power is 1 μW with a laser spot diameter of 2 μm. Two peaks (E_1 and E_2) originate from the two InAs quantum disks with varied thickness.

To test the uniformity of the samples, four NWs were then selected at different locations across the substrate to determine whether the optical characteristics were similar. The result shows that two dominant peaks (E_1 and E_2), which correspond to the two InAs quantum disks with different thicknesses; appear in all four NWs (Fig. 12.18d). The photoemissions of the two peaks, which are completely isolated, are tuned by controlling the InAs thickness through quantum confinement effect. The variation in the luminescence wavelength is ±10 nm for E_1 and ±15 nm for E_2. This is largely induced by the variation in the NW spacing caused by the position fluctuation of the indium seed particles in the open windows. The NW spacing and diameter have been shown to have a marked effect on the axial growth rate [71]. We believe that the variation in the NW spacing can be suppressed by further improving the site control of the nanometer-scale-diameter NWs in the micrometer-scale open windows.

12.6 Summary and Outlook

In summary, we show the growth, structural and optical properties of InAs and InP/InAs heterostructure NWs grown by the Au-free self-catalyzed VLS approach. We have achieved the heterostructure NWs by using an indium-particle-assisted VLS synthesis approach. The growth temperature as low as 320°C enables the formation of an atomically abrupt InP/InAs interface. The low growth temperature also enables us to grow multi-stacked InAs/InP NWs in the axial direction without any growth on the NW side face in the VPE mode. The high controllability of the growth technology ensures that the luminescence can be tailored by the thickness of InAs segment and cover the 1.3–1.5 μm telecommunication window range.

In contrast to the widely used top-down approach, we developed a bottom-up approach to form the site-defined InP/InAs NWs by successfully pairing bottom-up self-assembly with top-down micro-photolithography techniques. The growth procedure we used is a two-step ("seed and grow") process whereby indium adatoms are allowed to diffuse to the open windows of an exposed InP substrate and form an indium nanoparticle array under optimized temperature and window spacing conditions.

The site-defined InP NWs with a nanometer-scale diameter are then grown in the indium-particle-assisted VLS mode by using the indium nanoparticle array. To examine the growth feasibility of the InP/InAs heterostructure NWs with the method, we added two InAs segments with different thicknesses to the InP NW. The luminescence spots that originated from the InAs quantum disks exhibit the same pattern formed by the exposed InP windows that exist on the original substrate.

The high controllability of the synthesis approach described here in terms of thickness and diameter enable to realize state-of-the-art quantum devices in a NW. It is a very powerful approach for designing and synthesizing new nanostructures with high reproducibility allowing a whole new range of NW optoelectronic devices. Heterostructure NWs with high-purity materials and excellent optical characteristics will significantly improve the NW-based-device performance in electronics, photonics, and solar cells and provide new opportunities in fundamental physics research. Successfully pairing a controllable bottom-up growth technique with a top-down substrate preparation technique greatly improves the potential for mass-production and the widespread adoption of this technology. It will expand the application of NWs into many other areas that require large-area and low-cost NWs. Moreover, the fact that the approach is Au-free and has a growth temperature as low as 320–330°C makes it compatible with the CMOS process [72]. This in turn makes it possible to integrate various NW-based advanced functionalities into the CMOS process with the aim of moving "beyond CMOS" [73].

Acknowledgments

We thank Drs. M. D. Birowosuto and M. Takiguchi for their contribution in optical characterization, Dr. S. Sasaki for his contribution in electrical devices, Drs. M. Notomi and T. Sogawa for their discussions. We also thank Mr. C. Rainville, Ms. A. Salmon, Drs. E. Kuramochi, and T. Tamamura for their help in the preparation of patterned substrates, Dr. T. Mitate for his help in the TEM measurement. This work was supported by Scientific Research grant (No. 23310097, 15H05735, and 16H03821) from the Japanese Society for the Promotion of Science.

References

1. Wagner, R. S., and Ellis, W. C. (1964). Vapor-liquid-solid mechanism of single crystal growth, *Appl. Phys. Lett.*, **4**, pp. 89–90.
2. Bullis, W. M. (1966). Properties of gold in silicon, *Solid State Electron.*, **9**, pp. 143–168.
3. Sprokel, G. J., and Fairfield, J. M. (1965). Diffusion of gold into silicon crystals, *J. Electrochem. Soc.*, **112**, pp. 200–203.
4. Doh, Y. J., van Dam, J. A., Roest, A. L., Bakkers, E. P., Kouwenhoven, L. P., and De Franceschi, S. (2005). Tunable supercurrent through semiconductor nanowires, *Science*, **309**, pp. 272–275.
5. Mourik, V., Zuo, K., Frolov, S. M., Plissard, S. R., Bakkers, E. P., and Kouwenhoven, L. P. (2012). Signatures of Majorana fermions in hybrid superconductor-semiconductor nanowire devices, *Science*, **336**, pp. 1003–1007.
6. Yan, R. X., Gargas, D., and Yang, P. D. (2009). Nanowire photonics, *Nat. Photonics*, **3**, pp. 569–576.
7. Birowosuto, M. D., Yokoo, A., Zhang, G., Tateno, K., Kuramochi, E., Taniyama, H., Takiguchi, M., and Notomi, M. (2014). Movable high-Q nanoresonators realized by semiconductor nanowires on a Si photonic crystal platform, *Nat. Mater.*, **13**, pp. 279–285.
8. Hu, Y., Kuemmeth, F., Lieber, C. M., and Marcus, C. M. (2012). Hole spin relaxation in Ge-Si core-shell nanowire qubits, *Nat. Nanotechnol.*, **7**, pp. 47–50.
9. Qin, Y., Wang, X., and Wang, Z. L. (2008). Microfibre-nanowire hybrid structure for energy scavenging, *Nat. Commun.*, **451**, pp. 809–813.
10. Wallentin, J., Anttu, N., Asoli, D., Huffman, M., Åberg, I., Magnusson, M. H., Siefer, G., Fuss-Kailuweit, P., Dimroth, F., Witzigmann, B., Xu, H. Q., Samuelson, L., Deppert, K., and Borgström, M. T. (2013). InP nanowire array solar cells achieving 13.8% efficiency by exceeding the ray optics limit, *Science*, **339**, pp. 1057–1060.
11. Yazawa, M., Koguchi, M., and Hiruma, K. (1991). Heteroepitaxial ultrafine wire-like growth of InAs on GaAs substrates, *Appl. Phys. Lett.*, **58**, pp. 1080–1082.
12. Perea, D. E., Hemesath, E. R., Schwalbach, E. J., Lensch-Falk, J. L., Voorhees, P. W., and Lauhon, L. J. (2009). Direct measurement of dopant distribution in an individual vapour-liquid-solid nanowire, *Nat. Nanotechnol.*, **4**, pp. 315–319.

13. Putnam, M. C., Filler, M. A., Kayes, B. M., Kelzenberg, M. D., Guan, Y., Lewis, N. S., Eiler, J. M., and Atwater, H. A. (2008). Secondary ion mass spectrometry of vapor–liquid–solid grown, Au-catalyzed, Si wires, *Nano Lett.*, **8**, pp. 3109–3113.

14. Kelzenberg, M. D., Turner-Evans, D. B., Kayes, B. M., Filler, M. A., Putnam, M. C., Lewis, N. S., and Atwater, H. A. (2008). Photovoltaic measurements in single-nanowire silicon solar cells, *Nano Lett.*, **8**, pp. 710–714.

15. Putnam, M. C., Turner-Evans, D. B., Kelzenberg, M. D., Boettcher, S. W., Lewis, N. S., and Atwater, H. A. (2009). 10 μm minority-carrier diffusion lengths in Si wires synthesized by Cu-catalyzed vapor-liquid-solid growth, *Appl. Phys. Lett.*, **95**, p. 163116.

16. Breuer, S., Pfüller, C., Flissikowski, T., Brandt, O., Grahn, H. T., Geelhaar, L., and Riechert, H. (2011). Suitability of Au- and self-assisted GaAs nanowires for optoelectronic applications, *Nano Lett.*, **11**, pp. 1276–1279.

17. Motohisa, J., Noborisaka, J., Takeda, J., Inari, M., and Fukui, T. (2004). Catalyst-free selective-area MOVPE of semiconductor nanowires on (111)B oriented substrates, *J. Cryst. Growth*, **272**, pp. 180–185.

18. Noborisaka, J., Motohisa, J., and Fukui, T. (2005). Catalyst-free growth of GaAs nanowires by selective-area metalorganic vapor-phase epitaxy, *Appl. Phys. Lett.*, **86**, p. 213102.

19. Tomioka, K., Tanaka, T., Hara, S., Hiruma, K., and Fukui, T. (2011). III–V nanowires on Si substrate: Selective-area growth and device applications, *IEEE J. Sel. Topics Quantum Electron.*, **17**, pp. 1112–1129.

20. Kasahara, J., Kajiwara, K., and Yamada, T. (1977). GaAs whiskers grown by a thermal decomposition method, *J. Cryst. Growth*, **38**, pp. 23–28.

21. Novotny, C. J., and Yu, P. K. L. (2005). Vertically aligned, catalyst-free InP nanowires grown by metalorganic chemical vapor deposition, *Appl. Phys. Lett.*, **87**, p. 203111.

22. Mattila, M., Hakkarainen, T., and Lipsanen, H. (2007). Catalyst-free fabrication of InP and InP(N) nanowires by metalorganic vapor phase epitaxy, *J. Cryst. Growth*, **298**, pp. 640–643.

23. Jabeen, F., Grillo, V., Rubini, S., and Martelli, F. (2008). Self-catalyzed growth of GaAs nanowires on cleaved Si by molecular beam epitaxy, *Nanotechnology*, **19**, p. 275711.

24. Fontcuberta i Morral, A., Colombo, C., Abstreiter, G., Arbiol, J., and Morante, J. R. (2008). *Appl. Phys. Lett.*, **92**, p. 063112.

25. Pozuelo, M., Zhou, H., Lin, S., Lipman, S. A., Goorsky, M. S., Hicks, R. F., and Kodambaka, S. (2011). Self-catalyzed growth of InP/InSb axial nanowire heterostructures, *J. Cryst. Growth,* **329**, pp. 6–11.

26. Zhang, G., Tateno, K., Gotoh, H., and Sogawa, T. (2012). Vertically aligned InP nanowires grown via the self-assisted vapor–liquid–solid mode, *Appl. Phys. Express,* **5**, p. 055201.

27. Zhang, G., Sasaki, S., Tateno, K., Gotoh, H., and Sogawa, T. (2013). Au-free InAs nanowires grown in In-particle-assisted vapor-liquid-solid mode: Growth, structure, and electrical property, *AIP Adv.,* **3**, p. 052107.

28. Zhang, G., Tateno, K., Birowosuto, M. D., Notomi, M., Sogawa, T., and Gotoh, H. (2015). Controlled 1.1–1.6 μm luminescence in gold-free multi-stacked InAs/InP heterostructure nanowires, *Nanotechnology,* **26**, p. 115704.

29. Zhang, G., Rainville, C., Salmon, A., Takiguchi, M., Tateno, K., and Gotoh, H. (2015). Bridging the gap between the nanometer-scale bottom-up and micrometer-scale top-down approaches for site-defined InP/InAs nanowires, *ACS Nano,* **9**, pp. 10580–10589.

30. Zhang, Y., Aagesen, M., Holm, J. V., Jørgensen, H. I., Wu, J., and Liu, H. (2013). Self-catalyzed GaAsP nanowires grown on silicon substrates by solid-source molecular beam epitaxy, *Nano Lett.,* **13**, pp. 3897–3902.

31. Zhang, Y., Wu, J., Aagesen, M., Holm, J., Hatch, S., Tang, M., Huo, S., and Liu, H. (2014). Self-catalyzed ternary core–shell GaAsP nanowire arrays grown on patterned Si substrates by molecular beam epitaxy, *Nano Lett.,* **14**, pp. 4542–4547.

32. Mandl, B., Stangl, J., Hilner, E., Zakharov, A. A., Hillerich, K., Dey, A. W., Samuelson, L., Bauer, G. N., Deppert, K., and Mikkelsen, A. (2010). Growth mechanism of self-catalyzed group III–V nanowires, *Nano Lett.,* **10**, pp. 4443–4449.

33. Hertenberger, S., Rudolph, D., Bolte, S., Döblinger, M., Bichler, M., Spirkoska, D., Finley, J. J., Abstreiter, G., and Koblmüller, G. (2011). Absence of vapor-liquid-solid growth during molecular beam epitaxy of self-induced InAs nanowires on Si, *Appl. Phys. Lett.,* **98**, p. 123114.

34. Dimakis, E., Lähnemann, J., Jahn, U., Breuer, S., Hilse, M., Geelhaar, L., and Riechert, H. (2011). Self-assisted nucleation and vapor–solid growth of InAs nanowires on bare Si(111), *Cryst. Growth Des.,* **11**, pp. 4001–4008.

35. Grap, T., Rieger, T., Blömers, C., Schäpers, T., Grützmacher, D., and Lepsa, M. I. (2013). Self-catalyzed VLS grown InAs nanowires with twinning superlattices, *Nanotechnology,* **24**, p. 335601.

36. Gomes, U. P., Ercolani, D., Zannier, V., David, J., Gemmi, M., Beltram, F., and Sorba, L. (2016). Nucleation and growth mechanism of self-catalyzed InAs nanowires on silicon, *Nanotechnology,* **27**, p. 255601.

37. Wu, J., Ramsay, A., Sanchez, A., Zhang, Y., Kim, D., Brossard, F., Hu, X., Benamara, M., Ware, M. E., Mazur, Y. I., Salamo, G. J., Aagesen, M., Wang, Z., and Liu, H. (2016). Defect-free self-catalyzed GaAs/GaAsP nanowire quantum dots grown on silicon substrate, *Nano Lett.,* **16**, pp. 504–511.

38. Björk, M. T., Ohlsson, B. J., Sass, T., Persson, A. I., Thelander, C., Magnusson, M. H., Deppert, K., Wallenberg, L. R., and Samuelson, L. (2002). One-dimensional heterostructures in semiconductor nanowhiskers, *Appl. Phys. Lett.,* **80**, pp. 1058–1060.

39. Björk, M. T., Ohlsson, B. J., Thelander, C., Persson, A. I., Deppert, K., Wallenberg, L. R., and Samuelson, L. (2002). Nanowire resonant tunneling diodes, *Appl. Phys. Lett.,* **81**, pp. 4458–4460.

40. Tateno, K., Zhang, G., Gotoh, H., and Sogawa, T. (2012). VLS growth of alternating InAsP/InP heterostructure nanowires for multiple-quantum-dot structures, *Nano Lett.,* **12**, pp. 2888–2893.

41. Dick, K. A. (2008). A review of nanowire growth promoted by alloys and non-alloying elements with emphasis on Au-assisted III–V nanowires, *Prog. Cryst. Growth Charact. Mater.,* **54**, pp. 138–173.

42. Dick, K. A., Deppert, K., Samuelson, L., and Seifert, W. (2006). Optimization of Au-assisted InAs nanowires grown by MOVPE, *J. Cryst. Growth,* **297**, pp. 326–333.

43. Mandl, B., Stangl, J., Mårtensson, T., Mikkelsen, A., Eriksson, J., Karlsson, L. S., Bauer, G., Samuelson, L., and Seifert, W. (2006). Au-free epitaxial growth of InAs nanowires, *Nano Lett.,* **6**, pp. 1817–1821.

44. Lockwood, D. J., Yu, G., and Rowell, N. L. (2005). Optical phonon frequencies and damping in AlAs, GaP, GaAs, InP, InAs and InSb studied by oblique incidence infrared spectroscopy, *Solid State Commun.,* **136**, pp. 404–409.

45. Buchner, S., and Burstein, E. (1974). Raman scattering by wave-vector–dependent LO-phonon–plasmon modes in n-InAs, *Phys. Rev. Lett.,* **33**, pp. 908–911.

46. Xu, X. X., Yu, K. H., Wei, W., Peng, B., Huang, S. H., Chen, Z. H., and Shen, X. S. (2006). Raman scattering in InAs nanowires synthesized by a solvothermal route, *Appl. Phys. Lett.,* **89**, p. 253117.

47. Sasaki, S., Tateno, K., Zhang, G., Suominen, H., Harada, Y., Saito, S., Fujiwara, A., Sogawa, T., and Muraki, K. (2013). Encapsulated gate-all-around InAs nanowire field-effect transistors, *Appl. Phys. Lett.,* **103**, p. 213502.

48. Sasaki, S., Tateno, K., Zhang, G., Suominen, H., Harada, Y., Saito, S., Fujiwara, A., Sogawa, T., and Muraki, K. (2015). Self-aligned gate-all-around InAs/InP core–shell nanowire field-effect transistors, *Jpn. J. Appl. Phys.,* **54**, p. 04DN04.

49. Bryllert, T., Wernersson, L. E., Froberg, L. E., and Samuelson, L. (2006). Vertical high-mobility wrap-gated InAs nanowire transistor, *IEEE Electron Device Lett.,* **27**, pp. 323–325.

50. Dayeh, S. A., Soci, C., Yu, P. K. L., Yu, E. T., and Deli, W. (2007). Influence of surface states on the extraction of transport parameters from InAs nanowire field effect transistors, *Appl. Phys. Lett.,* **90**, pp. 162112–162111.

51. Jiang, X., Xiong, Q., Nam, S., Qian, F., Li, Y., and Lieber, C. M. (2007). InAs/InP radial nanowire heterostructures as high electron mobility devices, *Nano Lett.,* **7**, pp. 3214–3218.

52. Björk, M. T., Ohlsson, B. J., Sass, T., Persson, A. I., Thelander, C., Magnusson, M. H., Deppert, K., Wallenberg, L. R., and Samuelson, L. (2002). One-dimensional steeplechase for electrons realized, *Nano Lett.,* **2**, pp. 87–89.

53. Varela, M., Lupini, A. R., Benthem, K. V., Borisevich, A. Y., Chisholm, M. F., Shibata, N., Abe, E., and Pennycook, S. J. (2005). Materials characterization in the aberration-corrected scanning transmission electron microscopy, *Annu. Rev. Mater. Res.,* **35**, pp. 539–569.

54. Dick, K. A., Bolinsson, J., Borg, B. M., and Johansson, J. (2012). Controlling the abruptness of axial heterojunctions in III–V nanowires: Beyond the reservoir effect, *Nano Lett.,* **12**, pp. 3200–3206.

55. Yamakawa, I., Oga, R., Fujiwara, Y., Takeda, Y., and Nakamura, A. (2004). Atomic-scale observation of interfacial roughness and As–P exchange in InGaAs/InP multiple quantum wells, *Appl. Phys. Lett.,* **84**, pp. 4436–4438.

56. Streubel, K., Härle, V., Scholz, F., Bode, M., and Grundmann, M. (1992). Interfacial properties of very thin GaInAs/InP quantum well structures grown by metalorganic vapor phase epitaxy, *J. Appl. Phys.,* **71**, pp. 3300–3306.

57. Chu, M., Sun, Y. K., Aghoram, U., and Thompson, S. E. (2009). A solution for higher carrier mobility in nanoscale MOSFETs, *Annu. Rev. Mater. Res.,* **39**, pp. 203–229.

58. Adams, A. R. (2011). Strained-layer quantum-well lasers, *IEEE J. Sel. Top. Quantum Electron.*, **17**, pp. 1364–1373.

59. Ertekin, E., Greaney, P. A., Chrzan, D. C., and Sands, T. D. (2005). Equilibrium limits of coherency in strained nanowire heterostructures, *J. Appl. Phys.*, **97**, p. 114325.

60. Dalacu, D., Mnaymneh, K., Wu, X., Lapointe, J., Aers, G. C., Poole, P. J., and Williams, R. L. (2011). Selective-area vapor-liquid-solid growth of tunable InAsP quantum dots in nanowires, *Appl. Phys. Lett.*, **98**, p. 251101.

61. Dalacu, D., Mnaymneh, K., Lapointe, J., Wu, X., Poole, P. J., Bulgarini, G., Zwiller, V., and Reimer, M. E. (2012). Ultraclean emission from InAsP quantum dots in defect-free wurtzite InP nanowires, *Nano Lett.*, **12**, pp. 5919–5923.

62. Ikejiri, K., Noborisaka, J., Hara, S., Motohisa, J., and Fukui, T. (2007). Mechanism of catalyst-free growth of GaAs nanowires by selective area MOVPE, *J. Cryst. Growth,* **298**, pp. 616–619.

63. Paetzelt, H., Gottschalch, V., Bauer, J., Benndorf, G., and Wagner, G. (2008). Selective-area growth of GaAs and InAs nanowires-homo- and heteroepitaxy using SiNx templates, *J. Cryst. Growth,* **310**, pp. 5093–5097.

64. Schumann, T., Gotschke, T., Limbach, F., Stoica, T., and Calarco, R. (2011). Selective-area catalyst-free MBE growth of GaN nanowires using a patterned oxide layer, *Nanotechnology,* **22**, p. 095603.

65. Dayeh, S. A., Chen, P., Jing, Y., Yu, E. T., Lau, S. S., and Wang, D. (2008). Integration of vertical InAs nanowire arrays on insulator-on-silicon for electrical isolation, *Appl. Phys. Lett.*, **93**, p. 203109.

66. Tatebayashi, J., Ota, Y., Ishida, S., Nishioka, M., Iwamoto, S., and Arakawa, Y. (2012). Site-controlled formation of InAs/GaAs quantum-dot-in-nanowires for single photon emitters, *Appl. Phys. Lett.*, **100**, p. 263101.

67. Tomioka, K., Motohisa, J., Hara, S., and Fukui, T. (2008). Control of InAs nanowire growth directions on Si, *Nano Lett.*, **8**, pp. 3475–3480.

68. Plissard, S., Larrieu, G., Wallart, X., and Caroff, P. (2011). High yield of self-catalyzed GaAs nanowire arrays grown on silicon via gallium droplet positioning, *Nanotechnology,* **22**, p. 275602.

69. Ibach, H. (2006). *Physics of Surfaces and Interfaces* (Springer).

70. Pohl, U. (2013). *Epitaxy of Semiconductors* (Springer Berlin Heidelberg), pp. 171–224

71. Borgstrom, M. T., Immink, G., Ketelaars, B., Algra, R., and BakkersErik, P. A. M. (2007). Synergetic nanowire growth, *Nat. Nanotechnol.,* **2**, pp. 541–544.

72. Miller, D. A. B. (2009). Device requirements for optical interconnects to silicon chips, *Proc. IEEE,* **97**, pp. 1166–1185.

73. Bernstein, K., Cavin, R. K., Porod, W., Seabaugh, A., and Welser, J. (2010). Device and architecture outlook for beyond CMOS switches, *Proc. IEEE,* **98**, pp. 2169–2184.

Chapter 13

III-Nitride Nanowires and Their Laser, LED, and Photovoltaic Applications

Wei Guo,[a] Pallab Bhattacharya,[b] and Junseok Heo[c]

[a]*Department of Physics and Applied Physics, University of Massachusetts, Lowell, Massachusetts 01854, USA*
[b]*Department of Electrical Engineering and Computer Science University of Michigan, Ann Arbor, Michigan 48109, USA*
[c]*Department of Electrical and Computer Engineering Ajou University, Yeongtong-gu, Suwon 414-735, South Korea*

Wei_Guo@uml.edu

Semiconductor nanowire materials provide a unique opportunity for exploring novel physics phenomena at the nanoscale and they have enabled a new class of devices. Developments in nanowire growth technologies have led to the realization of III-nitride nanowires with precise control of composition, morphology, and electrical properties. In this chapter, we will review the recent development of III-nitride nanowire materials and nanowire growth and characteristics and discuss nanowire device applications. First, we will discuss the growth techniques of III-nitride nanowires by molecular beam epitaxy. Second, we will describe the characteristics of the III-nitride nanowires, where both the optical and electrical properties of the nanowires will be discussed. Finally, we will review the optoelectronics device applications of the In-nitride nanowires.

Novel Compound Semiconductor Nanowires: Materials, Devices, and Applications
Edited by Fumitaro Ishikawa and Irina Buyanova

ISBN 978-981-4745-76-5 (Hardcover), 978-1-315-36440-7 (eBook)
www.panstanford.com

13.1 Introduction

Over the past decades, nanowire materials have attracted enormous interest due to their unique mechanical, electronic, and optical properties [1–12]. An example is the more recently development of III-nitride (III-N) nanowires, which have generated more interest because of their favorable material properties and capability to achieve various device applications beyond traditional III-N materials [13–19]. For instance, due to the large lattice mismatch, it is extremely challenging to grow III-N films on Si substrates with good crystal quality [20, 21]. On the other hand, it has been demonstrated that high crystal quality III-N nanowires can be epitaxially grown on Si and promising device applications has been demonstrated [22]. Thus, the growth, characterizations, and device applications of III-N nanowires have emerged as important topics for research and development in the nitride community. Generally, III-N nanowires and nanowire heterostructures for these device applications can be grown in two ways: self-catalyst growth of III-N nanowires has been realized with the diameter of ~from 20 to 60 nm and randomly positioned; they can also be grown with metal catalyst, such as Au and Ni. The nanowire diameters in this case tend to better controlled. Due to compatibility with high-purity environment, catalyst-free growth of III-N nanowires has become the major focus in the III-N nanowire research. In addition, with the unique properties of III-N nanowires compared to their bulk counterparts, the recent development of InGaN nanowires has enabled several new classes of device applications [23, 24]. This chapter is devoted to the review of epitaxy, design, fabrication, and characteristics of III-N nanowires and their device applications. We focus on III-N nanowire light-emitting diodes (LEDs), laser, and photovoltaics applications.

13.2 Epitaxy Growth of III-AS Nanowires

Plasma-assisted molecular beam epitaxy (PA-MBE) systems are typical adopted as the major tool for the III-N nanowire growths, although it can have been reported that III-N nanowires have been successfully grown by metal-organic chemical vapor deposition (MOCVD) as well [11, 25, 26]. The crystalline III-N nanowires have been grown on various substrates as well, such as [111] and [100] Si, sapphire substrates, as well as amorphous layers, SiO_2 and SiN_x, the capability to be grown on different substrates opens large opportunities for various device applications. The choice of the Si substrates is more economical and compatible with modern CMOS industries. In this chapter, we will focus on the growth and characterization of III-N nanowires grown on Si substrates by MBE.

It has been largely demonstrated that vertically aligned III-N nanowires can be grown without introducing foreign metal catalyst by MBE. Even though the exact growth mechanism is still under debut, it is supported by several comprehensive investigations that the III-N nanowires grown on Si in a self-catalyst manner with the vapor-liquid-solid (VLS) growth mechanism, where the group V adatoms, such as Ga droplets, are serving as the catalyst to initialize the nanowire growth [16, 27]. A typical VLS growth process consists of four sequential steps: (1) transportation of the vapor phase precursor and adatoms, (2) precursor and adatom adsorption and desorption at the catalyst surface, (3) adatom diffusion across the liquid alloy catalyst, and (4) precipitation of solid crystals forming the nanowire morphology. Before the nanowire growth takes place, it is also essential to carefully prepare the substrate surface to achieve nanowire array growth with high reproducibility and good coverage. The Si substrates are first cleaned with acetone and isopropanol alcohol (IPA) to remove the organic contaminations. Then the Si surface is treated with diluted HF to remove the native SiO_2 layers before immediately loading into the growth chamber for nanowire growth. In the UHV growth chamber, the Si substrates are heated up to a temperature of 900°C to further remove the contaminations and surface oxide layers. Then, the substrate temperature is lowered to 800°C and a few monolayers of Ga are deposited with a Ga flux of 1.3×10^{-7} Torr in the absence of N. The growth temperature is calibrated by the (111) silicon reflective high-energy electron diffraction (RHEED) pattern transition from 1×1 to 7×7 pattern. Following that, the GaN nanowire growth is initiated at 800°C at a rate of 300 nm/h under N-rich conditions. The Ga flux is maintained at 1.3×10^{-7} Torr and the N flow rate is held constant at 1 sccm. The density of the GaN nanowires is well controlled by varying the Ga flux from 1.3×10^{-7} to 2×10^{-8} Torr when depositing the Ga droplets, the nanowire density can be significantly reduced from $\sim 5 \times 10^{10}$ to 1×10^{8} cm^{-2}. In order to explore the full advantages of III-N materials, InGaN nanowires and InGaN nanowire heterostructures with varied indium compositions are synthesized as well. When growing InGaN nanowires, the growth temperature is further reduced to 500 to 600°C to better incorporate the indium content and the In and Ga with varied flux combinations are supplied at the same time.

13.3 Characterizations of III-N Nanowires

In order to investigate the material and optical properties, the nanowires are examined by various material characterization methods, such as scanning electron microscope (SEM), transmission electron microscope (TEM), and photoluminescence (PL) measurements. Figure 13.1 compares the oblique view SEM images of the GaN nanowires grown with the Ga droplet deposition flux of 1.3×10^{-7} and 2×10^{-8} Torr, respectively. The GaN nanowires are vertically aligned with the average diameter of 20 to 60 nm. It is found that the nanowire density is well controlled by changing the Ga flux during the nanowire growth and the nanowire areal density varied from $\sim 5 \times 10^{10}$ to 1×10^{8} cm^{-2} is obtained. In particular, the nanowires with extreme low density have unique device applications to achieve single nanowire laser with microcavities [28].

Figure 13.1 SEM images of the GaN nanowires grown with the Ga droplet deposition flux of 1.3×10^{-7} Torr (a) and 2×10^{-8} Torr (b), respectively.

Figure 13.2a,b shows the TEM images of GaN nanowires, which show that they are structurally uniform, with a constant diameter of ~20 nm along the length and free of extended defects. The selected area diffraction (SAD) pattern in the inset reveals that the entire wire is single crystal with wurtzite structure and the *c*-plane is normal to the growth direction. In Fig. 13.2c, the cross-sectional TEM at the interface of the GaN nanowire and Si substrate shows that there is an amorphous SiN_x layer presented at the nanowire/Si interface. Bhattacharya et al. have shown that SiN_x is formed during the initial substrate heating step and playing an essential role of the nanowire initialization process [29]. Finally, the TEM image of Fig. 13.2d shows a GaN nanowire with InGaN dot-in-nanowire heterostructures where multiple layers of InGaN quantum dots with 2 nm height are self-aligned along the nanowire growth direction. The inset high-resolution TEM image depicts a smooth and dislocation-free interface between InGaN and GaN. It has been shown that a large number of device applications will be benefitted from the quantum confinement effect of the dot-in-nanowire structures [22, 30–32]. The microscopic characteristics of the nanowires clearly reveal the high crystal quality grown on Si substrates.

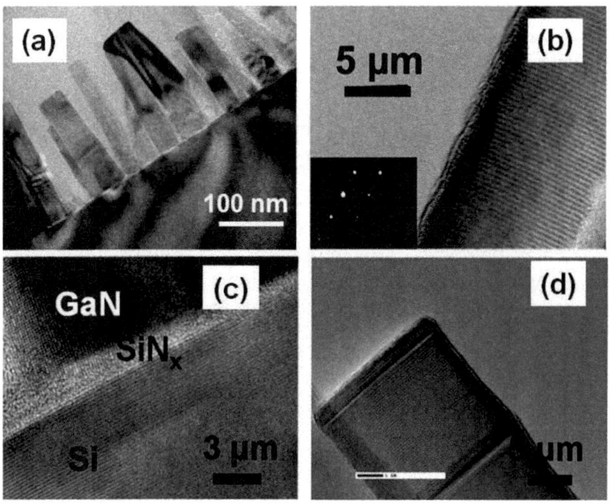

Figure 13.2 Cross-sectional TEM images of (a) single GaN nanowires; (b) nanowire/Si interface; inset: selective area diffraction (SAD) pattern of GaN nanowire; (c) cross-sectional TEM at the interface of the GaN nanowire and Si substrate; (d) TEM image of InGaN/GaN dot-in-nanowire structures.

As-grown nanowires are characterized by photoluminescence (PL) measurements. Figure 13.3a illustrates the normalized room temperature PL spectrum of the InGaN nanowire bundles with varied indium compositions, where the nanowires have average length of ~2 μm and the indium compositions are varied by changing the Ga/In flux ratio. It is found that the InGaN nanowire emission wavelength covers the entire visible wavelength, from 450 to 680 nm and the nanowires with longer emission wavelength exhibit broader emission linewidth, which is likely due to the increased alloy inhomogeneity with large indium compositions. Figure 13.3b illustrates the PL spectrum comparison of InGaN dot-in-nanowire and InGaN bulk nanowires with the same indium composition, where the emission blueshift of the dot-in-nanowire structures is due to the quantum confinement effect.

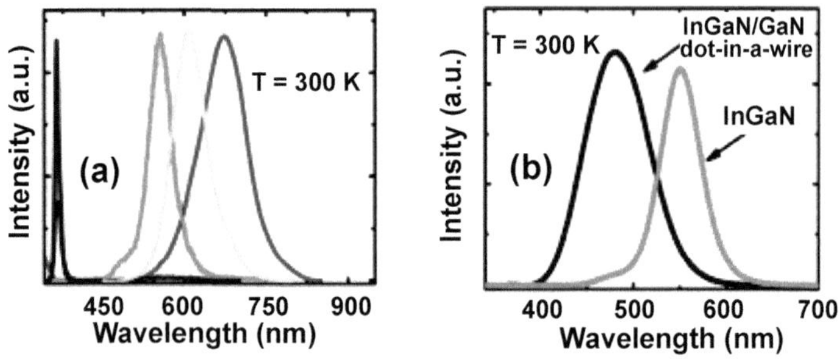

Figure 13.3 (a) Normalized room temperature PL spectrum of the InGaN nanowire bundles with varied indium compositions, where the nanowires have average length of ~2 μm and the indium compositions are varied by changing the Ga/In flux ratio; (b) PL spectrum comparison of InGaN dot-in-nanowire and InGaN bulk nanowires with the same indium composition, where the emission blueshift is due to the quantum confinement effect.

The carrier dynamics of the InGaN nanowires are measured by time-resolved photoluminescence measurements (TRPL), shown in Fig. 13.4a. The decay times are fitted by a stretched exponential model. For the peak energy (490 nm), the decay time is 1.12 ns and the stretching parameter is 0.74×. Nanoscale fluctuations of the In mole fraction are responsible for the non-monoexponential decay. Shown in Fig. 13.4b are the decay times as a function of emission energy, varying from 0.31 to 1.98 ns.

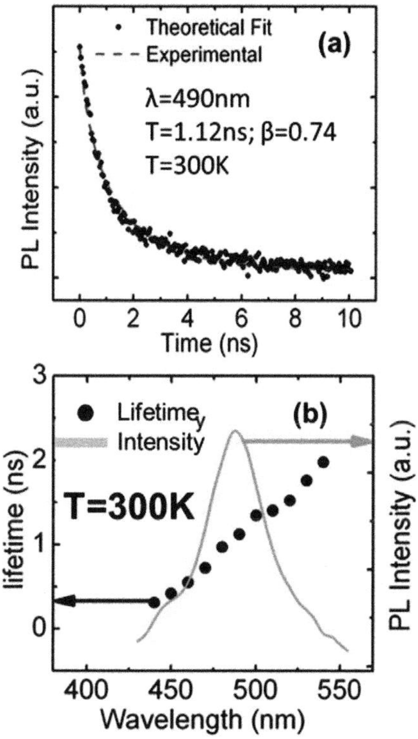

Figure 13.4 (a) Time-resolved photoluminescence spectrum of InGaN nanowire; (b) fitted carrier lifetime vs. emission wavelength.

Finally, in bulk InGaN LEDs, an efficiency *droop* observed in the devices at elevated injection levels largely hinders the further development of LEDs and solid-state lighting [33, 34]. As a result, the origin of the efficiency droop has been the centerpiece in the InGaN LED research community. The Auger recombination, carrier leakage and defect-related non-radiative recombination are the dominant arguments citing the efficiency droop [35–38]. The problem with ascertaining the most likely cause of droop is the large disparity between the theoretical and experimental value of the Auger coefficient, where the former is usually calculated with the assumption of defect-free bulk material. On the other hand, measurement of the Auger coefficient and fabrication of LEDs are done with InGaN materials with high defect density of 10^7–10^{10} cm^{-2}. It is quite possible that a defect-assisted Auger recombination process is attributed to the much higher value of *C*, and operational in the devices [35]. As a result, the defect-free nanowire provides a unique platform to characterize the intrinsic Auger process in InGaN materials.

In this context, we have measured for the first time the Auger coefficient in the InGaN nanowire materials. The Auger recombination in the nanowires is characterized by the combination of excitation-dependent and time-resolved PL measurements. Measurements are made with a frequency doubled mode-locked Ti:sapphire laser (pulse width 130 fs; repetition rate 80 MHz) with the photon energy adjusted to 3.1 eV. For the excitation-dependent TRPL measurements, the transient emission signal is measured by a time-correlated photon counting system equipped with a Hamamatsu high-speed photomultiplier tube. Figure 13.5a shows the excitation power-dependent photoluminescence spectra for the InGaN NW bundles. The emission peak at 500 nm and the linewidth (FWHM) of ~41.4 nm are obtained. It is an important feature in the spectra that the emission peak wavelength remains invariant with increasing excitation power, which has also been observed in the injection current-dependent electroluminescence (EL) in the InGaN nanowire LEDs devices [39]. This experimental observation implies that the InGaN nanowire materials exhibit negligible strain and strain-induced quantum confined Stark effect (QCSE). In addition, band-filling effects due to compositional inhomogeneity and alloy disorder are also absent in the InGaN nanowires, which can cause a redshift with increasing excitation power. The measured external quantum efficiency (EQE), as a function of carrier density, obtained from the excitation-dependent PL data, is shown in Fig. 13.5a. The carrier density in these plots is calculated from the incident power density. A feature that is immediately observed in both is that the efficiency droop is almost absent. The EQE (η_{ext}) is given by the following relation:

$$\eta_{\text{ext}} = \frac{Bn^2}{An + Bn + Cn^3}\eta, \tag{13.1}$$

where A and B are the Shockley–Read–Hall (SRH) coefficient and the radiative recombination coefficient, respectively and η is the extraction efficiency and is used here as a scaling factor to relate the measured EQE to the internal quantum efficiency (IQE), η_{int}. The coefficient B is related to η_{int} via the steady-state relation, $\eta_{\text{int}} = Bn^2/G$, where G is the carrier generation rate. The carrier density n can be derived from the measured TRPL data in the high excitation regime, using the relation: $n = G(t_1 - t_2)/(1 - L(t_1)/L(t_2))$, where $L(t)$ is the spectrally integrated photoluminescence intensity at time t, and t_1 and t_2 are two times in the initial decay of the time resolved PL signal. The generation rate G is related to the incident power P_{inc}, the absorption coefficient α and the photon energy by $G = P_{\text{inc}}(1 - R)\alpha/Ah\nu$ where A is the incident excitation area and R is the surface reflectivity of the nanowire samples. In order to measure the absorption coefficient of the nanowire samples, the InGaN nanowires are removed from the as-grown Si substrates and embedded in polydimethylsiloxane (PDMS) matrix. Wavelength-dependent transmission measurements are then made and $\alpha \cong 0.78 \times 10^{-5}\text{cm}^{-1}$ is measured for $\lambda = 500$ nm. The value of R is calculated to be ~0.04 by Fresnel's law in which the effective refractive index of the nanowire–air composite $n_{\text{InGaN effective}} = 1.5$ is calculated by the Maxwell–Garnett equation [40] with a nanowire fill factor of 40%. Thus, after fitting the values of G, n and η_{int} for each excitation power density, the values of B are calculated by using $B = \eta_{\text{int}}\, G/n^2$ and then fitted with $B = B_0/(1 + n/n^*)$ where B_0 and n^* are the fitting parameters. The values of A and C are then iteratively obtained by fitting η_{int} as a function of excitation power density with the measured PL and TRPL data.

It has been theoretically shown that at high excitation, the Auger recombination rate R_{Auger} attains a sub-cubic carrier density dependence $C = C_0/(1 + n/n^*)$ [41], where C_0 is the Auger coefficients at low injection. The fits to the measured EQE data with Eq. 13.1 are shown with the solid lines in Fig. 13.5b, where the square symbols represent the measured results. Finally, Fig. 13.5c shows the calculated values of the Auger coefficients C in the high excitation regime versus carrier density, and the Auger recombination coefficient at low excitation $C_0 = 6.1 \times 10^{-32}$ cm^{6}s^{-1} is obtained. The corresponding values of n^* are 1.23×10^{19}cm^{-3} is derived as well. The lower value of C_0 in the nanowires sample agree well with the data of David and Grundmann, where a smaller Auger coefficient is measured in InGaN quantum wells compared to those in bulk InGaN [42]. It is worth noting that in the nanowire samples the value of C_0 is more than an order of magnitude smaller than those measured in quantum wells and quantum dots emitting at ~400–500 nm [36]. Thus, it is concluded that in the nanowire samples with much smaller defect density compared to their bulk counterparts the measured Auger coefficient is much closer to the calculated values assuming an intraband Auger process.

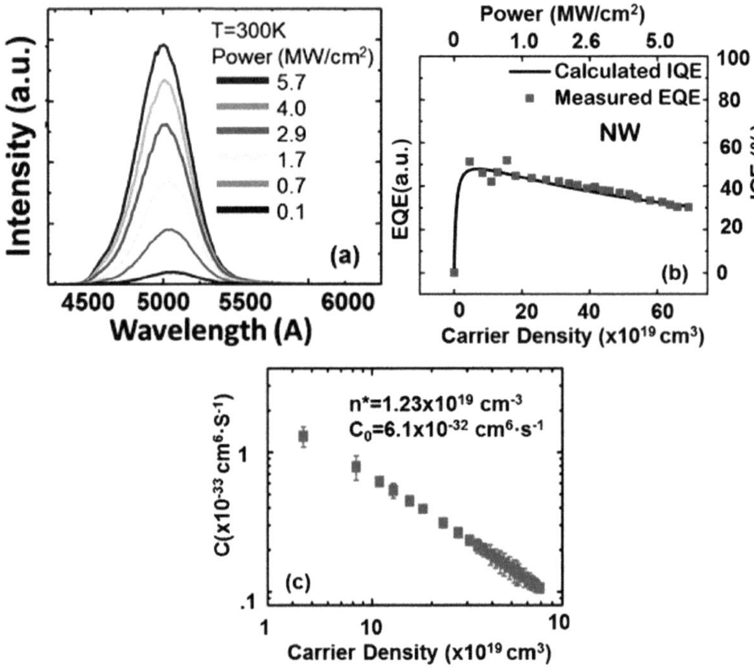

Figure 13.5 (a) Power-dependent PL spectra of InGaN nanowires; (b) EQE and IQE vs. carrier density and excitation power; (c) fitted Auger coefficient C vs. carrier density.

13.4 III-Nitride Nanowire LEDs and Lasers

13.4.1 III-Nitride Nanowire LEDs

The development of solid-state LEDs and laser diodes operating at visible wavelengths holds great promise for the development and miniaturization for a broad area of applications, such as full color mobile projectors, optical data storage, and media applications [43]. Besides being widely used in consumer electronics, high-brightness green LEDs and laser diodes are also required in polymer optical fiber–based optical data communications, it offers many advantages, such as low cost and large flexibility for plugin packing compared to other data communication solutions [44]. Solid-state visible lasers also play an important role in many medical applications, including body composition detectors, such as hemoglobin, human body fat and prostate-specific antigen (PSA) detections [45–47], dental photoinitiator [48], and therapy devices (e.g., sterilization and skin homeostasis) [49, 50]. However, the development of the bulk InGaN material-based visible LEDs and laser devices is largely hindered by the poor material quality. On the other hand, due to the unique material properties of III-N nanowires, nanowire LEDs for solid-state lighting have been largely viewed as the major device applications of InGaN nanowires. Due to the target applications, the large-area InGaN nanowire devices are still of great interest for mass production purpose. In addition to the demonstration purpose, nanowire LEDs have shown favorable properties compared to their QW counterparts. As one of the major challenges in solid-state lighting, the efficiency droop has been largely investigated in the nanowire LEDs. Bhattacharya et al. have for the first time investigated the Auger recombination process in nanowires LEDs and shown that the defect-assisted Auger process is likely the origin of the origin of efficiency droop in QW LEDs [51], and Mi et al. have improved the nanowire LED performance by managing the electron overflow [51] in the nanowire LEDs.

In this context, a top-down fabrication technology has been developed demonstrated electrically injected green and white InGaN nanowire LEDs and lasers on Si substrates. The nanowire device structure is schematically shown in Fig. 13.6a. Si-doped GaN nanowire with the length of 300 nm is first grown, followed by 300 nm undoped InGaN with varied In composition and 150 nm Mg-doped p-type GaN on top. The nanowires are planarized with a parylene insulating layer. From this point now, the nanowires are treated as bulk materials in the entire fabrication process. The device mesa with the dimension of 900 × 900 μm is defined by dry chlorine-based etching technique. Following that the mesa is covered with 5 nm/5 nm Ni/Au and 250 nm indium tin oxide (ITO) as the top ohmic contact to the p-GaN nanowires. Aluminum is deposited on the n-type Si to form the bottom electrode. In the InGaN nanowire active region designs, both the bulk InGaN nanowires and InGaN/GaN dot-in-nanowire structures are employed and both parylene and ITO are nearly transparent to visible light to ensure maximum light extraction efficiency. Figure 13.6b shows the measured light–current (L–I) characteristics of a green-emitting InGaN nanowire LED with the 300 nm InGaN nanowire active region design. The corresponding electroluminescence (EL) spectra are shown in the inset, where the injection current has been varied from 15 to 100 mA. There are several important features in the data of Fig. 13.6b. No saturation is observed in the L–I characteristics up to 100 mA. The emission peak is at 520 nm and the linewidth (FWHM) is ~50 nm (250 meV). In addition, InGaN nanowires with different indium compositions and InGaN/GaN disk-in-nanowire structures have been employed as the active region in the nanowire LEDs. Figure 13.6c shows the optical microscope images of nanowire LEDs with different emission wavelengths, it shows the large emission wavelength tunability of the InGaN nanowire LEDs.

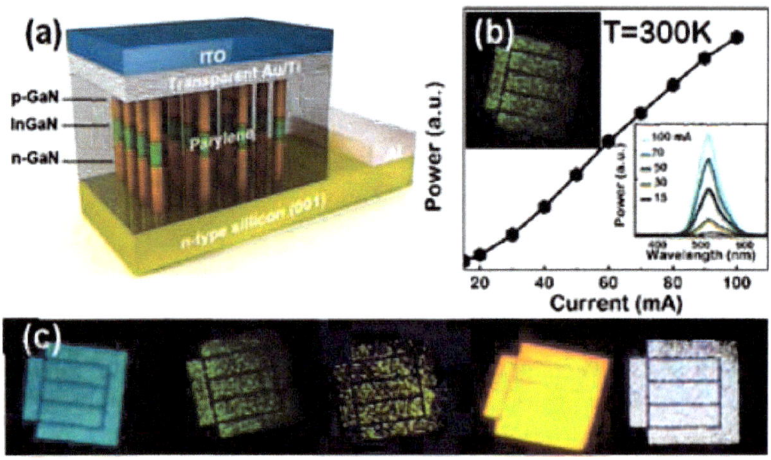

Figure 13.6 (a) Device structure of InGaN nanowire LEDs; (b) light–current characteristics of green InGaN nanowire LED; inset: current-dependent electroluminescence spectra; (c) optical microscope images of nanowire LEDs with different emission color.

13.4.2 III-Nitride Nanowire Lasers in Microcavities

Since the development of InGaN LEDs in the visible wavelength regime, it has been challenging to achieve high-performance

InGaN lasers due to the large defect density, especially in the green and red wavelength [52]. In this regard, due to the improved crystal quality, InGaN nanowire materials have been largely investigated for laser applications. Bhattacharya et al. and Mi et al. have shown electrically injected large-area edge emitting nanowire lasers covering the wavelength spectrum from UV and NIR [24, 53, 54]. In addition, single nanowires have been demonstrated where the Fabry–Pérot cavities and waveguides are obtained within the nanowires [11, 55–57]. However, due to the low facet reflectivity (<0.4) between semiconductor and air and waveguide confinement factor, these lasers require the nanowire to be very long (~40 μm) to achieve stimulated emission, which is not conducive to the realization of nanoscale lasers. In this context, huge efforts have been made to combine the semiconductor gain region with dielectric cavity to achieve laser in compact size [58]. In addition, such cavity design can enable a new class of lasers, such as GaN polariton lasers demonstrated by Bhattacharya et al. [23, 59, 60].

We demonstrate an optically pumped monolithic single GaN nanowire laser on silicon, with a two-dimensional (2D) photonic crystal resonant cavity and a sub-wavelength modal volume, operating at room temperature. The device heterostructure is schematically shown in Fig. 13.7a, where a single GaN nanowire is located at the center of a H_2 defect and surrounded by a 2D PC microcavity. The three-dimensional (3D) finite difference time domain (FDTD) method is used to design a H_2 defect PC microcavity made of TiO_2 and spin-on-glass (SOG) layers. The optical properties of the deposited SOG and TiO_2 materials are characterized by ellipsometry measurements. The measured optical refractive indices of SOG and TiO_2 at λ = 380 nm are 1.4 and 2.5, respectively. The PC with a triangular lattice of air holes of radius 40 nm and period 130 nm is employed. A small hollow that is formed around the nanowire in the center of the H_2 defect during device fabrication is assumed to be a partially filled hole with a diameter of 80 nm in the calculation. The PC cavity Q-factor, mode volume, and confinement factor Γ, of 750, ~0.003223 μm^3 (1.36$(\lambda/n)^3$), and 0.04, respectively, are obtained from the FDTD simulation. The modal gain estimated from the Q-factor is ~400 cm^{-1} and the corresponding optical gain at threshold is ~9000 cm^{-1}, which is achievable in GaN [61].

The device fabrication process is described as follows. In order to achieve a single nanowire in the PC cavity, the GaN nanowires grown on Si with low density ($\sim 10^7$ cm^{-2}) are employed to accommodate the PC cavity. The SOG and TiO_2 films with thickness of 380 nm and 120 nm, respectively, are deposited on the as-grown nanowire samples consequently, where the low index SOG layer provides the bottom cladding of the TiO_2 PC cavity and separates the nanowire emission from the absorbing Si substrates. Following that, the 2D PC microcavity is fabricated in the TiO_2 layer around the nanowire by electron beam lithography and dry etching techniques. The PC microcavity consists of a triangular lattice of air holes with period a = 120–130 nm and hole diameter d = 70–80 nm. The alignment between the single GaN nanowires and PC microcavity is achieved during the electron beam lithography process. Figure 13.7b shows the fabricated device with a single nanowire aligned in the center of the H_2 defect of the PC microcavity. As mentioned earlier, the non-conformal deposition of TiO_2 causes a small hollow near the nanowire and TiO_2 is also inevitably deposited on top of the nanowire. The nanowire diameter in the device is estimated to be 40 nm from SEM measurements.

The device is optically excited at room temperature with a Ti:sapphire pulsed laser at λ = 266 nm with a pulse duration of ~100 fs and a repetition rate of 80 MHz. Figure 13.7c shows the output spectrum recorded at 70 kW/cm^2 (below threshold), 143 kW/cm^2 (near threshold), and 477 kW/cm^2 (above threshold). At low pump power density (~70 kW/cm^2), a broad GaN bandedge emission with a full width at half maximum (FWHM) of ~10 nm and the small peak at λ = 370.4 nm are observed. The peak on the broad emission is a manifestation of the Purcell effect, which enhances spontaneous emission. The enhanced spontaneous emission peak becomes prominent near threshold (~143 kW/cm^2) and evolves into a coherent lasing peak above threshold. A lasing spectrum recorded at a high pump power density of 477 kW/cm^2 shows a narrow lasing peak at λ = 371.3 nm with a linewidth of ~1.5 nm. The redshift (~0.9 nm) of the peak wavelength is due to the combination of band-filling effects and device temperature change at high excitation. The output emission at low pump power density exhibits a fairly narrow peak (~4.5 nm) compared to a FWHM of GaN bandedge emission, known to be ~12 nm [11]. This narrow linewidth in spontaneous emission, as mentioned earlier, is due to the Purcell effect. A further linewidth narrowing is observed at approximately 120 kW/cm^2 and the linewidth is saturated at 1.5 nm at higher pump powers. The inset of Fig. 13.7c shows the high-resolution PL spectrum above the lasing threshold, where a non-Gaussian peak, sum of two Gaussian peaks, is observed; the fitted emission peaks have linewidths of 0.95 and 0.8 nm. We believe that the dual lasing peaks result from a lifting of the degeneracy of the dominant H_2 cavity modes due to imperfections in the fabrication process, which results in an off-center positioning of the GaN nanowire within the defect.

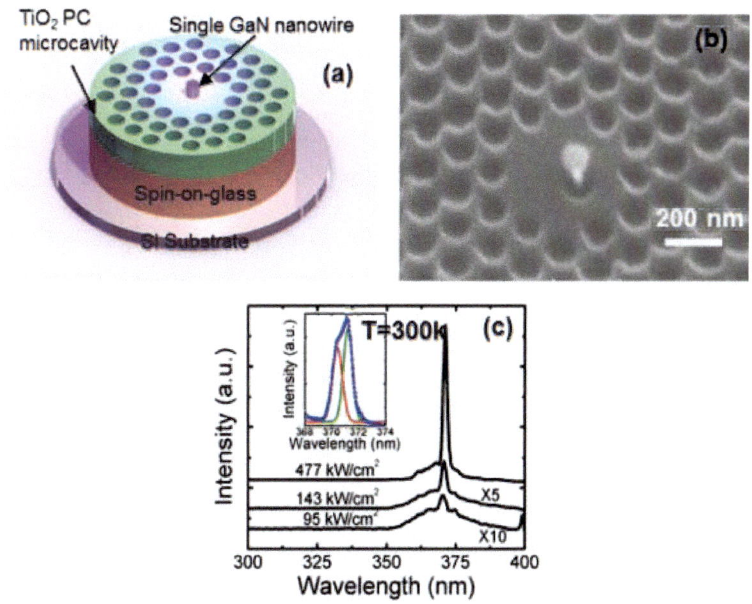

Figure 13.7 Schematic (a) and SEM image (b) of the single GaN nanowire PC microcavity lasers; (c) photoluminescence spectra at pump power densities of 95 kW/cm^2 (below threshold), 143 kW/cm^2 (near threshold) and 477 kW/cm^2 (above threshold). The inset shows the lasing spectrum recorded with high spectral resolution (blue circles), which is matched to the sum of two Gaussian peaks (green and red solid lines). The data indicates that the measured peak consists of two closely spaced transitions by splitting of the degenerate cavity mode.

13.5 III-Nitride Nanowires for Photovoltaic Applications

Due to the constantly increasing oil prices, there is an urgent need to develop high-efficiency solar cells to harvest the solar energy as the alternative energy source. In order to reduce the cost of solar cells, it is essential to significantly increase the conversion efficiency of photovoltaic materials and devices. In this context, progress has been made in design, demonstration, and understanding of solar cell devices with improved efficiency. Due to the large band gap, from 0.7 to 3.4 eV, (In)GaN materials have drawn huge interest in the research of high-efficiency solar cells. (In)GaN-based solar cell devices have the potential to absorb full sun spectrum more efficiently beyond the 2.2 eV band gap limit of the solar cells made from the traditional III–V alloys [62, 63]. However, the development has been impeded due to a variety of reasons: (1) large dislocation density in GaN template grown on foreign substrate; (2) difficulty in growing high-quality InGaN with indium contents >50% due to the large lattice mismatch between InN and GaN; and (3) the fact that indium contents tend to segregate into regions with high and low indium content (indium phase separation). Due to the large surface-to-volume ratio, InGaN nanowire material becomes an ideal candidate for photovoltaics applications. In addition, in PV devices, an antireflection coating (ARC) layer for solar cells is essential to reduce the surface reflection loss [64]. This layer usually contains dielectric materials with small effective refractive index to reduce the Fresnel reflection at the air/device interface [65]. Recently, one-dimension nanoscale structures such as nanowire, nanotube, and nanorods have been studied as the potential candidates for wide-angle and broadband antireflective coatings [66–69]. As a result, this is another inherent advantage of employing nanowire materials for photovoltaic applications.

13.5.1 Characterization of GaN Nanowire Anti-Reflection Coating (ARC)

We have demonstrated a novel antireflection hierarchical structure containing GaN nanowires grown on textured Si substrates. The novel ARC structure exhibits low reflectance over a wide-angle and broadband. To grow the hierarchical ARC structure, inverted pyramid pattern is first created on (001) Si substrates by KOH crystallographic wet etching with SiO_2 as the etching mask [70]. After removing the SiO_2 mask, the GaN nanowires are grown on the textured Si substrates. Figure 13.8 shows a cross-sectional SEM image of GaN nanowires conformally grown on Si substrate with inverted pyramid pattern, the nanowires are successfully grown on both {111} and {100} facets with similar density, average diameter and length of ~10^{11} cm^{-2}, 50 nm, and 1 μm, respectively.

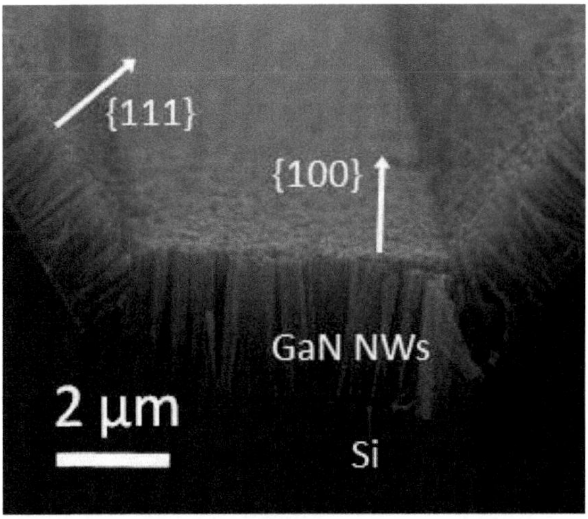

Figure 13.8 Cross-sectional SEM image of GaN nanowires conformally grown on Si substrate with inverted pyramid pattern.

The reflectance spectra of the hierarchical GaN nanowire structures grown on textured Si substrates at different incident angles are shown in Fig. 13.9a. It is found that the reflectance of the nanowire hierarchical ARC structure remains low, ~2%, in a wide range of incident angles and broadband in the solar spectrum, from 20° to 70° and 400 to 1500 nm, respectively. And Fig. 13.9b compares the reflectance spectra at 20° incident of 500 nm GaN film on Si, GaN nanowires grown (001) Si substrate and GaN nanowire hierarchical structure. It is found that the nanowire hierarchical structure can effectively reduce the surface reflection by a factor of 15 compared to the GaN film on Si samples. The reflection reduction in the nanowire hierarchical structure is explained in two points. The reduced effective refractive index in the GaN nanowire layer can reduce the Fresnel reflection and the inverted pyramid structure can increase the effective surface reflection area by bouncing the incident beam between the two facets before escaping from the surface. It is noticed that the inverted pyramid is usually employed in the Si-based solar cells to increase the absorption.

In summary, the GaN nanowires grown on textured Si substrates can effectively reduce the surface reflection. Therefore, (In)GaN nanowires can absorb the sunlight and reduce the reflection simultaneously and there is no need to invest an additional antireflection coating in such devices. This is important in the context of reducing device fabrication complexity and cost.

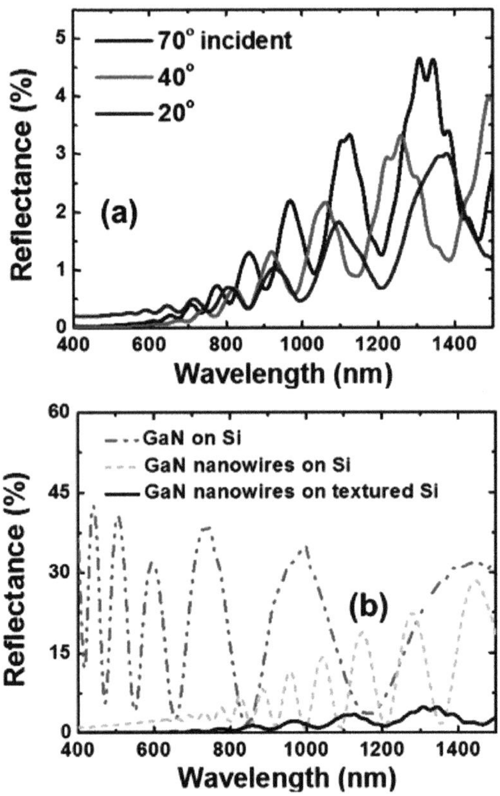

Figure 13.9 (a) Reflection spectra of GaN nanowire hierarchical ARC at 20°, 40°, and 70° incident; (b) Reflection spectra of GaN on Si, GaN nanowires on Si, and GaN nanowires on textured Si.

13.5.2 InGaN Nanowire Solar Cells

The nanowire solar cell device structure is schematically shown in Fig. 13.10a. The device is fabricated by top-down fabrication techniques developed by us. P-GaN/i-InGaN/n-InGaN nanowires with the total length of 1.5 μm are first grown on *n*-type (001) silicon substrates. After planarization by parylene, transparent Ni/Au/ITO p-ohmic and Al n-ohmic contacts are deposited on p-GaN nanowires and Si substrate, respectively. It is worthy to note that in this fabrication process the back metal contact is employed to avoid the mesa etching step, since the nanowires are electronically isolated in the lateral direction. This is of importance to simplify the device fabrication complexity and cost. Figure 13.10b shows the *I–V* characteristics of the proposed InGaN nanowire devices under 1× sun (AM 1.5G), where the open-circuit voltage, V_{oc}, of 0.72 V, short-circuit current density, J_{sc}, of 0.6 mA/cm^2, and external efficiency of ~0.25% are measured in InGaN nanowire solar cells. The nanowires employed in this study have a band gap of 1.8 eV measured from a separated PL study. The device has the size of 900 × 900 μm^2. The measured open circuit voltage is larger than what has been achieved in Si nanowire solar cells (~0.3 V) and is comparable to the number obtained in single InGaN/GaN core-shell nanowire solar cell devices reported by Dong et al. after taking the band gap difference into account [71, 72]. Furthermore, as illustrated in the inset of Fig. 13.10b, the power-dependent V_{oc} and J_{sc} measurements show that the InGaN nanowire solar cells have unusually strong dependency on the incident power compared to other bulk solar cell devices, where V_{oc} and J_{sc} have logarithmic and linear dependence on the incident power, respectively [73]. And this novel phenomenon is quantitatively explained by the nanoscale structure of the nanowires. Calarco et al. have conducted the photoconductivity measurements of single GaN nanowires and shown that the surface depletion layer thickness will be reduced with the increasing number of photogenerated carriers [74]. Therefore, in the nanowire solar cell structures, increased incident power can reduce the depletion layer thickness and increase the width of carrier conduction path in the nanowires, as a result, the measured V_{oc} and J_{sc} will increase non-linearly with the incident power. This novel phenomenon in nanowire solar cells shows that the surface depletion layers will not reduce the device performance and the nanowire solar cells devices are suitable for the high-efficiency concentrated photovoltaic (CPV) applications. In summary, this work has clearly shown that InGaN nanowire PV devices are a promising candidate in the high-efficiency and low-cost PV industry. By further improving the structure design and device fabrication and incorporating multi junction cells, the InGaN nanowire PV can become a serious counterpart in the PV industry.

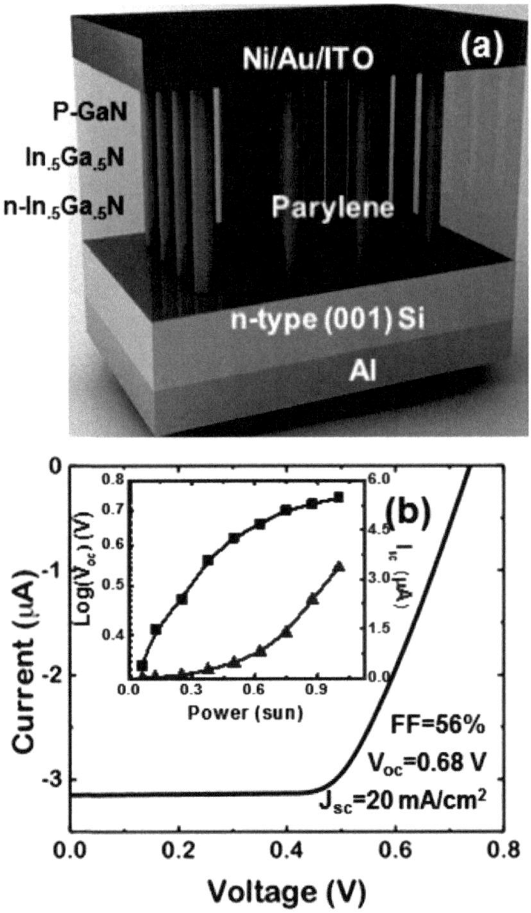

Figure 13.10 (a) Schematic of InGaN nanowire solar cell; (b) *I–V* characteristics of InGaN nanowire solar cells; inset: measured V_{oc} and I_{sc} vs. Power.

13.6 Conclusions

The III-nitride nanowire array has been successfully grown on Si substrates, and the nanowires exhibit favorable material properties for various device applications as well as a defect-free material platform to study the fundamental physics in nitride materials and devices. Nanowire LEDs have largely demonstrated promising results to revolutionize the solid-state lighting industry. Due to unique nanoscale, nanowire materials have been utilized in nanoscale laser applications, where single GaN nanowire lasers have been demonstrated. Finally, III-nitride nanowires have also been employed for photovoltaic applications, and the early demonstration of nanowire PV devices has shown unique properties and performance.

References

1. Jensen, L. E., Björk, M. T., Jeppesen, S., Persson, A. I., Ohlsson, B. J., and Samuelson, L. (2004). Role of surface diffusion in chemical beam epitaxy of InAs nanowires, *Nano Lett.*, 4(10), pp. 1961–1964.
2. Dick, K. A., Deppert, K., Mårtensson, T., Mandl, B., Samuelson, L., and Seifert, W. (2005). Failure of the vapor-liquid-solid mechanism in Au-assisted MOVPE growth of InAs nanowires, *Nano Letters*, 5(4), pp. 761–764.
3. Bryllert, T., Wernersson, L.-E., Froberg, L., and Samuelson, L.: Vertical high-mobility wrap-gated InAs nanowire transistor, *IEEE Electron Device Letters*, 2006, 27(5), pp. 323–325.
4. Mandl, B., Stangl, J., Mårtensson, T., Mikkelsen, A., Eriksson, J., Karlsson, L. S., Bauer, G., Samuelson, L., and Seifert, W.: Au-free epitaxial growth of InAs nanowires, *Nano Letters*, 2006, 6,(8), pp. 1817–1821.
5. Wei, W., Bao, X.-Y., Soci, C., Ding, Y., Wang, Z.-L., and Wang, D.: Direct heteroepitaxy of vertical InAs nanowires on Si substrates for broad band photovoltaics and photodetection, *Nano Letters*, 2009, 9(8), pp. 2926–2934.
6. Fröberg, L., Seifert, W., and Johansson, J.: Diameter-dependent growth rate of InAs nanowires, *Physical Review B*, 2007, 76(15), p. 153401.
7. Dick, K. A., Thelander, C., Samuelson, L., and Caroff, P.: Crystal phase engineering in single InAs nanowires, *Nano Letters*, 2010, 10(9), pp. 3494–3499.

8. Tchernycheva, M., Travers, L., Patriarche, G., Glas, F., Harmand, J.-C., Cirlin, G. E., and Dubrovskii, V. G.: Au-assisted molecular beam epitaxy of InAs nanowires: Growth and theoretical analysis, *Journal of Applied Physics*, 2007, 102(9), p. 094313.
9. Zhou, X., Dayeh, S. A., Aplin, D., Wang, D., and Yu, E. T.: Direct observation of ballistic and drift carrier transport regimes in InAs nanowires, *Applied Physics Letters*, 2006, 89(5), p. 053113.
10. Duan, X., and Lieber, C. M.: Laser-assisted catalytic growth of single crystal GaN nanowires, *Journal of the American Chemical Society*, 2000, 122(1), pp. 188–189.
11. Gradečak, S., Qian, F., Li, Y., Park, H.-G., and Lieber, C. M.: GaN nanowire lasers with low lasing thresholds, *Applied Physics Letters*, 2005, 87(17), p. 173111.
12. Caroff, P., Wagner, J. B., Dick, K. A., Nilsson, H. A., Jeppsson, M., Deppert, K., Samuelson, L., Wallenberg, L., and Wernersson, L. E.: High-quality InAs/InSb nanowire heterostructures grown by metal-organic vapor-phase epitaxy, *Small*, 2008, 4(7), pp. 878–882.
13. Hersee, S. D., Sun, X., and Wang, X.: The controlled growth of GaN nanowires, *Nano Letters*, 2006, 6(8), pp. 1808–1811.
14. Kuykendall, T., Pauzauskie, P., Lee, S., Zhang, Y., Goldberger, J., and Yang, P.: Metalorganic chemical vapor deposition route to GaN nanowires with triangular cross sections, *Nano Letters*, 2003, 3(8), pp. 1063–1066.
15. Songmuang, R., Landré, O., and Daudin, B.: From nucleation to growth of catalyst-free GaN nanowires on thin AlN buffer layer, *Applied Physics Letters*, 2007, 91(25), p. 251902.
16. Calarco, R., Meijers, R. J., Debnath, R. K., Stoica, T., Sutter, E., and Lüth, H.: Nucleation and growth of GaN nanowires on Si (111) performed by molecular beam epitaxy, *Nano Letters*, 2007, 7(8), pp. 2248–2251.
17. Bertness, K., Roshko, A., Sanford, N., Barker, J., and Davydov, A.: Spontaneously grown GaN and AlGaN nanowires, *Journal of Crystal Growth*, 2006, 287(2), pp. 522–527.
18. Li, Q., and Wang, G. T.: Strain influenced indium composition distribution in GaN/InGaN core-shell nanowires, *Applied Physics Letters*, 2010, 97(18), p. 181107.
19. Yoshizawa, M., Kikuchi, A., Mori, M., Fujita, N., and Kishino, K.: Growth of self-organized GaN nanostructures on Al_2O_3 (0001) by RF-radical source molecular beam epitaxy, *Japanese Journal of Applied Physics*, 1997, 36(4B), p. L459.

20. Feltin, E., Beaumont, B., Laügt, M., De Mierry, P., Vennegues, P., Lahreche, H., Leroux, M., and Gibart, P.: Stress control in GaN grown on silicon (111) by metalorganic vapor phase epitaxy, *Applied Physics Letters*, 2001, 79(20), pp. 3230–3232.

21. Katona, T., Craven, M., Fini, P., Speck, J., and DenBaars, S.: Observation of crystallographic wing tilt in cantilever epitaxy of GaN on silicon carbide and silicon (111) substrates, *Applied Physics Letters*, 2001, 79(18), pp. 2907–2909.

22. Nguyen, H. P. T., Cui, K., Zhang, S., Djavid, M., Korinek, A., Botton, G. A., and Mi, Z.: Controlling electron overflow in phosphor-free InGaN/GaN nanowire white light-emitting diodes, *Nano Letters*, 2012, 12(3), pp. 1317–1323.

23. Das, A., Heo, J., Jankowski, M., Guo, W., Zhang, L., Deng, H., and Bhattacharya, P.: Room temperature ultralow threshold GaN nanowire polariton laser, *Physical Review Letters*, 2011, 107(6), p. 066405.

24. Frost, T., Jahangir, S., Stark, E., Deshpande, S., Hazari, A., Zhao, C., Ooi, B. S., and Bhattacharya, P.: Monolithic electrically injected nanowire array edge-emitting laser on (001) silicon, *Nano Letters*, 2014, 14(8), pp. 4535–4541.

25. Lee, S. K., Choi, H. J., Pauzauskie, P., Yang, P., Cho, N. K., Park, H. D., Suh, E. K., Lim, K. Y., and Lee, H. J.: Gallium nitride nanowires with a metal initiated metal–organic chemical vapor deposition (MOCVD) approach, *Physica Status Solidi (b)*, 2004, 241(12), pp. 2775–2778.

26. Wang, G. T., Talin, A. A., Werder, D. J., Creighton, J. R., Lai, E., Anderson, R. J., and Arslan, I.: Highly aligned, template-free growth and characterization of vertical GaN nanowires on sapphire by metal–organic chemical vapour deposition, *Nanotechnology*, 2006, 17(23), p. 5773.

27. Bertness, K., Sanford, N., Barker, J., Schlager, J., Roshko, A., Davydov, A., and Levin, I.: Catalyst-free growth of GaN nanowires, *Journal of Electronic Materials*, 2006, 35(4), pp. 576–580.

28. Heo, J., Guo, W., and Bhattacharya, P.: Monolithic single GaN nanowire laser with photonic crystal microcavity on silicon, *Applied Physics Letters*, 2011, 98(2), p. 021110.

29. Grossklaus, K. A., Banerjee, A., Jahangir, S., Bhattacharya, P., and Millunchick, J. M.: Misorientation defects in coalesced self-catalyzed GaN nanowires, *Journal of Crystal Growth*, 2013, 371, pp. 142–147.

30. Guo, W., Banerjee, A., Bhattacharya, P., and Ooi, B. S.: InGaN/GaN disk-in-nanowire white light emitting diodes on (001) silicon, *Applied Physics Letters*, 2011, 98(19), p. 193102.

31. Jahangir, S., Schimpke, T., Strassburg, M., Grossklaus, K. A., Millunchick, J. M., and Bhattacharya, P.: Red-emitting (nm) In 0.51 Ga 0.49 N/GaN disk-in-nanowire light emitting diodes on silicon, *IEEE Journal of Quantum Electronics*, 2014, 50(7), pp. 530–537.

32. Yalavarthi, K., Chimalgi, V., and Ahmed, S.: How important is nonlinear piezoelectricity in wurtzite GaN/InN/GaN quantum disk-in-nanowire LED structures?, *Optical and Quantum Electronics*, 2014, 46(7), pp. 925–933.

33. Pimputkar, S., Speck, J. S., DenBaars, S. P., and Nakamura, S.: Prospects for LED lighting, *Nature Photonics*, 2009, 3(4), pp. 180–182.

34. Feezell, D. F., Speck, J. S., DenBaars, S. P., and Nakamura, S.: Semipolar InGaN/GaN light-emitting diodes for high-efficiency solid-state lighting, *Journal of Display Technology*, 2013, 9(4), pp. 190–198.

35. Kioupakis, E., Rinke, P., Delaney, K. T., and Van de Walle, C. G.: Indirect Auger recombination as a cause of efficiency droop in nitride light-emitting diodes, *Applied Physics Letters*, 2011, 98(16), p. 161107.

36. Zhang, M., Bhattacharya, P., and Guo, W.: InGaN/GaN self-organized quantum dot green light emitting diodes with reduced efficiency droop, *Applied Physics Letters*, 2010, 97(1), p. 011103.

37. Bhattacharya, P., Zhang, M., and Hinckley, J.: Tunnel injection $In_{0.\ 25}Ga_{0}$. 75N/GaN quantum dot light-emitting diodes, *Applied Physics Letters*, 2010, 97(25), p. 1107.

38. Kim, M.-H., Schubert, M. F., Dai, Q., Kim, J. K., Schubert, E. F., Piprek, J., and Park, Y.: Origin of efficiency droop in GaN-based light-emitting diodes, *Applied Physics Letters*, 2007, 91(18), p. 183507.

39. Guo, W., Zhang, M., Banerjee, A., and Bhattacharya, P.: Catalyst-free InGaN/GaN nanowire light emitting diodes grown on (001) silicon by molecular beam epitaxy, *Nano Letters*, 2010, 10(9), pp. 3355–3359.

40. Sipe, J., and Boyd, R. W.: Nonlinear susceptibility of composite optical materials in the Maxwell Garnett model, *Physical Review A*, 1992, 46(3), p. 1614.

41. Hader, J., Moloney, J. V., and Koch, S. W.: Beyond the ABC: carrier recombination in semiconductor lasers, (International Society for Optics and Photonics, 2006 ed.), pp. 61151T-61151T-61157.

42. David, A., and Grundmann, M. J.: Influence of polarization fields on carrier lifetime and recombination rates in InGaN-based light-emitting diodes, *Applied Physics Letters*, 2010, 97(3), p. 033501.

43. Ohta, H., DenBaars, S. P., and Nakamura, S.: Future of group-III nitride semiconductor green laser diodes [Invited], *J. Opt. Soc. Am. B*, 2010, 27(11), pp. B45–B49.

44. Haupt, M., and Fischer, U. H. P.: WDM over POF: The inexpensive way to break through the limitation of bandwidth of standard POF communication, *Proc. SPIE* 6478, 64780I (2007).

45. Jeon, K. J., Kim, S.-J., Park, K. K., Kim, J.-W., and Yoon, G.: Noninvasive total hemoglobin measurement, *Journal of Biomedical Optics*, 2002, 7(1), pp. 45–50.

46. Conway, J. M., Norris, K. H., and Bodwell, C. E.: A new approach for the estimation of body composition: Infrared interactance, *American Journal of Clinical Nutrition*, 1984, 40(6), pp. 1123–1130.

47. Ahn, C.-G., Ah, C. S., Kim, T.-Y., Park, C. W., Yang, J.-H., Kim, A., and Sung, G. Y.: Photosensitive biosensor array system using optical addressing without an addressing circuit on array biochips, *Applied Physics Letters*, 2010, 97(10), p. 103703.

48. Walsh, L. J.: The current status of laser applications in dentistry, *Australian Dental Journal*, 2003, 48(3), pp. 146–155.

49. Hamamoto, A., Mori, M., Takahashi, A., Nakano, M., Wakikawa, N., Akutagawa, M., Ikehara, T., Nakaya, Y., and Kinouchi, Y.: New water disinfection system using UVA light-emitting diodes, *Journal of Applied Microbiology*, 2007, 103(6), pp. 2291–2298.

50. Babilas, P., Landthaler, M., and Szeimies, R. M.: Photodynamic therapy in dermatology, *European Journal of Dermatology*: EJD, 2006, 16(4), pp. 340–348.

51. Guo, W., Zhang, M., Bhattacharya, P., and Heo, J.: Auger recombination in III-nitride nanowires and its effect on nanowire light-emitting diode characteristics, *Nano Letters*, 2011, 11(4), pp. 1434–1438.

52. Nakamura, S., Senoh, M., Nagahama, S.-I., Iwasa, N., Yamada, T., Matsushita, T., Kiyoku, H., and Sugimoto, Y.: InGaN-based multi-quantum-well-structure laser diodes, *Japanese Journal of Applied Physics*, 1996, 35(1B), p. L74.

53. Hazari, A., Aiello, A., Ng, T.-K., Ooi, B. S., and Bhattacharya, P.: III-nitride disk-in-nanowire 1.2 μm monolithic diode laser on (001) silicon, *Applied Physics Letters*, 2015, 107(19), p. 191107.

54. Li, K., Liu, X., Wang, Q., Zhao, S., and Mi, Z.: Ultralow-threshold electrically injected AlGaN nanowire ultraviolet lasers on Si operating at low temperature, *Nature Nanotechnology*, 2015, 10(2), pp. 140–144.

55. Johnson, J. C., Choi, H.-J., Knutsen, K. P., Schaller, R. D., Yang, P., and Saykally, R. J.: Single gallium nitride nanowire lasers, *Nature Materials*, 2002, 1(2), pp. 106–110.

56. Duan, X., Huang, Y., Agarwal, R., and Lieber, C. M.: Single-nanowire electrically driven lasers, *Nature*, 2003, 421(6920), pp. 241–245.

57. Qian, F., Li, Y., Gradečak, S., Park, H.-G., Dong, Y., Ding, Y., Wang, Z. L., and Lieber, C. M.: Multi-quantum-well nanowire heterostructures for wavelength-controlled lasers, *Nature Materials*, 2008, 7(9), pp. 701–706.

58. Oulton, R. F., Sorger, V. J., Zentgraf, T., Ma, R.-M., Gladden, C., Dai, L., Bartal, G., and Zhang, X.: Plasmon lasers at deep subwavelength scale, *Nature*, 2009, 461(7264), pp. 629–632.

59. Bhattacharya, P., Xiao, B., Das, A., Bhowmick, S., and Heo, J.: Solid state electrically injected exciton-polariton laser, *Physical Review Letters*, 2013, 110(20), p. 206403.

60. Bhattacharya, P., Frost, T., Deshpande, S., Baten, M. Z., Hazari, A., and Das, A.: Room temperature electrically injected polariton laser, *Physical Review Letters*, 2014, 112(23), p. 236802.

61. Domen, K., Kondo, K., Kuramata, A., and Tanahashi, T.: Gain analysis for surface emission by optical pumping of wurtzite GaN, *Applied Physics Letters*, 1996, 69(1), pp. 94–96.

62. Jani, O., Ferguson, I., Honsberg, C., and Kurtz, S.: Design and characterization of GaN/InGaN solar cells, *Applied Physics Letters*, 2007, 91(13), p. 132117.

63. Dahal, R., Pantha, B., Li, J., Lin, J. Y., and Jiang, H. X.: InGaN/GaN multiple quantum well solar cells with long operating wavelengths, *Applied Physics Letters*, 2009, 94(6), p. 063505.

64. Green, M. A.: Solar cells: Operating principles, technology, and system applications (1982).

64. Green, M. A.: Solar Cells: Operating Principles, Technology, and System Applications (Prentice-Hall series in solid state physical electronics), Prentice Hall (1982).

65. Zhao, J., and Green, M. A.: Optimized antireflection coatings for high-efficiency silicon solar cells, *Electron Devices, IEEE Transactions on*, 1991, 38(8), pp. 1925–1934.

66. Lee, Y.-J., Ruby, D. S., Peters, D. W., McKenzie, B. B., and Hsu, J. W. P.: ZnO nanostructures as efficient antireflection layers in solar cells, *Nano Letters*, 2008, 8(5), pp. 1501–1505.

67. Hu, L., and Chen, G.: Analysis of optical absorption in silicon nanowire arrays for photovoltaic applications, *Nano Letters*, 2007, 7(11), pp. 3249–3252.

68. Chhajed, S., Poxson, D. J., Yan, X., Cho, J., Schubert, E. F., Welser, R. E., Sood, A. K., and Kim, J. K.: Nanostructured multilayer tailored-refractive-index antireflection coating for glass with broadband and omnidirectional characteristics, *Applied Physics Express*, 4 (Copyright © 2011 The Japan Society of Applied Physics), p. 052503.

69. Park, C. Y., Lim, J. M., Yu, J. S., and Lee, Y. T.: Structural and antireflective characteristics of catalyst-free GaN nanostructures on GaN/sapphire template for solar cell applications, *Applied Physics Letters*, 2010, 96(15), pp. 151909–151903.

70. Zhao, J., Wang, A., Green, M. A., and Ferrazza, F.: 19.8% efficient "honeycomb" textured multicrystalline and 24.4% monocrystalline silicon solar cells, *Applied Physics Letters*, 1998, 73(14), pp. 1991–1993.

71. Dong, Y., Tian, B., Kempa, T. J., and Lieber, C. M.: Coaxial group III–nitride nanowire photovoltaics, *Nano Letters*, 2009, 9(5), pp. 2183–2187.

72. Tian, B., Zheng, X., Kempa, T. J., Fang, Y., Yu, N., Yu, G., Huang, J., and Lieber, C. M.: Coaxial silicon nanowires as solar cells and nanoelectronic power sources, *Nature*, 2007, 449(7164), pp. 885–889.

73. Fonash, S. J.: Solar cell device physics (1981. 1981).

74. Calarco, R., Marso, M., Richter, T., Aykanat, A. I., Meijers, R., V. D. Hart, A., Stoica, T., and Luth, H.: Size-dependent photoconductivity in MBE-grown GaN-Nanowires, *Nano Letters*, 2005, 5(5), pp. 981–984.

Chapter 14

III–V Nanowires: Transistor and Photovoltaic Applications

Katsuhiro Tomioka, Junichi Motohisa, and Takashi Fukui

Graduate School of Information Science and Technology, and Research Center for Integrated Quantum Electronics (RCIQE), Hokkaido University, Kita14, Nishi9, Sapporo 060-8628, Japan

tomioka@rciqe.hokudai.ac.jp

III–V nanowires (NWs) are promising building blocks for future electronic and optical devices because they have the advantage of having a core–shell structure that can be used to build three-dimensional (3D) devices not limited to being thin films. Position- and size-controlled NWs are very important for exploiting these benefits. In this chapter, we report on selective-area growth of III–V NWs and recent progress on NW-based transistor and photovoltaic devices.

14.1 Introduction

Silicon (Si)-based electronics and photovoltaic devices are becoming increasingly common tools in our life. Our descendants several centuries ago might therefore call the times we live in today the "Silicon Age." It was an inherently natural choice for us to use

Novel Compound Semiconductor Nanowires: Materials, Devices, and Applications
Edited by Fumitaro Ishikawa and Irina Buyanova

ISBN 978-981-4745-76-5 (Hardcover), 978-1-315-36440-7 (eBook)
www.panstanford.com

such abundant materials, and exploit the properties of Si in devices. For electronics, in particular, the invention of field-effect transistors (FETs) and their miniaturization have accelerated the evolution of uses of Si. One need only consider that the number of the FETs in devices such as laptops and smartphones will exceed the total number of neurons in all human brains within the next 20 years. Moreover, all one has to do is look out on the vast planes, desert, and urban landscape of Japan, and you can easily see Si-based photovoltaic panels. However, despite the growing prevalence of such devices, it is worth asking whether the Si Age will continue to evolve forever. The answer is that we cannot expect it to, since the limits of Si-based electronics and photovoltaics are on horizon. More inventions will be required to prolong the technology node and make further progress.

The slowdown in evolution that we currently face is ultimately a consequence of power consumption and power supply limitations. In the case of Si-based integrated circuits (ICs), the tremendous number of FETs integrated on a chip consume a lot of power. A main goal of future ICs is thus to lower the power consumption of the FETs while enhancing their performance. Electronic devices will soon consume over 30% of the energy produced worldwide; the search is on for improved power-supply technology, e.g., solar power generation with energy harvesting that can be ideally incorporated into individual electronic devices. In this regard, the size of photovoltaic devices should be reduced while their energy efficiency is increased.

The semiconductor NWs have attracted much attention as possible building blocks for future electronic and optical devices. In particular, research on NWs has been growing since the early 2000s [1] and advent of the vapor–liquid–solid (VLS) [2] method, which makes it easy to synthesize very thin whisker structures. NW materials can form unique axial and radial heterostructures, and a number of state-of-the-art NW devices have been demonstrated [3–5]. In order to make the most of these benefits, it is necessary to control the NWs as to their size and position in a device. Here, the VLS method is not suitable for such position-controlled synthesis, and the metal catalysts it used form deep impurities that degrade device performance [6, 7, 132, 133, 153, 154, 160]. Selective-area growth (SAG), on the other hand, is more suitable for forming both size- and position-controlled NWs, regardless of its use of metal catalysts. This chapter describes electrical and photovoltaic device applications of III–V NWs and selective-area growth of NWs.

14.2 Transistor Applications

The NW-channels can be formed by taking a top-down approach using etching or a bottom-up approach using NW synthesis. In this chapter, we will focus on NWs synthesized using the bottom-up approach. As of June 2016, there have been about 400 reports on transistor using NWs built by taking the bottom-up approach. Channel materials for transistor applications are based on Si, Ge, III–V compound semiconductors, nitrides, ZnO, II-VI compound semiconductors, oxides, SiC, Se, and B. Reports on Si NW-channel devices are the most numerous, with papers on ZnO and III–V NWs coming second and third. The NW architectures using these channels are categorized into those with lateral NW-FETs and those with vertical NW-FETs, as shown in Fig. 14.1. Lateral FETs use a laid-down single NW grown by VLS or SAG and have three-terminals (source, gate, drain) fabricated by electron-beam lithography. The gate architecture is usually Ω-shaped. On the other hand, vertical FETs use as-grown vertical NWs as a channel. These vertical three-terminal devices are fabricated by using a low-dielectric insulating layer and etch-back procedure. The gate architecture is basically a surrounding-gate (wrapped-gate, or gate-all-around) structure.

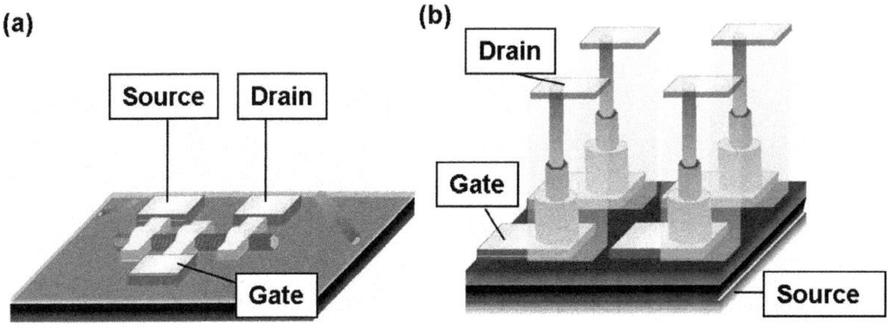

Figure 14.1 Illustrations of NW-FETs: (a) a lateral FET and (b) a vertical FET.

14.2.1 Lateral NW FETs

The initial studies dating from the early 2000s were on the lateral FET structure using a single or a few NWs because EB lithography available at the time simplified the fabrication process. The structures were used for characterizations of electronic properties of NW materials and for explorations of functionality in NWs rather than in studies for device applications and improvements in device performance.

With regard to device applications, there have been some demonstrations of basic logic circuits and sensor devices (ZnO and oxide-based NW materials are suitable for sensors). As for logic applications, Cui and Huang et al. were the first demonstrate basic logic circuits (NAND, NOR and so on) crossing single Si NWs in 2001 [9, 10]. In the early research, two Si NWs were used as cross-bar architecture, where one NW serves as a gate electrode and the other serves as a channel. Late came reports of both an n-channel and a p-channel using single Si NW and CdS nanoribbon [11], while more recently, programmable circuits using these cross-bar NW-FET logic structure have been demonstrated [12]. Basic logic circuits of a cross-bar array using oxide-related NWs such as ZnO and In_2O_3 were also demonstrated as driving current-sources for organic light-emitting diodes (OLED) on flexible substrate [13]. Moreover, inverter circuits were fabricated using a single ZnO NW [14].

As for sensor applications, gas- [15], pH- [16], molecular- [17], and bio-sensors [18, 19] have been demonstrated. These sensors take advantage of the surface area afforded by the NW-geometry and change the charge potential of the NW-channel by terminating and decorating the NW-surfaces with adduct atoms. In addition, Si NWs have been used in bio-sensing applications such as monitoring small voltages in neurons and recording/decoding of neural signals [19–23]. As for gas-sensors, Fan et al. were the first to use ZnO NWs sensing oxygen-gas [15]. Later, Kang et al. reported a GaP NW-based gas sensor [24]. There have been many reports on oxide-related NWs devices, because the charge potential of oxide-related NWs is sensitive to surface termination with atoms and molecules. A humidity sensor was recently demonstrated [25].

As for bio-sensors using NWs, Cui et al. were the first to demonstrate that the conductance of Si-NWs varied with the molecular-termination [15]. After that, Si NW arrays were used to monitor signals from neurons of mouse brains [19]. Later, sensing, recording, and decoding of neural signals were reported by combining Si-NW arrays with nano-probes [23].

Now let us focus on the use of lateral NW-FETs as electrical switching devices. Here the most important consideration is whether the conventional MOSFETs and technology nodes can be replaced with lateral NW-FETs. In this regard, lateral NW-FETs, which are typically several micrometers long in plane, are at a disadvantage; they are incompatible with the conventional integration process in which Si-FETs have reached to the 7 nm -scale node. Although it may be possible to integrate many n-channel and p-channel FETs inside the same NW, it seems unlikely that lateral NW-FETs can be used to go beyond the current technology node. Despite the drawback, though, research on lateral NW-FETs has led to areas of important technological progress.

In particular, p-channel operation of lateral NW-FETs was first demonstrated in 2003 [26]. Not long after, Zheng et al. demonstrated n-channel lateral NW-FET by using phosphorus-doped Si NWs [27]. These developments were followed by a demonstration of n/p-channel FETs inside a single Si NW [28] and a method of fabricating two FETs inside single NW [29]. The first demonstration of a lateral FET using Si NWs was reported by Cui et al. in 2001 [9]. Then, in 2003, Wang et al. demonstrated a lateral Ge NW-FET and examined its p-channel switching properties [30]. As for lateral FETs using III–V NWs, cross-bar logics circuits using InP NWs were demonstrated by Duan et al., [4] and, subsequently, lateral FETs using InAs and GaP NWs were reported [24, 31]. As for the nitride-based NWs, Huang et al. reported back-gate lateral FETs using VLS-grown GaN NWs in 2002 [32].

Significant progress in lateral NW-FETs was made with the introduction of a junction-less FET (JL-FET) structure by using a silicide source/drain region [33]. Silicidation is an important technology for obtaining low contact resistance in the conventional Si-CMOS process. A JL-FET, in which the channel region consisted only of a Si-NW, was made by using a Ni-based silicidation and alloy process. Although the NW JL-FET is credited to Conlinge et al. in 2010 [33], similar structures using NWs were demonstrated as early as 2000. For example, Wu et al. fabricated Si NW-based JL-FETs by forming NiSi alloy in the source/drain regions through exploitation of the solid diffusion mechanism [34]. Liow et al. demonstrated a Ge NW-based JL-FET in 2008 [35]. In the case of Ge NWs, NiGe alloy has been used in the source/drain regions [36]. Regarding III–V NWs, NW-based JL-FETs were fabricated through exploitation of the solid diffusion mechanism in Ni/InAs and Ni/InGaAs alloys [37, 38].

14.2.2 Vertical NW FETs

The primary goal of vertical NW-FETs is replacements for conventional planar metal-oxide-semiconductor FETs (MOSFETs) and CMOS technology in large-scale integrated circuits (LSIs). Continued miniaturization of FETs faces serious issues with power consumption because of the inherently large off-state leakage current and short-channel effect. In this regard, to reduce power dissipation sufficiently for gate length (L_G) of the 7 nm node, Si-based CMOS technologies are expected to have a multigate architecture [39–43], new channel materials [44–46] and transport mechanisms [46–49].

The multigate architecture, as represented by the a fin-gate [39], in which the channel is biased by a gate wraps around three sides of the channel, is already being used in state-of-the-art LSI chips to control the gate voltage and maintain scalability. Sooner or later, the multigate architecture will be replaced with a gate-all-around (GAA) [40] or surrounding-gate transistor (SGT) architecture [41–43], in which the gate metal wraps around the whole surface of channel in order to provide the best electrostatic control of the gate. The resulting gate structure can suppress the short-channel effect and the off-state leakage current and shrink the area occupied by the FETs in LSIs. Moreover, III–V compound semiconductors (III–Vs) and Ge are promising channel materials because their field-effect mobility is much greater than that of strained SiGe. Channels made from these materials have higher ON-state current at lower supply voltages. Thus, the combination of the multigate architecture and these channel materials may open the way to the development of high-performance FETs with low power consumption. In fact, high-performance InGaAs FinFETs [50, 51], GAA FETs [52, 53], Ge recess-gated FETs [54, 55] and Ge FinFETs [56] have recently been reported. GAA and SGT architectures have NW channel morphologies; in particular, the SGTs (the vertical NW-FETs described in this book chapter) use vertical NW-channels.

Not as many vertical NW-FETs as lateral NW-FETs have been reported because the device fabrication is complicated by the 3D processing of the as-grown vertical NW-channels. The first Si-based SGT with wrapped-gate structure was demonstrated by Masuoka et al. in 1988 [41]. At the time, Si wire was made by top-down etching because bottom-up approaches, such as VLS, were immature. SGTs using VLS-grown NWs have been demonstrated since the early 2000s. The first demonstration, in 2003, used CuSCN NWs [57]; it was followed by demonstrations of ZnO NW-based SGTs [58]. VLS-grown Si and InAs NW-based SGTs were reported in 2006 [59, 60]. Later, Svensson et al. demonstrate complementary integration of InAs NW-SGT and GaSb NW SGT for logic CMOS circuit [61]. Recently, high-performance III–V SGTs were demonstrated for RF application and CMOS applications [62–65]. And Tomioka et al. demonstrated high-performance SGTs using modulation-doped InGaAs NW channels (NW-based high electron-mobility transistor) on Si [66, 67].

Among the NW materials, III–V elements are the most promising materials for SGT applications. III–V compound semiconductors are expected to be fast n-channel materials because of their higher electron mobility compared with bulk silicon and their small electron effective mass. Additionally, these materials have band-engineering capabilities that allow several heterostructures, e.g., high-electron-mobility transistors (HEMTs), to be fabricated. Nanometer-scale transistors made from III–V materials have achieved much higher injection velocities (I_{on} is determined by the injection velocity) than those of Si MOSFETs under low drain-source voltages (V_{DS}). Of these III–V compound semiconductors, $In_xGa_{1-x}As$ offers promising n-channels on Si because the Schottky barrier height (SBH) at the MIS interface can be adjusted by changing the In composition [68]. Many investigations have been devoted to In-rich InGaAs (In-content > 70%), because its higher electron mobility and moderate SBH.

Besides the potential of this material for future n-channels, a III–V multigate FinFET using $In_{0.7}Ga_{0.3}As$ on Si substrates has recently been reported [50]. This trend of using InGaAs channels with a fin-gate architecture will lead to lateral GAA [52, 53] or vertical surrounding-gate transistors (SGTs) [62–67] of future low-power CMOS technologies.

Considering the recent progress in Si-CMOS technology, co-integration of n-channel III–V and p-channel Ge or III–V material will be important in future CMOS technology. Moreover, the area occupied by n-channel FETs should be smaller than that of the p-channel FETs for current density matching. When the miniaturization for p-channel FETs reaches the 7 nm-node, the integration area of the n-channel FETs should be smaller than the 7 nm scale. III–V devices made this small will not work as conventional FETs because the quantum confinement effect and depletion layer influence the electrostatic gate controlling at this scale. Instead, using a vertical architecture can shrink the area occupied by III- V-based MOSFETs and thereby avoid the above issues. For example, integration of vertical FET with a 20 nm diameter corresponds to that of a 5 nm-scale planar FET [41, 69].

Despite the technological importance of the SGT architecture, there have been issues related to heterogeneous integration of III–V NWs on Si platforms and 3D processing for making vertical SGT structure using III–V NW channels. Heterogeneous integration of size-controlled and position-controlled one-dimensional III–V NWs on Si will be very important for integration of vertical III–V SGTs on Si. For heterogeneous integration of III–Vs on Si platforms, these issues include mismatches of lattice constants, thermal expansion coefficients, and polarity. In particular, the polarity mismatch results in both vertical and tilted NW growth (this issue will be discussed in the next section). Recently, novel crystal growth methods have enabled direct integration of III–V NWs on Si [70, 71], and SGT applications have been demonstrated [72–75].

Table 14.1 summarizes the benchmarks of the early demonstrations of NW-based vertical and lateral FETs. Figure 14.2 compares the trend of Si-CMOS technology as predicted by International Technology Roadmaps for Semiconductors (ITRS, which was closed in 2016) with results for III–V NW-based SGTs. In this figure, the gate length of the SGTs is normalized by the thickness of the gate metal.

Table 14.1 FET performance benchmarks

Materials	Sub.	Struc.	d_{NW} (nm)	L_G (μm)	I_{DS} (mA/μm)@ V_{DS} = 1.00 V	I_{DS} (mA/μm)@ V_{DS} = 0.50 V	I_{OFF} (nA/μm)	SS (mV/dec)	Ref.
InGaAs CMS	Si	V	160	0.2	1.3	0.45	10 3	75	66
InAs	Si	V	100	0.3	0.083	0.025	3.3	320	75
Si		L	5	0.2	1.7	0.06	0.2	63	74
InAs	Si	V	60	0.05	0.006	0.002	4	300	73
Si	Si	V	4	0.5	10–5	NA	0.02	120	59
InAs	InAs	V	40	0.05	0.07	0.05	0.4	75	72
InAs	InAs	V	80	1.0	NA	0.008	5	100	60
ZnO	SiC	V	35	0.2	0.001	NA	0.05	130	58
GaN CMS		L	100	1	0.03	0.01	10-4	68	155
Ge/Si		L	18	0.19	0.53	0.32	5.26	100	34
InAs		L	50	2	0.03	0.01	NA	NA	156
InAs/InP CS		L	20	0.17	NA	3.2	0.29	260	157
InGaAs CMS	Si	V	60	0.15	1.6	0.46	0.02	70	67
InGaAs	InP	L	20–35	20–80	1.0	0.45 – 0.60	0.1	63	158
InAs	Si	V	13	0.23	1.7	0.90	$> 10^3$	180	63
InAs		L	12–15	0.266	NA	0.314	100	68	64
InGaAs	InP	V	60	0.100	NA	~ 0.100	0.01	68	65
InAs	Ge	V	70	0.200	2.0	0.65	0.01	78	159

V, vertical SGT; L, Lateral GAA.

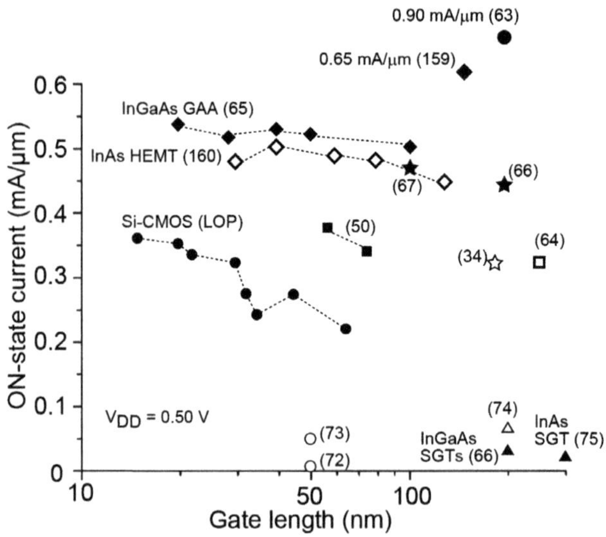

Figure 14.2 On-state current versus gate-length for NW-FETs, III–V MOSFETs, and Si-MOSFET. V_{DD} is 0.50 V.

The table lists demonstrations in which NW-FETs perform better than Si-FinFETs. The switching properties of conventional FETs are dictated by a quantity known as the subthreshold slope (SS), which cannot be lower than the theoretical limit of 60 mV/decade of current at room temperature; this means that a 60 mV increase in voltage results in a ten-fold increase in switching current. SS is inversely proportional to the rate at which the current flow through the FETs increases toward its ON state with increasing supply voltage. Hence, an SS which close to 60 mV/decade results in a lower supply voltage for FETs. Practical Si MOSFETs have SS of 70–100 mV/dec. On the other hand, III–V MOSFETs have had poor SS because of surface Fermi level pinning and a large interface-state defect density (D_{it}) across the gate-oxide/III–V interface. Recent progress of atomic-layer deposition (ALD) has improved D_{it} and the gate-stacks for III–V MOS structures, and nearly ideal SS close to 60 mV/decade have been demonstrated for III–V MOSFETs [53, 67]. A III–V NW-SGT with a nearly ideal SS was reported.

Figure 14.2 plots the on-state current versus gate length for representative MOSFETs and SGTs. The on-state current for the Si-MOSFETs can be categorized as high-performance (HP), low operating power (LOP), and low-standby power (LSTP). HP MOSFETs are used in servers and data centers. LOP MOSFETs are intended for laptop and mobile devices, while LSTP MOSFETs are for mobile device and sensor devices. The LOP MOSFETs have been operated at a V_{DD} of 0.50 V. The figure also plots the fastest planar HEMTs using InAs channels and InGaAs FinFETs. The on-state current for the NW-based SGTs are below 0.1 mA/μm, which is lower than that of Si-MOSFETs of the 16 nm scaling node, indicating that the SGT has inherent issues with the on-state current. This is because the 3D processing of making a vertical SGT structure has inherently high parasitic resistances because of process-induced damage and poor contact technology. Additional current boosting technologies such as HEMT and strain are needed in order to make improvements to the on-state current.

Figure 14.2 indicates that the SGTs that use core–shell (CS) or core–multishell (CMS) structure outperform Si-MOSFETs. These structures achieve high on-state currents by using modulation doping and surface passivation. Recently, InGaAs NW-based HEMT-type SGTs on Si platforms were demonstrated to have significantly high on-state current at a V_{DD} of 0.50 V [66, 67]. They were made by forming a modulation doped multishell layer around the InGaAs NW-sidewalls. These SGTs are discussed in the next section, but the point that should be emphasized here is that there is plenty of room for improving the device parameters of SGTs, i.e., the gate length, gate-oxide, channel-length, and NW-diameter. These SGTs are potential replacements for conventional Si CMOS. Thus, current enhancement by suppressing the off-state leakage current at a V_{DD} of 0.50 V and outperforming state-of-the-art Si MOSFETs will be important milestones for future switching devices to reach. One of the key approaches to these milestones is the utilization of modulation doping on the small NWs.

In the next section, we will discuss the heterogeneous integration of vertical III–V NWs and describe progress on vertical SGTs using vertical III–V NW-channels on Si and a typical modulation-doping CMS structure.

14.3 Selective-Area Growth of III–V NWs

The typical growth method of NWs is based on the VLS mechanism. Wagner and Ellis first reported the existence of this mechanism in 1964 in relation of formation of Si whiskers [2]. The mechanism uses catalysts and their liquid phase underneath the metal particles for crystallization.

Another approach to growing NWs is catalyst-free SAG. This approach uses partially masked templates with lithographically defined opening patterns (Fig. 14.3a) at specific positions. The masked templates are made from amorphous or metal films, such as SiO_2 and SiNx. Polygonal nanostructures surrounded by several facets can be formed inside the openings, because the crystal growth proceeds through the faceting growth mechanism. The use of III–V (111)A or B substrates enables formation of vertically aligned hexagonal pillars surrounded by {–110} vertical sidewalls. Such "position-controlled" NWs have been made using III–V compound semiconductors, such as GaAs [76–79], InP [80, 81], InAs [82], InGaAs [83–86], and GaAsP [87], as well as nitrides [88] and oxides [89]. As for core–shell NWs, position-controlled AlGaAs/GaAs [90], InAs/InP [91] and GaAsP/GaAs [92] core–shell NW, and InP/InAs/InP [93] CMS NWs have been fabricated by using selective-area metal-organic vapor phase epitaxy (SA-MOVPE). In addition, position-controlled growth of InGaAs/GaAs [94, 95], InAsP/InP [96], and AlGaAs/GaAs axial heterostructure NWs [97] have been investigated. Usually III–V NWs tend to grow in the <111>B or <111>A directions. For example, GaAs and InAs NWs preferentially grow in the <111>B direction, so, vertically-aligned GaAs and InAs NWs can be grown on a (111)B (Fig. 14.3b).

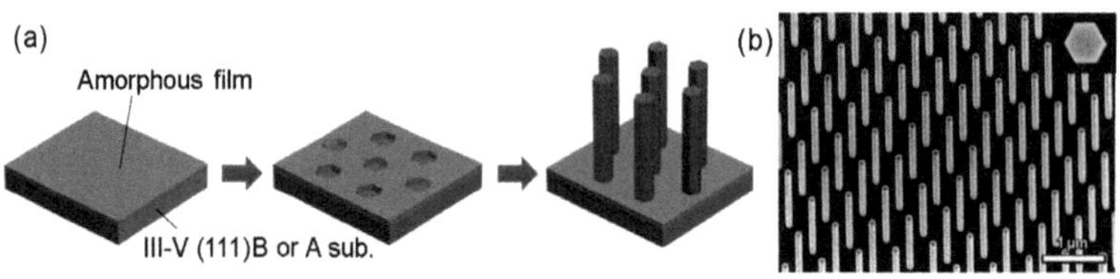

Figure 14.3 (a) Schematic NW-fabrication using SAG. (b) A SEM image of GaAs NW-array on GaAs(111)B substrate. The Inset shows a plan-view of the GaAs NW.

14.3.1 Formation of Core–Shell NWs by SAG

The interesting feature of SAG for NWs is that the lateral-over growth (LOG) mode can be switched to the NW growth-mode (or vice versa) by controlling the growth temperature (T_G) or V/III ratio. For example, low T_G enhances the LOG of GaAs without NW growth after the GaAs NW growth at the optimum T_G [78]. The diameter of the GaAs NW becomes larger than the mask opening without NW growth. This is because the growth rate of the {1–10} sidewall increases as a result of the suppression of desorption reactions. This phenomenon occurs in other materials as well. Its use in the growth of heterostructures enables formation of core–shell or CMS NWs such as GaAs/AlGaAs core–shell NWs [90–93, 98].

The surface states of NWs increase in number as compared with planar III–V materials because the NWs have a comparatively larger surface area. The surface states degrade the optical properties of III–V NWs and device performance through non-radiative recombination process. The CS structures effectively passivate the surface states of NWs. In particular, the shell layer is more stable at atmosphere compared with NWs that have been subjected to chemical treatments such as sulfur passivation. It has been shown that GaAs/AlGaAs core–shell NWs improve photoluminescence (PL) intensity by up to 490 times compared with bare NWs through a reduction in the surface states density [98].

The above solution means that the large surface area and small diameter of CS or CMS NWs are obvious geometrical advantage. Let us consider a core–shell NW with a diameter of 200 nm and a height of 3 μm. The junction area is 1.8 μm^2 when the NW has a radial p-n junction on the NW-sidewalls. Supposing we integrate NWs with a pitch of 400 nm in a 50 μm square (the surface area is 2.5×10^3 μm^2), the total junction area of the NW array will be 2.8×10^4 μm^2. The junction area of the NW array in this case is approximately ten times larger than that of a planar LED. Supposing further that device performance of a single NW LED is close to that of a planar LED, the chip area of the LED can be reduced to a tenth area that of the planar device. Photovoltaic applications can hence take advantage of these geometrical benefits. Moreover, integrating such NWs on Si substrates should ideally reduce manufacturing costs to 1/100th that of planar LED. Thus, making the most of the geometrical advantages of NWs should lead to new high-brightness and low-cost light sources. Moreover, NWs can be used to create new functionalities because double-heterostructures (DH) can be formed on the sidewalls.

14.3.2 Integration of Vertical III–V NWs on Si by SAG

Recent progress in NW-synthesis using the VLS method [2] and SAG has enabled integration of III–V NWs with Si substrates. Much of the attention has focused on the use NWs as building blocks for future electronics and photonics because they can be used as fast channels in vertical FETs on Si wafers and have optical applications such as NW-LEDs and photodetectors on Si platforms.

Since Märtensson's first report on As/P-related III–Vs NWs on Si in 2004 [70], more than two hundred papers have been published on III–V NW growth on Si substrates. It should be noted as well that nitride-related NWs on Si were described by Kishino et al. in 1997 [99]. Almost all of the reports have focused on synthesis or integration of III–V NWs on Si. In particular, GaAs, InP, InAs, GaSb, and ternary-alloy NWs have been grown on Si by using the VLS, catalyst-free, and SAG methods (see references in Refs. [100, 101]). III–V NW-based devices such as LEDs [102, 103], and SGTs integrated on Si have recently been demonstrated [66, 73, 75, 104–107] and novel devices using III–V NW/Si heterojunctions, such as solar cells [108], Esaki tunnel diodes [109–113], and tunnel FETs [67, 113–117] have been reported. In heteroepitaxy of III–V NWs, one must consider not only the conventional problems of III–Vs/Si integration but also the following four problems: (a) positioning of the NW sites, (b) polarity of III–V NWs, (c) unintentional doping from the Si substrate, (d) misfit dislocation at the heterointerface.

The III-As NWs tend to grow in the <111>B directions. Thus, vertically aligned III-As NWs can be grown on a III–V(111)B substrate. On a III–V(111)A surface, the III-As NWs grow in the other three equivalent <111>B directions. Unlike III–V semiconductors, the <111> direction of group-IV semiconductors, on the other hand, has no such polarities, i.e., A or B. Thus, equivalent surface orientations and directions always occur on a Si(111) surface and these equivalencies form anti-phase domains in the case of direct growth of III–Vs on Si. Instead of anti-phase domain formation, equivalent growth directions always occur in III–V NW/Si integrations. That is, the III–V NWs grow in vertical <111> and three equivalent tilted <111> directions at the same time on Si(111). The differences result from either the co-existence of (111)A and B surfaces that are from the termination of group-III or V atoms on the Si(111) surface during SAG. These equivalent growth directions should be suppressed in order to align III–V NWs vertically on Si.

Selective-area growth initiates from an atomically flat surface without catalysts. The issue regarding the SAG of III–V NWs on Si involves the surface reconstruction of the Si, the SiO_2/Si interface and the incorporation process of group-III/V atoms on the Si(111) surface. The methods for dealing with these issues depend on the growth material because the chemisorption processes of group-III atoms on Si surfaces and the bonding strength between group-III atoms and Si differ at various temperatures. If we neglect these basic issues, vertically aligned III–V NWs will not form on the Si surface.

Various reconstructions of Si(111) surfaces have been observed in an ultrahigh-vacuum (UHV). c2×1, 7×7, and 1×1 reconstructed surfaces form during heating. Metastable reconstructed $n \times n$ ($n = 1, \sqrt{3}, 2$), c2×4, and c2×8 surfaces form at temperatures from 500 to 830°C, and a 1×1 reconstructed structure forms above 830°C or below 430°C [118–121] by cooling procedure. Note that the growth temperature windows for most III–V semiconductors are within the range in which metastable reconstructions are formed. These surface reconstructions suppress the adsorption and nucleation processes of III–V materials because the resulting surfaces are very stable and there are no dangling bonds for adsorption. A promising way of avoiding metastable surface reconstructions is to cool the Si(111) surface to 400°C in an H_2 ambient, because the 1×1 reconstructed surface formed at a higher temperature can regenerate at around 400°C. Native oxides also suppress the nucleation process of III–Vs. Moreover, these oxides result in unexpected NW growth on Si [108]. Thermal cleaning at temperatures above 900°C can remove native oxides and form a Si(111) 1×1 reconstructed surface because it decomposes the SiO_2 and also alters the surface reconstruction of Si(111). These phenomena are the basic processes for cleaning Si(111) in order to control the growth directions of III–V NWs on Si(111) in SAG.

The SiO_2/Si interface has Si^{1+}, Si^{2+}, and Si^{3+} chemical structures [122, 123]. After the native oxides have been removed at high temperature, the Si(111) surface should be reconstructed into a 1×1 surface, and group-III or -V atoms terminated or incorporated on the Si(111) 1×1 surface should be precisely controlled to form a (111)B surface to align III–V NWs vertically. NWs can be grown vertically by simply forming an (111)B-oriented surface on the Si(111) surface. As shown in Figs. 14.4a and 4(b), once As-incorporating Si^{3+} and/or In-terminated Si^{1+} have formed on the Si surface, only vertical III–V NWs should grow on the Si(111) substrates. Conversely, the growth directions of the III–V NWs can be controlled by optimizing the initial surface. Group-V atoms should be replaced with the outermost Si atoms of the 1×1 reconstructed surface to form an As-incorporating Si^{3+} surface because it is equivalent to a V-atom terminated Si^{3+} surface and a (111)B-oriented surface. It was found that a Si(111): As 1×1 reconstructed surface formed at low temperature in an As ambient [124]. The method by which these (111)B-oriented surfaces form differs for each III–V/Si system because these processes strongly depend on the bonding strength between Group-III and Si atoms.

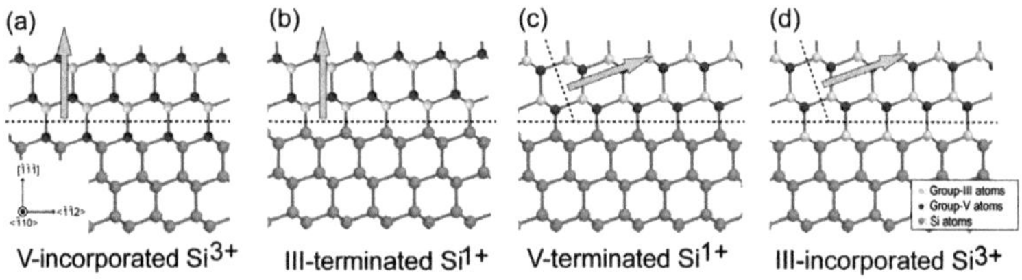

Figure 14.4 Schematics diagrams of chemical structures: (a) group-V-incorporating Si^{3+} structure, (b) group-III-terminated Si^{1+} surface, (c) group-V-terminated Si^{1+} surface, and (d) group-III-incorporating Si^{3+} structure. These are viewed from the <–110> direction. Arrows indicate directions of growth for III–V NWs.

So far, we have described the growth of vertical III-As NWs on Si [66, 71, 98, 199, 101, 103]. Now let us review some of the typical results for InAs, GaAs, and InGaAs NWs on Si(111). In particular let us examine the integration of InAs NWs grown on Si by using SAG in Fig. 14.5a.

An important consideration in aligning vertical III–V NWs on Si is the formation of an As-adsorbed Si(111) 1 × 1 surface before the NW growth. This surface corresponds to the (111)B orientation, and it can be formed at temperatures below 430°C in an As ambient. Here, after the thermal cleaning, the Si surface can be cooled to 400°C in the H_2 ambient. Then, the supply of AsH_3 gas can be resumed. After this procedure, almost the entire Si(111) surface changes into an As-adsorbed Si(111) 1×1 surface, which is equivalent to a (111)B-polar surface. However, dangling-bonds of Si(111) still remain because of the imperfect formation of the As-adsorbed Si(111) 1×1 surface. To deal with this problem, flow-rate modulation epitaxy (FME) at low temperatures can be performed on the As-adsorbed Si(111) 1×1 surface.

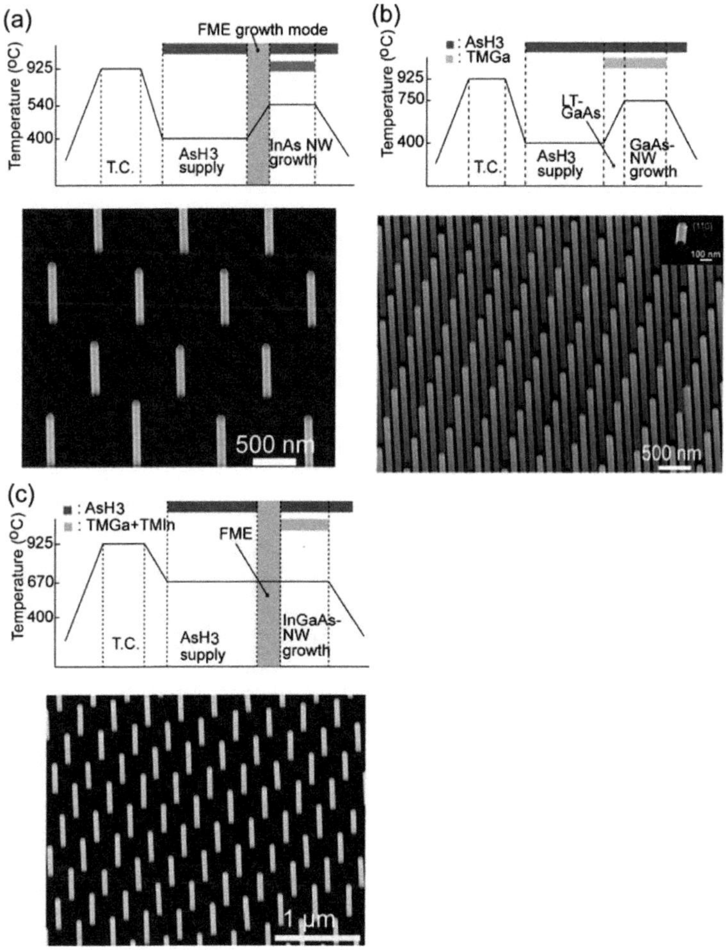

Figure 14.5 Formation of vertical III–V NWs on Si(111): (a) InAs NW on Si and growth sequence, (b) GaAs NWs on Si (Ref. 98, Copyright (2010) Institute of Physics Science), and (c) InGaAs NWs on Si (Ref. 100, Copyright (2011) IEEE).

Flow-rate modulation epitaxy basically increases surface migration of adatoms at low temperature [125]; it is an effective way of controlling the growth direction of InAs NWs, because both terminations of the In atoms to the As-adsorbed Si(111) 1×1 surface and the dangling-bonds of the Si(111) surface form (111)B surfaces. An arbitrary number of FME-cycles alternating between TMIn and AsH_3 supplies are performed at 400°C. Completely vertical InAs NWs on Si(111) can be obtained under optimal conditions, as illustrated in Fig. 14.5a. The InAs NWs depicted in the image grew only in the openings in the mask and were oriented perpendicular to the Si(111) surface.

Figure 14.5c shows that SAG can successfully integrate vertical InGaAs NWs on Si(111). In this case, the growth sequence is different from that of InAs and GaAs NWs in that there is no AsH_3 treatment at low temperature. This is to suppress a formation of In/Ga alloy that would otherwise occur during the low-temperature buffer-layer growth and FME. Here, the melting point of In/Ga alloy is lower than In and enhances Si-Ga/In intermixing at low temperature, which means that temperature is an important consideration in ternary III–V NW growth. The growth sequence has to be carefully chosen for each individual NW, not each individual NW material. Thun, the growth sequence should be determined for each InGaAs ternary NW material such that the solid-phase of the NWs has the desired In/Ga ratio.

InAs/undoped InAlAs/Si-doped InAlAs CMS NWs have been investigated in attempts to passivate the surface of InAs NWs. An undoped-InAlAs/Si-doped InAlAs multilayer is grown on the sidewalls of the InAs NWs by using the lateral over growth mode of selective-area MOVPE (SA-MOVPE) (Fig. 14.6a) after growth of the InAs NWs on Si (Fig. 14.6b). Fig. 14.6c shows the resultant growth on Si(111).The diameter of the CMS NWs during the InAlAs growth increased from 80 to 140 nm, while the average height of the InAs/InAlAs CMS NWs remained constant at 1.25 μm after the formation of the InAs/InAlAs CMS structures. This means that the InAlAs layers grew only radially in the <–110> directions. The thicknesses of the n-InAlAs and undoped-InAlAs shell layer were 23 and 7 nm, respectively.

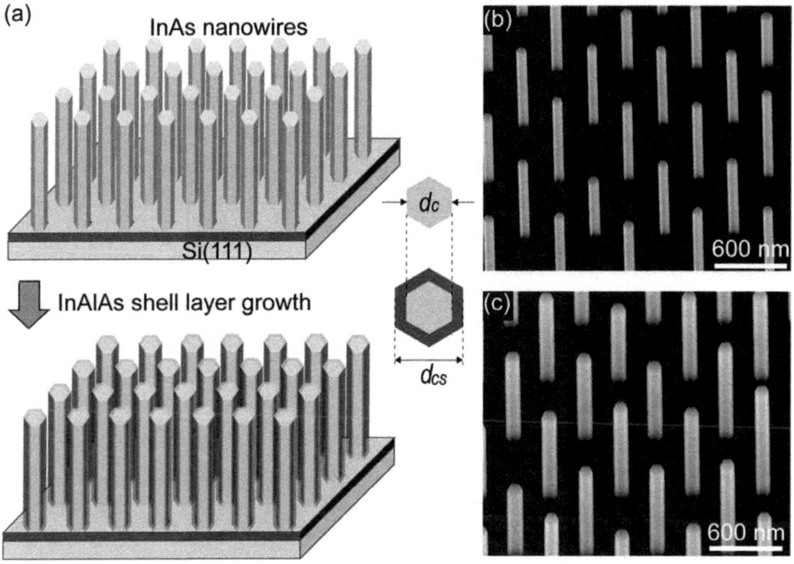

Figure 14.6 (a) Illustration of InAs NW and InAs/InAlAs CMS NW growth, (b) 30°-tilted SEM image view showing resultant growth of InAs NWs on Si, and (c) 30°-tilted SEM image view showing growth of InAs/InAlAs CMS NWs on Si.

14.4 Vertical NW SGTS

The first demonstration of SGTs was reported by Takato et al. [41, 42]. The main advantage of the vertical surrounding-gate structure is superior electrostatic gate-bias control in the channel due to the wrapped gate metal, and it enables the occupied area to be reduced so as to allow denser integration. Additionally, certain scaling parameters in the effective-oxide thickness (EOT) of the gate-oxide and gate length can be simply determined by the deposition thickness and process thickness because the integration dimensions change into the vertical orientation. SGTs, however, still face challenges of high parasitic capacitance and resistance in high-frequency device applications and 3D device processes. Here, we describe device-process flow for making III–V NW-SGTs on Si.

14.4.1 Device Process-Flow of SGTs

First, let us consider a typical device process-flow of SGT. In this case, we use some numbers of lithography-defined NWs for vertical SGTs as the multichannels with a high-*k*/metal. The details are described in some reports [75, 105]. The schematic of the device structure is shown in Fig. 14.7. The entire surface of the NWs is covered with a high-*k* dielectric film formed by atomic layer deposition (ALD) (Fig. 14.7b). Then, a RF sputtering and photolithography are used to form the gate metal and contact pads (Fig. 14.7c). In this case, the gate-metal is deposited on both the top surface and the sidewalls of the NWs. Next, a low-*k* benzocyclobutene (BCB) resin (Dow Chemical CYCLOTENE) is spin-coated and etched back to the desired thickness (approximately 200 nm in this case) by reactive-ion etching (RIE) (Fig. 14.7d). After the RIE, wet etching is used to remove the metal and high-*k* dielectric remaining on the top portions of the NWs (Fig. 14.7d). At this point, the BCB layer acts as a protection layer for the wet etching, whose thickness defines the gate length, L_G. After that, the devices are spin-coated again with BCB to electrically isolate the gate from the top drain electrode (Fig. 14.7f), and then etched back by using RIE to expose only the top part of the NW. Drain and source metal are then evaporated on the top portion of the vertical III–V NWs and the back side of the Si (Fig. 14.7g). Finally, the gate contact pads are exposed by using RIE (Fig. 14.7h).

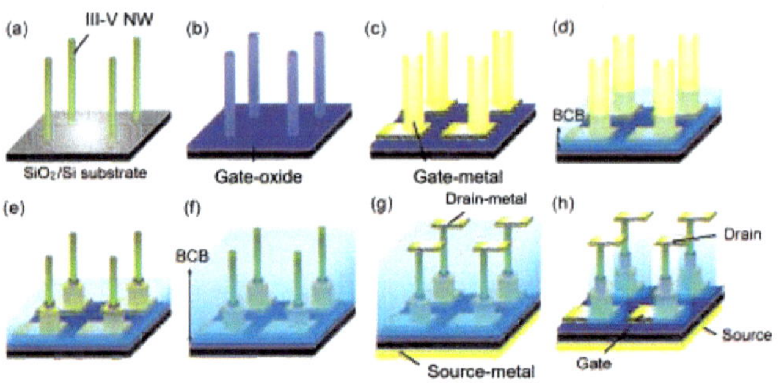

Figure 14.7 Device fabrication processes: (a) SA-MOVPE growth of NWs, (b) atomic layer deposition, (c) surrounding-gate metal sputtering, (d) etching stopper layer formation, (e) selective etching of metal and high-*k*, (f) electrical separation layer formation, (g) drain and source metal evaporation, and (h) Gate contacting-pads exposure (from Ref. 75, Copyright (2012) The Japan Society of Applied Physics).

14.4.2 InGaAs SGTs

Through the device procedure described in Section 14.4.1, the structure of an SGT device using vertical III–V NWs on Si is illustrated in Fig. 14.8a [66, 105] and Fig. 14.8b depicts growth of an InGaAs NW-array on Si. In the corresponding experiment, the In composition was 70%, and a gate-first process and etch-back procedures were used for making the SGT device. Low-*k* insulating polymer (benzocyclobutene, BCB) was used for the interlayer dielectric of the vertical SGT. Ten InGaAs NWs were formed, each InGaAs NW was composed of an undoped InGaAs (carrier concentration: 1×10^{16} cm^{-3})/Si-doped InGaAs (carrier concentration: 3×10^{18} cm^{-3}) axial junction. The height of the undoped InGaAs segment corresponded to the channel length (L_{ch}), which in this case was 50 nm. After the InGaAs NW-channels had been formed on Si(111), the NWs were treated with an alkaline solution and a buffered hydrofluoric acid (BHF) solution to etch the native oxides of the InGaAs NWs. Then, the NWs were covered with high-*k* HfAlO film by using atomic layer deposition. Then, the tungsten (W) film was sputtered on the NW array covered with the gate oxide. The thickness of the W was 150 nm. After the gate stack formed, the W was lithographically defined as a gate-pad by reactive ion etching (RIE). Then, the sample was spin-coated with BCB and etched back by RIE with a CF_4/O_2 mixture to simultaneously etch the BCB, W, and HfAlO gate oxide. After the RIE, the NWs were spin-coated with BCB, and then etched back by RIE to isolate the gate and drain metals. Finally, the drain and source metals were deposited on top of the NWs and back-surface of the Si substrate. The drain metal was Ni/Ge/Au/Ni/Au and the source metal was an Ni/Au multilayer. After the series of device-process flow, the sample was annealed at 400°C in N_2 to form Ohmic contacts and laminate the $HfAlO_x$ structure.

As for gate-stacking of the SGTs, a self-cleaning effect by trimethylaluminum (TMAl) pulsed deposition [126] is performed, and then HfAlO film is deposited. The oxide is based on a composite structure of one monolayer (ML) of Al_2O_3 and four MLs of HfO_2, and this multilayer is laid down in each deposition cycle. The thickness of the gate-oxide film ranged from 4 to 20 nm. The composite film formed alloy laminate HfAlO films. According to the integrated intensity of the spectra, the hafnium content was estimated to be about 80% and the Al content was about 20%. The dielectric constant of the $Hf_{0.8}Al_{0.2}O$ was estimated to be 20.4 by using the Maxwell-Garnet effective medium approximation [127]. The capacitance–voltage (*C–V*) properties and gate-leakage (I_G) characteristics of the InGaAs NW-SGTs are measured in order to characterize the quality of the InGaAs NW/HfAlO. The drain metal was grounded and the source metal was open to avoid the contribution of the parasitic capacitance from the Si substrate, mask template, and BCB interlayer. Figure 14.9a is a representative *C–V* curve at 100 kHz. There is no hysteresis, which means that the HfAlO had no charge traps, and Fig. 14.9b has no frequency dispersion. The I_G–V_G curve indicates I_G was moderately low for an EOT of 0.75 nm.

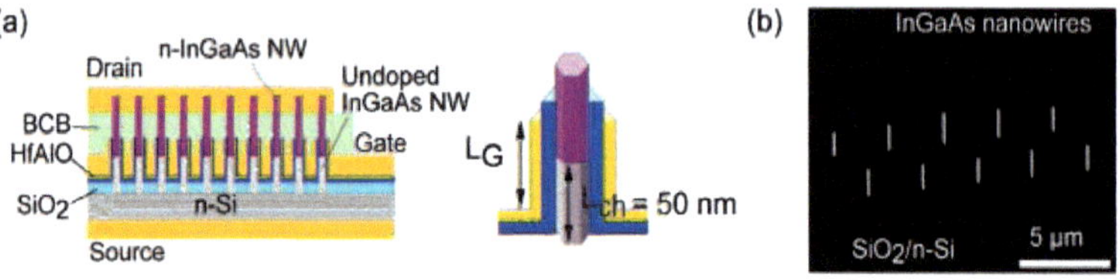

Figure 14.8 (a) Schematics of InGaAs NW-SGT on Si substrate and, (b) representative SEM image of InGaAs NW-array.

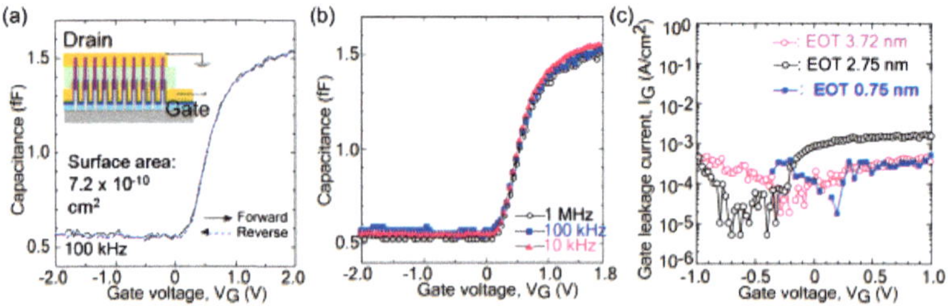

Figure 14.9 (a) *C–V* properties at 100 kHz. Oxide thickness was 14 nm. (b) Frequency dispersion of *C–V* for InGaAs NW-SGT structure. (c) I_G–V_G characteristics with variations in EOT.

Figure 14.10 shows the transfer and output characteristics for the InGaAs NW-SGTs on Si. Carrier transport is along the <111> direction and it occurs on {–110} surfaces. The EOT is 0.75 nm. The measured current was divided by the number of NWs and normalized by the gate perimeter (250 nm). The maximum drain current I_D is 75 μA/μm (at V_{DS} = 0.50 V), minimum subthreshold slope (SS) is 78 mV/decade, and the average subthreshold swing is about 90 mV/dec with a V_G window of 0.5 V. The on–off ratio, I_{on}/I_{off}, is approximately 10^5, while I_{off} is 10^{-9} A/μm at V_{DS} = 0.50 V. The drain-induced barrier lowering (DIBL) is 68 mV/V. The moderately low I_{off} is owing to the discontinuity in the band structure across the InGaAs/Si. Figure 14.11a plots the I_D–V_{DS} curve, indicating asymmetrical properties. Under a positive V_{DS}, carrier transport occurs regardless of the potential across the junction in the conduction band, but the carrier feels a potential under the negative V_{DS}. The potential barrier seems to form at the InGaAs/Si interface, as shown in Fig. 14.11b, and it depends on the diameter of the heterointerface of the InGaAs NW/Si and the number of misfit dislocations. In addition, the potential barrier affects the series resistance slightly. The complex behavior arising from the potential barrier involves a trade-off between I_{off} and R_{on}. Although the characteristics depicted here are not so quantitative, it can be seen that the InGaAs NW-SGT on Si with low I_{off} reaching < pA/μm always has a high R_{on}. The switching behavior of the SS properties is close to that of recent InGaAs-based multigate FinFETs [50] and gate-all-around FETs (GAA-FETs) [52, 53]. The SS of the InGaAs NW-SGT remains almost constant to variations in the EOT [67]. This is because the interface effect density (D_{it}) across the InGaAs NW and HfAlO is almost constant. Figure 14.11c depicts the variations in D_{it} estimated from the subthreshold slopes. The D_{it} ranges from 1.8 to 3.9 × 10^{12} cm^{-2}eV^{-1}.

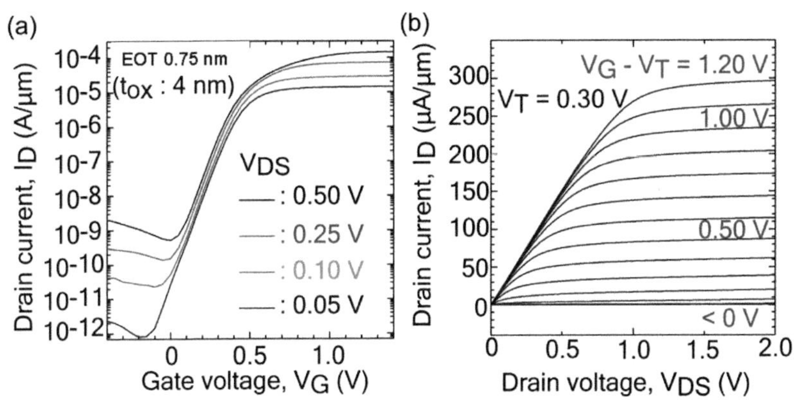

Figure 14.10 Switching performance of InGaAs NW-SGTs on Si with EOT of 0.70 nm: (a) transfer curve and (b) output characteristics.

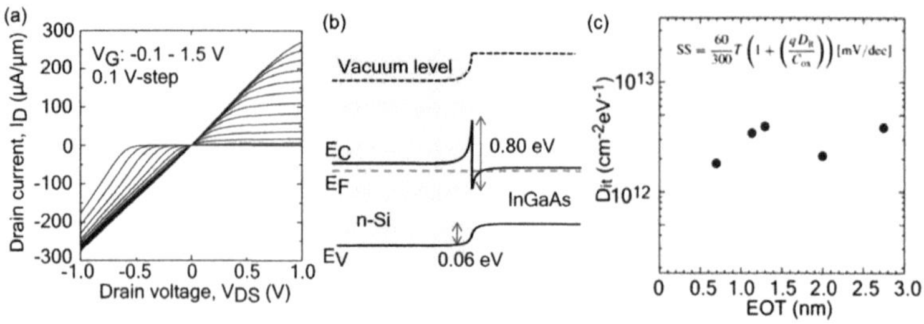

Figure 14.11 (a) Representative output characteristics of InGaAs NW-SGT, (b) band-diagram of InGaAs NW/n-Si substrate and (c) D_{it} estimated from subthreshold slope with variations in EOT.

14.4.3 Modulation-Doped NW-Channels on Si

Although the switching properties of the InGaAs NW-SGTs are better than those of other III–V MOSFETs, I_{on} is still lower than that of the benchmark stated by the International technology roadmap for semiconductors (ITRS). The EOT scaling in Fig. 14.9 makes no contribution to enhancement of I_{on}. There are two main reasons for the constant I_{on}. The first is the large D_{it}, which causes a depletion layer. A thinner EOT forms a stronger inversion layer, but the large D_{it} fixes the formation of the inversion layer. The second is the large diameter of the NWs. The inversion layer forms near the surface of the NW-channel, and the large diameter results in a weak inversion. Very thin InGaAs NW-channels, 20 nm in diameter, have recently been found to form a strong body inversion [20]. Therefore, a smaller NW-diameter is predicted to enhance I_{on}. Another approach to enhancing I_{on} exploits the distinctive properties of selective-area growth that enables the formation of a core–multishell structure allowing passivation and band-engineering of HEMT structures. The use of radial growth modes in selective-area growth have been used to form functional heterostructures on NW-sidewalls, including LEDs [103], solar cells, and HEMTs [66].

Figure 14.12a illustrates a core–multishell NW-HEMT structure consisting of InP/InAlAs/δ-doped layer/InAlAs/InGaAs multishell layers as a modulation doped structure. This complex structure is to obtain high I_{on} and to maintain better gate-control of InGaAs NW-SGTs. The InP layer serves as a barrier layer, creating spatially separate regions of carrier confinements under various

V_G. The InAlAs/δ-doped layer/InAlAs is the source of modulation doping. The outer InGaAs serves as a capping layer to maintain SS < 100 mV/decade. The main advantage of the multilayer is in the ability to stack many δ-doping layers to form multi quantum well structure. Even if the surface of the NW sidewalls is rough, these multi δ-doping layers uniformly create 2DEG inside the NW channel. The solution to the one-dimensional Poisson–Schrödinger equation reveals that this structure confines the carrier inside the InGaAs NW-channel under positive V_G (Fig. 14.12b) [66]. Figure 14.12c shows a high-angle annular dark field STEM (HAADF-STEM) image of cross section of a grown InGaAs/InP/InAlAs/δ-doped layer/InAlAs/InGaAs core–multishell NW. The core InGaAs is 68 nm in diameter, and the thicknesses of the InP, InAlAs/δ-doped layer/InAlAs, and outer InGaAs are 4, 20, and 2 nm, respectively. The NW is 120 nm in diameter. Interestingly, dark contrast can be observed at the corner of the InAlAs shell layer indicating Al segregation occurs at the corner. The Al segregation forms Al-rich InAlAs. This Al-rich InAlAs splits the cylindrical HEMT structure into six-sided HEMT layers. Although a three-dimensional simulation of an AlGaAs/GaAs NW-HEMT revealed carrier confinement at the corner of a core NW [128], the Al-rich InAlAs was thought to act as a barrier to carrier confinement at the corner of the NW.

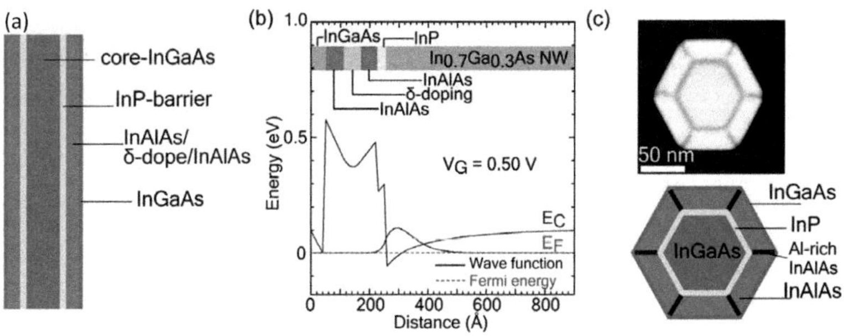

Figure 14.12 (a) Illustration of InGaAs/InP/InAlAs/InGaAs core–multishell NW, (b) one-dimensional simulation of band structure at V_G = 0.50 V, and (c) HAADF-STEM image of core–multishell NW and schematic cross section.

Figure 14.13 shows the I_D–V_G and I_G–V_G curves at a V_{DS} of 0.50 V. The core InGaAs NW is 90 nm in this case, and the thicknesses of the InP, InAlAs/δ-doped layer/InAlAs, and outer InGaAs are 2.6, 16.5 and 5 nm. The SGT demonstrates an n-type enhancement mode with a threshold voltage of 0.38 V. The I_{on}/I_{off} ratio is approximately 10^8. I_{off} is below 10 pA/μm. The SS was 75 mV/dec, and DIBL is 35 mV/V. The use of the HEMT layers enhances I_{on} and the transconductance (G_m). I_{on} is 0.45 mA/μm at a ($V_{DS} = V_G - V_T$) of 0.50 V, and G_m reaches 1.42 mS/μm at a V_{DS} of 0.50 V [66]. Figure 14.13a plots the field effect mobility as a function of V_G for InGaAs NWs, InGaAs/InAlAs core–shell NWs, the InGaAs/InP/InAlAs/InGaAs core–multishell with a diameter of 90 nm and a thicker HEMT layer, and the core–multishell NW. The moderately thin multishell layer acts as a good modulation doped layer and enhances the field effect mobility. The field-effect mobility offered by the core–multishell layer (7850 cm²/Vs) is approximately ten-times that of Si-MOSFETs [66].

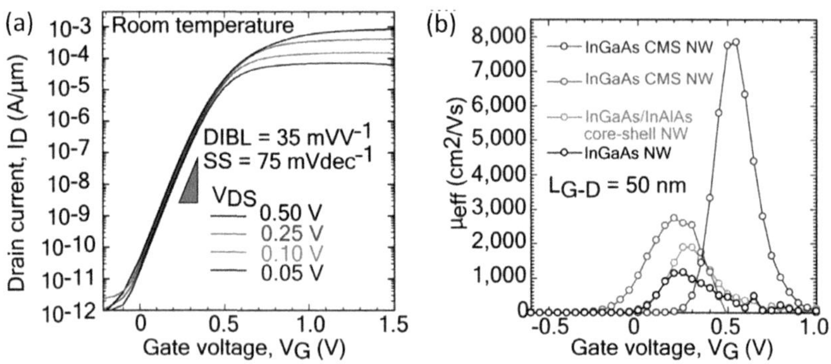

Figure 14.13 (a) Illustration of InGaAs/InP/InAlAs/InGaAs core–multishell NW (Ref. 66, Copyright (2012) Nature Publishing Group), and (b) One-dimensional simulation of band structure under $V_G = 0.50$ V.

Further scaling of the InGaAs NWs and the thickness of the HEMT layer are needed to make high-performance InGaAs NW SGTs on Si-LSIs. The thin HEMT layer (Fig. 14.13c) is being investigated and so far it has been revealed that such thin layers enhance I_{on} and G_m. Although further calculations using three-dimensional Poisson–Schrödinger equations are required to determine the distribution of 2DEG inside the NWs, NWs with a diameter of < 10 nm with a total shell thickness of 10 nm have been predicted to work as HEMTs beyond the limitation of the quantum size effect.

14.5 Photovoltaic Devices

The rapidly increasing global demand for energy with our population has led to an interest in using renewable energy sources such as solar energy to avoid wasting fossil fuel resources. Photovoltaic applications using nanotechnologies and light-trapping effects are being explored as a way of reducing the costs of amount of materials used in solar cells (SCs). III–V materials are promising for solar cells because high-energy efficiency can be expected from their tandem structure. However, these materials are expensive enough to impede widespread commercialization of the terrestrial application of NW-based SC. The use of III–V NW arrays may be a way of reducing costs and materials while maintaining the power conversion efficiency. This is because the NW array can be transferred to an inexpensive foreign substrate. Moreover, through recycling of expensive III–V substrates, the materials used to produce NW SCs can be reduced to 0.2% that for conventional planar III–V SCs. Here, an InP NW-array SC with an energy efficiency above 13% was recently demonstrated [129]. The required III–V materials to generate 1 W of power can be reduced by more than two orders of the magnitude as compared to that of Si-based SC panels thanks to the high absorption coefficients of III–V materials.

The NW-SC has a diffraction limit in absorption in classical optics, and the maximum light absorption is proportional to the coverage of the NW-array relative to the whole illuminated surface. A periodic NW array can potentially enhance optical absorption by reducing optical reflectance [130–132]. In particular, a size- and position-optimized NW array can achieve an overall sunlight absorption level that exceeds the ray-optics light-trapping absorption limit over a broad range of incidence angles because of its near-field-optically enhanced absorption and anti-reflection effect [129, 131]. Even if the NWs are less than 200 nm in diameter, the NW-array SC can increase their short circuit current (I_{SC}). Numerical simulations predict that NW arrays with very low surface coverage (~20%) would have a J_{SC} as high as that of thin-film SC, that is, energy efficiency would be as high as that of planar SCs without having to use as much material [131].

The most distinctive feature of NW materials for SC applications is the CS structure. It is predicted that a CS structure with a radial p-n junction, which is aligned parallel to the incident light direction, improves the collection efficiency of photo-generated electrons and holes throughout the junction area.Moreover, as explained above, this architecture saves natural resources when compared with conventional thin-film SC. Although many photovoltaic applications of NWs have been reported [134–138], comparatively few periodically aligned NWs with uniform CS structures have been fabricated. Here, Goto et al. fabricated highly uniform InP NWs with radial p-n junctions and demonstrated InP NW-array SCs on a p-InP substrate in 2009 [139].

Nanowire materials have a large surface area and surface Fermi-level pinning. Surface Fermi-level pinning degrades J_{SC} by enhancing non-radiative recombination centers; however, surface passivation can alleviate the pinning problem and lead to a open-circuit voltage (V_{OC}). For example, Wallentine et al., demonstrated InP NW-array SC with an energy efficiency of 13.8% by using array with 180 nm-diameter NWs and surface coverage of 18%. The J_{SC} of the SC reached 83%, a world record for planar InP SCs, exceeds the V_{OC} against the planar InP SCs [129]. Moreover, the CS NWs with a radial p-n junction enlarge optical absorption area in the NW-array structure.

In this section, we describe the recent progress on NW-array SCs and the benefits of their NW geometry and optical functionality. We also describes SCs using III–V NW array grown by SAG.

14.5.1 Benefits of III–V NW Materials for SC Applications

A solar cell has four dominant photovoltaic processes: (a) light absorption, (b) generation of photo-generated carriers, (c) separation of the photo-generated carriers, and (d) collection of the photo-generated carriers into a current-flow. The NW-SC has benefits in terms of light absorption and separation of carriers; these are distinctive features of the SC applications. NW-SCs can be categorized as having two different structures; one is an axial p-n junction SC in which the p-n junction is formed inside the NWs (Fig. 14.14b); the other is a radial p-n junction SC in which the p-n junction is formed on the NW-sidewalls (Fig. 14.14c).

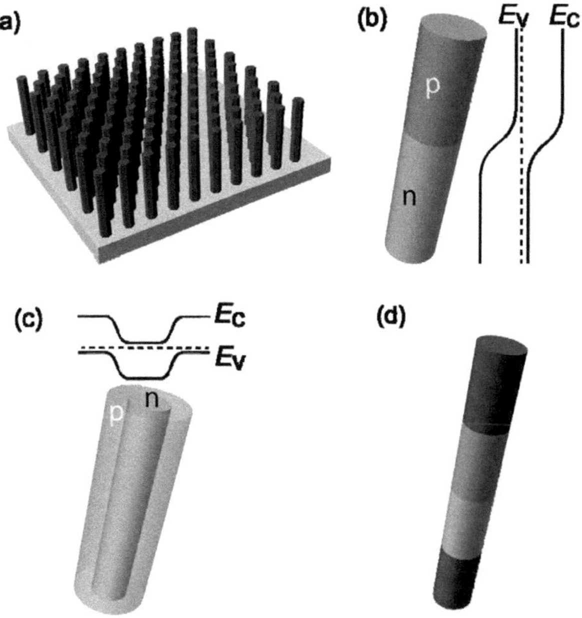

Figure 14.14 (a) Schematic diagram of NW array structure, (b) NW with axial p-n junction, (c) NW with radial p-n junction, and (d) axial multijunction NW with different band gaps.

A NW-array with a regular period shows enhanced light absorption [131]. Kupec et al. predicted the light absorption in solar spectrum (wavelength of 300–900 nm) would be virtually 100% for an array 180 nm in diameter and 360 nm in period. The important point here is that the array does not require any anti-reflection film. Huang et al. simulated a simple periodic array of Si NWs and found that the structure can suppress reflection of light. Here, reflection of both p-polarized and s-polarized light can be suppressed for a wide range of angles of incidence through the moth-eye effect [140]. These improvements in light absorption result from decreased reflection due to a reduction in the effective index, which decreases with distance from the substrate, and from the increased optical path length due to internal light scattering.

Conventional III–V SCs have a heavily doped n-type emitter layer several dozen nanometer thick on a shallow doped p-type base layer with several micrometers thick in order to enlarge the diffusion length of minority carriers. In this case, short-wavelength light with a higher absorption coefficient will penetrate the shallow part, and the photo-generated minority carriers near the p-n junction can easily arrive at the junction. On the other hand, long-wavelength light has a large penetration depth, so photo-generated minority carriers in the deep region will be lost owing to the short diffusion length (or in some cases the fast recombination velocity) and recombination process. Therefore, in the case of long-wavelength light, high-quality III–V materials are required for enhancing quantum efficiency. On the other hand, NW-SCs with a radial p-n junction, which is shown in Fig. 14.14c, can separate the photo-generated carriers inside the radial p-n junction. The separation length of the minority carriers does not depend on the penetration depth of the incident light and can be the whole-diameter of the NW. Thus, the carrier diffusion length has less influence on the power conversion efficiency of NW as compared with planer SCs [141].

Multijunction tandem SCs composed of several cells each having different band gaps can potentially exceed 50% power conversion efficiency because four-junction (4J) tandem structure effectively covers the whole solar spectrum. III–V materials are feasible for multijunction tandem SCs because the energy gap can be changed by varying the ternary material and composition. Up to the time of this writing, the highest power conversion efficiency for 3J-tandem SCs is reported to be 43.5% [142] under concentrated sun light, and 4J and 5J tandem SCs are expected to exceed this figure. In this regard, lattice-matched heterojunction is required for further improvement of the power conversion efficiency. However, it is very difficult to obtain high-quality layer growth without misfit dislocations for 4J or 5J tandem SCs In the case of planar structures, there is no good pair of materials for such multijunction SCs.

On the other hand, the small footprints of NWs can overcome the problem of misfit dislocations for the higher lattice mismatched system. Glas reported that the small footprint of NW allows lattice strain without lattice relaxation and the critical thickness can be increased by decreasing the NW diameter [143]. For example, no misfit dislocations are formed in 50 nm-diameter NWs with as much as a 3%-lattice mismatched system. The geometrical advantages that NWs have for multijunction NW SCs can be exploited in the various multijunction combinations regardless of the lattice mismatch (Fig. 14.14d).

As mentioned above, NW-arrays have a light-trapping effect and required no anti-reflection films and they would have resource and cost savings. Moreover, almost all III–V materials have a direct band gap that results in a high absorption coefficient at visible wavelengths. That means SC films with a few micrometers thick sufficiently absorb incident light, and thus, resource-saving SC panels can be realized by combining the peeling-off method which will be described later. Table 14.2 lists the sized of SC panels needed for supplying 5 MW (enough power for about 1700 households). Si-based SC panel amounting to 9.8t of Si are required for the supplying that much power; on the other hand, NW-based SCs, especially InP NW-arrays, can reduce the amount to 127th part of the Si-panel, i.e., 71 kg. This figure may be able to be reduced to only 32 kg by optimizing the NW-SC structure in the future. Moreover, concentrated solar power generation can also reduce the amount. Resource savings will be an important factor in the commercialization of III–V based SC. They will also lead to cost savings.

Table 14.2 Amount of semiconductor materials used for power generation (5 MW)

	Power conversion efficiency (%)	Standard quality of material (mg/W)	Amount used for 5 MW (kg)
c-Si (bulk) (comm.)	25.6	1819	9080
c-Si thin (lab.)	20.1	515	2575
InP NW array (lab.)	6.35	1	71
4J NW + 500 sun	50.0	0.042	0.21

14.5.2 Progress in NW SCs

The previous section described the benefits of the NW SC structure. In this section, we review the progress in NW-SCs. Table 14.3 summarizes the performance of representative III–V NW-SCs.

LaPierre et al. demonstrated NW-SCs consisting of a GaAs CS NW-array grown using an MBE-based VLS method. Although they used ITO transparent metal, the power conversion efficiency of the SC remained below 1% because of the non-uniformity of the array and low V_{OC} with poor rectification [137]. While, Huffaker et al. also fabricated a GaAs CS NW-array SC with transparent electrodes and achieved a power conversion efficiency of 2.54% [144]. They used the SAG method for forming the array, and the catalyst-free and size-controlled synthesis seemed to improve performance. Additionally, a single GaAs CS NW with a p-i-n junction SC [134] and InAs NW integrated on Si was utilized in an SC application to absorb long wavelengths region [108]. Figure 14.15a shows the device structure and performance of this InP NW-array SC. The size-controlled InP NW array was formed with an SiO_2 mask. It had well size-controlled NW array shows moderately large filling factor (FF) and a power conversion efficiency of 3.37%. InP is best material for a single-junction SCs because the band-gap is optimum for absorbing visible range of solar spectrum. Furthermore, this material has good Ohmic contact with transparent electrodes such as indium tin oxide (ITO); thus, InP-related CS NWs would be compatible with such ITO films. Recently, Yoshimura et al. demonstrated an InP/AlInP CS NW-array SC with improving the power conversion efficiency of 6.35% [145] (Fig. 14.15b).

Table 14.3 Summary of NW-based SC performances

Mat.	Struc.	Junction	Growth	V_{OC} (V)	J_{SC} (mA/cm^2)	FF(%)	η(%)	Ref.
GaAs	Array	p-n RA	MBE	0.20	15.5	26.7	0.8	137
	Array	p-n RA	MBE	0.14	29.4	26.5	1.1	150
	Array	p-n RA	MOVPE	0.39	17.6	37	2.5	144
(on Si)	Array	p-i-n RA	MBE	0.39	18.2	46.5	3.3	161
	Array	p-i-n RA	MOVPE	0.50	12.7	65	4.0	162
	Single	p-i-n RD	MBE	NA	NA	65	4.5	134
	Array	n-p RA	MOVPE	0.44	24.3	62	6.6	163
	Array	n-i-p RA	MOVPE	0.57	18.9	69	7.4	164
	Array	n-i-p AX	MOVPE	0.57	21.1	63.7	7.6	165
	Array	p-i-n RA	MOVPE	0.91	21.3	79.2	15.3	148
InAs	Array	p-n AX	MOVPE	NA	NA	NA	0.8	108
InP	Single	p-n AX	MOVPE	0.69	8.25	58	3.3	152
	Array	p-n RA	MOVPE	0.43	13.7	57	3.4	139
	Array	p-n RA	MOVPE	0.67	11.1	59	4.2	151
	Array	p-i-n RD	MOVPE	0.50	15.7	67.7	5.3	166

Mat.	Struc.	Junction	Growth	V_{OC} (V)	J_{SC} (mA/cm^2)	FF(%)	η(%)	Ref.
	Array	p-n RA	MOVPE	0.46	23.4	60	6.4	145
	Array	p-n AX	MOVPE	0.73	21	73	11.1	149
	Array	p-i-n AX	MOVPE	0.78	24.6	72	13.8	129
InGaAs	Array	p-n RD	MOVPE	0.26	17.2	55.3	2.5	167
	Array	p-i-n AX	MOVPE	0.54	18.2	72.1	7.1	168
GaN	Single	p-i-n RD	MOVPE	1.00	0.39	56	0.2	146
	Array	p-n AX	VPE	0.95	7.6	38	2.7	147
GaAs/Si	Array	Axial	MOVPE	0.96	20.6	57.8	11.4	169

RD, radial junction; AX, axial junction; RA, radial and axial junction.

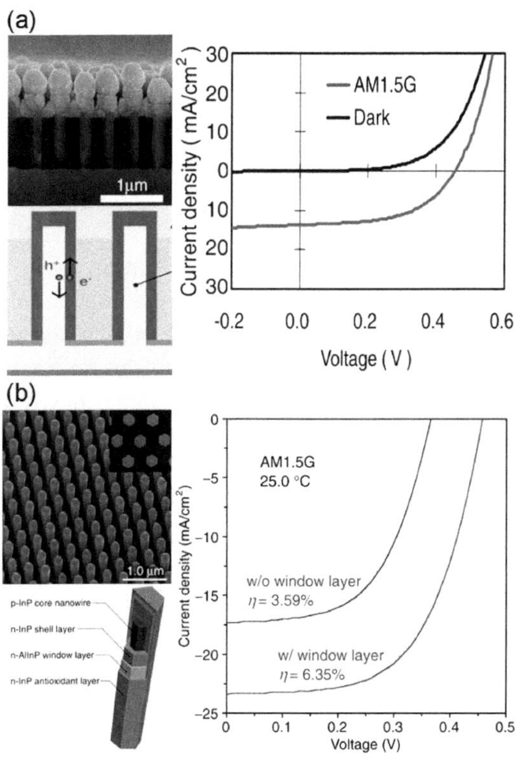

Figure 14.15 (a) InP core–shell (Ref. 139, Copyright (2008) The Japan Society of Applied Physics) and (b) InP/AlInP core–multishell NW array solar cells (Ref. 145, Copyright (2013) The Japan Society of Applied Physics).

Nitride-based NWs, which cover the whole range of solar spectrum by changing the contents of the ternary materials, can potentially be used in multijunction SC applications. However, the materials have serious crystal defects because even a small difference in alloy composition results in a large lattice mismatch. So far, it has been difficult to make high-performance planar GaN-based SCs. The NW geometry, especially its footprint, could be used to suppress the generation of defects and improve SC performance. Recently, GaN CMS NWs with intrinsic InGaN layers were used in SC application to control the energy band gap [146]. Tang et al. fabricated a novel SC using Mg-doped GaN NW array integrated with n-Si. The n-Si/p-GaN heterojunction act as an SC and showed a low dark current and good rectifying properties. The power conversion efficiency was 2.73% [147].

Borgström et al. demonstrated the highest power conversion efficiency among III–V NW-based SCs developed so far (13.8%). They used InP axial-junction NW array and passivated the surface by using PECVD-SiO_2 films to suppress surface recombination process [129]. They showed record high J_{SC} and V_{OC}. Recently, the group broke the record of power conversion efficiency by using GaAs NW array SC. Aberg et al. demonstrated GaAs NW-array SC with the efficiency of 15.3% [148]. Later, Cui et al. demonstrated a similar NW-SC structure that had a power conversion efficiency above 10% [149]. The research on III–V NW SCs seems to have reached the second stage. While there is as yet no clear answer as to which architecture (axial or radial) has higher power conversion efficiency, we can look forward to further evolution of these NW-SCs.

14.5.3 Future Prospects in III–V NW-SCs

The highest power conversion efficiency for the III–V thin film SCs has already achieved 28.8% for single-junction and 38.8% for multijunction architecture under AM1.5G. On the other hand, the power conversion efficiency of III–V NW-based SCs remains around 10% to 15%.

The primary issue in the degradation of efficiency is enhancement of recombination process due to the large surface area of the NW-array. The complicated p-n junction of the axial- and radial-junctions is responsible for the low efficiency. The surface recombination process can be suppressed by installing a reflection layer against diffusing minority carriers and by growing a passivation layer with a low surface recombination velocity.

Yoshimura et al. reported that an AlInP window layer acts as a passivation layer in the InP NW-based SCs [145]. They improved the power conversion efficiency while suppressing surface recombination. Figure 14.16 shows the improvement in internal quantum efficiency (IQE) at longer wavelengths that is had by using such a window layer. Although AlInP/InP is a lattice-mismatched system, they grew thin AlInP less than the critical thickness of AlInP/InP lattice mismatch, in order to degrade the crystal quality. Lapierre et al. used sulfur-passivation on GaAs NW-SCs and improved the power conversion efficiency by decreasing the surface recombination velocity [150].

Various materials option can be used for making multijunction SCs by using lattice relaxation and pseudomorphic-growth. Fukui et al. proposed axial 4J NW SCs [151]; these are expected to have a power conversion efficiency of 50%. Samuelson et al. demonstrated InP NW tunnel diodes integrated on a Si substrate [152]; these diodes could be building blocks for monolithic integrated multijunction NW-SCs.

In addition, advances in the characterization of device parameters such as the carrier concentration will be required because the conventional characterizations such as SIMS and Hall measurements have difficulty characterizing thin NWs quantitatively. In particular, quantitative characterization of dopants in NWs will be essential for optimizing NW-SC architecture.

The III–V SCs using GaAs and InP are promising SC for high power conversion efficiency because of their flexible material compositions and the capability of their lattice-mismatched systems to be used in multijunction structures. However, cost is a serious issue hampering the practicality of such III–V SCs since the materials themselves are so much more expensive than earth-abundant Si. Moreover, synthesis of NWs such as VLS and SAG methods basically are high cost process because of they use metal particles and masked substrates. In this regard, the peeling-off method and recycling of masked substrates may be able to reduce the cost of mass-produced SCs.

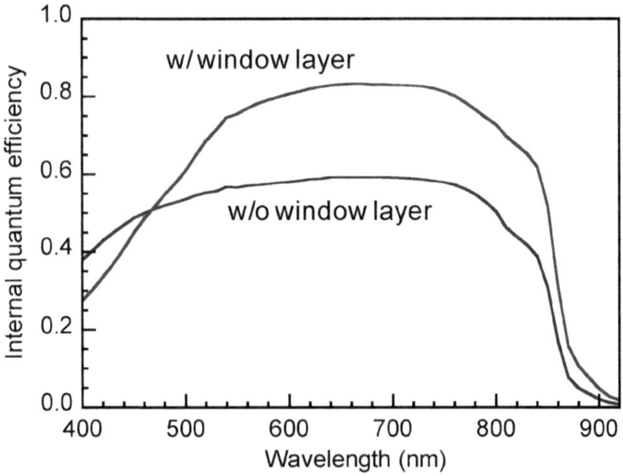

Figure 14.16 Internal quantum efficiency characteristics of InP core–shell NW array solar cells with and without an AlInP window layer (Ref. 145, Copyright (2013) The Japan Society of Applied Physics).

Peeling-off methods using flexible PDMD or transparent polymers have been demonstrated by several research groups. Anttu et al. and Chen et al. characterized the optical properties of the peeled-off flexible film with an InP NW-array in it. They also demonstrated recycling of the host masked substrates in NW synthesis. Figure 14.17a illustrates a typical peeling-off method for making a NW-array. Once the film has been peeled off, the host substrate can be used again for another NW-synthesis again. Figure 14.17b illustrates recycling of the mask substrate used in making GaAs NWs. The GaAs NWs are regrown on a recycled masked substrate after the peeling-off process. The regrown GaAs NWs have a uniform size that is the same as that of first growth of GaAs NWs. Thus, recycling the host masked substrate may be another way to decrease material costs.

The benefit of flexible NW-SCs offer resource savings as well as cost savings. The used resources can be decreased by about two orders of magnitude because flexible devices do not have host substrates. Unlike the earth-abundant Si, III–V materials with rare metals are relatively limited and expensive resource. The peeling-off and flexible device procedure could thus be the core process of future III–V NW-SCs. Moreover, flexible III–V NW SCs have a backside electrode that reflects incident light inside the cell layer. Thus, flexible and lightweight solar panels made from the SCs should exhibit both enhanced light confinement and high power conversion efficiency. These advantage can be available for space-satellite applications and megawatt-class power generation plants.

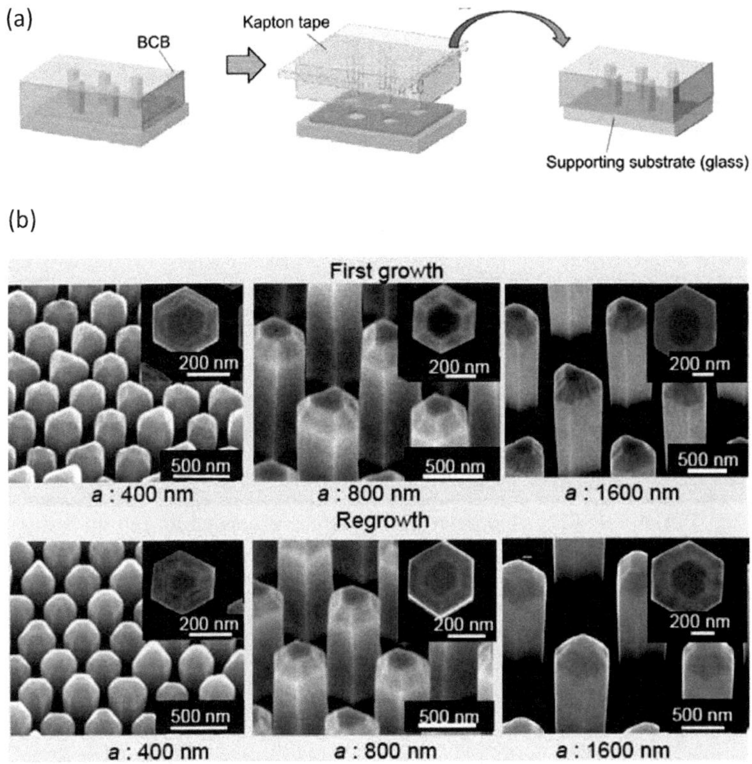

Figure 14.17 (a) Peel-off method of NW array and (b) SEM images of first growth and regrowth of NWs for a same substrate.

Acknowledgments

The authors would like to thank Drs. Masatoshi Yoshimura, Eiji Nakai, Keitaro Ikejiri, and Mrs. Tomotaka Tanaka, Takahito Endo, Muyi Chen, and Yoshihiro Hiraya for their fruitful discussions and experiments described in this chapter. This work was financially supported by a Grant-in-Aid for Scientific Research from the Ministry of Education, Culture, Sports, Science and Technology (MEXT) and the Japan Science and Technology Agency (JST) PRESTO program.

References

1. Yang, P., Yan, R., and Fardy, M. (2010) Semiconductor nanowire: What's next?, *Nano Lett.*, **10**, pp. 1529–1539.
2. Wagner, R. S., and Ellis, W. C. (1964) Vapor-liquid-solid mechanism of single crystal growth, *Appl. Phys. Lett.*, **4**, p. 89.
3. Yan, H., Wu, Y., Kind, H., Weber, E., Russo, R., and Yang, P. (2001) Room-temperature ultraviolet nanowire nanolasers, *Science*, **292**, p. 1897.
4. Duan, X., Huang, Y., Cui, Y., Wang, J., and Lieber, C. M. (2001) Indium phosphide nanowires as building blocks for nanoscale electronic and optoelectronic devices, *Nature*, **409**. p. 66.
5. Gundiksen, M. S., Yang, S. I., Lieber, C. M., and Kelly, J. W. (2002) Growth of nanowire superlattice structures for nanoscale photonics and electronics, *Nature*, **415**, p. 617.
6. Bulllis, W. M. (1966) Properties of gold in silicon, *Solid-State Elec.*, **9**, p. 143.
7. Perea, D. E., Hemesath, E. R., Schwalbach, E. J., Lensch-Falk, J. L., Voorhees, P. W., and Lauhon, L. J. (2009) Direct measurement of dopant distribution in an individual vapor-liquid-solid nanowire, *Nat. Nanotech.*, **4**, p. 315.
8. Burke, J. E. (1964). Scattering of surface waves on an infinitely deep fluid, *J. Math. Phys.*, **6**, pp. 805–819.
9. Cui, Y., and Lieber, C. M. (2001) Functional nanoscale electronics devices assembled using silicon nanowire building blocks, *Science*, **291**, p. 851.
10. Huang, Y., Duan, X., Wei, Q., and Lieber, C. M. (2001) Directed assembly of one-dimensional nanostructures into functional networks, *Science*, **291**, p. 630.
11. Duan, X., Niu, C., Sahi, V., Chen, J., Parce, J. W., Empedocles, S., and Goldman, J. L. (2003) High-performance thin-film transistors using semiconductor nanowires and nanoribbons, *Nature*, **425**, p. 274.
12. Yan, H., Choe, H. S., Nam, S. W., Hu, Y., Das, S., Klemic, J. F., Ellenbogen, J. C., and Lieber, C. M. (2011) Programmable nanowire circuits for nanoprocessors, *Nature*, **470**, p. 240.
13. Park, W. I., Kim, J. S., Yi, G.-C., Lee, H.-J. (2005) ZnO nanorod logic circuits, *Adv. Mater.*, **17**, p.1393.
14. Roy, S., and Gao, Z. (2010) Direct-write fabrication of a nanoscale digital logic element on a single nanowire, *Nanotechnology*, **21**, p. 245306.

15. Fan, Z., Wang, D., Chang P.-C., Tseng, W.-Y., and Lu, J. G. (2004) ZnO nanowire field-effect transistor and oxygen sensing property, *Appl. Phys. Lett.*, **85**, p. 5923.

16. Patolsky, F., Zheng, G., and Lieber, C. M. (2006) Nanowire-based biosensors, *Anal. Chem.*, **78**, p. 4260.

17. Cui, Y., Zhong, Z., Wang, D., Wang, W. U., Lieber, C. M. (2003) High performance silicon nanowire field effect transistors, *Nano Lett.*, **3**, pp 149.

18. Patolsky, F., Zheng, G., and Lieber, C. M. (2006) Fabrication of silicon nanowire devices for ultrasensitive, label-freem real-time detection of biological and chemical species, *Nat. Protocols*, **1**, p. 1711.

19. Patolsky, F., Timko, B. P., Yu, G., Fang, Y., Greytak, A. B., Zheng, G., and Lieber, C. M. (2006) Detection, simulation, and inhibition of neural signals with high-density nanowire transistor arrays, *Science*, **313**, p. 1100.

20. Timko, B. P., Cohen-Karni, T., Qing, Q., Tian, B., and Lieber, C. M. (2010) Design and implementation of functional nanoelectronic interfaces with biomolecules, cells, and tissue using nanowire device arrays, *IEEE Trans. Nanotech.*, **9**, p. 269.

21. Cohen-Karni, T., Casanova, D., Cahoon, J., Qing, Q., Bell, D., and Lieber, C. M. (2012) Synthetically-encoded ultrashort-channel nanowire transistors for fast, point-like cellular signal detection, *Nano Lett.*, **12**, p. 2639.

22. Xie, P., Xiong, Q., Fang, Y., Qing, Q., and Lieber, C. M. (2012) Local electrical potential detection of DNA by nanowire-nanopore sensors, *Nat. Nanotech.*, **7**, p. 119.

23. Duan, X., Gao, R., Xie, P., Cohen-Karni, T., Qing, Q., Choe, H. S., Tian, B., Jiang, Z., and Lieber, C. M. (2012) Intracellular recordings of action potentials by an extracellular nanoscale field-effect transistor, *Nat. Nanotech.*, **7**, p. 174.

24. Kang, D., Ko, J.-H., Bae, E., Hyun, J., Park, W., Kim, B.-K., Kim, J.-J., and Lee, C. (2004) Ambient air effects on electrical characteristics of GaP nanowire transistor, *J. Appl. Phys.* **96**, p. 7574.

25. Jiang, P., Jie, J., Yu, Y., Wang, Z., Xie, C., Zhang, X., Wu, C., Wang, L., Zhu, Z., and Luo, L. (2012) Aluminum-doped n-type ZnS nanowries as high-performance UV and humidity, *J. Mater. Chem.* **22**, p. 6856.

26. McAlpine, M. C., Friedman, R. S., Jin, S., Lin, K.-H., Wnag, W. U., and Lieber, C. M. (2003) High-performance nanowire electronics and photonics on glass and plastic substrates, *Nano Lett.*, **3**, p. 1531.

27. Zheng, G., Lu, W., Jin, S., and Lieber, C. M. (2004) Synthesis and fabrication of high-performance n-type silicon nanowire transistors, *Adv. Mater.*, **16**, p. 1890.

28. Sheriff, B. A., Wang, D., Heath, J. R., and Kurtin, J. N. (2008) Complementary symmetry nanowire logic circuits: Experimental demonstrations and in silico optimization, *ACS Nano*, **2**, p. 1789.

29. Shan, Y., Ashok, S., and Fonash, S. J. (2007) Unipolar accumulation-type transistor configuration implemented using Si nanowires, *Appl. Phys. Lett.,* **91**, p. 093518.

30. Wang, D., Wang, Q., Javey, A., Tu, R., Dai, H., Kim, H., McIntyre, P. C., Krishnamohan, T., and Saraswat, K. C. (2003) Germanium nanowire field-effect transistors with SiO_2 and high-k HfO_2 dielectrics, *Appl. Phys. Lett.*, **83**, p. 2432.

31. Thelander, C., Bjork, M. T., Larsson, M. W., Hansen, A. E., Wallenberg, L. R., and Samuelson, L. (2004) Electron transport in InAs nanowires and heterostructure nanowire devices, *Solid State Commun.*, **131**, p. 573.

32. Huang, Y., Duan, X., Cui, Y., and Lieber, C. M. (2002) Gallium nitride nanowire nanodevices, *Nano Lett.*, **2**, p. 101.

33. Colingem J.-P., Lee, C.-W., Afzalian, A., Akhavan, N. D., Yan, R., Ferain, I., Razavi, P., O'Neill, B., Blake, A., White, M., Kelleher, A.-M., McCarthy, B., and Murphy, R. (2010) Nanowire transistors without junctions, *Nat. Nanotech.*, **5**, p. 225.

34. Wu, Y., Xiang, J., Yang, C., Lu, W., and Lieber, C. M. (2004) Single-crystal metallic nanowires and metal/semiconductor nanowire heterostructures, *Nature*, **430**, p. 61.

35. Liow, T.-Y., Tan, K.-M., Lee, R. T. P., Zhu, M., Tan, B. L.–H., Balasubramanian, N., and Yeo, Y.-C. (2008) Germanium source and drain stressors for ultrathin-body and nanowire field-effect transistors, *IEEE Elec. Dev. Lett.*, **29**, p. 808.

36. Tang, J., Wang, C.-Y., Xiu, F., Hong, A. J., Chen, S., Wang, M., Zeng, C., Yang, H.-J., Tuan, H.-Y., Tsai, C.-J., Chen, L. J., and Wang, K. L. (2010) Single-crystalline $Ni_2Ge/Ge/Ni_2Ge$ nanowire heterostructure transistors, *Nanotechnology*, **21**, p. 505704.

37. Chueh, Y.-L., Ford, A. C., Ho, J. C., Jacobson, Z. A., Fan, Z., Chen, C.-Y., Chou, L.-J., and Javey, A. (2008) Formation and characterization of NixInAs/InAs nanowire heterostructures by solid source reaction, *Nano Lett.*, **8**, p. 4528.

38. Shen, L. F., Yip, S. P., Yang, Z.-X., Fang, M., Hung, T. K., Pun, E. Y. B., and Ho. J. C. (2015) High-performance wrap-gated InGaAs nanowire field-effect transistors with sputtered dielectric, *Sci. Rep.*, **5**, p. 16871.

39. Huang, X., Lee, W.-C., Kuo, C., Hisamoto, D., Chang, L., Kedzierski, J., Anderson, E., Tekeuchi, H., Choi, Y.-K., Asano, K., Subramanian, V., King, T.-J., Boker, J., and Hu. C. (1999) Sub 50 nm finfet: PMOS, *IEEE IEDM Tech. Dig.*, p. 67.

40. Ferain, I., Colinge, C. A., and Conlinge, J.-P. (2011) Multigate transistors as the future of classical metal-oxide-semiconductor field-effect transistors, *Nature*, **479**, p. 310.

41. Takato, H., Sunouchi, K., Okabe, N., Nitayama, A., Hieda, K., Horiguchi, F., and Masuoka, F. (1991) High performance CMOS surrounding gate transistor (SGT) for ultra high density LSIs, *IEEE IEDM Tech. Dig.*, p. 222 (1988).

42. Takato, H., Sunouchi, K., Okabe, N., Nitayama, A., Hieda, K., Horiguchi, F., and Masuoka, F. (1991) Impact of surrounding gate transistor (SGT) for ultra-high-density LSI's, *IEEE Trans. Elec. Dev.*, **38**, p. 573.

43. Nitayama, A., Takato, H., Okabe, N., Sunouchi, K., Hieda, K., Horiguchi, F., and Masuoka, F. (1991) Multi-pillar surrounding gate transistor (M-SGT) for compact and high-speed circuits, *IEEE Trans. Elec. Dev.*, **38**, p. 579.

44. del Alamo, J. A. (2011) Nanometre-scale electronics with III-V compound semiconductors, *Nature*, **479**, p. 317.

45. Pillarisetty, R. (2011) Academic and industry research progress in germanium nanodevices, *Nature*, **479**, p. 324.

46. Riel, H., Wernesson, L.-E., Hong, M., and del Alamo, J. A. (2014) III-V compound semiconductor transistors—from planar to nanowire structures, *MRS Bull.*, **39**, p. 668.

47. Seabaugh, A. C., and Zhang, Q. (2010) Low-voltage tunnel transistors for beyond CMOS logic, *Proc. IEEE*, **98**, p. 2095.

48. Ionescu, A. M., and Riel, H. (2011) Tunnel field-effect transistors as energy-efficient electronic switches, *Nature*, **479**, p. 329.

49. Pott, V., Kam, H., Nathanael, R., Jeon, J., Alon, E., and Liu, T.-J. K. (2010) Mechanical computing redux: Relays for integrated circuit applications, *Proc. IEEE.*, **98**, p. 2076.

50. Radosavljevic, M., Dewey, G., Basu, D., Boardman, J., Chu-Kung, B., Fastenau, J. M., Kabehie, S., Kavalieros, J., Le, V., Liu, W. K., Lubyshev, D., Metz, M., Millard, K., Mukherjee, N., Pan, L., Pillarisetty, R., Rachmady, W., Shah, U., Then, H. W., Chau, R. (2011) *IEEE IEDM Tech. Dig.*, p. 765.

51. Wu, Y. Q., Wang, R. S., Shen, T., Gu, J. J., and Ye, P. D. (2009) First experimental demonstration of 100 nm inversion-mode InGaAs FinFET through damage-free sidewall etching, *IEEE IEDM Tech. Dig.* p. 331.

52. Gu, J. J., Liu, Y. Q., Colby, R., Gordon, R. G., and Ye, P. D. (2011) First experimental demonstration of gate-all-around III-V MOSFETs by top-down approach, *IEEE IEDM Tech. Dig.* p. 769.

53. Gu. J. J., Wang, X., Wu, H., Gordon, R. G., and Ye, P. D. (2013) Variablity improvement by interface passivation and EOT scaling of InGaAs nanowire MOSFETs, *IEEE Elec. Dev. Lett.*, **34**, p. 608.

54. Wu, H., Si, M., Lin, D., Zhang, J., and Ye, P. D. (2014) Ge CMOS: Breakthroughs of nFETs (I_{max} = 714 mA.mm, g_{max} = 590 mS/mm) by recessed channel and S/D, *IEEE VLSI symp. Tech. Dig.*, p. 76.

55. Wu, H., Luo, W., Si, M., Zhang, J., Zhou, H., and Ye, P. D. (2014) Deep sub-100 nm Ge CMOS devices on Si with the recessed S/D and channel, *IEEE IEDM Tech. Dig.*, p. 426.

56. Wu, H., Luo, W., Zhou, H., Si, M., Zhang, J., and Ye, P. D. (2015) First experimental demonstration of Ge 3D FinFET CMOS circuits, p. T59.

57. Chen, J., and Könenkamp, R. (2003) Vertical nanowire transistor in flexible polymer foil, *Appl. Phys. Lett.*, **82**, p. 4782.

58. Ng, H. T., Han, J., Yamada, T., Nguyen, P., Chen, Y. P., and Meyyappan, M. (2004) Single crystal nanowire vertical surround-gate field-effect transistor, *Nano Lett.*, **4**, p. 1247.

59. Goldberger, J., Hochabaum, A. I., Fan, R., and Yang, P. (2006) Silicon vertically integrated nanowire field effect transistor, *Nano Lett.*, **6**, p. 973.

60. Bryllert, T., Wernersson, L.-E., Fröberg, L. E., and Samuelson, L. (2006) Vertical high-mobility wrap-gated InAs nanowire transistor, *IEEE Elec. Dev. Lett.*, **27**, p. 323.

61. Svensson, J., Dey, A. W., Jacobsson, D. J., and Wernersson, L.-E. (2015) III-V nanowire complementary metal-oxide semiconductor transistors monolithically integrated on Si, *Nano Lett.*, **15**, p. 7898.

62. Johansson, S., Memisevic, E., Wernersson, L.-E., Lind, E. (2014) High-frequency gate-all-around vertical InAs nanowire MOSFETs on Si substrates, *IEEE Elec. Dev. Lett.*, **35**, p. 518.

63. Persson, K.-M., Berg, M., Borg, B., Wu, J., Johansson, S., Svensson, J., Jansson, K., Lind, E., and Wernersson, L.-E. (2013) Extrinsic and intrinsic performance of vertical nanowire MOSFETs on Si substrates, *IEEE Trans. Elec. Dev.*, **60**, p. 2761.

64. Vasen, T., Ramvall, P., Afzalian, A., Thelander, C., Dick, K. A., Holland, M., Doornbos, G., Wand, S. W., Oxland, R., Velliantis, G., van Dal, M. J. H., Duriez, B., Ramirez, J.–R., Droopad, R., Wernersson, L. E., Samuelson, L., Chen, T.–K., Yeo, Y.-C., and Passlack, M. (2016) InAs nanowire GAA n-MOSFETs with 12–15 nm diameter, *IEEE VLSI Symp.*, p. 160.

65. Ramesh, S., Ivanov, T., Camerotto, E., Sun, N., Franco, J., Sibaja-Hernandez, A., Rooyackers, R., Alian, A., Loo, J., Veloso, A., Milenin, A., Lin, D., Favia, P., Bender, H., Collert, N., Thean, V.-Y., and Meyer, K. D. (2016) Top-down InGaAs nanowire and fin vertical FETs with record performance, *IEEE VLSI Symp. Tech. Dig.*, p. 164.

66. Tomioka, K., Yoshimura, M., and Fukui, T. (2012) A III-V nanowire channel on silicon for high-performance vertical transistors, *Nature*, **488**, p. 189.

67. Tomioka, K., Yoshimura, M., Nakai, E., Ishizaka, F., and Fukui, T. (2013) Integration of III-V nanowires on Si: From high-performance vertical FET to steep-slope switch, *IEEE IEDM Tech. Dig.* p. 88.

68. Wieder, H. H. (2003) Surface and interface barriers of $In_xGa_{1-x}As$ binary and ternary alloys, *J. Vac. Sci. Technol. B*, **21**, p. 1915.

69. Yakimets, D., Eneman, G., Schuddinck, P., Bao, T. H., Bardon, M. G., Raghavan, P., Veloso, A., Collaert, N., Mercha, A., Verkest, D., Thean, A. V.-Y., and Meyer, K. D. (2015) Vertical GAAFETs for the ultimate CMOS scaling, *IEEE Trans. Elec. Dev.*, **62**, p.1433.

70. Mårtensson, T., Patrik, C., Svensson, T., Wacasar, B. A., Larsson, M. W., Seifert, W., Deppert, K., Gustafsson, A., Wallenberg, L. R., and Samuelson, L. (2004) Epitaxial III-V nanowires on silicon, *Nano Lett.*, **4**, p. 1987.

71. Tomioka, K., Motohisa, J., Hara, S., and Fukui, T. (2008) Control of InAs nanowire growth directions on Si, *Nano Lett.*, **8**, p. 3475.

72. Thelander, C., Rehnstedt, C., Froberg, L. E., Lind, E., Mårtensson, T., Caroff, P., Lowgren, T., Ohlsson, B. J., Samuelson, L., and Wernersson, L.-E. (2008) Development of a vertical wrap-gated InAs FET, *IEEE Trans. Elec. Dev.*, **55**, p. 3030.

73. Rehnstedt, C., Mårtensson, T., Thelander, C., Samuelson, T., and Wernersson, L.-E. (2008) Vertical InAs nanowire wrap gate transistors on Si substrates, *IEEE Trans. Elec. Dev.*, **55**, p. 3037.

74. Li, X., Chen, Z., Shen, N., Sakar, D., Singh, N., Banerjee, K., Lo, G.-Q., Kwong, D.-L. (2011) Vertically stacked and independently controlled twin-gate MOSFETs on a single Si nanowire, *IEEE Elec. Dev. Lett.*, **32**, p. 1492.

75. Tanaka, T., Tomioka, K., Hara, S., Motohisa, J., Sano, E., and Fukui, T. (2010) Vertical surrounding gate transistors using single InAs nanowires grown on Si substrates, *Appl. Phys. Exp.*, **3**, p. 025003.

76. Motohisa, J., Noborisaka, J., Takeda, J., Inari, M., and Fukui, T. (2004) Catalyst-free selective-area MOVPE of semiconductor nanowires on (111)B oriented substrates, *J. Cryst. Growth*, **272**, p. 180.

77. Noborisaka, J., Motohisa, J., and Fukui, T. (2005) Catalyst-free growth of GaAs nanowires by selective-area metalorganic vapor-phase epitaxy, *Appl. Phys. Lett.*, **86**, p. 213102.

78. Ikejiri, K., Noborisaka, J., Hara, S., Motohisa, J., and Fukui, T. (2007) Mechanism of catalyst-free growth of GaAs nanowires by selective area MOVPE, *J. Cryst. Growth*, **298**, p. 616.

79. Ikejiri, K., Sato, T., Yoshida, H., Hiruma, K., Motohisa, J., Hara, S., and Fukui, T. (2008) Growth characteristics of GaAs nanowires obtained by selective area metal-organic vapour-phase epitaxy, *Nanotechnology*, **19**, p. 265604.

80. Mohan, P., Motohisa, J., and Fukui, T. (2005) Controlled growth of highly uniform, axial/radial direction-defined, individually addressable InP nanowire arrays, *Nanotechnology*, **16**, p. 2903.

81. Kitauchi, Y., Kobayashi, Y., Tomioka, K., Hara, S., Hiruma, K., and Fukui, T. (2010) Structural transition in indium phosphide nanowires, *Nano Lett.*, **10**, p. 1699.

82. Tomioka, K., Mohan, P., Noborisaka, J., Hara, S., Motohisa, J., and Fukui, T. (2007) Growth of highly uniform InAs nanowire arrays by selective-area MOVPE, *J. Cryst. Growth* **298**, p. 644.

83. Akabori, M., Takeda, J., Motohisa, J., and Fukui, T. (2003) InGaAs nano-pillar array formation on partially masked InP(111)B by selective area metal-organic vapour phase epitaxial growth for two-dimensional photonic crystal application, *Nanotechnology*, **14**, p. 1071.

84. Sato, T., Motohisa, J., Noborisaka, J., Hara, S., and Fukui, T. (2008) Growth of InGaAs nanowires by selective-area metalorganic vapor phase epitaxy, *J. Cryst. Growth*, **310**, p. 2359.

85. Sato, T., Kobayashi, Y., Motohisa, J., Hara, S., and Fukui, T. (2008) SA-MOVPE of InGaAs nanowires and their compositions studied by micro-PL measurement, *J. Cryst. Growth*, **310**, p. 5111.

86. Yoshimura, M., Tomioka, K., Hiruma, K., Hara, S., Motohisa, J., and Fukui, T. (2010) Growth and characterization of InGaAs nanowires formed on GaAs (111)B by selective-area metal organic vapor phase epitaxy, *Jpn. J. Appl. Phys.*, **49**, p. 04DH08.

87. Fujisawa, S., Sato, T., Hara, S., Motohisa, J., Hiruma, K., and Fukui, T. (2011) Growth and characterization of a GaAs quantum well buried in GaAsP/GaAs vertical heterostructure nanowire by selective-area metal organic vapor phase epitaxy, *Jpn. J. Appl. Phys.*, **50**, p. 04DH03.

88. Sekiguchi, H., Kishino, K., and Kikuchi, A. (2010) Emission color control from blue to red with nanocolumn diameter of InGaN/GaN nanocolumn arrays grown on same substrate, *Appl. Phys. Lett.*, **96**, p. 231104.

89. Kim, Y.-J., Lee, C.-H., Hong, Y. J., Yi, G.-C., Kim, S. S., Cheong, H. (2006) Controlled selective growth of ZnO nanorod and microrod arrays on Si substrate by a wet chemical method, *Appl. Phys. Lett.*, **89**, p. 163128.

90. Noborisaka, J., Motohisa, J., Hara, S., and Fukui, T. (2005) Fabrication and characterization of freestanding GaAs/AlGaAs core-shell nanowires and AlGaAs nanotubes by using selective-area metalorganic vapor phase epitaxy, *Appl. Phys. Lett.*, **87**, p. 093109.

91. Mohan, P., Motohisa, J., and Fukui, T. (2006) Realization of conductive InAs nanotubes based on lattice-mismatched InP/InAs core-shell nanowires, *Appl. Phys. Lett.*, **88**, p. 013110.

92. Hua, B., Motohisa, J., Kobayashi, Y., Hara, S., and Fukui, T. (2009) Single GaAs/GaAsP coaxial core-shell nanowire laser, *Nano Lett.*, **9**, p. 112.

93. Mohan, P., Motohisa, J., and Fukui, T. (2006) Fabrication of InP/InAs/InP core-multishell heterostructure nanowires by selective area metalorganic vapor phase epitaxy, *Appl. Phys. Lett.*, **88**, p. 133105.

94. Yang, L., Motohisa, J., Takeda, J., Tomioka, K., and Fukui, T. (2007) Selective-area growth of hexagonal nanopillars with single InGaAs/GaAs quantum wells on GaAs (111)B substrate and their temperature-dependent photoluminescence, *Nanotechnology*, **18**, p. 105302.

95. Shapiro, J. N., Lin, A., Wong, P. S., Scofield, A. C., Tu, C., Senanayake, P. N., Mariani, G., Liang, B. L., Huffaker, D. L. (2010) InGaAs heterostructure formation in catalyst-free GaAs nanopillars by selective-area metal-organic vapor phase epitaxy, *Appl. Phys. Lett.*, **97**, p. 243102.

96. Sasakura, H., Kumano, K., Suemune, I., Mothisa, J., Kobayashi, Y., van Kouwen, M., Tomioka, K., Fukui, T., Akopian, N., and Zwiller, V. (2009) Exciton coherence in clean single InP/InAsP/InP nanowire quantum dots emitting in infra-red measured by Fourier spectroscopy, *J. Phys. Conf. Ser.*, **193**, p. 012132.

97. Hayashida, A., Sato, T., Hara, S., Motohisa, J., Hiruma, K., and Fukui, T. (2010) Fabrication and characterization of GaAs quantum well buried in AlGaAs/GaAs heterostructure nanowires, *J. Cryst. Growth*, **312**, p. 3592.

98. Tomioka, K., Kobayashi, Y., Motohisa, J., Hara, S., and Fukui, T. (2009) Selective-area growth of vertically aligned GaAs and GaAs/AlGaAs core-shell nanowires on Si(111) substrate, *Nanotechnology*, **20**, p. 145302.

99. Kikuchi, A., kawai, M., Tada, M., and Kishino, K. (2004) InGaN/GaN multiple quantum disk nanocolumn light-emitting diodes grown on (111) Si substrate, *Jpn. J. Appl. Phys.*, **43**, p. L1524.

100. Tomioka, K., Tanaka, T., Hara, S., Hiruma, K., and Fukui, T. (2011) III-V nanowires on Si substrate: selective-area growth and device applications, *IEEE J. Select. Top. Quant. Elec.*, **17**, p. 1112.

101. Tomioka, K., and Fukui, T. (2014) Recent progress in integration of III-V nanowire transistors on Si substrate by selective-area growth, *J. Phys. D: Appl. Phys.* **47**, p. 394001.

102. Chuang, L. C., Sedwick, F. G., Chen, R., Ko, W. S. Moewe, M., Ng, K. W., Tran, T.-T. D., and Chang-Hasnain, C. (2011) GaAs-based nanoneedle light emitting diode and avalanche photodiode monolithically integrated on a silicon substrate, *Nano Lett.*, **11**, p. 385.

103. Tomioka, K., Motohisa, J., Hara, S., Hiruma, K., and Fukui, T. (2010) GaAs/AlGaAs core multishell nanowire-based light-emitting diodes on Si, *Nano Lett.* **10**, p. 1639.

104. Wernersson, L.-E., Thelander, C., Lind, E., and Samuelson, L. (2010) III-V nanowires—extending a narrowing road, *Proc. IEEE*, **98**, p. 2047.

105. Tomioka, K., Yoshimura, M., and Fukui, T. (2011) Vertical $In_{0.7}Ga_{0.3}As$ nanowire surrounding-gate transistors with high-k gate dielectric on Si substrate, *IEEE IEDM Tech. Dig.* p. 773 (2011).

106. Chalamestani, S. G., Johansson, S., Bor, B. M., Lind, E., Dick, K. A., and Wernersson, L.-E. (2012) Uniform and position-controlled InAs nanowires on 2" Si substrates got transistors applications, *Nanotechnology*, **23**, p. 015302.

107. Johansson, S., Egard, M., Ghalamestani, S. G., Bor, B. M., Berg, M., Wernersson, L.-E., and Lind, E. (2012) RF characterization of vertical InAs nanowire wrap-gate transistors integrated on Si substrates, *IEEE Trans. Micro. Theory Tech.*, **59**, p. 2733.

108. Wei, W., Bao, X.–Y., Soci, C., Ding, Y., Wand, Z. L., and Wang, D. (2009) Direct heteroepitaxy of vertical InAs nanowries on Si substrates for broad band photovoltaics and photodetector, *Nano Lett.*, **9**, p. 2926.

109. Bjork, M. T., Schmid, H., Bessire, C. D., Moselund, K. E., Ghoneim, H., Karg, S., Lortscher, E., and Riel, H. (2010) Si-InAs heterjunction Esaki tunnel diodes with high current densities, *Appl. Phys. Lett.*, **97**, p. 163501.

110. Bessire, C. D., Bjork, M. T., Schmid, H., Schenk, A., Ruters, K. B., and Riel, H. (2011) Trap-assisted tunneling in Si-InAs nanowire heterojunction tunnel diodes, *Nano Lett.*, **11**, p. 4195.

111. Yang, T., Hertenberger, S., Morkotter, S., Abstreiter, G., and Koblmuller, G. (2012) Size, composition, and doping effects on In(Ga)As nanowire/Si tunnel diodes probed by conductive atomic force microscopy, *Appl. Phys. Lett.*, **101**, p. 233102.

112. Koblmüller, G., and Abstreiter, G. (2014) Growth and properties of InGaAs nanowires on silicon, *Phys. State Solid. PRL*, **8**, p. 11.

113. Riel, H., Moslund, K. E., Bessire, C. D., Bjork, M. T., Schenk, A., Ghoneim, H., and Schmid, H. (2012) InAs-Si heterojunction nanowire tunnel diodes and tunnel FETs, *IEEE IEDM Tech. Dig.* p. 391.

114. Tomioka, K., and Fukui, T. (2012) Tunnel field-effect transistor using InAs nanowire/Si heterojunction, *Appl. Phys. Lett.*, **98**, p. 083114.

115. Moselund, K. E., Schmid, H., Bessire, C., Bjork, M. T., Ghoneim, H., Riel, H. 2012, InAs-Si nanowire heterojunction tunnel FETs, *IEEE Elec. Dev. Lett.*, **33**, p. 1453.

116. Tomioka, K., and Fukui, T. (2014) Current increment of tunnel field-effect transistor using InGaAs nanowire/Si heterojunction by scaling of channel length, *Appl. Phys. Lett.*, **104**, p. 073507.

117. Tomioka, K., Yoshimura, M., and Fukui, T. (2013) Sub 60 mV/decade switch using an InAs nanowire-Si heterojunction and turn-on voltage shift with a pulsed doping technique, *Nano Lett.*, **13**, p. 5822.

118. Takayanagi, K., Tanishiro, Y., Takahashi, S., and Takahashi, M. (1985) Structure analysis of Si(111)-7×7 reconstructed surface by transmission electron diffraction, *Surf. Sci.*, **164**, p. 367.

119. Hoshino, T., Kokubun, K., Fujiwara, H., Kumamoto, K., Ishimaru, T., and Ohdomari, I. (1995) Critical domain size of the 7×7 structure for nucleation and growth on Si(111) quenched surfaces, *Phys. Rev. Lett.*, **75**, p. 2372.

120. Hoshino, M., Shigeta, Y., Ogawa, K., and Homma, Y. (1996) STM and SEM studies on the character of triangular Si(111)-7×7 domains formed in quenched Si(111) surface, *Surf. Sci.*, **365**, p. 29.

121. Chiba, M., Tanishiro, Y., Monida, H., and Yagi, K. (1999) Scanning tunneling microscopy studies on the reversible phase transition between metastable structures of Si(11)-c2×8 and "1×1", *Surf. Sci.*, **441**, p. 179.

122. Himspel, F. J., McFeely, F. R., Taleb-Ibrahimi, A., Yarmoff, J. A., and Hollinger, G. (1988) Microscopic structure of the SiO_2/Si interface, *Phys. Rev. B*, **38**, p. 6084.

123. Hattori, T., Aiba, T., Iijima, E., Okube, Y., Nohira, H., Tate, N., and Katayama, M. (1996) Initial stage of oxidation of hydrogen-terminated silicon surfaces, *Appl. Surf. Sci.*, **104–105**, p. 323.

124. Olmstead, M. A., Bringans, R. D., Uhrberg, R. I. G., and Bachrach, R. Z. (1986) Arsenic overlayer on Si(111): Removal of surface reconstruction, *Phys. Rev. B*, **34**, p. 6041.

125. Kobayashi, N., Makimoto, T., and Horikoshi, Y. (1985) Flow-rate modulation epitaxy of GaAs, *Jpn. J. Appl. Phys.*, **24**, p. 962.

126. Hinkle, C. L., Sonnet, A. M., Vogel, E. M., McDonnell, S., Hughes, G. J., Milojevic, M., Lee, B., Aguirre-Tostado, F. S., Choi, K. J., Kim, H. C., Kim, J., and Wallace, M. (2008) GaAs interfacial self-cleaning by atomic layer deposition, *Appl. Phys. Lett.*, **92**, p. 071901.

127. Tuck, C. (1999) *Effective Medium Theory*, 1st ed. (Oxford).

128. Nextnano3, (http://www.nextnano.de/nextnano3/tutorial/2Dtutorial_core_shell_circle_hexagon.htm).

129. Wallentin, J., Anttu, N., Asoli, D., Huffman, M., Aberg, I., Magunusson, M. H., Siefer, G., Fuss-Kailuweit, P., Dimroth, F., Witzgmann, B., Zu, H. Q., Samuelson, L., Deppert, K., and Borgström, M. T. (2013) InP nanowire array solar cells achieving 13.8% efficiency by exceeding the ray optics limit, *Science*, **339**, p. 1057.

130. Hu, L., and Chen, G. (2007) Analysis of optical absorption in silicon nanowire arrays for photovoltaic applications, *Nano Lett.*, **7**, p. 3249.

131. Kupec, J., Stoop, R. L., and Witzigmann, B. (2010) Light absorption and emission in nanowire array solar cells, *Opt. Exp.*, **18**, p. 27589.

132. Burgos, S. P., de Waele, R., Polman, A., and Atwater, H. A. (2010) A single-layer wide-angle negative-index metamaterials at visible frequencies, *Nature Mater.*, **9**, p. 407.

133. Wen, L., Zhao, Z., Li, Z., Shen, Y., Guo, H., and Wang, Y. (2011) Theoretical analysis and modeling of light trapping in high efficiency GaAs nanowire array solar cells, *Appl. Phys. Lett.*, **99**, p. 143116.

134. Colombo, C., Heib, M., Gratzel, M., and i Morral, A. F. (2009) Gallium arsenide p-i-n radial structures for photovoltaic applications, *Appl. Phys. Lett.*, **94**, p. 173108.

135. Kelzenberg, M. D., Turner-Evans, D. B., Kayes, B. M., Filler, M. A., Putnam, M. C., Lewis, N. S., and Atwater, H. A. (2008) Photovoltaic measurements in single-nanowire silicon solar cells, *Nano Lett.*, **8**, p. 710.

136. Tian, B., Zheng, X., Kempa, T. J., Fand, Y., Yu, N., Yu, G., Huang, J., and Lieber, C. M. (2007) Coaxial silicon nanowires as solar cells and nanoelectronics power sources, *Nature*, **499**, p. 885.

137. Czaban, J. A., Thompson, D. A., and LaPierre, R. R. (2009) GaAs core-shell nanowires for photovoltaic applications, *Nano Lett.*, **9**, p. 148.

138. Borström, M. T., Ohlsson, B. J., Sass, T., Persson, A. I., Thelander, C., Magunusson, M. H., Deppert, K., Wallenberg, L. R., and Samuelson, L. (2011) Nanowire with promise for photovoltaics, *IEEE J. Selec. Quant. Elec.* **17**, p. 1051.

139. Goto, H., Nosaki, K., Tomioka, K., Hara, S., Hiruma, K., Motohisa, J., and Fukui, T. (2008) Growth of core-shell InP nanowires for photovoltaic application by selective-area metal organic vapor phase epitaxy, *Appl. Phys. Exp.*, **2**, p. 035004.

140. Huang, Y.-F., Chattopadhyay, S., Jen, Y.-J., Peng, C.-Y., Liu, T.-A., Hsu, Y.-K., Pan, C.-L., Lo, H.-C., Hsu, C.-H., Chang, Y.-H., Lee, C.-S., Chen, K.-H., and Chen, L.-C. (2007) Improved broadband and quasi omnidirectional anti-reflection properties with biomimetic silicon nanostructures, *Nature Nanotech.*, **2**, p. 770.

141. Kayes, B. M., Atwater, H. A., and Lewis, N. S. (2005) Comparison of the device physics principles of planar and radial p-n junction nanorod solar cells, *J. Appl. Phys.*, **97**, p. 114302.

142. Bremner, S. P., Levy, M. Y., and Honsberg, C. B. (2008) Analysisi of tandem solar cell efficiencies under AM1.5G spectrum using a rapid fluc calculation method, *Prog. Photovol. Res. App.*, **8**, p. 2638.

143. Glas, F. (2006) Critical dimensions for the plastic relaxation of strained axial heterostructures in free-standing nanowires, *Phys. Rev. B*, **74**, p. 12302(R).

144. Mriani, G., Wong, P. S., Katzenmeyers, A. M., Leonard, F., Shapiro, J., and Huffaker, D. L. (2011) Patterned radial GaAs nanopiller solar cells, *Nano Lett.*, **11**, p. 2490.

145. Yoshimura, M., Nakai, E., Tomioka, K., and Fukui, T. (2013) Indium phosphide core-shell nanowire array solar cells with lattice-mismatched window layer, *Appl. Phys. Exp.*, **6**, p. 052301.

146. Dong, Y., Tian, B., Kempa, T. J., and Lieber, C. M. (2009) Coaxial group III-nitride nanowire photovoltaics, *Nano Lett.*, **9**, p. 2183.

147. Tang, Y. B., Chen, Z. H., Song, H. S., Lee, C. S., Cong, H. T., Cheng, H. M., Zhang, W. J., Bello, I., and Lee, S. T. (2008) Vertically aligned p-type single-crystalline GaN nanorod arrays on n-type Si for heterojunction photovoltaic cells, *Nano Lett.*, **8**, p. 4191.

148. Åberg, I., Vescovi, G., Asoli, D., Naseem, U., Gilboy, J. P., Sundvall, C., Dahlgren, A., Svensson, K. E., Anttu, N., Björk, M. T., and Samuelson, L. (2016) A GaAs nanowire array solar cell with 15.3% efficiency at 1 sun, *IEEE J. Photovol.*, **6**, p. 185.

149. Cui, Y., Wang, J., Plissard, S. R., Cavalli, A., Vu, T. T. T., van Veldhoven, R. P. J., Gao, L., Trainor, M., Verheijen, M. A., Haverkort, J. E. M., and Bakkars, E. P. A. M. (2013) Efficiency enhancement of InP nanowire solar cells by surface cleaning, *Nano Lett.*, **13**, p. 4113.

150. Tajik, N., Peng, Z., Kuyanov, P., and LaPierre, R. R. (2011) Sulfur passivation and contact methods for GaAs nanowire solar cells, *Nanotechnology*, **22**, p. 225402.

151. Fukui, T., Yoshimura, M., Nakai, E., and Tomioka, K. (2012) Position-controlled III-V compound semiconductor nanowire solar cells by selective-area metal-organic vapor phase epitaxy, *AMBIO*, **41** (Supplement 2), p. 119.

152. Heurlin, M., Wicket, P., Fält, S., Borgström, M. T., Deppert, K., Samuelson, L., and Magnusson, M. H. (2011) Axial InP nanowire tandem junction grown on a silicon substrate, *Nano Lett.*, **11**, p. 2028.

153. Anttu, N., Abrand, A., Asoli, D., Heurlin, M., Åberg, I., Samuelson, L., and Borgström, M. (2014) Absorption of light in InP nanowire arrays, *Nano Res.*, **7**, p. 816.

154. Chen, M., Nakai, E., Tomioka, T., and Fukui, T. (2015) Application of free-standing InP nanowire arrays and their optical properties for resource-saving solar cell, *Appl. Phys. Exp.*, **8**, p. 012301.

155. Li, Y., Xiang, J., Qian, F., Gradecak, S., Wu, Y., Yan, H., Blom, D. A., and Lieber, C. M. (2006) Dopant-free GaN/AlN/ALGaN radial nanowire heterostructures as high electron mobility transistor, *Nano Lett.*, **6**, p. 1468.

156. Do, Q.-T., Blekker, K., Regolin, I., Prost, W., and Tegude, F. J. (2007) High transconductance MISFET with a single InAs nanowire channel, *IEEE Elec. Dev. Lett.*, **28**, p. 682.

157. Zhang, X., Lew, K.-K., Nimmatoori, P., Redwing, J. M., and Dickey, E. C. (2007) Diameter-dependent composition of vapor-liquid-solid grown $Si_{1-x}Ge_x$ nanowires, *Nano Lett.*, **7**, p. 3241.

158. Gu, J. J., Wang, Z. W., Wu, H., Shao, J., Neal, A. T., Manfra, M. J., Gordon, R. G., and Ye, P. D. (2012) 20–80 nm channel length InGaAs gate-all-around nanowire MOSFETs with EOT = 1.2 nm and lowest SS = 63 mV/dec, *IEEE IEDM Tech. Dig.* p. 633.

159. Tomioka, K., Ishizaka, F., and Fukui, T. (2015) Selective-area growth of InAs nanowires on Ge and vertical transistor application, *Nano Lett.*, **15**, p. 7253.

160. Kim, D.-H., del Alamo, J. A., Chen, P., Ha, W., Urtega, M., and Brar, B. (2010) 50 nm E-mode $In_{0.7}Ga_{0.3}As$ PHEMTs on 100 mm InP substrate with f_{max} > 1 THz, *IEEE IEDM Tech. Dig.*, p. 692.

161. Boulanger, J. P., Chia, A. C. E., Wood, B., Yazdi, S., Kasama, T., Aagesen, M., and LaPierre, R. R. (2016) Characterization of a Ga-assisted GaAs nanowire array solar cell on Si substrate, *IEEE J. Photovol.*, **6**, p. 661.

162. Nakai, E., Yoshimura, M., Tomioka, K., and Fukui, T. (2013) GaAs/InGaAs core-multishell nanowire-array-based solar cells, *Jpn. J. Appl. Phys.*, **52**, p. 055002.

163. Mariani, G., Scofield, A. C., Hung, C.-H., and Huffaker, D. L. (2013) GaAs nanopillar-array solar cells employing in situ surface passivation, *Nat. Commun.*, **4**, p. 1497.

164. Mariani, G., Zhou, Z., Scofield, A., and Huffaker, D. L. (2013) Direct-bandgap epitaxial core-multishell nanopillar photovoltaics featuring subwavelength optical concentrators, *Nano Lett.*, **13**, p. 1632.

165. Yao. M., Huang. N., Cong, S., Chi, C., Seyedi, M. A., Lin, Y., Cao, Y., Povinelli, M. L., Dapkus, P. D., and Zhou, C. (2014) GaAs nanowire array solar cells with axial p-i-n junctions, *Nano Lett.*, **14**, p. 3293.

166. Nowzari, A., Heurin, M., Jain, V., Strom, K., Hesseinnia, A., Anttu, N., Borgström, M. T. (2015) A comparative study of absorption in vertically and laterally oriented InP core-shell nanowire photovoltaics devices, *Nano Lett.*, **15**, p. 1809.

167. Mohseni, P. K., Behnam, A., Wood, J. D., Zhao, X., Yu, K. J., Wang, N. C., Rockett, A., Rogers, J. A., Lyding, J. W., Pop, E., and Li, X. (2014) Monolithic III-V nanowire solar cells on graphene via direct van der Waals epitaxy, *Adv. Mater.*, **26**, p. 3755.

168. Nakai, E., Chen, M., Yoshimura, M., Tomioka, K., and Fukui, T. (2015) InGaAs axial-junction nanowire-array solar cells, *Jpn. J. Appl. Phys.*, **54**, p. 015201.

169. Yao, M., Cong, S., Arab, S., Huang, N., Povinelli, M. L., Cronin, S. B., Dapkus, P. D., and Zhou, C. (2015) Tandem solar cells using GaAs nanowires on Si: Design, fabrication, and observation of voltage addition, *Nano Lett.*, **15**, p. 2424.

Index